MASTERING ELECTRICITY

Dedication

To Professor Joseph B. Aidala

A distinguished pioneer in electrical engineering technology

Professor Aidala was the founder and former chairman of the Electrical and Computer Engineering Technology Department at Queensborough Community College. For 31 years, his energy, work, and leadership created one of the largest and finest programs in the country. It has served as a model to other educational institutions throughout the nation. His enthusiasm and dedication to teaching has launched thousands of students, including the authors, on to productive careers in electronics.

MASTERING ELECTRICITY

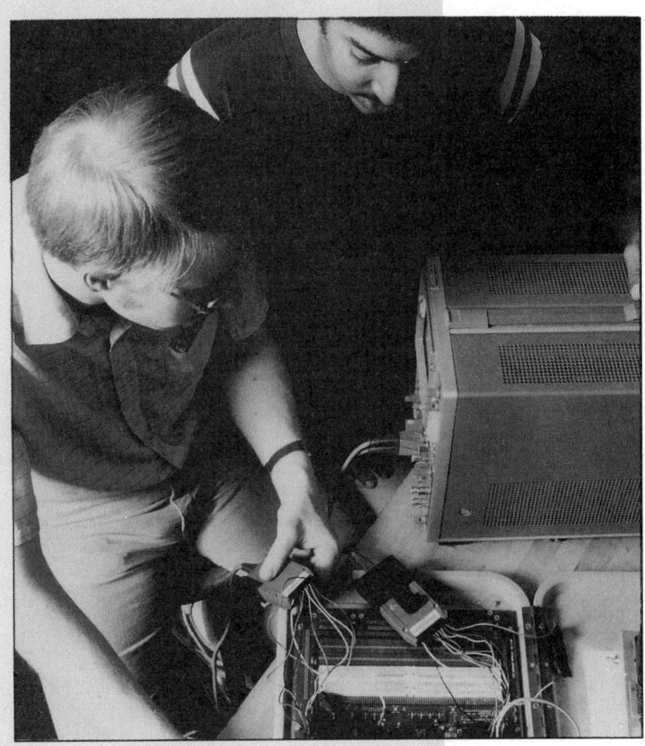

Courtesy of DeVry Institutes

Stuart Asser, PE

Queensborough Community College of the City University of New York

Vincent Stigliano, PE

Queensborough Community College of the City University of New York

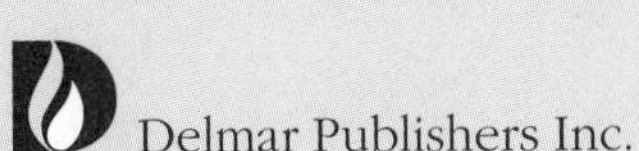

Delmar Publishers Inc.

I(T)P™

NOTICE TO THE READER

Publisher does not warrant or guarantee any of the products described herein or perform any independent analysis in connection with any of the product information contained herein. Publisher does not assume, and expressly disclaims, any obligation to obtain and include information other than that provided to it by the manufacturer.

The reader is expressly warned to consider and adopt all safety precautions that might be indicated by the activities described herein and to avoid all potential hazards. By following the instructions contained herein, the reader willingly assumes all risks in connection with such instructions.

The publisher makes no representations or warranties of any kind, including but not limited to, the warranties of fitness for particular purpose or merchantability, nor are any such representations implied with respect to the material set forth herein, and the publisher takes no responsibility with respect to such material. The publisher shall not be liable for any special, consequential or exemplary damages resulting, in whole or in part, from the readers' use of, or reliance upon, this material.

Cover illustration by Larry Hamill

Delmar Staff
Administrative Editor: Wendy Welch
Developmental Editor: Mary Clyne
Project Editor: Eleanor Isenhart
Senior Production Supervisor: Karen Leet
Cover Design Coordinator: Lisa L. Pauly
Text Design Coordinator: Megan DeSantis

For information address Delmar Publishers Inc.
3 Columbia Circle, Box 15-015
Albany, New York 12212-5015

Printed in the United States of America
Published simultaneously in Canada
by Nelson Canada,
a division of the Thomson Corporation

10 9 8 7 6 5 4 3 2 1 XXX 99 98 97 96 95 94 93

Library of Congress Cataloging-in-Publication Data:
Asser, Stuart.
Mastering electricity / Stuart Asser. Vincent Stigliano.
p. cm.
Includes index.
ISBN 0-8273-4604-2
1. Electric circuits. I. Stigliano, Vincent. II. Title
TK454.A78 1994 92-26598
621.3—dc20 CIP

CONTENTS

PREFACE

The electronics industry has been growing in geometric proportions for many years. That growth would not have been possible without the bright young minds that have emerged from our nation's schools. The next several years will place a great burden on our educational institutions as the need for technicians, engineers, computer specialists, service personnel, and salespeople continues to grow.

The fundamental principles and theories of electricity have not changed, but our society has changed. The preparation and background of today's schools, students, and teachers is far different and diverse from what it was twenty years ago. This textbook addresses these changes. It is designed to appeal to both student and faculty needs. It will also prepare students for the rapidly changing electrical and electronic industry.

Mastering Electricity is an introductory text for students who are beginning their study of electric circuits. No previous knowledge of electricity or electronics is assumed. The student should have a basic understanding of arithmetic and algebra, but the required knowledge of mathematics is integrated and explained throughout the text. The mathematics level in the illustrated examples and homework exercises is intentionally simple. The student can therefore focus on electric theory without getting entangled in the computations. Furthermore, all illustrated examples are presented using a detailed step-by-step method that includes the use of standard units. The illustrated examples and homework exercises become progressively more challenging as the use of practical examples and circuit values are introduced.

The reading level is as simple and clear as possible. A list of key terms is presented at the beginning of each chapter. When a key term is first introduced, it is defined and printed in boldface type. Furthermore, the Lab Manual contains crossword puzzle exercises that utilize the key terms found in each chapter. This is intended to provide an interesting and entertaining way of building the student's technical vocabulary. Thus each chapter presents the key terms, defines the key terms, and tests the understanding of the key terms. In effect, each chapter becomes a self-contained technical glossary.

The text follows a consistent format and pedagogy. Each chapter contains performance objectives, key terms, introduction, self-tests, sequentially numbered equations and illustrated examples, *T*ech *T*idbits—T^2 (or clarifying notes and safety points), a related practical section entitled *T*ech *T*ips and *T*roubleshooting—T^3, a summary, and homework exercises arranged by chapter section. Many illustrated examples, charts, diagrams, photographs, and tables are used to present the material.

Chapter 1 is an introduction to the study of electricity. It presents a brief history but concentrates on safety and the use of hand tools. Chapters 2 and 3 introduce the fundamental concepts and units of electricity and the necessary mathematics, including the use of the scientific calculator. Chapters 4-7 are devoted to the study of direct current circuits and components. Chapter 8 discusses basic magnetism and is intended as a transition for the student into the study of alternating current. Chapters 9-12 are devoted to the study of alternating current (sinusoidal) circuits and components. Chapter 13 focuses on power, motors, and generators in both AC and DC applications. Finally, Chapter 14 introduces the student to the use of the computer as a problem-solving tool for electric circuits. It presents in detail the use of modern computer simulation programs.

The text is modular to allow instructors to customize its use to the individual needs of their students. For example, for a more practical, hands-on coverage of electricity, more theoretical and mathematical topics like calculating the resistance of wire, Thevenin's theorem, Norton's theorem, source conversion, superposition theorem, loop analysis, RLC circuits, and resonance can be omitted or presented qualitatively. Instructors conducting more theoretical programs, and particularly articulation programs, will find that the mathematical topics are presented in a clear and easy manner useful for quantitative instruction.

ACKNOWLEDGMENTS

The authors gratefully acknowledge the following reviewers for sharing their knowledge and expertise with us in this project.

William D. Amos, St. Helen's High School, Oregon

John Brotherman, Douglas MacArthur High School, Texas

George Carr, Lancaster High School, Ohio

Mark Cotner, Central Oklahoma A.V.T.S., Sepulpa, Oklahoma

Ron Felty, Ashland State Vocational-Technical School, Kentucky

Ronald Poor, Southwestern Community College, North Carolina

Dave Roth, Ferguson/Florissant P.R.O.B. Center, Missouri

Michael R. Sanderson, DeVry Institute of Technology, Ohio

John Stackhouse, I.T.T., Aurora, Colorado

Richard Thomson, Northwestern Electronics Institute, Minnesota

We are also indebted to Diane Asser for typing and reviewing the original manuscript and for her many helpful suggestions.

Authors and teachers must also learn their craft. Our learning began at Queensborough Community College of the City University of New York. We would like to express our deepest gratitude and respect to our colleagues and teachers at Queensborough Community College, particularly Joseph B. Aidala for teaching, guiding, and inspiring us. Professor Aidala's detailed review of the manuscript and many clarifying suggestions have greatly enhanced and improved the text. We can never thank him enough. Finally, the authors especially thank their families for their patience, encouragement, and love during the writing of this book.

Stuart M. Asser

Vincent J. Stigliano

CHAPTER 1

INTRODUCTION TO ELECTRICITY

Courtesy of DeVry Institutes

KEY TERMS

amber rod
electromagnetic induction
electron
electron theory
heat sink
Leyden Jar
long nose pliers
nutdriver
screwdriver
soldering iron
voltaic cell
wire cutter
wire stripper

OBJECTIVES

After completing this chapter, the student should be able to:

- identify uses of electricity.
- outline the early history of electricity.
- discuss the general rules for safety in the electronics laboratory.
- discuss and identify the basic hand tools used in the electronics laboratory.

1.1 INTRODUCTION

Electricity influences our daily lives in thousands of ways. It is difficult to imagine life today without the presence of electricity. Electricity is used in homes, offices, industry, and schools. In homes, electricity is used to operate appliances such as washing machines, televisions, radios, and refrigerators. Offices use electricity to operate devices such as computers, typewriters, and telecommunications equipment. Industry utilizes electricity to operate many different types of manufacturing machinery including robots. Today's schools use electricity to power audiovisual and other types of learning equipment that train students for our high technology society. Without electricity, the world we know today would not be possible.

Electron theory—describes the behavior of electrons.

Although electricity cannot be seen, the effects of electricity are surely visible. The effects of electricity are caused by tiny particles called **electrons.** When electrons move, things begin to happen. The rules that describe the behavior of electrons are known as the **electron theory.** The electron theory, along with being the basis for the design and operation of electric and electronic equipment, is also used to investigate the very nature of life and the universe.

1.2 EARLY HISTORY

Leyden Jar—first capacitor, a storage device for electric charge.

The age of electricity can be traced back at least 2,500 years. The Greek philosopher, Thales of Miletus (600 B.C.), demonstrated the ability to pick up paper and straw with an **amber rod** (fossilized resin) that was rubbed with a cloth. The same effect can be accomplished today by rubbing a glass rod and attracting the hair on your head.

The Greek word for amber is *elektron*. William Gilbert, an English scientist, applied the English spelling *electrics* to materials he found that behaved like amber. Gilbert's studies were published in a thesis called *De Magnete* in 1600. He used words such as *electric force* and *electric attraction,* which have become modern terms.

Conventional current theory—plus to minus.

Over the next 100 years, there were many observations of electric and magnetic phenomena. In 1745, Pieter van Musschenbroek developed the **Leyden Jar** (the first capacitor) for the storage of electric charge. The modern study of electricity begins in 1746 with Benjamin Franklin's experiments. His studies led to the development of a practical condenser or capacitor for the storage of static electricity. He was the first to identify lightning as electricity with his famous kite experiment in 1752. Franklin later developed his fluid theory of electricity, which defined the direction of electric current. This later became known as *conventional current theory*. Conventional current theory assumes that electricity flows from plus to minus. Electron current theory assumes that electricity flows from minus to plus. These theories are explained in detail in Chapter 2.

Electron current theory—minus to plus.

Voltaic cell—first battery.

In 1784, Charles Coulomb demonstrated the relationship of the forces between electric charges. This became known as Coulomb's law. Professor Luigi Galvani performed experiments in 1791 demonstrating that electricity is present in animals. The first **voltaic cell** or battery that could produce electricity through chemical action was developed by the Italian engineer, Alessandro Volta in 1799. In the early 1800s, men like Hans Christian Oersted of Sweden and Andre Ampere of France studied the relationship between electricity and magnetism. In 1826, Georg Ohm observed the fundamental relationship of electricity known as Ohm's law. In 1831, the English scientist, Michael Faraday, developed the theory of **electromagnetic induction** whereby energy can be transmitted between two coils of wire magnetically without any physical connection. Electromagnetic theory was further developed by James Clark Maxwell of Scotland. Maxwell's equations showed us that electromagnetic waves travel through the air at the speed of light. Maxwell's work led to developments by Heinrich Hertz and Guglielmo Marconi in the transmission of electromagnetic energy. This led to the development of radio, television, and the electronics we know today.

Electro-magnetic Induction—transmitting electrical energy through a magnetic field without any physical connections.

✓ Self-Test 1

1. Thales of Miletus demonstrated the ability to pick up paper and straw with an AMBER ROD that was rubbed with a cloth.
2. Pieter van Musschenbroek developed the first capacitor known as the LEYDEN JAR.
3. Alessandro Volta developed the first battery known as VOLTAIC CELL.

Electricity can kill. Safety is everyone's responsibility.

1.3 ELECTRIC SAFETY

Safety is everyone's responsibility. Whether you are on the job, in a laboratory, or in the classroom, you are responsible for your own safety and for the safety of other people around you. Electricity can kill. It can also cause fires, damage to equipment, and injury. Therefore, electricity must be respected. Safe work with electricity requires a careful, well-thought-out approach to each job or task. The safe operation and handling of test equipment and hand tools is a necessary requirement for each individual working with electricity.

Workers and students who operate, maintain, and experiment with electricity always face the risk of electric shock. The best safeguard is common sense, care, and the following list of proven safety rules. Electric shock can affect the body in many ways. As little as one hundred millionths (0.000100) ampere can cause pain. Less than ten thousandths (0.010) ampere can cause muscular paralysis and can possibly be fatal. Remember that the body is largely water and minerals, which means it is a good conductor of electricity. The only insulation you have is your skin, which can become wet and will then be a conductor of electricity.

Electric shock (see Figure 2-10)
pain = 0.0001 ampere
paralysis = 0.010 ampere
death = 0.1 ampere

Many people think it takes a great amount of voltage to cause a fatal electric shock. We have all heard stories of people being electrocuted by high voltages. Some of these stories may be true, but the fact remains that most electrical shocks come from small amounts of electricity. Death from electric shock happens most often from the standard 120-volt AC, 60-hertz power line. The result is an instantaneous, violent type of paralysis. Smaller amounts of electricity may have less violent effects but can still be harmful.

When an emergency occurs, it is often accompanied by panic. This can cause the mind and muscles to become paralyzed. Safety education teaches us what to do during an emergency and helps to minimize anxiety. First aid courses teach us how to handle an accident. Sometimes a wrong action can be worse than no action at all.

In the case of an electric shock, the first thing to do is remove the victim from the current path. If the power switch is readily accessible, turn it off. Always know where the power switch and the emergency power cut-off switches are located. If you cannot disconnect the power, the person can be detached from the energy source by using an insulator like a piece of dry wood. Disconnecting a person from an energy source can be dangerous. The rescuer can become another path for current and also receive an electric shock.

Since the time of Benjamin Franklin, it has become apparent that electricity can be dangerous. In fact, shortly after Franklin's experiment, a Russian, a French, and a British physicist were killed trying to duplicate the kite experiment. We should therefore conclude that good safety procedures are important to all who use and work with electricity.

T² *tech tidbit*

Public law 91-596, commonly known as the Occupational Safety and Health Act (OSHA), sets standards for health and safety in industry.

The following general rules of laboratory safety apply to anyone who works with tools, power equipment, and electricity.

1. Always plan ahead. Keep your work area clean, dry, and neat.
2. Don't fool around while in the laboratory. Injuries can be caused through thoughtless acts.
3. Always get your instructor's permission before using tools and equipment.
4. Have your instructor approve your work before starting any project or turning on any electricity.
5. Read and familiarize yourself with the manual or instructions pertaining to any piece of equipment or tool you are using.
6. Familiarize yourself with the location and operation of all safety equipment, including emergency power cut off switches, eye wash stations, fire extinguishers, fire alarm boxes, first aid kits, and so forth.

Water and foam can conduct electricity.

7. Never use water- or foam-type fire extinguishers on electrical fires. Carbon dioxide (CO_2)- or halon-type fire extinguishers are preferred. Fire extinguishers are rated by the type of fire they are designed to control. Always use the proper type of fire extinguisher or the fire may spread.
8. Inspect all power tools and equipment before turning them on. Be certain all line cords are intact and have three-prong safety plugs. Or use two-prong plugs on double-insulated equipment to protect against shock hazards.
9. Never defeat any safety devices such as circuit breakers, fuses, and cut-off switches.
10. Power line voltages should be insulated from ground by an isolation transformer to reduce shock hazards.
11. Do not stand in water when making electrical measurements or working with electrical equipment.
12. Measure voltages with one hand behind your back. Connect only one lead at a time.
13. Always wear appropriate safety equipment such as safety glasses and goggles when working with soldering irons, rotary equipment, and hammers.
14. Report all injuries at once to your instructor no matter how small.
15. Use proper methods when handling or lifting objects. Lift with your legs and not with your back.

16. Do not distract others while they are working.
17. Never leave machines or electrical equipment running while unattended.
18. Always wear proper and sensible clothing while working in the laboratory. Particular care must be taken when using rotary machinery. Loose clothing and long hair should be tucked away or firmly tied.
19. Always use the proper tool for the job. Use tools correctly and only if they are in proper working condition.
20. Certain devices, such as TV picture tubes and large value capacitors, can store large amounts of electric energy. Never work with these devices until you have been trained in the proper safety procedures.
21. Never assume when working with electricity. Be certain things are the way you believe them to be. For example, never assume the power switch is turned off.
22. Never work while under the influence of medication, alcohol, drugs, or when you are tired.
23. Always strive to avoid accidents. Take your time and know exactly what you are doing. Don't rush, run, or throw things.
24. If an accident should occur, immediately shut off the power, report the accident to the instructor, and render first aid if you are qualified.

✓ Self Test 2

1. Safety is __________ responsibility.
2. Electric current as low as __________ amperes can be fatal.
3. Never use __________ fire extinguishers on an electric fire.
4. In case of an accident, you should immediately __________, __________, and render first aid if you are qualified.

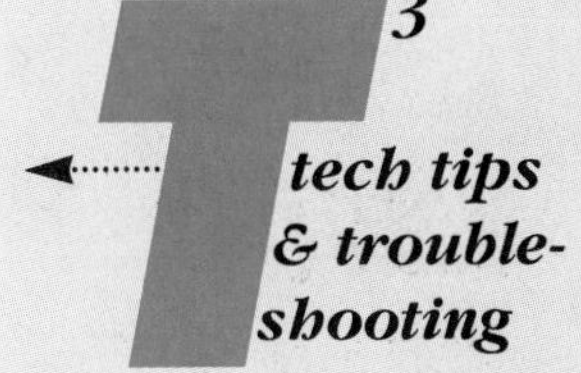

T3 HAND TOOLS

Applying electric theory to practice requires the use of many different types of tools. Every engineer and technician should have a basic knowledge of the use of common hand tools. Figure 1-1 illustrates a typical technician's tool kit. Purchasing a good set of tools can be expensive. However, a good quality set of tools can last a lifetime.

The following discussion summarizes a number of basic hand tools required when working on electric circuits:

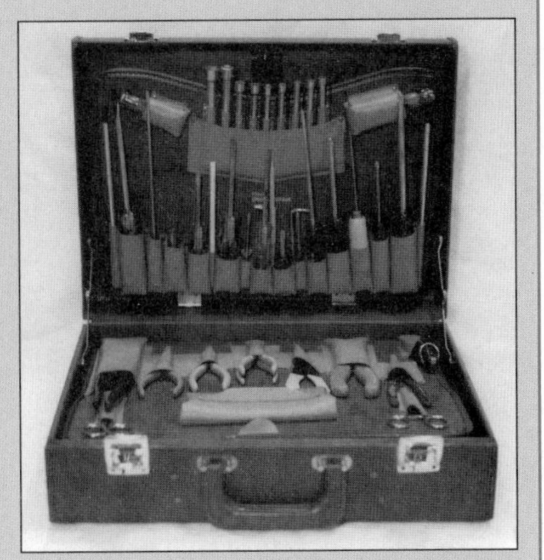

Figure 1-1 Technician's tool kit

Long nose pliers—for holding, picking up, and heat sinking.

Wire cutters are classified by the angle of cut.

T² tech tidbit

When using wire strippers, be careful not to nick or cut the wire.

1. **Long nose pliers** are one of the most useful tools in the workplace. They are used for picking up and holding small components. They can also be used as **heat sinks** when assembling or repairing electric circuits and devices. A heat sink is something that conducts heat away from an electric component or device. Heat sinks are often used for thermal protection during soldering. When using long nose pliers as a heat sink, the handles may be held together with a rubber band. This allows the jaws of the pliers to act as a heat sink while freeing up your hands for soldering or other work.

   When buying long nose pliers, make sure the jaws of the pliers touch firmly together along their serrated edge. The handles should be insulated to protect from electric shock hazards. Often the handles appear to be insulated but are actually designed for holding convenience. Some long nose pliers have a wire stripper and wire cutter feature built into the jaws. Long nose pliers are also called *needle nose pliers*. Figure 1-2 is a photo of three sizes of long nose pliers.

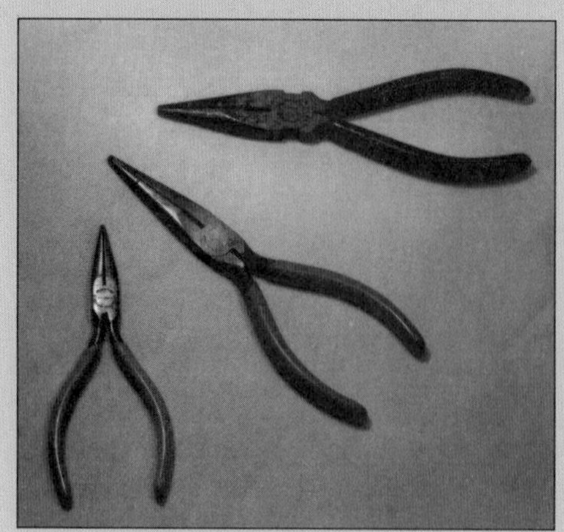

Figure 1-2 Long nose pliers

2. **Wire cutters** are used to cut wire and trim leads of circuit components. There are many different types and sizes of wire cutters. The most common type is the diagonal cutter. Wire cutters can be classified by the angle of cut. For example, 45°, 90°, and so forth.

   When buying wire cutters, check for sturdiness of the tool, alignment of the jaws, and insulated handles. When the jaws are closed, no light should pass through the cutting edges. Figure 1-3 is a photo of three sizes of wire cutters.

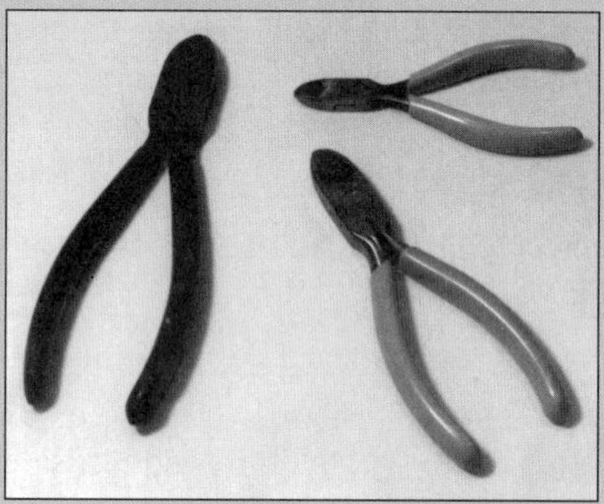

Figure 1-3 Wire cutters

3. **Wire strippers** are used to remove insulation from electric wire so that connections can be made. Wire strippers come in two basic types: manual and automatic. The manual wire stripper works with a scissorlike action, and the automatic wire stripper uses a crimping action. Most wire strippers have settings for different sizes of wire.

   When buying wire strippers, check for sturdy construction and a strong, sharp stripping edge. Make sure the handles fit your hand comfortably. When stripping insulation from a wire, care should be taken to assure that the wire is not nicked or cut. This can weaken the strength of the wire. Figure 1-4 shows a manual and an automatic wire stripper.

Figure 1-4 Wire strippers

4. **Screwdrivers** are used for inserting and removing screws. They are also used to make adjustments and can be used when dismounting desoldered components. Screwdrivers are generally classified by the type of head, size of the head, and length of the shank (shaft). The three most common head styles are the slotted-, phillips-, and torx-type screwdrivers. A typical tool kit should contain several different sizes of each type, because the correct size screwdriver head must be used or damage to the screw head may occur.

   When buying screwdrivers, check for a sturdy construction and a comfortable length handle. The insulated handle should be large enough for your hand. Screwdrivers are also available with interchangeable heads. Figure 1-5 illustrates a number of different screwdrivers.

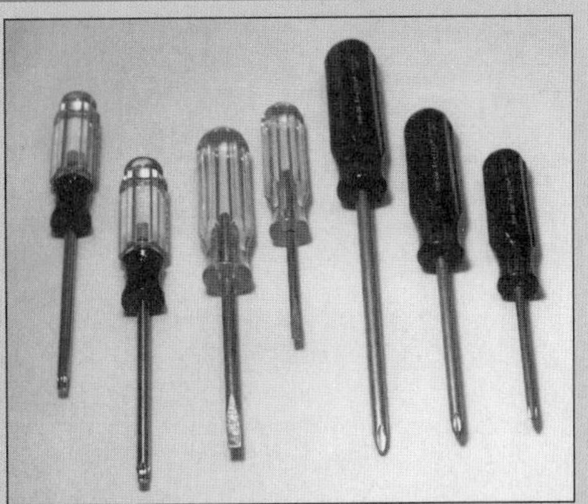

Figure 1-5 Screwdrivers

5. **Nutdrivers** are used to remove hex nuts from and install hex nuts onto screws and bolts. They are often more convenient than using wrenches or pliers. Nutdrivers come in different sizes to fit different size hex nuts. They are also available with interchangeable heads.

   When buying nutdrivers, check for sturdy construction and a comfortable length handle. Figure 1-6 shows a number of different size nutdrivers.

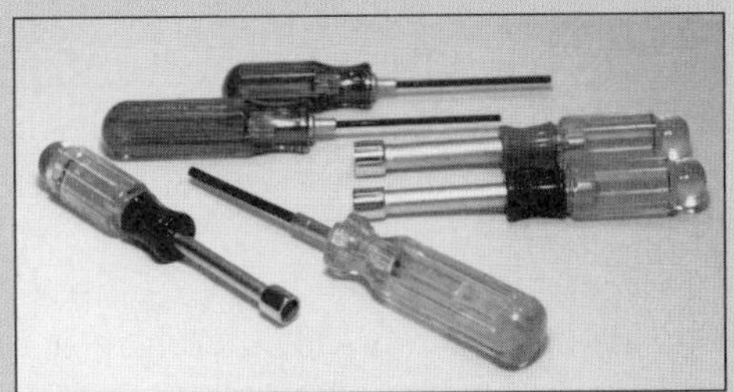

Figure 1-6 Nutdrivers

6. **Soldering irons** are used to apply solder to electrical connectors. Electricity is used to heat an element in the soldering iron's tip. The heat produced melts the solder and causes the connection to be bonded. The tip of the soldering iron should be at least 100°F above the melting

Screwdrivers are classified by the type of head, size of the head, and length of the shank.

Nutdrivers are also available with ratchet handles.

Twenty-five to 40 watt soldering irons are a good size to use on most electronic circuit work.

tech tidbit

When soldering delicate components, care must be taken not to overheat the component.

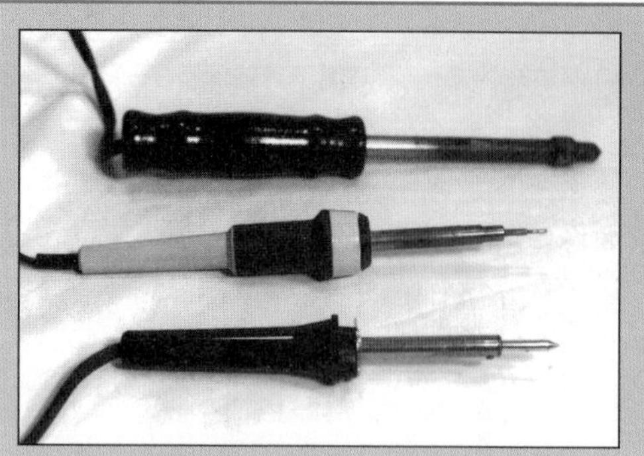

Figure 1-7 **Soldering irons**

temperature of the solder. For most printed circuit board work, a tip temperature of 650°F is desirable.

Soldering irons are available in different sizes. The tip of the iron comes in different shapes to accommodate different soldering jobs. Figure 1-7 shows three typical soldering irons.

✓ Self-Test 3

1. Name five basic hand tools that are used when working on electric circuits. LONG NOSE PLIERS, WIRE CUTTERS, WIRE STRIPPERS, SCREW DRIVER, SOLDERING IRONS
2. The most common type of wire cutter is called the DIAGONAL CUTTER.
3. Name the three most common styles of screwdrivers. SLOTTED, PHILIPS, TORX
4. When soldering delicate components, care must be taken not to OVERHEAT COMPONENT

SUMMARY

Electricity influences our lives in many ways. It is used in homes, offices, industry, and schools. The effects of electricity are caused by tiny particles called electrons. The rules that describe the behavior of electricity are known as the electron theory. The age of electricity can be traced back to 600 B.C. when it was demonstrated that paper and straw could be picked up with an amber rod that was rubbed with a cloth.

Good safety procedures are important to all who use and work with electricity. Electricity can kill, therefore it must be respected and good safety habits must be practiced. As little as ten thousandths (0.010) ampere can cause muscular paralysis and can be fatal. In case of electric shock, the first thing to do is to remove the victim from the current path. Disconnecting a person from an energy source can be dangerous. The rescuer can become another path for current and also receive an electric shock. Always follow the general rules for laboratory safety.

Applying electric theory to practice requires the use of many different types of tools. Every engineer and technician should have a basic knowledge of the proper use of common hand tools. This chapter discussed the standard tool kit including long nose pliers, wire cutters, wire strippers, screwdrivers, nutdrivers, and soldering irons.

EXERCISES

Section 1.3

Select the best correct answer for each question.

1. Which one of the following safety statements is *not* true?
   (A) You must have your instructor's approval before starting to work.
   (B) A small cut need not be reported.
   (C) If tools are not in proper working condition, do not use them.
   (D) After turning off a machine, stay by it until it is completely stopped.
2. Which one of the following safety statements is true?
   (A) Safety is everyone's responsibility.
   (B) Electricity is not dangerous.
   (C) There is no risk of electric shock when operating electric equipment on a wooden workbench.
   (D) The human body will not conduct electricity.
3. Before working in any laboratory or shop you should always
   (A) do all your homework.
   (B) familiarize yourself with all safety equipment.
   (C) get all the tools you need for the job.
   (D) turn on all rotating equipment.
4. When soldering and drilling, you should always
   (A) hold your work with pliers.
   (B) have a screwdriver handy.
   (C) talk to your friend about the job.
   (D) wear safety glasses or goggles.
5. Which one of the following safety statements is *not* true?
   (A) Power line voltages should be isolated from ground.
   (B) Do not stand in water when making electrical measurements.
   (C) Always use two hands when measuring voltage.
   (D) Report all injuries to your instructor.
6. Which of the following safety statements is *not* true?
   (A) When lifting heavy objects, use your back.
   (B) Do not disturb others when they are working.
   (C) Never leave machinery running unattended.
   (D) Always wear suitable clothing when working in the lab.

7. The technician should
   (A) use the proper tool for the job.
   (B) work when tired.
   (C) assume that the electricity is off.
   (D) override a safety device.
8. The technician should always
   (A) wear rings and loose clothing in the laboratory.
   (B) rush to finish a job on time.
   (C) read and be familiar with equipment operating manuals.
   (D) work in poorly lighted areas.
9. Accidents can be caused by
   (A) following instructions.
   (B) paying attention.
   (C) running in the lab.
   (D) reading safety rules.
10. When electricity travels through the human body
    (A) it does not hurt.
    (B) 1 ampere must be present to cause pain.
    (C) 10 amperes are necessary to cause muscular paralysis.
    (D) as little as 0.01 ampere can be fatal.
11. See the Lab Manual for the crossword puzzle for Chapter 1.

CHAPTER 2

BASIC UNITS OF ELECTRICITY

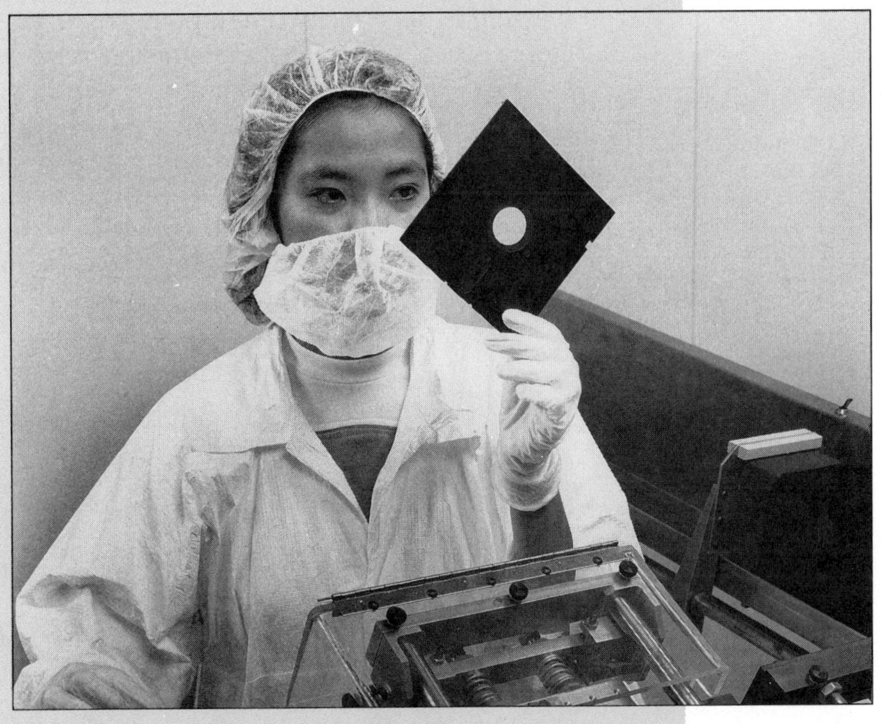

Courtesy of Eastman Kodak Company

KEY TERMS

ampere (A)
atoms
charge
conductance
conductor
coulomb (C)
current
efficiency
electricity
electromotive force (EMF)
electron
electrostatic
energy
flux
flux density
horsepower (HP)
insulators
ion
joule (J)
kilowatt-hour (KWh)
magnets
matter
neutron
newton (N)
nucleus
ohm (Ω)
permeability
poles
power
proton
resistance
retentivity
siemens (S)
static electricity
volt (V)
voltage
watt (W)
work

OBJECTIVES

After completing this chapter, the student should be able to:

- define energy.
- state the units of energy.
- describe the composition of matter.
- define electric charge.
- define the coulomb.
- define voltage, current, and resistance.
- calculate voltage, current, and resistance.
- define power and efficiency.
- calculate power and efficiency.
- calculate the cost of electricity.
- define materials and properties of magnets and magnetism.

2.1 INTRODUCTION

Energy is defined as the ability to do work. Some kind of external source of energy is required to cause water to flow into a pipe. For example, a pump can supply the pressure needed to push water through a pipe. The pump's pressure is often measured in pounds per square inch. It follows that the higher the pressure, the greater the amount of water that will flow through a certain size pipe. Similarly, some kind of external source of electric energy is required to cause electrons to flow in a wire or electric circuit. The unit of electric pressure is the **volt (V)**, which is sometimes referred to as **electromotive force (EMF)**. For example, a battery with a rating of 6 volts will force electricity through a wire to light a bulb (Figure 2-1). If the battery voltage is increased to 12 volts, more electricity will flow in the wire and the light will become brighter.

T² tech tidbit

Volt—unit of electric pressure.

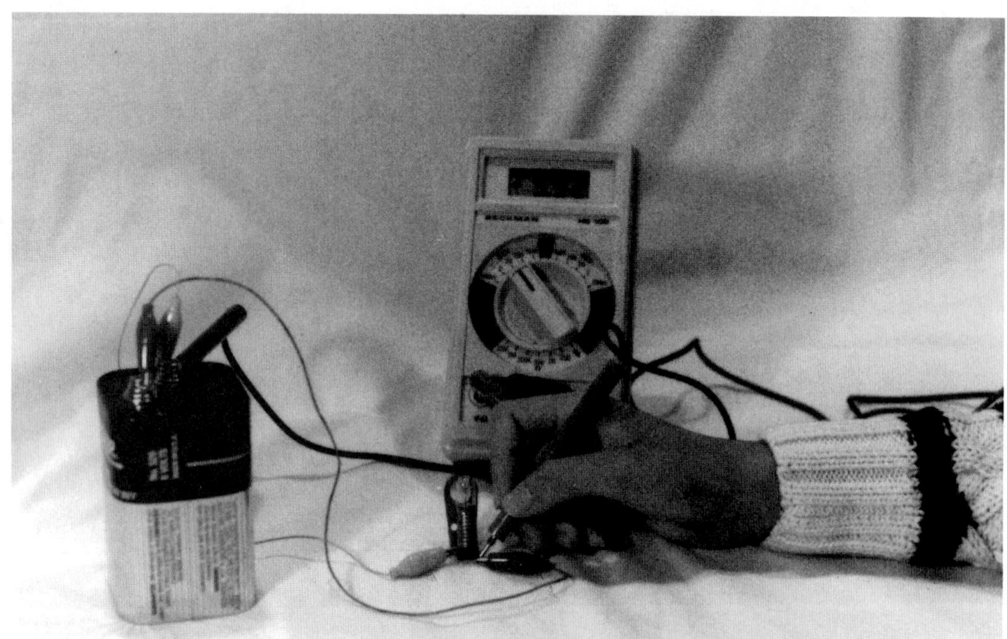

Figure 2–1 **Measuring voltage**

As with any type of energy, electric energy cannot be created or destroyed, only changed in form. For example, the energy from a waterfall in Niagara Falls can be used to turn a generator and make electricity. The electricity can be sent through copper wires to New York City where it can be used to operate appliances and lights, provide heat or air conditioning, and perform hundreds of other functions for our society. Thus, the energy from the waterfall was converted to electric energy and then to other forms of mechanical, light, and heat energy, but never destroyed.

2.2 UNITS OF ENERGY

The basic unit of energy or work is the **joule (J)**. The joule represents a very small quantity. For example, it would take hundreds of thousands of joules to light the average light bulb. The amount of energy used by a typical light bulb in one day will total millions of joules of energy.

Joule—unit of energy or work.

In a mechanical system, **work** is equal to force times distance. Mathematically this can be expressed as:

$$W = F \times D$$

Equation 2-1

where F = mechanical force measured in newtons (N) (approximately 0.225 pounds)

D = distance measured in meters (approximately 39.4 inches)

W = work measured in newton-meters or joules

$W = F \times D$

$F = \frac{W}{D}$

$D = \frac{W}{F}$

EXAMPLE 2-1

A force of 100 newtons is required to lift a computer (load) 3 meters. Calculate the work performed.

Solution:

$$W = F \times D$$
$$= 100 \text{ newtons} \times 3 \text{ meters}$$
$$= 300 \text{ newton-meters}$$
$$= 300 \text{ joules}$$

Matter is anything that occupies space and has mass.

2.3 COMPOSITION OF MATTER

In the study of electricity, it is necessary to understand why electric energy exists. All material is composed of **matter**. Thus matter can be considered as anything that occupies space and has mass. Looking deeper into the structure of matter, all matter can be thought of as being composed of millions of tiny particles called **atoms**. Atoms are the basic building blocks of all matter in our universe. They are the smallest recognizable part of an element. Within these atoms are the forces that cause electricity to exist. The atom can be broken down into its subatomic parts, which are called electrons, protons, and neutrons, as shown in Figure 2-2.

Atoms are composed of electrons, protons, and neutrons.

Actually atoms have a complex structure that is somewhat analogous to our solar system. In Figure 2-2A, we see the schematic diagram of the simplest atom, hydrogen. The center part of the atom is called the **nucleus**. The nucleus of the hydrogen atom contains one particle called a **proton**. The proton is defined as having a positive (+) electric charge. Revolving around the nucleus, just as the planets revolve around the sun, is a tiny particle called an **electron**. The electron is defined as having a negative (–) electric charge.

Nucleus—center of an atom.

Proton—positive electric charge.

In Figure 2-2B, we see the schematic diagram of a helium atom. Notice that the nucleus of the helium atom has another subatomic particle called the **neutron**. A neutron is defined as having no electrical charge. Neutrons do,

Electron—negative electric charge.

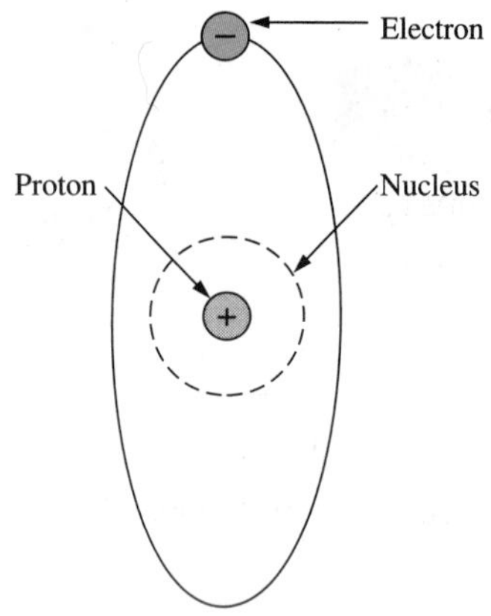

(A) Schematic diagram of a hydrogen atom

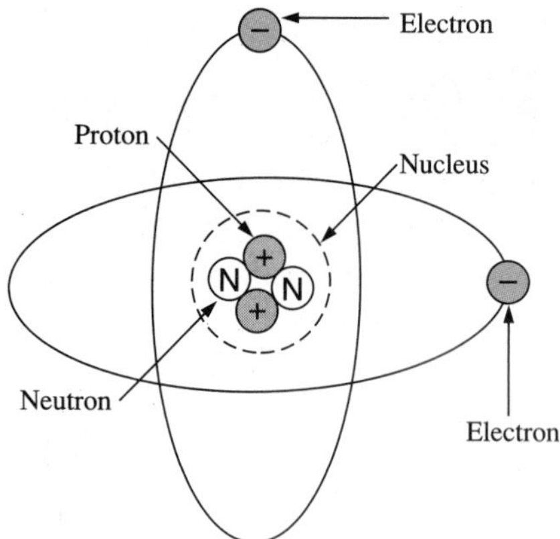

(B) Schematic diagram of a helium atom

Figure 2–2 The atom

Neutron—no electric charge.

however, add mass to an atom. Each of the elements in nature is made up of a unique combination of protons, neutrons, and electrons. It should be noted that the electrons orbit the atom in three-dimensional space. Each electron in an atom has its own orbit, which is determined by the complex interaction between all of the particles. Most of the structure of the atom is empty space. That is, if the nucleus of an atom was considered to be the size of a dime, the closest orbiting electron would be over a quarter of a mile away. When the number of electrons equals the number of protons, the atom is said to be neutral. It is neutral because the magnitude of the positive charge of a proton is equal and opposite in polarity to the negative charge of an electron. If an atom loses or gains one or more electrons, it is called an **ion**. An atom that has lost an electron is called a **positive ion**. An atom that has gained an electron is called a **negative ion**. Table 2-1 summarizes the characteristics of subatomic particles.

Ion—a charged atom (+ or –).

When an electron in the outer orbit becomes dislodged from an atom it is known as a **free electron**. Free electrons can be created by light, heat, chemical reaction, and so forth. These free electrons are responsible for the phenomenon known as **electricity**. Electricity is then the flow of these free electrons. Generally the fewer the number of free electrons in the outer shell of an atom, the better the atom will conduct electricity. The greater the number of electrons in the outer shell of an atom, the harder the electrons are to free. This makes for a poor conductor. Thus, materials with few free electrons are called

Electricity is the flow of free electrons.

Table 2-1 Characterisitics of Subatomic Particles

Particle	Charge	Mass (grams)	Relative Weight
proton	+1	1.672×10^{-24}	1
neutron	0	1.672×10^{-24}	1
electron	–1	9.11×10^{-28}	1/1836

Note: 10^{-24} = 0.000000000000000000000001. Working with very small and very large numbers will be discussed in Chapter 3.

insulators. Materials with many free electrons are called **conductors**. In order for electricity to flow through a conductor, we need an external source of energy to attract and repel the free electrons.

tech tidbit

Insulator—few free electrons.

Conductor—many free electrons.

✓ Self-Test 1

1. The basic unit of energy is the Joule.
2. The basic building block of all matter is the Atoms.
3. Identify the subatomic particles that are found in the nucleus of an atom. PROTONS + NEUTRONS
4. The electron has a NEGITIVE electrical charge.
5. An atom that has lost an electron is called a POSITIVE ION.
6. Materials with few free electrons are called INSULATORS.
7. The relative weight of an electron to a proton is ______.

2.4 ELECTRICAL CHARGES

When one substance has an excess of electrons and another has a deficiency of electrons an electric **charge** exists between these two materials. Insulators such as glass and paper have the ability to separate these charges and are called **dielectrics**. This stored electrical charge is said to be static or not in motion and is known as **static electricity**. Static electricity can be made to flow when in the presence of a conductor. For example, if you take a glass rod and rub it vigorously with a silk cloth, the glass rod will lose electrons and become positively charged. The silk cloth will gain electrons and become negatively charged. If the glass rod is placed near a metal door knob, the charge or static electricity will arc or flow between the two materials. Another example of static electricity is lightning. Lightning is caused by the movement of charge between the clouds and the earth's surface.

Dielectrics are insulators.

Consider another interesting fact about static electricity. If you can see a spark leave your body (when you have been charged with static electricity), your body was charged to more than 10,000 volts. If you hear a snap, but cannot see the discharge, your body was charged to more than 5,000 volts. So, with all this voltage, why doesn't a person become seriously injured? Because there is no appreciable current. Recall from Chapter 1 that it is the current that causes harm.

To arc a 1-inch air gap, 20,000 volts in dry air are required.

It should be noted that the effects **electrostatic** charges have on each other are very important to the study of electricity. Coulomb's law states that objects with like charges will repel or move away from each other and objects with unlike charges will attract or move together toward each other, as illustrated in Figure 2-3.

Like charges repel, unlike charges attract.

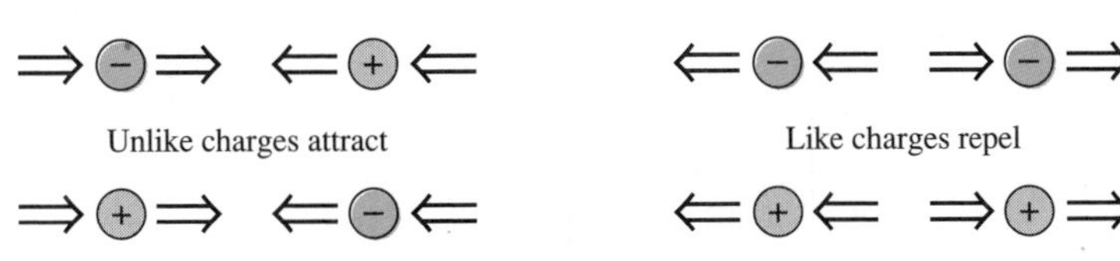

Figure 2-3 **Interaction of charges**

The basic unit of electric charge is the **coulomb (C)**. It was named after Charles Augustin de Coulomb. One coulomb of charge is defined as equal to the total charge of 6.25×10^{18} electrons (6.25 billion billion). This unit was developed because the size and charge of a single electron is very small. Therefore, the coulomb is a more convenient unit to use, just like miles are more convenient than inches when we deal with very long distances.

2.5 DEFINING POTENTIAL DIFFERENCE

If you lift a lead ball from the ground to the top of a building, you will perform an amount of work that is equal to the weight of the ball times the height of the building. This work is stored in the position of the ball and in the form of potential energy. Similarly, when you move an electric charge from one position to another, you will also perform work that is stored as potential energy. In both cases, the potential energy will be returned in the form of kinetic energy (energy of motion) when the ball is allowed to drop back to the ground or the electric charge is allowed to return to its original point or origin, as shown in Figure 2-4.

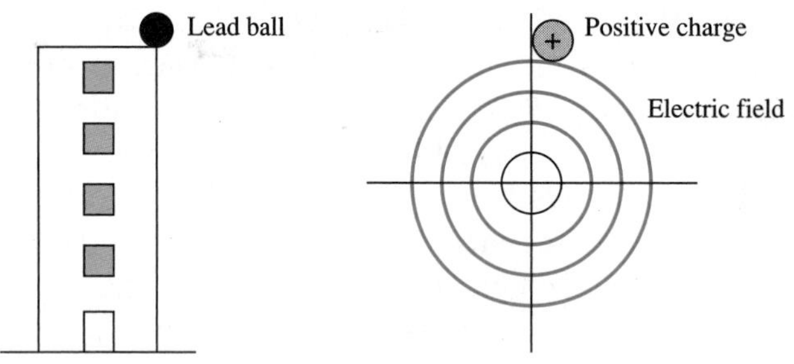

Figure 2–4 **Potential energy**

Often, however, we are interested in the work required to move an object from one point to another (W_{AB}.) For example, in Figure 2-5, the lead ball on the top of building A has a certain amount of potential energy with respect to the ground. If the lead ball on top of building A were moved to the top of building B, it will now have a greater amount of potential energy with respect to the ground due to its increased height. However, the energy required to

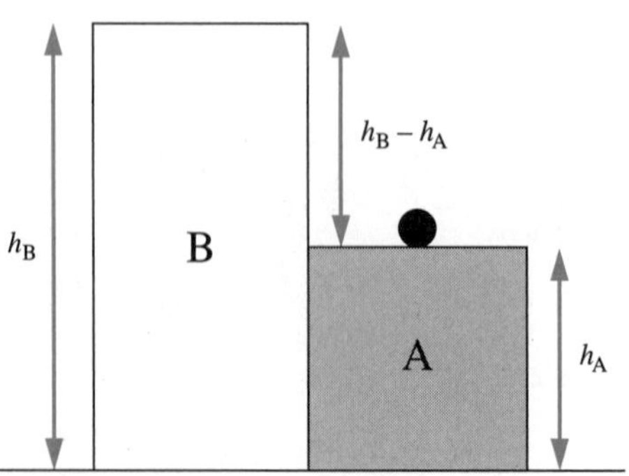

Figure 2–5 **Potential energy**

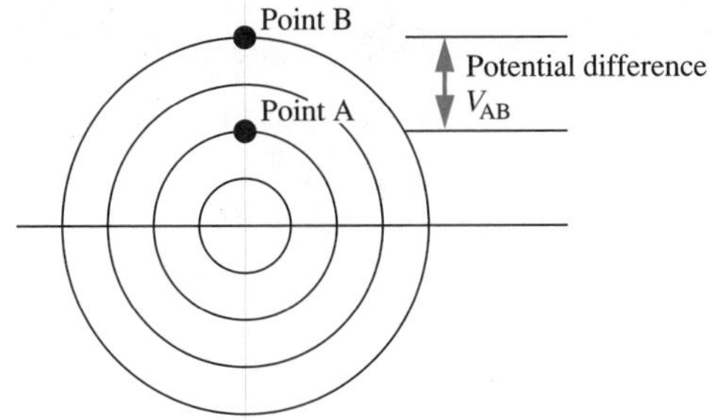

Figure 2–6 **Potential difference between two points**

move the lead ball from the top of building A to the top of building B is equal to the weight of the ball times the difference in the total heights of the two buildings. Therefore, the potential energy, *PE,* is

$$PE_{AB} = weight \times (height_B - height_A)$$

Similarly, the amount of energy required to move an electric charge from one point to another is equal to the work required to move the charge from point A to point B, as shown in Figure 2-6. This is called the potential difference or **voltage** between the two points. It was named after the Italian physicist Count Alessandro Volta.

Voltage—the potential difference between two points.

By definition, a potential difference of 1 volt is said to exist between point A and point B, when 1 joule of work is required to move 1 coulomb of charge from point A to point B. This is expressed mathematically as,

Equation 2-2

$$V_{AB} = \frac{W_{AB}}{Q}$$

where V_{AB} = the potential difference in volts

W_{AB} = the work required to move from point A to point B in joules

Q = the amount of charge moved in coulombs

$$V_{AB} = \frac{W_{AB}}{Q}$$

$$Q = \frac{W_{AB}}{V_{AB}}$$

$$W_{AB} = Q \bullet V_{AB}$$

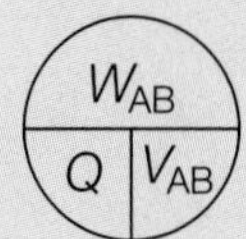

EXAMPLE 2-2

What is the potential difference between points A and B if 42 joules of energy are required to move 7 coulombs of charge from point A to point B?

Solution:

$$V_{AB} = \frac{W_{AB}}{Q}$$

$$= \frac{42 \text{ joules}}{7 \text{ coulombs}}$$

$$= 6 \text{ volts}$$

EXAMPLE 2-3

How much energy does a 9-volt battery use to move 30 coulombs of charge through a circuit?

Solution:

$$V_{AB} = \frac{W_{AB}}{Q}$$

$$W_{AB} = (V_{AB}) \cdot (Q)$$

$$W_{AB} = (9 \text{ volts}) \cdot (30 \text{ coulombs})$$

$$W_{AB} = 270 \text{ volt-coulombs}$$

$$W_{AB} = 270 \text{ joules}$$

✓ Self-Test 2

1. What causes lightning? MOVEMENT of CHARGE BETWEEN Clouds & EARTH
2. Objects with unlike charges will ATTRACT each other.
3. The basic unit of electric charge is the COULOMB.
4. The unit of potential difference is the VOLTAGE.
5. What is the potential difference required to move 5 coulombs of electric charge if 10 joules of energy are used? 2 VOLTS

T² tech tidbit

Electric current is the rate of flow of electrons or electric charge.

2.6 CHARGE IN MOTION—CURRENT

When a potential difference exists across a conductor, the free electrons in the conductor are caused to move or flow. This movement of free electrons or electric charge is known as electric **current**, as shown in Figure 2-7.

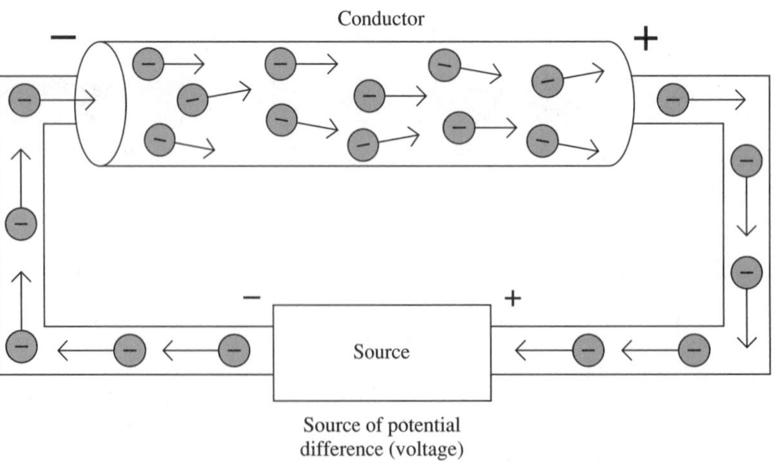

Figure 2–7 ***Electric current***

Recall that the electrostatic charge of an electron is extremely small. Consequently, a larger and more practical unit for charge, the coulomb, is used. As previously explained, 1 coulomb of charge is equal to the charge carried by 6.25×10^{18} electrons (more than 6 billion billion electrons).

tech tidbit

Ampere—unit of electric current.

Current is then the rate of the flow of electric charge in a circuit. The unit of electric current is the **ampere (A)**, which was named after the French scientist Andre M. Ampere. When 1 coulomb of charge flows past a point in 1 second of time, this is defined as 1 ampere of electric current. Mathematically this is expressed as,

$$I = \frac{Q}{t}$$

Equation 2-3

where I = the current in amperes

Q = the charge in coulombs

t = the time in seconds

tech tidbit

$$I = \frac{Q}{t}$$

$$t = \frac{Q}{I}$$

$$Q = I \cdot t$$

EXAMPLE 2-4

Determine the current in a circuit if 600 coulombs of charge passes a point in 30 seconds.

Solution:

$$I = \frac{Q}{t}$$

$$= \frac{600 \text{ coulombs}}{30 \text{ seconds}}$$

$$= 20 \text{ amperes}$$

EXAMPLE 2-5

A current of 4 amperes passes through a wire for a period of 2 hours. What is the total charge transferred?

Solution:

$$I = \frac{Q}{t}$$

$$Q = I \cdot t$$

$$Q = (4 \text{ amperes})(2 \text{ ~~hours~~} \times 60 \text{ ~~minutes/hour~~} \times 60 \text{ seconds/~~minute~~})$$

$$= (4 \text{ amperes})(7{,}200 \text{ seconds})$$

$$= 28{,}800 \text{ coulombs}$$

Note: t must be in seconds. Therefore, in the preceding example, we had to convert 2 hours to 7,200 seconds.

2.7 RESISTANCE—THE OPPOSITION

tech tidbit

Resistance is the opposition to current.

Resistance is the opposition to the flow of electric charge or current. It results in the transformation of electric energy into heat. All materials in nature have resistance. Conversely, the ease in which charge is allowed to move through a material is known as **conductance**. The symbol for resistance is R and the symbol for conductance is G.

tech tidbit

The resistor symbol R —\/\/\/—

The unit of resistance is the ohm (Ω). The symbol used is the Greek letter *omega*. It was named in honor of George Simon Ohm. The unit of conductance is **siemens (S),** or **mho (℧),** which is ohm spelled backward. Conductance was renamed in honor of Ernst von Siemens.

One ohm is defined as the resistance of a conductor when 1 ampere of current passes through the conductor with a potential difference of 1 volt across the conductor. It is sometimes more convenient to use conductance instead of resistance. Resistance and conductance are reciprocals. Mathematically,

Equation 2-4 ▶

$$G = \frac{1}{R} \quad \text{or} \quad R = \frac{1}{G}$$

where R = the resistance in ohms

G = the conductance in siemens or mhos

Unit of resistance is ohm (Ω).

✓ Self-Test 3

1. The rate of flow of electric charge is called CURRENT.
2. When 1 coulomb of charge flows past a point in 1 second, this is defined as 1 AMP.
3. Determine the current if 1,000 coulombs of charge passes a point in 25 seconds. 40 AMPS
4. The opposition to current is RESISTANCE.
5. The unit of conductance is mho (℧).

tech tidbit

Unit of conductance is siemens (S) or mho (℧).

2.8 POWER AND EFFICIENCY

tech tidbit

Power is the rate of doing work.

Power, whether electrical or mechanical, is defined as the rate of doing work. Mathematically, power is expressed as work per unit time or

Equation 2-5 ▶

$$P = \frac{W}{t}$$

where P = the power in joules per second or watts

W = the work or energy in joules

t = time in seconds

$$P = \frac{W}{t}$$

$$t = \frac{W}{P}$$

$$W = P \bullet t$$

W / P | t

The unit of power, the **watt (W)**, is named after the Scottish inventor of the steam engine, James Watt. The watt is a more practical unit for electric power. Engineers seldom use joules per second in real-world applications.

EXAMPLE 2-6

What is the power rating of a light bulb that converts 600 joules of energy in 10 seconds?

Solution:

$$P = \frac{W}{t}$$

$$= \frac{600 \text{ joules}}{10 \text{ seconds}}$$

$$= 60 \text{ joules/second}$$

$$= 60 \text{ watts}$$

Unit of power is the watt (W). Household appliances are usually rated in watts.

EXAMPLE 2-7

How much energy does it take to solder a connection using a 25 watt soldering iron for 5 seconds?

Solution:

$$P = \frac{W}{t}$$

$$W = P \bullet t$$

$$= (25 \text{ watts})(5 \text{ seconds})$$

$$= 125 \text{ watt-seconds}$$

$$= 125 \text{ joules}$$

Large amounts of power are usually expressed in kilowatts or thousands of watts. One kilowatt equals 1×10^3 watts or 1,000 watts. When you pay your electric bill, you are charged for the work performed or energy used and not the power. Thus as already stated,

One kilowatt equals 1,000 watts.

$$W = P \bullet t$$

$$\text{joules} = \text{watts} \bullet \text{seconds}$$

The watt-second or joule is a very small unit. Electric power companies charge their customers in the more convenient unit of kilowatt-hours. Mathematically,

$$\text{Electric energy} = \text{power} \times \text{time}$$

or

$$\text{Electric energy} = \text{kilowatts} \times \text{hours (KWh)}$$

or

$$\text{Electric energy} = \frac{\text{watts} \times \text{hours}}{1{,}000}$$

◀ Equation 2-6

EXAMPLE 2-8

How much energy is consumed by a 100-watt light bulb in one day? In one year?

Solution:

$$\begin{aligned}\text{Electric energy} &= \frac{\text{power} \times \text{time}}{1{,}000}\\ &= \frac{(100 \text{ watts})(24 \text{ hours/day})}{1{,}000}\\ &= \frac{2{,}400 \text{ watt-hours/day}}{1{,}000}\\ &= 2.4 \text{ kilowatt-hours/day}\\ &= 2.4 \text{ KWh/day}\end{aligned}$$

$$\begin{aligned}\text{Electric energy} &= \frac{\text{power} \times \text{time}}{1{,}000}\\ &= \frac{(100 \text{ watts})(24 \text{ hours/}\cancel{\text{day}})(365 \text{ }\cancel{\text{days}}\text{/year}}{1{,}000}\\ &= \frac{876{,}000 \text{ watt-hours/year}}{1{,}000}\\ &= 876 \text{ kilowatt-hours/year}\\ &= 876 \text{ KWh/year}\end{aligned}$$

To calculate your electric bill, you must multiply the total number of kilowatt-hours of energy consumed by the cost of electricity per kilowatt-hour. Mathematically,

Equation 2-7 ▶

$$\text{Cost of electricity} = \text{KWh} \times \text{rate}$$

EXAMPLE 2-9

Find the cost of operating a 300-watt television set for 4 hours at 15 cents per kilowatt-hour.

Solution:

$$\begin{aligned}\text{Electric energy} &= \frac{\text{power} \times \text{time}}{1{,}000}\\ &= \frac{(300 \text{ watts})(4 \text{ hours})}{1{,}000}\\ &= \frac{1{,}200 \text{ watt-hours}}{1{,}000}\\ &= 1.2 \text{ KWh}\end{aligned}$$

$$
\begin{aligned}
\text{Cost of electricity} &= \text{KWh} \times \text{rate} \\
&= (1.2\ \cancel{\text{KWh}}) \times (15\ \text{cents}/\cancel{\text{KWh}}) \\
&= 18\ \text{cents} \\
&= \$0.18
\end{aligned}
$$

The New York City power company, Consolidated Edison, presently charges about 14 cents per kilowatt-hour of electrical energy. This rate is one of the highest in the United States. It is interesting to note that at any instant in time the power delivered by Consolidated Edison ranges from 7,000 to 9,000 megawatts (7,000,000 to 9,000,000 kilowatts).

Table 2-2 summarizes the power consumption of a number of electric devices.

Table 2-2 Appliance Power Consumption

Device	Approximate Power Consumption (watts)
coffee maker	850
television set	300
toaster	1,200
washing machine	500
electric heater	1,500
air conditioner (10,000 BTU)	1,200
heating pad	50
electric blanket	150
stereo/radio	250
home computer	300
audiocassette recorder	15
frying pan	1,300
electric shaver	10
hair dryer	1,500
clothes dryer	2,400

The term **horsepower (HP)** is frequently used to describe the power of electric motors. One horsepower is equal to 746 electrical watts. Therefore, to convert horsepower to electrical watts, we must multiply the horsepower rating by 746.

One horsepower equals 746 watts.

EXAMPLE 2-10

Find the cost of operating a 2-horsepower pump motor for 6 hours at 15 cents per kilowatt-hour.

Solution:

$$\begin{aligned}\text{Electric power} &= \text{horsepower} \times 746 \text{ watts/horsepower}\\ &= (2\ \cancel{\text{horsepower}}) \times (746 \text{ watts}/\cancel{\text{horsepower}})\\ &= 1{,}492 \text{ watts}\end{aligned}$$

$$\begin{aligned}\text{Electric energy} &= \frac{\text{power} \times \text{time}}{1{,}000}\\ &= \frac{(1{,}492 \text{ watts})(6 \text{ hours})}{1{,}000}\\ &= \frac{8{,}952 \text{ watts-hours}}{1{,}000}\\ &= 8.952 \text{ KWh}\end{aligned}$$

$$\begin{aligned}\text{Cost of energy} &= \text{KWh} \times \text{rate}\\ &= (8.952\ \cancel{\text{KWh}})(15 \text{ cents}/\cancel{\text{KWh}})\\ &= 134.28 \text{ cents}\\ &= \$1.34\end{aligned}$$

As previously stated, energy cannot be created or destroyed but only changed in form. However, no conversion process is 100% efficient. For example, a 100-watt incandescent light bulb will produce about 80 watts of heat energy and 20 watts of light energy. Therefore, we can see that an incandescent light bulb is a fairly efficient heat source and an inefficient light source. **Efficiency** is usually expressed as the power out of a device divided by the power supplied to the device. Mathematically,

Equation 2-8 ▶

$$\%\ \text{efficiency} = \frac{P_{out}}{P_{in}} \times 100\%$$

where the power out and the power in *must* be expressed in the same units of power.

EXAMPLE 2-11

Find the efficiency of a 100-watt light bulb that produces 20 watts of light energy.

Solution:

$$\% \text{ efficiency} = \frac{P_{out}}{P_{in}} \times 100\%$$

$$= \frac{20 \text{ watts}}{100 \text{ watts}} \times 100\%$$

$$= 0.2 \times 100\%$$

$$= 20\%$$

EXAMPLE 2-12

Find the efficiency of a 1-horsepower motor that is supplied with 1,000 watts of power.

Solution:

$$\% \text{ efficiency} = \frac{P_{out}}{P_{in}} \times 100\%$$

$$= \frac{(1 \text{ horsepower})(746 \text{ watts/horsepower})}{1{,}000 \text{ watts}} \times 100\%$$

$$= \frac{746}{1{,}000} \times 100\%$$

$$= 0.746 \times 100\%$$

$$= 74.6\%$$

2.9 MAGNETISM

Magnets are materials that are capable of attracting iron and other ferromagnetic substances. A **natural magnet** is a mineral-like lodestone (a black ore mineral) that exhibits the properties of magnetic attraction. Materials like iron and steel can be artificially magnetized by using the natural magnet or an electric field. In 1820, Hans Christian Oersted, a Danish scientist, discovered the relationship between magnetism and electricity. Oersted observed through experiments with a compass and a current-carrying wire that an electric current is always surrounded by a magnetic field. The strength of the field varies *directly* with the amount of current in the wire and *inversely* with the distance from the wire, as shown in Figure 2-8.

Natural magnets come from lodestone.

Magnets made of hard steel tend to retain their magnetic qualities for a long time. They are said to have a high **retentivity**. Magnets made of soft iron can only retain their magnetic qualities for a short time. Soft iron magnets are said to have a low retentivity. The ease in which a material can be magnetized is referred to as its **permeability**. The ends of a magnet are known as the **poles**. Magnetic forces at the poles are the strongest because the magnetism is concentrated at the poles, as shown in Figure 2-9. Magnetism is measured in

Poles—the ends of a magnet.

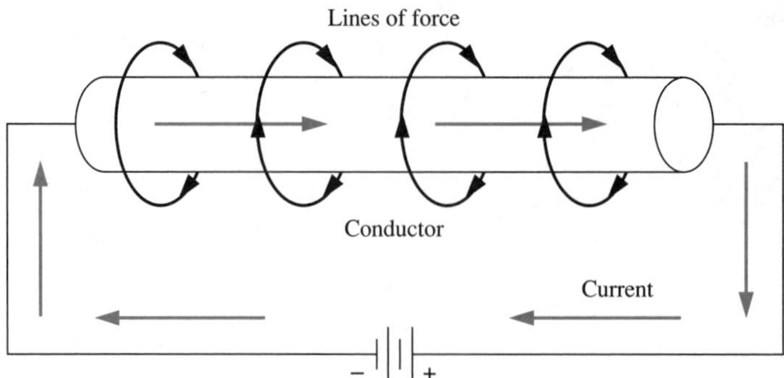

Figure 2–8 **Magnetic field caused by current around a conductor**

magnetic lines known as magnetic **flux**. A single magnetic line of force is equal to *one Maxwell* of magnetic flux. The amount of magnetic lines or flux per unit area is known as the **flux density**.

tech tidbit

Flux—number of magnetic lines.

Flux density—the number of magnetic lines per unit area.

Unlike magnetic poles attract, like magnetic poles repel.

The north pole of a magnet is defined as the pole that points approximately toward the geographic north pole. It is the pole from which the magnetic lines come out, as shown in Figure 2-9. The south pole of a magnet is defined as the pole that points approximately toward the geographic south pole. It is the pole into which the magnetic lines enter, as shown in Figure 2-9. *Unlike* magnetic poles tend to *attract* each other, while *like* magnetic poles tend to *repel* each other.

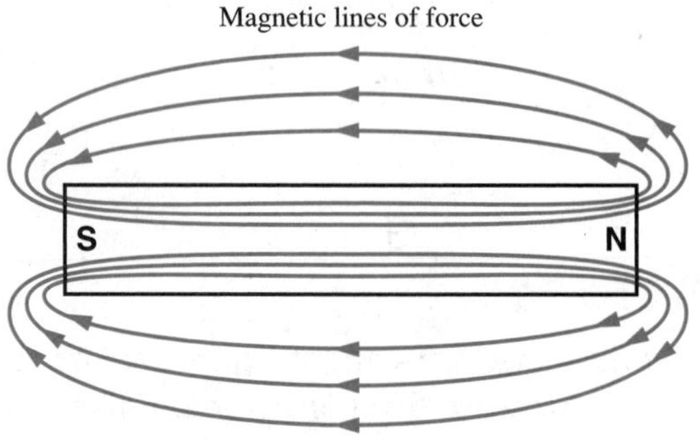

Figure 2–9 **Magnetic lines of force for a magnet**

Self-Test 4

1. Power is defined as THE RATE OF DOING WORK.
2. The unit of power is WATT - JOULES.
3. Does leaving your television on for 12 hours or leaving a 100-watt light bulb on for 24 hours consume more power?.
4. One kilowatt equals 1000 watts.
5. One horsepower is equal to 746 electrical watts.
6. Like magnetic poles tend to REPEL each other.

T^3 AVOIDING ELECTRIC SHOCK

While pure water is not a good conductor of electricity, most water contains impurities and will allow electric charge to flow. Therefore, we say that "water and electricity should *never* be mixed." Moisture increases the danger of electric shock.

When working with electricity, you must exercise safety precautions to prevent current from passing through your body. The higher the voltage, the greater the danger of electric shock. If we think of the human body as an electric circuit, then the amount of charge that can flow (current) depends on the amount of voltage and the resistance of the body. Normally the human body is not a good conductor because of the skin. However, while your skin is a good insulator, your body contains a lot of water that is a good conductor. Generally speaking, the resistance of the body is approximately 50,000 ohms. However, if the skin gets wet, the body's resistance can drop drastically. Also, holding a conductor tighter will lower body resistance. The lower the resistance, the greater the charge that can flow (current) through the body. As little as 10 millionths (0.000010) ampere can cause electric shock. Ten thousandths (0.010) ampere could be fatal because it may cause muscular paralysis. Figure 2-10 shows what can occur when different amounts of current pass through the body.

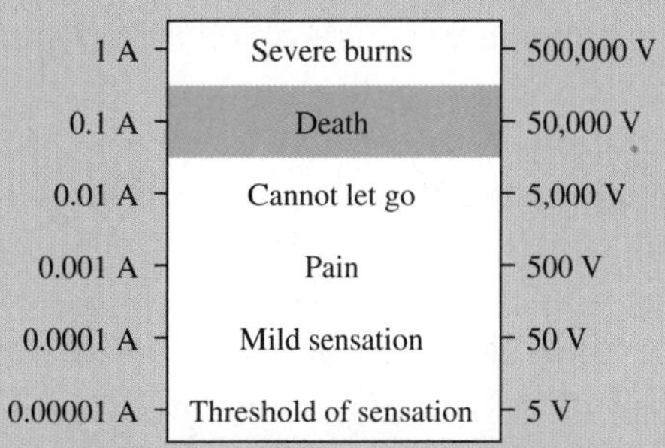

Figure 2–10 **Effect of electric current on the body (based on 50,000 ohms of body resistance)**

A good safety rule is to keep yourself insulated from earth ground. It is also a good idea to work with one hand. In this way, you are less likely to complete a circuit and allow current to pass through your entire body. Furthermore, never work with the power turned on if it is not completely necessary.

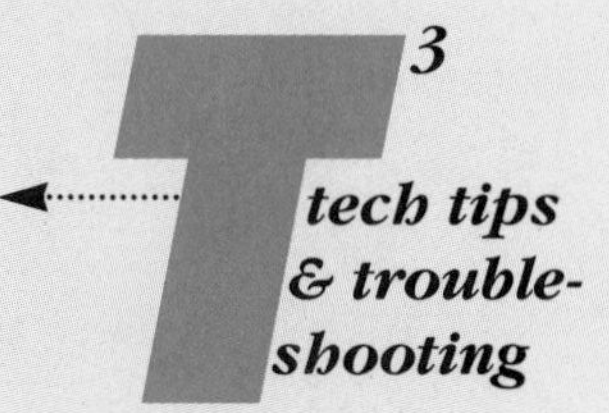

Moisture increases the danger of electric shock.

SAFETY TIP: Keep yourself insulated from earth ground.

SAFETY TIP: Make electrical measurements with one hand.

SAFETY TIP: Never assume the power is turned off.

SUMMARY

Energy is defined as the ability to do work. The basic unit of energy is the joule. Matter can be considered as anything that occupies space and has mass. Matter is composed of tiny particles called atoms. The center part of the atom is called the nucleus. The nucleus contains subatomic particles called neutrons and protons. Particles called electrons orbit around the nucleus. When electrons in the outer orbit become dislodged from an atom, they are called free electrons. Free electrons are responsible for the phenomenon known as electricity.

When a substance has an excess of electrons and another substance has a deficiency of electrons, an electric charge is said to exist between the two materials. The movement of electric charge per unit of time is called current. The unit of current is the ampere. The amount of energy required to move electric charge from one point to another is called voltage. The unit of voltage is the volt. Resistance is the opposition to the flow of electric charge or current. The unit of resistance is the ohm. The opposite of resistance is conductance. The unit of conductance is the siemens. Power is defined as the rate of doing work. The unit of electrical power is the watt. Electric energy is often measured in kilowatt-hours. Efficiency is defined as the ratio of the power out of a device to the power supplied to the device.

Magnets are materials that are capable of attracting iron and other ferromagnetic substances. The ends of a magnet are called the poles. Unlike magnetic poles tend to attract each other, and like magnetic poles tend to repel each other. Magnetism is measured in magnetic lines known as flux.

FORMULAS

$$W = F \bullet D \qquad F = \frac{W}{D} \qquad D = \frac{W}{F}$$

$$E_{AB} = \frac{W_{AB}}{Q} \qquad W_{AB} = E_{AB} \bullet Q \qquad Q = \frac{W_{AB}}{E_{AB}}$$

$$I = \frac{Q}{t} \qquad Q = I \bullet t \qquad t = \frac{Q}{I}$$

$$P = \frac{W}{t} \qquad W = P \bullet t \qquad t = \frac{W}{P}$$

$$G = \frac{1}{R} \qquad R = \frac{1}{G}$$

$$\text{Electric energy (KWh)} = \frac{P \bullet t}{1{,}000}$$

$$\text{Cost of electricity} = \text{KWh} \bullet \text{rate}$$

$$\%\ \text{efficiency} = \frac{P_{out}}{P_{in}} \times 100\%$$

$$1\ \text{coulomb} = 6.25 \times 10^{18}\ \text{electrons}$$

$$1\ \text{horsepower} = 746\ \text{watts}$$

EXERCISES

Section 2.2

1. A force of 50 newtons is required to lift a radio (load) 2 meters. Calculate the work performed.
2. Find the distance a car will move if 60,000 joules of energy are used when a force of 10,000 newtons is applied.

Section 2.5

3. What is the potential difference between two points if 150 joules of energy are required to move 25 coulombs of charge?
4. Find the energy used by a 12-volt battery delivering 1,000 coulombs of charge.
5. How many coulombs of charge will be moved by a 6-volt battery using 1,800 joules of energy?
6. Determine the energy expended when moving a charge of 500 coulombs through a potential difference of 60 volts.

Section 2.6

7. Determine the current in a circuit if 750 coulombs of charge pass a point in 50 seconds.
8. Find the current in amperes when 24 coulombs pass a point every 0.06 seconds.
9. Find the current in a circuit if 1.5 coulombs of charge pass a point in 5 minutes.
10. How long does it take 0.006 coulomb of charge to pass a point in a circuit when the current is 0.005 ampere.
11. A current of 7 amperes exists in a wire for 2 minutes. What is the total charge transferred?
12. Will a circuit breaker rated at 15 amperes open when 1,500 coulombs of charge pass through it in 1.5 minutes?

Section 2.7

13. Find the conductance of a 100-ohm resistor.
14. Find the conductance of a 2,200-ohm resistor.
15. Find the resistance of a wire whose conductance is 0.15 siemens.

Section 2.8

16. What is the power rating of a radio that converts 30 joules of energy in 5 seconds?
17. What is the power rating of an air conditioner that converts 432,000 joules of energy in 3 minutes?
18. How much energy does it take to operate a 150-watt light bulb for 2 hours?
19. If the power to a device is 120 joules per second, how long will it take to deliver 1,800 joules?

20. How much energy is consumed by a 1,500-watt heater in 7 hours?
21. Find the cost of operating a 500-watt microwave oven for 4 hours at 12 cents per kilowatt-hour.
22. Find the cost of operating a 0.5 horsepower sump pump for 12 hours at 12 cents per kilowatt-hour.
23. Find the *total cost* of operating the following devices at 12 cents per kilowatt-hour:

    (A) 300-watt TV for 4 hours

    (B) 1,000-watt dishwasher for 2 hours

    (C) 60-watt light for 10 hours

    (D) 90-watt stereo for 6 hours
24. Find the efficiency of a 2.5-horsepower (power output) motor that requires 3,000 watts of input power.
25. A 2-horsepower motor (power output) operates at 90% efficiency. Find the input power required to operate the motor.
26. See Lab Manual for the crossword puzzle for Chapter 2.

CHAPTER 3

WORKING WITH NUMBERS

Courtesy of Hewlett Packard

OBJECTIVES

After completing this chapter, the student should be able to:

- demonstrate mathematical techniques used by engineers and technicians.
- convert between conventional, scientific, and engineering notation.
- use engineering prefixes.
- perform arithmetic operations in conventional, scientific, and engineering notation.
- use a scientific calculator to perform mathematical operations.

KEY TERMS

base number
engineering notation
engineering prefix
exponent
exponential notation
powers of ten
scientific notation

3.1 INTRODUCTION

Mathematics—the international language of science.

Mathematics is the international language of science and technology. For example, engineers in Japan can read and understand mathematical equations written by engineers from the United States even though they do not speak the same language. It is vital to begin your study of electricity with a firm understanding of numbers. Therefore, you must be able to work confidently with numbers.

The study of electricity often requires the use of some very large and some very small numbers. Numbers like 20 million (20,000,000) or five millionths (0.000005) are common. As you can see, these numbers have a lot of zeros. When working with numbers with many zeros, it is easy to make mistakes. For example, if we multiply the previous two numbers,

$$\begin{array}{r} 20{,}000{,}000 \\ \times\ 0.000005 \\ \hline 100.000000 \end{array}$$

it would be very easy to lose or gain a zero. It is common to lose count of zeros when working with the type of calculations performed by electronics students. In order to avoid this kind of error, a method known as **scientific notation** was developed. This chapter introduces some of the basic mathematical techniques that engineers around the world use to make it easier to work with large and small numbers.

Scientific Notation—based on the powers of ten.

3.2 POWERS OF TEN

Scientific notation is based on the **powers of ten** theory. A power of ten is simply the number of times the number ten is multiplied by itself. For example:

$$10^2 = 10 \times 10 = 100$$
$$\therefore 10^2 = 100$$

Count the number of zeros to find the power of ten.

Here the number ten is raised to the second power, which means it is equal to 10 multiplied by 10, which equals 100. In this example, ten is known as the **base number** and two is known as the **exponent** or power of the base number. This is similar to **exponential notation.** A quick way to determine the correct power of ten is to simply count the number of zeros to the right of the one. The number of zeros becomes the power of ten. As another example, consider

$$10^3 = 10 \times 10 \times 10 = 1{,}000$$
$$\therefore 10^3 = 1{,}000$$

EXAMPLE 3-1

Convert the following numbers to powers of ten:

(A) 10 (B) 10,000 (C) 1,000,000

Solution:

(A) $10 = 10^1$ Note there is only one zero after the one.

(B) $10{,}000 = 10^4$ Note there are four zeros after the one.

(C) $1{,}000{,}000 = 10^6$ Note there are six zeros after the one.

Very small numbers (numbers whose value is less than one) use negative powers of ten. For example,

$$10^{-2} = \frac{1}{10^2} = \frac{1}{100} = 0.01$$

$$\therefore\ 10^{-2} = 0.01$$

Here the number ten is raised to the minus two power, which means it is equal to 1 divided by 10^2 or 1/100. Note that in mathematics $10^{-2} = 1/10^2$. As another example, consider

$$10^{-3} = \frac{1}{10^3} = \frac{1}{1{,}000} = 0.001$$

$$\therefore\ 10^{-3} = 0.001$$

$$10^{-n} = \frac{1}{10^n}$$

The method of converting numbers that are *less than one* to powers of ten is similar to the method used for large numbers that are greater than one. To convert numbers that are less than one into powers of ten, simply move the decimal point to the right until the decimal point is immediately to the right of the first number or first significant digit. The number of places the decimal point is moved becomes the negative power of ten.

EXAMPLE 3-2

Convert the following numbers to powers of ten:

(A) 0.1 (B) 0.0001 (C) 0.000001

Solution:

(A) $0.1 \Rightarrow 0.1. \Rightarrow 1.0 \times 10^{-1} = 10^{-1}$

1 place right

(B) $0.0001 \Rightarrow 0.0001. \Rightarrow 1.0 \times 10^{-4} = 10^{-4}$

4 places right

(C) $0.000001 \Rightarrow 0.000001. \Rightarrow 1.0 \times 10^{-6} = 10^{-6}$

6 places right

Note that in the previous example, the number 1 may be written as 1.0. This is done to minimize errors. Table 3-1 illustrates the representing of numbers as powers of ten.

It should be noted that 10^0 is equal to 1. In fact, in mathematics any number (except zero and infinity) raised to the zero power is equal to 1. For example:

$$2^0 = 1 \qquad 17^0 = 1 \qquad x^0 = 1$$

Any number raised to the zero power is equal to 1.

Table 3-1 Powers of Ten

$1 = 10^0$	
$10 = 10^1$	$0.1 = 10^{-1}$
$100 = 10^2$	$0.01 = 10^{-2}$
$1{,}000 = 10^3$	$0.001 = 10^{-3}$
$10{,}000 = 10^4$	$0.0001 = 10^{-4}$
$100{,}000 = 10^5$	$0.00001 = 10^{-5}$
$1{,}000{,}000 = 10^6$	$0.000001 = 10^{-6}$

✓ Self-Test 1

Convert the following numbers to powers of ten:

1. 100
2. 10,000,000
3. 0.01
4. 0.00001
5. 1.0

3.3 SCIENTIFIC NOTATION

Scientific notation expresses numbers as a power of ten.

Scientific notation is a method of expressing any number as a power of ten. For example:

$$200 = 2 \times 100 = 2 \times 10^2$$
$$= 2.0 \times 10^2$$

To perform the conversion into scientific notation, the original number (200) is divided by some power of ten. Note that the previous example can also be performed as follows:

$$200 = 20 \times 10 = 20 \times 10^1$$
$$= 20.0 \times 10^1$$

In scientific notation, there is only one significant digit to the left of the decimal point.

It should be noted that the first part of any number expressed in scientific notation, should be a number between 1.00 and $9.\overline{99}$. (Note that the bar indicates a repeating number.) Therefore, the number 200 should be written as 2.0×10^2. An easy way to convert a number into scientific notation is to simply move the decimal point until a number between 1.00 and $9.\overline{99}$ is obtained. The number of places that the decimal point is moved becomes the power of ten. For example:

$$3{,}700 \Rightarrow 3.700.0 \Rightarrow 3.7 \times 10^3$$

3 places left

Moving the decimal point three places to the left results in 3.7 (a number between 1.00 and 9.$\overline{99}$).

$$91{,}300 \Rightarrow 9.1300.0 \Rightarrow 9.13 \times 10^4$$

4 places left

EXAMPLE 3-3

Convert the following numbers into scientific notation:

(A) 2,200 (B) 56,000 (C) 3,256,000

Solution:

(A) $2{,}200 \Rightarrow 2.200.0 \Rightarrow 2.2 \times 10^3$

3 places left

(B) $56{,}000 \Rightarrow 5.6000.0 \Rightarrow 5.6 \times 10^4$

4 places left

(C) $3{,}256{,}000 \Rightarrow 3.256000.0 \Rightarrow 3.256 \times 10^6$

6 places left

tech tidbit

Numbers greater than 1—move decimal point left. Numbers less than 1—move decimal point right.

Small numbers that are less than one can be converted into scientific notation using a very similar technique. The only difference is that we will have to move the decimal point to the right. The exponent of the base ten will now be negative. For example:

$$0.0033 \Rightarrow 0.003.3 \Rightarrow 3.3 \times 10^{-3}$$

3 places right

$$0.0000781 \Rightarrow 0.00007.81 \Rightarrow 7.81 \times 10^{-5}$$

5 places right

tech tidbit

Exponent holds the decimal place.

Note that when the decimal point is moved to the right, the exponent becomes negative. Thus, it is said, the exponent holds the decimal place.

EXAMPLE 3-4

Convert the following numbers into scientific notation.

(A) 0.025 (B) 0.00047 (C) 0.0000027

Solution:

(A) $0.025 \Rightarrow 0.02.5 \Rightarrow 2.5 \times 10^{-2}$

2 places right

(B) $0.00047 \Rightarrow 0.0004.7 \Rightarrow 4.7 \times 10^{-4}$

4 places right

(C) $0.0000027 \Rightarrow 0.000002.7 \Rightarrow 2.7 \times 10^{-6}$

6 places right

Self-Test 2

Convert the following numbers into scientific notation:

1. 8,900
2. 47,000
3. 0.0033
4. 0.000081

3.4 CONVERTING SCIENTIFIC NOTATION TO CONVENTIONAL NOTATION

The method of converting numbers in scientific notation back into conventional notation (no powers of ten) is to simply reverse the process. For example:

$$2.0 \times 10^3 \Rightarrow \underbrace{2.000.}_{\text{3 places right}} \Rightarrow 2{,}000$$

$$4.0 \times 10^{-2} \Rightarrow \underbrace{0.04.0}_{\text{2 places left}} \Rightarrow 0.04$$

In the first example, we reversed the process by moving the decimal point three places to the right. In the second example, we reversed the process by moving the decimal point two places to the left.

EXAMPLE 3-5

Convert the following numbers from scientific form to conventional form:

(A) 2.5×10^2

(B) 8.9×10^3

(C) 4.9×10^4

(D) 2.5×10^{-2}

(E) 8.9×10^{-3}

(F) 4.9×10^{-4}

Solution:

(A) $2.5 \times 10^2 \Rightarrow \underbrace{2.50.}_{\text{2 places right}} \Rightarrow 250$

(B) $8.9 \times 10^3 \Rightarrow \underbrace{8.900.}_{\text{3 places right}} \Rightarrow 8{,}900$

(C) $4.9 \times 10^4 \Rightarrow 4.9000. \Rightarrow 49{,}000$
4 places right

(D) $2.5 \times 10^{-2} \Rightarrow 0.02.5 \Rightarrow 0.025$
2 places left

(E) $8.9 \times 10^{-3} \Rightarrow 0.008.9 \Rightarrow 0.0089$
3 places left

(F) $4.9 \times 10^{-4} \Rightarrow 0.0004.9 \Rightarrow 0.00049$
4 places left

3.5 ENGINEERING NOTATION

Engineering Notation—the powers of ten are grouped in multiples of three.

Engineering notation can be considered a special case of scientific notation where the powers of ten are grouped in *multiples of three*. The powers of ten are given names known as **engineering prefixes.** Table 3-2 illustrates the engineering prefixes for the most common powers of ten used in electricity as well as other areas of engineering.

Note that uppercase letters are used here as symbols to indicate large or positive numbers. Lowercase letters are used as symbols to indicate small or negative exponent numbers. To illustrate the use of engineering notation, consider the following examples:

Table 3-2 Engineering Prefixes

Powers of Ten	Prefix	Symbol
10^{12}	Tera	T
10^{9}	Giga	G
10^{6}	Mega	M
10^{3}	Kilo	K
10^{0}	Units	—
10^{-3}	milli	m
10^{-6}	micro	μ
10^{-9}	nano	n
10^{-12}	pico	p

microsecond, Kilovolts, Megahertz, picofarads.

1,000 volts = 1.0×10^3 volts = 1 Kilovolt = 1 KV

0.001 ampere = 1.0×10^{-3} ampere = 1 milliampere = 1 mA

1,000,000 hertz = 1.0×10^6 hertz = 1 Megahertz = 1 MHz

To convert numbers into engineering notation, we follow the same rules as when converting to scientific notation except we only group in multiples of three. For example:

$$33{,}000 \Rightarrow 33.000.0 \Rightarrow 33 \times 10^3 = 33\ \text{K}$$

3 places left

$$560{,}000 \Rightarrow 560.000.0 \Rightarrow 560 \times 10^3 = 560\ \text{K}$$

3 places left

OR

$$560{,}000 \Rightarrow .560000.0 \Rightarrow 0.56 \times 10^6 = 0.56\ \text{M}$$

6 places left

$$0.00034 \Rightarrow 0.000.34 \Rightarrow 0.34 \times 10^{-3} = 0.34\ \text{m}$$

3 places right

OR

$$0.00034 \Rightarrow 0.000340. \Rightarrow 340 \times 10^{-6} = 340\ \mu$$

6 places right

T^2 tech tidbit

Keep your prefixes as consistent as possible.

In all cases the decimal point is always moved in multiples of three places. This allows for the use of the most commonly used engineering prefixes. Note that there is more than one way to express a number in engineering notation. Generally speaking, it is a good idea to keep your prefixes consistent. That is, if every quantity is in Kilo, then make your conversions to Kilo.

EXAMPLE 3-6

Express the following numbers in scientific and in engineering notation:

(A) 3,300 (B) 0.0007 (C) 67,890,000,000

Solution:

(A) $3{,}300 \Rightarrow 3.300.0 \Rightarrow 3.3 \times 10^3 = 3.3\ \text{K}$

3 places left

Note that scientific and engineering notation are the same.

(B) $0.0007 \Rightarrow 0.0007. \Rightarrow 7.0 \times 10^{-4}$ (scientific notation)

4 places right

$0.0007 \Rightarrow 0.000.7 \Rightarrow 0.7 \times 10^{-3} = 0.7\ \text{milli} = 0.7\ \text{m}$ (engineering notation)

3 places right

(C) $67{,}890{,}000{,}000 \Rightarrow 6.7890000000.0 \Rightarrow 6.789 \times 10^{10}$ (scientific notation)

10 places left

$67{,}890{,}000{,}000 \Rightarrow 67.890000000.0 \Rightarrow 67.89 \times 10^9 = 67.89\ \text{Giga} = 67.89\ \text{G}$

9 places left (engineering notation)

✓ Self-Test 3

Express the following numbers in scientific and engineering notation:

1. 24,000
2. 68,000,000
3. 0.044
4. 0.000015

3.6 MULTIPLICATION

Multiplication—add exponents.

Multiplying numbers that are in scientific form is easy. Generally speaking, the method is to add the exponents of the powers of ten and multiply the number part of the expressions. For example,

$$(5 \times 10^3)(3 \times 10^4) = (5 \times 3)(10^{(3+4)})$$
$$= 15 \times 10^{(3+4)}$$
$$= 15 \times 10^7$$

Note that we multiply the number parts 5 and 3. Next we add the exponents 3 and 4. The general rule is expressed mathematically as:

$$(A \times 10^X)(B \times 10^Y) = \mathrm{A} \bullet \mathrm{B} \times 10^{(X+Y)}$$

EXAMPLE 3-7

Multiply the following numbers:

(A) $(1.5 \times 10^6)(3 \times 10^3)$

(B) $(0.5 \times 10^{-5})(4 \times 10^2)$

(C) $(7 \times 10^{-2})(3 \times 10^{-3})$

Solution:

(A) $(1.5 \times 10^6)(3 \times 10^3) = 4.5 \times 10^{(6+3)}$
$= 4.5 \times 10^9$

(B) $(0.5 \times 10^{-5})(4 \times 10^2) = 2.0 \times 10^{(-5+2)}$
$= 2.0 \times 10^{-3}$

(C) $(7 \times 10^{-2})(3 \times 10^{-3}) = 21.0 \times 10^{[-2+(-3)]}$
$= 21.0 \times 10^{-5}$

3.7 DIVISION

Division—subtract exponents.

The division process is the opposite of the multiplication process. The exponents of the powers of ten are subtracted and the number parts of the expression are divided. Consider the following example:

$$\frac{9 \times 10^6}{3 \times 10^4} = \left(\frac{9}{3}\right) \times 10^{(6-4)}$$

$$= 3 \times 10^2$$

Note that the number part of the numerator, 9, is divided by the number part of the denominator, 3. The exponent in the denominator is subtracted from the exponent in the numerator. The general rule is expressed mathematically as:

$$\frac{A \times 10^X}{B \times 10^Y} = \left(\frac{A}{B}\right) \times 10^{(X-Y)}$$

Special care must be taken when the denominator contains a negative exponent. Consider the following example:

$$\frac{9 \times 10^6}{3 \times 10^{-4}} = \left(\frac{9}{3}\right) \times 10^{[6-(-4)]}$$

$$= 3 \times 10^{(6+4)}$$

$$= 3 \times 10^{10}$$

$+ \bullet +$ or $- \bullet - =$ positive.

$+ \bullet -$ or $- \bullet + =$ negative.

Note that the powers of ten result in a 6 minus a negative 4. A negative times a negative equals a positive, so this results in 6 + 4 or 10. When multiplying or dividing like signs ($+ \bullet +$ or $- \bullet -$) the result is *always* positive. When multiplying or dividing unlike signs ($+ \bullet -$ or $- \bullet +$) the result is *always* negative.

EXAMPLE 3-8

Divide the following numbers:

(A) $\dfrac{21.0 \times 10^7}{7.0 \times 10^3}$

(B) $\dfrac{18.0 \times 10^{-5}}{3.0 \times 10^3}$

(C) $\dfrac{14.0 \times 10^{-6}}{3.0 \times 10^3}$

Solution:

(A) $$\frac{21.0 \times 10^7}{7.0 \times 10^3} = \left(\frac{21.0}{7.0}\right) \times 10^{(7-3)}$$

$$= 3.0 \times 10^4$$

(B) $$\frac{18.0 \times 10^{-5}}{3.0 \times 10^3} = \left(\frac{18.0}{3.0}\right) \times 10^{(-5-3)}$$

$$= 6.0 \times 10^{-8}$$

(C) $$\frac{14.0 \times 10^{-6}}{2.0 \times 10^{-2}} = \left(\frac{14.0}{2.0}\right) \times 10^{[-6-(-2)]}$$

$$= 7.0 \times 10^{(-6+2)}$$

$$= 7.0 \times 10^{-4}$$

✓ Self-Test 4

Solve the following problems using scientific notation:

1. $(6 \times 10^2)(5 \times 10^4)$
2. $(1.1 \times 10^{-3})(7 \times 10^3)$
3. $\dfrac{3.6 \times 10^6}{1.2 \times 10^2}$
4. $\dfrac{15 \times 10^4}{5 \times 10^{-2}}$

3.8 EXPONENTIATION

The general rule for raising a power of ten to another exponent is:

$$(10^a)^b = 10^{(a \cdot b)} = 10^{ab}$$

Exponentiation—multiply exponents.

For example:

$$(10^2)^3 = 10^{2 \cdot 3} = 10^6$$
$$(10^2)^{-3} = 10^{2 \cdot (-3)} = 10^{-6}$$
$$(10^{-2})^{-3} = 10^{-2 \cdot (-3)} = 10^6$$

If the number is in scientific form, then the number part is also raised to the power. For example:

$$(3 \times 10^2)^3 = (3)^3 \times 10^{2 \cdot 3} = 27 \times 10^6$$
$$(5 \times 10^3)^{-2} = (5)^{-2} \times 10^{3 \cdot (-2)} = \frac{1}{5^2} \times 10^{-6} = 0.04 \times 10^{-6}$$

EXAMPLE 3-9

Perform the following exponentiations:

(A) $(5 \times 10^4)^2$

(B) $(2 \times 10^3)^{-2}$

(C) $(6 \times 10^{-5})^3$

Solution:

(A) $(5 \times 10^4)^2 = (5)^2 \times 10^{4 \cdot 2} = 25 \times 10^8$

(B) $(2 \times 10^3)^{-2} = (2)^{-2} \times 10^{3 \cdot (-2)} = \frac{1}{2^2} \times 10^{-6} = \frac{1}{4} \times 10^{-6} = 0.25 \times 10^{-6}$

(C) $(6 \times 10^{-5})^3 = (6)^3 \times 10^{(-5 \cdot 3)} = 216 \times 10^{-15}$

Addition/Subtraction—exponents must be the same.

3.9 ADDITION AND SUBTRACTION

In scientific notation, addition and subtraction can only be performed when the exponents or powers of ten of the numbers are all the same. For example:

$$(5 \times 10^3) + (6 \times 10^3) = 11 \times 10^3$$

$$(8 \times 10^{-2}) + (12 \times 10^{-2}) = 20 \times 10^{-2}$$

$$(9 \times 10^4) - (6 \times 10^4) = 3 \times 10^4$$

$$(12 \times 10^{-5}) - (8 \times 10^{-5}) = 4 \times 10^{-5}$$

Note that in all of these examples, the powers of ten are the same. If they were not the same, we could not perform the addition or subtraction operation. However, we can change a number so that the power of ten becomes the same as the other number in the problem. For example:

$$(50 \times 10^2) + (20 \times 10^3)$$

$$= (5 \times 10^1 \times 10^2) + (20 \times 10^3)$$

$$= (5 \times 10^{(1+2)}) + (20 \times 10^3)$$

$$= (5 \times 10^3) + (20 \times 10^3)$$

$$= 25 \times 10^3$$

3.10 PREFIX CONVERSIONS

Numbers can be converted from one prefix to another. This is done by moving the decimal point. For example, we know that

1 dollar = 100 cents

3.5 dollars = 350 cents

450 cents = 4.5 dollars

Two simple rules help us when performing prefix conversions.

1. If you want to convert a larger unit to a smaller unit, you should get a greater quantity of the smaller unit. Therefore, we must multiply by some factor.
2. If you want to convert a smaller unit to a larger unit, you should get a smaller quantity of the larger unit. Therefore, we must divide by some factor.

Larger to smaller rule—multiply by the factor. Smaller to larger rule—divide by the factor.

For example, to convert 16 dollars to cents, we realize that dollars is a larger unit than cents. Therefore by rule 1, we must multiply by the factor of 100. Thus, to convert dollars to cents:

$$16 \text{ dollars} \rightarrow 16\ (100) = 1{,}600 \text{ cents}$$

To convert 520 cents to dollars, using rule 2,

$$520 \text{ cents} \rightarrow 520 \div 100 = 5.2 \text{ dollars}$$

The trick is to recognize the factor. With dollars and cents, the factor will always be 100. Engineering prefixes are always in multiples of 3. Therefore, factors of 1,000 (10^3), 1,000,000 (10^6), 1,000,000,000 (10^9), and so forth will always result. Referring to Table 3-2 for adjacent engineering prefixes, the factor will always be 1,000. Skipping one prefix (Mega to tera) results in a factor of 1,000,000.

EXAMPLE 3-10

Perform the following prefix conversions:

(A) 8.0 milli to micro

(B) 500 Kilo to Mega

(C) 10 Mega to Kilo

Solution:

(A) milli to micro → large to small (Rule 1)

8.0 milli → 8.0(1,000) = 8,000 micro

(B) Kilo to Mega → small to large (Rule 2)

500 Kilo → 500 ÷ 1,000 = 0.5 Mega

(C) Mega to Kilo → large to small (Rule 1)

10 Mega → 10(1,000) = 10,000 Kilo

✓ Self-Test 5

Solve the following problems using scientific notation:

1. $(10^{-3})^3$
2. $(4 \times 10^3)^2$

Perform the following prefix conversions:

3. 1.2 Mega to Kilo
4. 0.5 milli to micro

T[3] tech tips & trouble-shooting

T³ SCIENTIFIC CALCULATORS

One of the most useful tools for a technician is the scientific calculator. Advances in solid state electronics have allowed the packaging of many advanced features into a low-cost, hand-held scientific calculator. The use of this type of calculator has become universal in the scientific and technological communities. Figure 3-1 illustrates two common calculators.

Figure 3-1 Scientific calculators

All scientific calculators allow you to directly enter data in scientific notation form. This is usually accomplished through the use of the *EXP* OR *EE* key. Some calculators use the term *EXP,* which stands for *exponential*. Other calculators use the term *EE,* which stands for *enter exponent.*

To enter the number 5×10^3 into a calculator in scientific form, press the *5* key, next press the *EXP* key, then press the three key, as shown in Figure 3-2A.

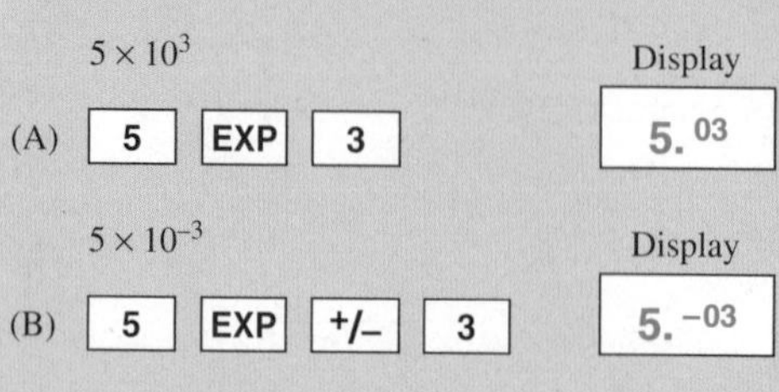

Figure 3-2 Scientific notation

Notice that when the *EXP* key is depressed, two small zeros appear in the upper righthand corner of the display. These characters indicate the exponent of the power of ten.

To enter the number 5×10^{-3}, follow the sequence indicated in Figure 3-2B. Notice the use of the +/– key to change the sign of the exponent.

On some calculators, it is necessary to be in the scientific mode before calculations can be performed in scientific notation. This is accomplished through the use of a mode key. On some calculators, this key is labeled *FSE,* which stands for *fixed, scientific, engineering notation* modes. Every time the *FSE* key is depressed, the calculator will change its mode of operation.

Figure 3-3 illustrates a number of calculations performed in scientific mode. Note that when the calculator is used in scientific mode, it will always display the answer in scientific form. That is, only one digit will be displayed to the left of the decimal point.

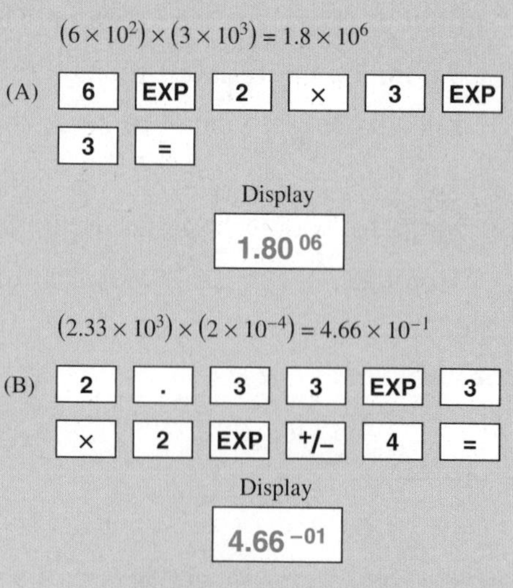

Figure 3-3 Scientific calculations (continued on page 45)

T[2] tech tidbit

For scientific notation use the *EXP* or *EE* key.

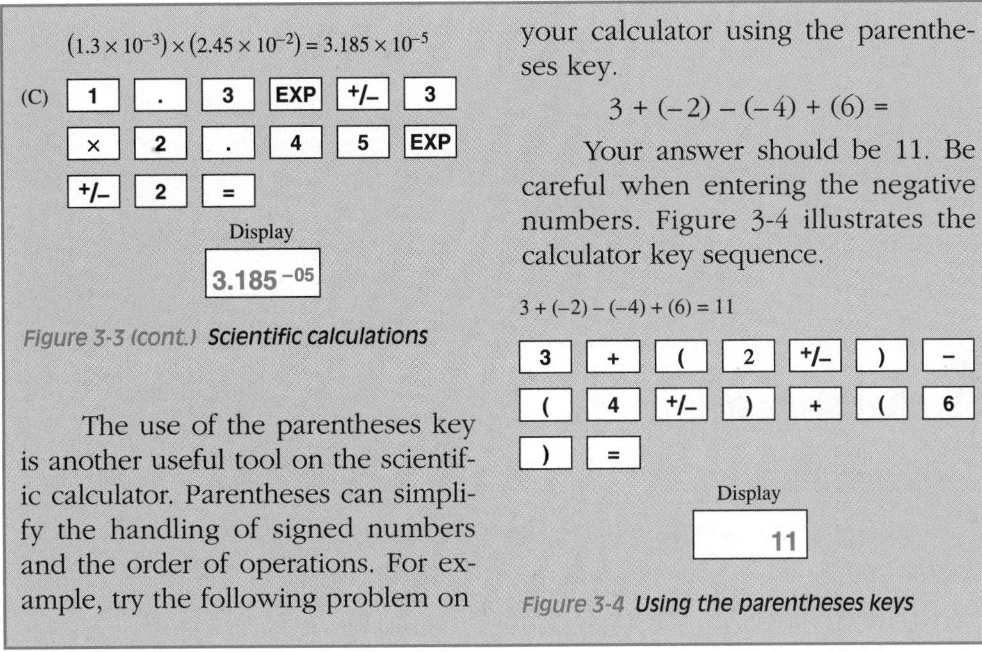

Figure 3-3 (cont.) **Scientific calculations**

The use of the parentheses key is another useful tool on the scientific calculator. Parentheses can simplify the handling of signed numbers and the order of operations. For example, try the following problem on your calculator using the parentheses key.

$$3 + (-2) - (-4) + (6) =$$

Your answer should be 11. Be careful when entering the negative numbers. Figure 3-4 illustrates the calculator key sequence.

Figure 3-4 **Using the parentheses keys**

SUMMARY

The study of electricity often requires the use of some very large and some very small numbers. This chapter introduced the basic mathematical techniques that are used to make working with very large and very small numbers easier.

Scientific notation is based on the powers of ten theory. The powers of ten are given names known as engineering prefixes. Engineering prefixes also help us to work with very large and small numbers. It is important to become comfortable and proficient in the use of the mathematics for electricity. However, do not be concerned that the study of electricity will require you to have a thorough understanding of advanced mathematical methods. Although the study of electricity does require the frequent use of basic algebra, advanced mathematics is not required to understand and apply the basic laws and formulas to electric circuits.

FORMULAS

$$(A \times 10^X) \cdot (B \times 10^Y) = A \cdot B \times 10^{(X+Y)}$$

$$\frac{A \times 10^X}{B \times 10^Y} = \left(\frac{A}{B}\right) \times 10^{(X-Y)}$$

$$(10^a)^b = 10^{(a \cdot b)}$$

$$10^{-a} = \frac{1}{10^a}$$

Rules for multiplying or dividing signs:

$$(+ \cdot +) \text{ or } (- \cdot -) = +$$

$$\left(\frac{+}{+}\right) \text{ or } \left(\frac{-}{-}\right) = +$$

$$(+ \cdot -) \text{ or } (- \cdot +) = -$$

$$\left(\frac{+}{-}\right) \text{ or } \left(\frac{-}{+}\right) = -$$

EXERCISES

Section 3.2

1. Express the following numbers as powers of ten:
   (A) 100 (C) 0.01
   (B) 10,000 (D) 0.0001
2. Express the following numbers as powers of ten:
   (A) 1 (C) 0.001
   (B) 1,000,000 (D) 0.000001
3. Express the following powers of ten in conventional form:
   (A) 10^3 (C) 10^{-1}
   (B) 10^5 (D) 10^{-5}

Section 3.3

4. Convert the following numbers into scientific notation form:
   (A) 170 (C) 0.04
   (B) 3,300 (D) 0.0025
5. Convert the following numbers into scientific notation form:
   (A) 56,000 (C) 0.000009
   (B) 91,000,000 (D) 0.000039
6. Convert the following numbers into scientific notation form:
   (A) 76,500 (C) 0.00739
   (B) 2,390,000 (D) 0.0000000825

Section 3.4

7. Convert the following scientific notation numbers into conventional form:
   (A) 7.6×10^2 (C) 3.8×10^{-2}
   (B) 2.2×10^3 (D) 4.6×10^{-4}
8. Convert the following scientific notation numbers into conventional form:
   (A) 4.02×10^3 (C) 6.023×10^{-1}
   (B) 8.608×10^5 (D) 9.17×10^{-5}

Section 3.5

9. Express the following numbers in engineering notation form:

(A) 1,200 (C) 0.015

(B) 62,000,000 (D) 0.000033

10. Express the following numbers in engineering notation form:

(A) 470 (C) 0.0000027

(B) 10,500 (D) 0.00024

11. Express the following numbers in engineering notation form:

(A) 1.5×10^4 (C) 2.3×10^{-2}

(B) 6.8×10^5 (D) 5.6×10^{-4}

Section 3.6

12. Perform each of the following operations and express the result as a power of ten:

(A) (10)(100) (C) (0.01)(0.1)

(B) (1,000)(1,000) (D) (0.001)(0.01)

13. Perform each of the following operations and express the result as a power of ten:

(A) $(4.0 \times 10^2)(2.0 \times 10^3)$ (C) $(8.2 \times 10^3)(3.5 \times 10^4)$

(B) $(2.3 \times 10^4)(4.2 \times 10^2)$ (D) $(5.7 \times 10^3)(3.2 \times 10^6)$

14. Perform each of the following operations and express the result as a power of ten:

(A) $\dfrac{10{,}000}{100}$ (C) $\dfrac{1{,}000}{10{,}000}$

(B) $\dfrac{100{,}000}{10{,}000}$ (D) $\dfrac{10}{1{,}000}$

15. Perform each of the following operations and express the result as a power of ten:

(A) $\dfrac{9.6 \times 10^4}{3.2 \times 10^3}$ (C) $\dfrac{5.0 \times 10^2}{1.25 \times 10^{-5}}$

(B) $\dfrac{8.5 \times 10^6}{3.4 \times 10^2}$ (D) $\dfrac{6.4 \times 10^{-3}}{3.2 \times 10^{-2}}$

Section 3.8

16. Perform each of the following operations and express the result as a power of ten.

(A) $(10^3)^2$ (C) $(10^{-3})^4$

(B) $(10{,}000)^4$ (D) $(10^{-2})^{-3}$

17. Perform each of the following operations and express the result as a power of ten:

(A) $\dfrac{(10^2)(10^3)^2}{100}$

(B) $\dfrac{(0.1)(10^{-2})^2}{1{,}000}$

(C) $\dfrac{(300)^2(0.001)}{90}$

(D) $\dfrac{(10^5)^2(10^3)^4}{10^{-8}}$

Section 3.9

18. Perform each of the following operations and express the result as a power of ten.

(A) $(8 \times 10^3) + (4 \times 10^3)$

(B) $(6 \times 10^{-2}) + (8 \times 10^{-2})$

(C) $(4.5 \times 10^4) - (2 \times 10^4)$

(D) $(12 \times 10^6) - (80 \times 10^5)$

Section 3.10

19. Perform the following prefix conversions:

(A) 5.0 milli = ________ micro

(B) 200 Kilo = ________ Mega

(C) 6.5 Mega = ________ Kilo

(D) 500 micro = ________ milli

Section T^3

20. Perform the following operations on your calculator and express your answer in scientific notation to three decimal places.

(A) $(652)(247)(159) =$

(B) $(3 \times 10^3)(900)(0.001) =$

(C) $\dfrac{(3 \times 10^2)^3(0.0001)}{9 \times 10^{-3}} =$

(D) $\dfrac{(75)(5.6 \times 10^{-4})}{107} =$

21. See Lab Manual for the crossword puzzle for Chapter 3.

CHAPTER 4

ELECTRICAL COMPONENTS AND CIRCUITS

Courtesy of Cleveland Institute of Electronics, Inc.

KEY TERMS

ampere-hour (Ah)
battery
capacitor
cell
circuit breaker
cold cranking ampere (CCA)
conductor
conventional current
double-pole, double-throw
electric circuit
electrolyte
electron current
energy source
farad (F)
fuse
henry (H)
inductor
insulator
load
nominal
ohm (Ω)
plates
pole
reserve capacity minutes
resistor color code
schematic diagram
semiconductor
series aiding
series opposing
single-pole, single-throw
switch
temperature coefficient
theromocouple
tolerance
transducers

OBJECTIVES

After completing this chapter, the student should be able to:

- define the units of ohms, volts, amperes, watts, henries, and farads.
- identify electric circuit components.
- interpret and draw schematic diagrams.
- list examples of conductors, semiconductors, and insulators.
- differentiate between electron and conventional current.
- identify sources of electric energy.
- color code resistors.
- determine the resistance of wire.
- read and interpret meter scales.
- measure voltage, current, and resistance with a volt-ohm-milliammeter (VOM).

4.1 INTRODUCTION

Now that we have a basic understanding of the fundamental properties of electricity, we can begin to study the physical devices that allow us to use electricity. The basic electrical components form the building blocks of all electric circuits. We see them in radios, televisions, computers, airplanes, automobiles, and almost everywhere we look. Without these electrical components, our modern electric world would not exist. In this chapter, we will learn about many of the basic electrical components and how they can be connected to form an electric circuit.

4.2 CURRENT IN AN ELECTRIC CIRCUIT

tech tidbit

Electric circuits consist of energy sources, loads, and conductors.

In order for current to exist, there must be a path or circuit. An **electric circuit** in its simplest form consists of an energy source, a load, and conductors that connect everything together. An **energy source** is something that separates electrons from their atoms. Devices like batteries, generators, and electronic power supplies are all examples of energy sources. A **load** is something that uses and

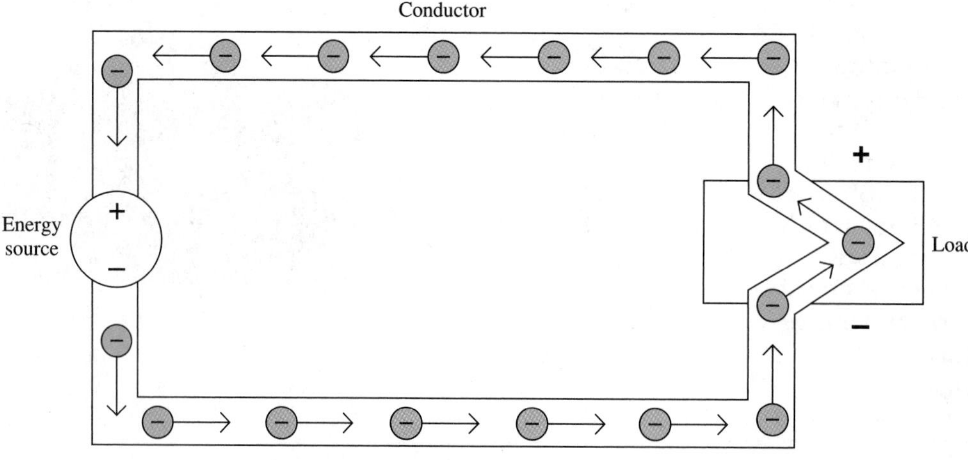

(A) Electron current

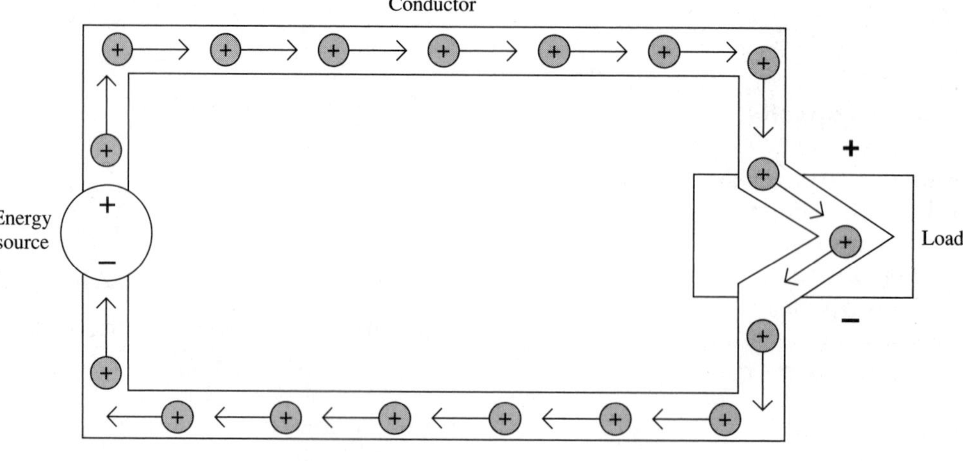

(B) Conventional current

Figure 4-1 **Electric current**

converts energy. For example, lights, motors, and radios are all electric loads. **Conductors** control the movement of charge (current) in a circuit. A good conductor allows charge (current) to flow easily through an electric circuit. Metallic materials such as gold, silver, and copper are examples of good conductors. **Semiconductors** allow a limited movement of charge (current) in a circuit. Materials such as silicon and germanium are examples of semiconductors. **Insulators** allow very little current movement of charge in an electric circuit. Materials such as wood, rubber, and glass are examples of different types of insulators. Figure 4-1A illustrates the flow of electrons in an electric circuit. Electrons leave the negative terminal of the energy source and flow through the load (performing work) as they are attracted back to the positive terminal of the energy source.

In the early days of electronics, scientists did not know which way the electric charge was flowing. Benjamin Franklin decided to adopt the convention that charge flows from plus (+) to minus (–) through the load. This convention is still widely used today and is referred to as **conventional current,** as shown in Figure 4-1B.

Benjamin Franklin was wrong. Modern scientists have proven that electrons, having less mass than protons, are the particles of motion. That is, electrons actually flow from minus to plus through the load.

This book uses the **electron current** standard and assumes that charge always flows from minus to plus through the load, as shown in Figure 4-1A. Keep in mind that what we are talking about is the direction in which we *assume* the electrons are flowing. A light bulb will light if the electrons are assumed to flow clockwise or counterclockwise. It is important to select a convention and be consistent.

Figure 4-2 illustrates the electrical diagram of a circuit, which is called a **schematic diagram.** Note the use of symbols to represent each device in the circuit. The light bulb is symbolized by a resistor, R_L (where the subscript L denotes load), which represents the electric resistance of the light bulb. The voltage source in this case is a battery, V, which is represented by a series of long and short lines. The long line represents the positive (+) terminal of the battery and the short line represents the negative (–) terminal. The switch, SW, is symbolized by two circles indicating contact points and a line representing the switch lever. When the switch lever is moved in the direction indicated by the arrow, the switch is said to be closed. This completes the circuit allowing current to light the light bulb.

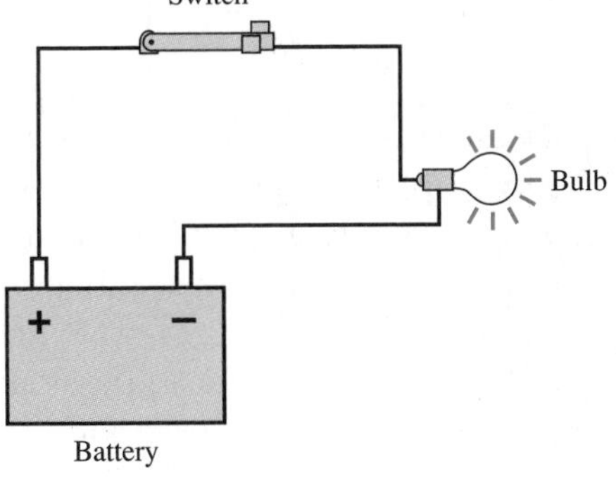

(A) Wiring diagram

SW
R_L
+
–
V

(B) Schematic diagram

Figure 4-2 Electric circuit

Good conductors have one or two electrons in their outer orbits.

Semiconductors have four electrons in their outer orbits.

Good insulators have seven or eight electrons in their outer orbits.

Conventional current + to –.

Electron current – to +.

A schematic diagram uses symbols to represent electric components.

✓ Self-Test 1

1. _______________ separates electrons from their atoms.
2. Conductors control the movement of _______________ in a circuit.
3. The standard that assumes that charge always flows from minus to plus is called _______________.
4. Draw the schematic symbol for a resistive load, a battery, and a switch.

4.3 VOLTAGE SOURCES

tech tidbit 2

Transducers convert energy.

Energy can neither be created nor destroyed, so it follows that electric energy must come from some other form of energy. The primary sources of energy come from chemical, mechanical, light, heat, and pressure. Devices that convert one type of energy to another type of energy are called **transducers.** For example, a loud speaker converts electric energy to sound energy.

Chemical Sources

Electricity can be produced chemically by inserting two dissimilar metals such as zinc and copper called **plates** into a chemical conducting solution called an **electrolyte,** as shown in Figure 4-3.

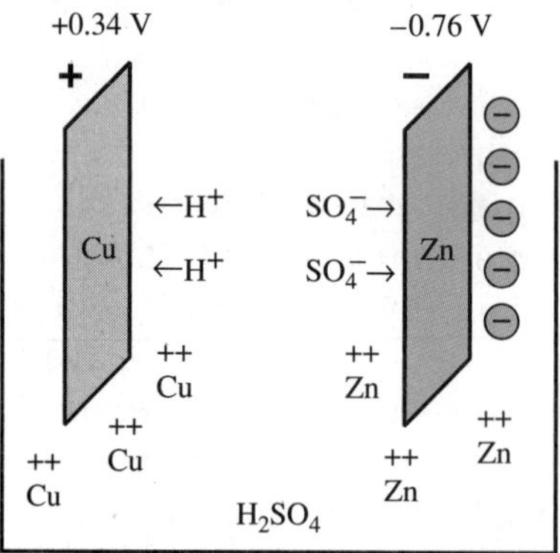

Figure 4-3 ***Voltaic or primary cell***

The solution slowly dissolves the zinc plates causing positively charged zinc ions to go into the solution, leaving the zinc plate negatively charged. This action repeats itself for the copper plate. However, because copper is less reactive, it becomes less negatively charged and therefore becomes the positive plate. The potential developed between the copper plate (+0.34 volt) and the zinc plate (−0.76 volt) is the difference between the individual potentials. That is,

$$+0.34\ \text{V} - (-0.76\ \text{V}) = 1.1\ \text{V}$$

The arrangement shown in Figure 4-3 is known as a primary or voltaic cell. It is named after its inventor, Italian scientist, Alessandra Volta. A secondary cell is a cell that can be recharged. An example of a secondary cell is the nickel-cadmium cell (Ni-Cad). It should be noted that primary cells generally cannot be recharged. Figure 4-4 shows a number of different types of commercially available batteries.

tech tidbit

Primary voltage cells *cannot* be recharged. Secondary voltage cells *can* be recharged.

Figure 4-4 ***Commercially avaliable betteries***

When a group of cells are connected together to provide either a greater voltage or current capacity, they are known as a **battery.** An example is the lead-acid automobile battery. Lead-acid automobile batteries are usually made up of six **cells** connected together. Each cell has a voltage of 2.1 volts when fully charged. Figure 4-5 illustrates a single battery cell circuit and the equivalent schematic diagram. Note that the short line is the negative terminal and the long line is the positive terminal.

tech tidbit

A battery is a group of cells connected together.

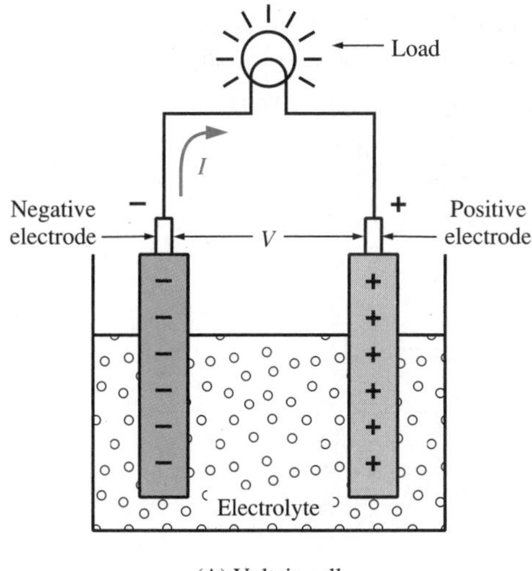

(A) Voltaic cell

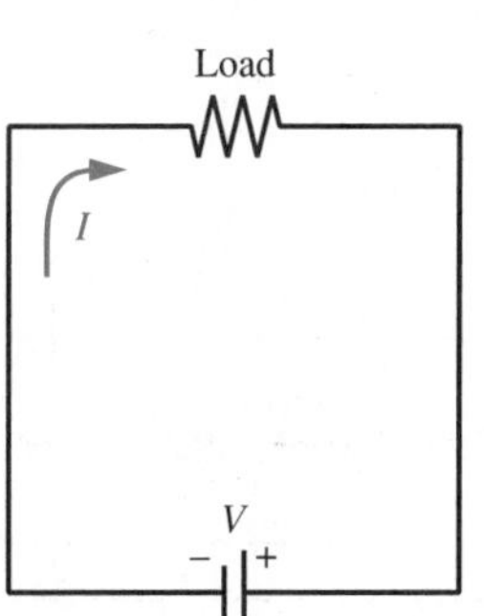

(B) Voltaic cell schematic diagram

Figure 4-5 ***Voltaic cell***

Cells and batteries can be connected together to increase the voltage or the current in a circuit. This is commonly seen in the standard household flashlight where two D cells are connected together.

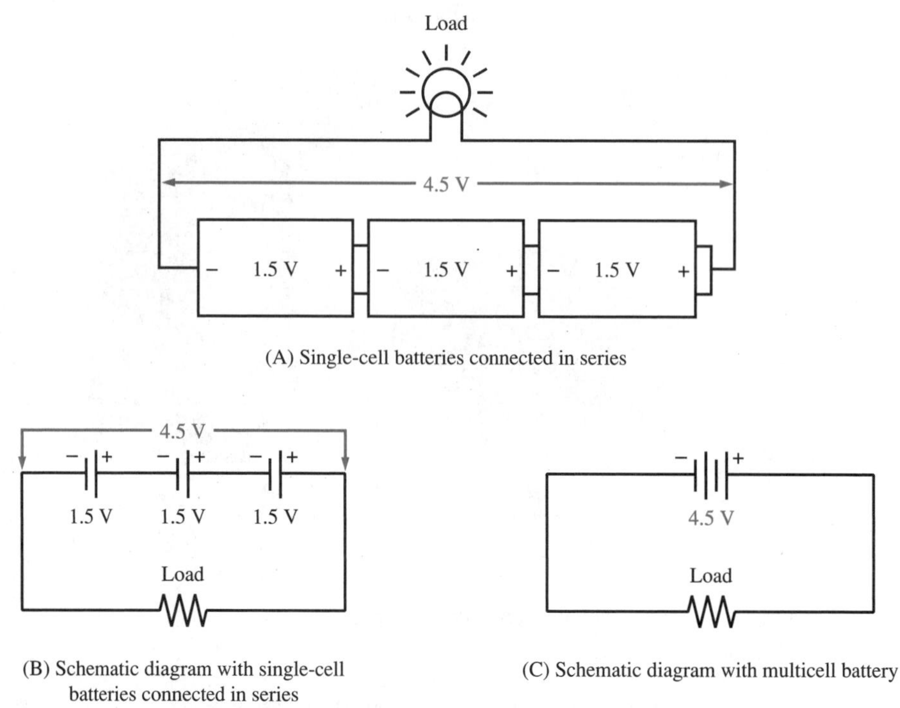

Figure 4-6 ***Connecting batteries in series***

2 tech tidbit

Series aiding—connection polarities are opposite. Series opposing—connection polarities are the same.

When batteries are connected together in a series, as shown in Figure 4-6A, the total voltage will be either the sum or the difference depending on how they are connected. When the connection polarities are opposite, the total voltage will be the sum. This is known as **series aiding** and is shown in Figure 4-6B. When the connection polarities are the same, the total voltage will be the difference. This is known as **series opposing** and is shown in Figure 4-7. It should be noted that series opposing connections have little practical use in modern electric circuits. The concept of connecting voltage sources together and determining the total voltage is further discussed in the next chapter.

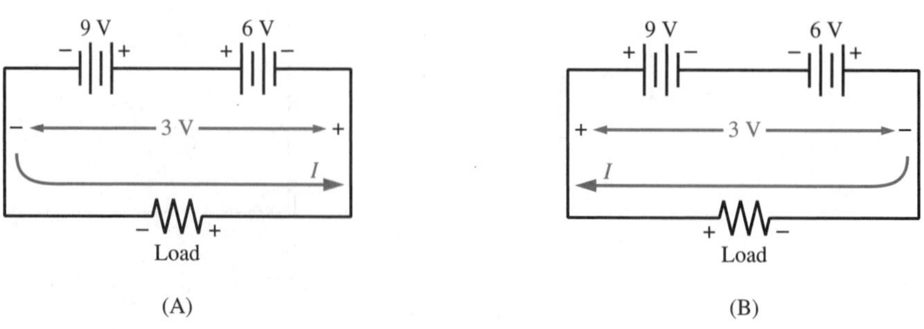

Figure 4-7 ***Series opposing batteries***

Batteries are sometimes given an **ampere-hour** rating in addition to a voltage rating.

The ampere-hour rating of a battery is equal to the current (I) the battery can deliver multiplied by the time (t) the battery can theoretically supply the current. Thus,

$$Ah = I \bullet t$$

◀ Equation 4-1

where Ah = the ampere-hour rating of the battery

I = the current in amperes

t = the time in hours

Note: Smaller batteries are often rated in milliampere-hours (mAh).

For comparative purposes, the theoretical life of a battery can be determined, if the ampere-hour rating of the battery is known, by using the following equation:

$$LIFE = \frac{Ah}{I}$$

◀ Equation 4-2

where $LIFE$ = the life of the battery in hours (when it is completely discharged)

Ah = the ampere-hour rating of the battery

I = the current leaving the battery

EXAMPLE 4-1

What is the life of an 80 ampere-hour battery supplying 5 amperes to a circuit?

Solution:

$$LIFE = \frac{Ah}{I}$$

$$= \frac{80 \text{ ampere-hours}}{5 \text{ amperes}}$$

$$= 16 \text{ hours}$$

It follows that the larger the ampere-hours of a battery, the longer the life of the battery will be. Thus, ampere-hour ratings become very important when selecting and purchasing a battery.

A good automobile battery has a 60–80 ampere-hour rating.

It should be noted that today many automobile battery manufacturers rate their batteries in units called **cold cranking amperes** (CCA) and **reserve capacity minutes**. Cold cranking amperes is a measure of battery's ability to start an engine. The rating is in amperes. It specifies the number of amperes a battery will deliver for 30 seconds at 0°F before the terminal voltage drops to a predetermined level (7.2 volts for a 12-volt battery). This measure is based on the reasoning that a vehicle should be able to start in 30 seconds of cranking if all other starting components are in good condition.

Reserve capacity minutes is a measure of how long a battery can produce power on a continuous basis. This rating tells you how many minutes at 80°F with a 25-ampere drain a fully charged battery will supply power. The test continues until the voltage of the battery drops to a predetermined level (10.5 volts for a 12-volt battery). This rating has become very important as onboard computer electronics have been added to new cars.

Mechanical Sources

Electricity can be generated mechanically by the movement of a conductor in a magnetic field, as shown in Figure 4-8.

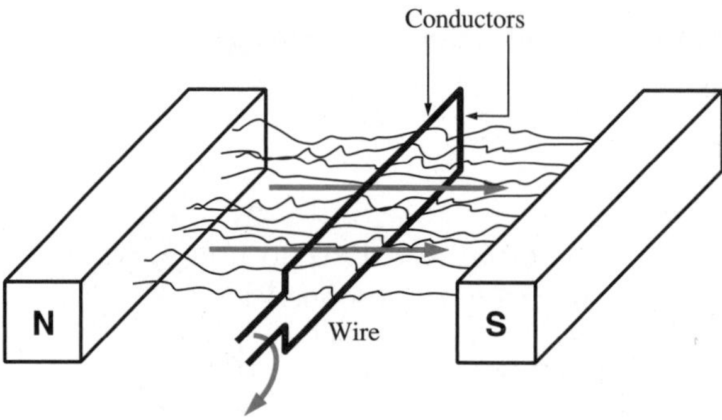

Figure 4-8 **Mechanical source**

This phenomena, which is known as induction, was discovered by Michael Faraday. When a wire is moved so it cuts the magnetic lines of force, a voltage will be induced into the wire. The amount of voltage generated is determined by:

1. the strength of the magnetic field,
2. the speed and angle that the conductor cuts the magnetic lines of force, and
3. the length of the conductor.

The induced voltage will increase as the speed, strength of the field, or size and length of the wire in the magnetic field are increased. Rotating machines operating under this principle are called electric generators. Generators convert mechanical energy of rotation to electric energy.

Light Sources

Light can be used to generate electricity. Certain materials like selenium, silicon, and cadmium will release electrons when excited by light. This release of electrons is known as the *photoelectric effect*.

Photoelectric effect—the release of electrons caused by light.

Today, solar cells like the one shown in Figure 4-9 can be used to power calculators, motors, satellites, and space vehicles. To produce large amounts of electric energy, large solar cells are required. Someday, it may be practical to power our homes with electricity using solar collector panels on the roofs of our homes.

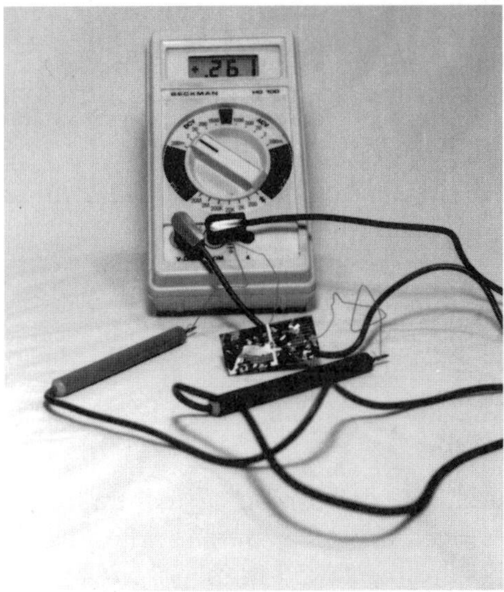

Figure 4-9 Solar cell

Temperature Sources

Devices that convert heat energy into electric energy are called **thermocouples.** Whenever two dissimilar metals are connected together to form a junction, as shown in Figure 4-10, a thermocouple will be formed.

tech tidbit

Thermocouples convert heat energy to electric energy.

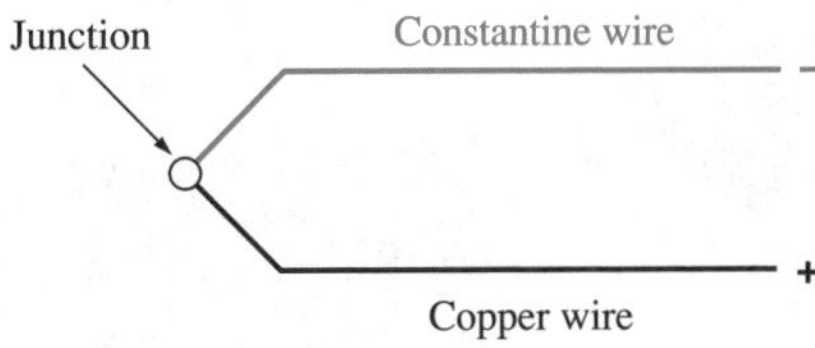

Figure 4-10 Temperature sources

When the junction is heated, electrons tend to move away from the heat. Different metals are used, so a difference in potential is formed at the open ends of the thermocouple wires. Thermocouples are most often used in measurement and control applications. The amount of energy created from a single junction is very small, making thermocouples impractical as a source of energy.

Pressure Sources

Electricity may be generated from pressure by compressing certain types of crystal materials. This process is known as the *piezoelectric effect.* When a pressure is exerted on a crystal material like quartz, a potential difference is created between the upper and lower surfaces of the materials. The greater the pressure exerted, the greater the potential difference (voltage).

tech tidbit

Piezoelectric effect—converts pressure to voltage.

One of the most common applications of the piezoelectric effect is the crystal microphone. Sound energy exerts a force on the microphone crystal to generate a voltage. Another application is the phonograph cartridge in which grooves on a record exert forces on a crystal. Other uses include pressure sensors for measurement and control and timing applications in clocks and other electronic pieces of equipment.

✓ Self-Test 2

1. A _______________ converts energy from one form to another.
2. Name five primary sources of electrical energy.
3. A group of cells connected together is called a _______________.
4. An example of a secondary cell is _______________.
5. A device that converts light energy into electric energy is the _______________.

4.4 RESISTORS

Resistors are rated in ohms and watts.

As previously stated, the opposition of resistors to current is measured in **ohms.** Large value resistors are measured in kilohms or thousands of ohms. Larger values of resistors are measured in megohms or millions of ohms. For example, a 2-kilohm (abbreviated 2 KΩ) resistor is equivalent to 2×10^3 or 2,000 ohms. A 3-megohm (abbreviated 3 MΩ) resistor is equivalent to 3×10^6 or 3,000,000 ohms. Figure 4-11 shows a number of different types of common resistors found in electronic circuits today.

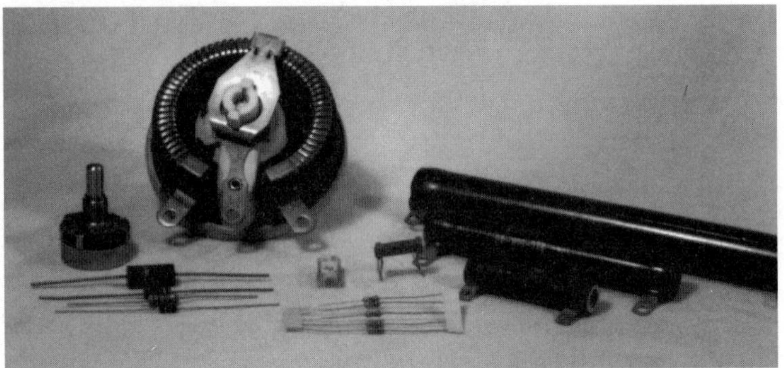

Figure 4-11 ***Resistors***

Resistors come in a wide range of values, determined by the number of ohms. The other important characteristic of a resistor is its power rating. That is, the amount of power or heat that it can safely dissipate. The power rating of a resistor is stated in watts. It is directly proportional to its physical size. Thus, the larger the power rating, the larger the physical size of the resistor.

Resistors are grouped according to their construction and whether they are of fixed or variable value. A fixed resistor is one that does not change value. For example, a 100-ohm resistor. A variable resistor is one that changes its value. This is usually done by turning a shaft that is connected to the resistor. For example, the volume control on a radio may vary between 0 ohms and 10 kilohms as the shaft is rotated. Variable resistors are often constructed from a coil of wire. A wiper or brush is rotated over the wire to cause a change in the resistance, as shown in Figure 4-12.

Potentiometers have three terminals. Rheostats have two terminals.

Variable resistors are also known as potentiometers when they have three terminals (Figure 4-12A). Two-terminal variable resistors are known as rheostats, as shown in Figure 4-12B.

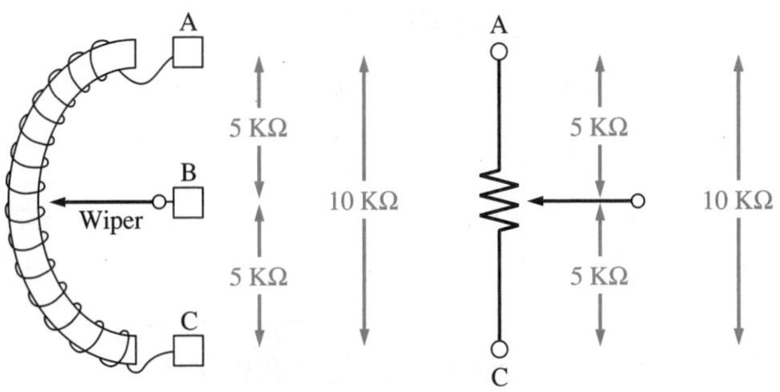

(A) Potentiometer and symbol

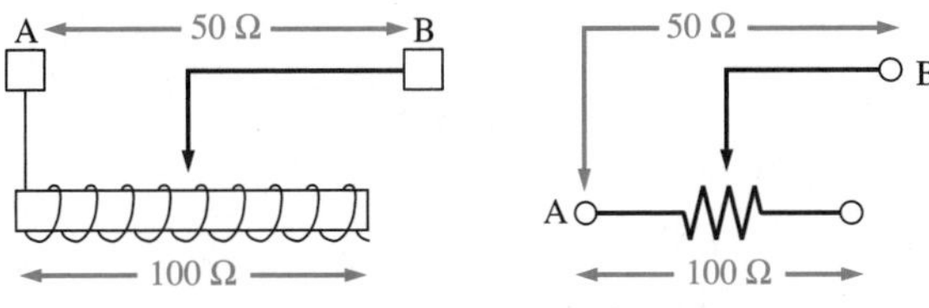

(B) Rheostat and symbol

Figure 4-12 ***Potentiometers and rheostats***

Other types of variable resistors are controlled by temperature or the voltage across them. *Thermistors* are resistors whose temperature controls their resistive value. When the temperature of a thermistor increases, the resistance of the thermistor will usually decrease. Thermistors are very useful in electric circuits because a small change in temperature will result in a large change in resistance.

Varistors are voltage dependent, nonlinear resistors. They are generally used to protect circuits from sudden increases in voltage. The voltage across a varistor will not exceed a certain predetermined amount. This is because the resistance of the varistor will change with changes in voltage.

Varistors are often used in TVs for circuit protection and safety.

Resistors are constructed from many different materials. The three most common types are carbon composition, wire wound, and thin film resistors.

The carbon composition resistor is constructed of finely powdered carbon mixed with a binding material. The ohms value is determined by the ratio of carbon to binding material. It should be noted that the physical size does not have an effect on the resistance value. However, the power rating is directly proportional to the physical size of the resistor. Carbon composition resistors are the most common type of resistor used in electric circuits today.

Wire wound resistors are constructed of many turns of wire. The resistor value is determined by the wire material alloy (i.e., copper-nickel) and the length of the coil of wire. The power rating is determined by the thickness of the diameter of the wire. Wire wound resistors are generally used for high power applications and may exceed 1,000 watts.

Thin film resistors are constructed of a thin film of resistive material on a nonconducting substrate or base like ceramic. They are often found in solid state applications when low power and small size is required. Thin film resistors are very stable and rugged. They do not change resistance value significantly with changes in temperature.

Color coding is used to identify the value of resistors.

4.5 RESISTOR COLOR CODE

Electronics is an international industry with people who speak many different languages, so a method known as the **resistor color code** was developed to identify resistive components. Color coding is one of the most popular methods of identifying the value of small carbon resistors. The basis of the color coding system is to use colors to represent numerical values, as shown in Table 4-1.

Table 4-1 Resistor Color Code

Color	Significant Digit Bands 1, 2	Multiplier Band 3	Tolerance (%) Band 4
Blank	—	—	20
Silver	—	10^{-2}	10
Gold	—	10^{-1}	5
Black	0	$10^0 = 1$	1
Brown	1	$10^1 = 10$	2
Red	2	$10^2 = 100$	
Orange	3	10^3	
Yellow	4	10^4	
Green	5	10^5	
Blue	6	10^6	
Violet	7	10^7	
Gray	8	10^8	
White	9	10^9	

Band 1 represents the first significant digit of the resistor value. Band 2 represents the second significant digit of the resistor value. Band 3 is called the multiplier band and band 4 is called the tolerance band.

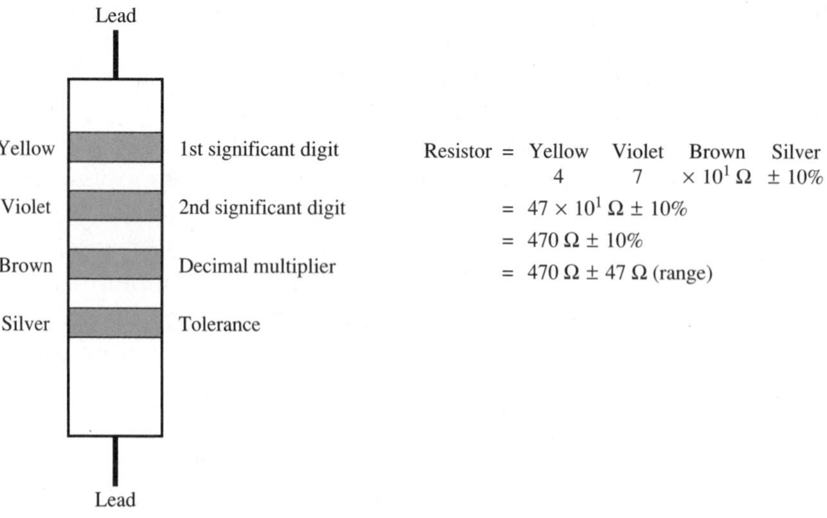

Figure 4-13 Resistor color code

Figure 4-13 demonstrates the method of converting the resistor color code to the resistor value in ohms. Band 1, which represents the first significant digit, is always the band closest to one of the ends of the resistor. Reading from top to bottom in Figure 4-13, the first band is yellow, which represents the number 4. Band 2 is violet, which represents the number 7. Therefore, the first two significant digits are 4 and 7 or 47 respectively. The third band, which is called the multiplier band, is brown. Referring to Table 4-1 we see that this tells us to multiply the two significant digits (47) by 10^1 or 10. Thus the value of the resistor is 47×10^1 ohms or 470 ohms. This is called the **nominal** or ideal value of the resistor. The fourth band is called the **tolerance** band. It tells us how close the actual resistor value is to the nominal value. In this case the fourth band is silver, which from Table 4-1 is equivalent to + or – 10%. This means that the actual resistor value lies somewhere in the range of 470 + 10% and 470 – 10% ohms. In actual numbers this equals 470 + 47 and 470 – 47 ohms, which ranges from 517 to 423 ohms.

Nominal is the ideal or color code value.

Table 4-2 illustrates several more examples of determining resistor values using the color code.

Table 4-2 Color Code Examples

Color Code						Range	
Band 1	Band 2	Band 3	Band 4	Nominal (Ω)	Tolerance (%)	Max. (Ω)	Min. (Ω)
Brown	Black	Red	Gold	1,000	±5	1,050	950
White	Brown	Brown	Silver	910	±10	100	819
Orange	Orange	Orange	Silver	33,000	±10	36,300	29,700
Gray	Red	Yellow	Gold	820,000	±5	861,000	779,000
Brown	Black	Gold	No band	1	±20	1.2	0.8

Resistors are generally valued in the units of kilohms and megohms. One kilohm is equal to 1,000 ohms and 1 megohm is equal to 1,000,000 ohms. A resistor whose value is 2,200 ohms can be called a 2.2-kilohm or 2.2-KΩ resistor. A resistor whose value is 1,500,000 ohms can be called a 1.5-megohm or 1.5-MΩ resistor. These conversions are easily performed by dividing the nominal resistor value by 1,000 for conversion to kilohms and by 1,000,000 for conversion to megohms.

Resistors are often rated in kilohms and megohms.

It is important to be able to reverse the color code process. That is to determine the colors that identify a certain value resistor. For example, what are the colors for a 5.6-kilohm resistor? A 5.6-kilohm resistor is equal to 5.6 × 1,000 ohms or 5,600 ohms. Thus the first significant digit is 5 or the color green. The second significant digit is 6 or the color blue, and the multiplier is 100 or 10^2, which is red. The colors for bands 1, 2, and 3 are then green, blue, and red respectively.

EXAMPLE 4-2

Determine the resistor color code for a 6.8-megohm resistor.

Solution:

$$
\begin{aligned}
\text{Resistor} &= 6.8 \text{ megohms} \\
&= 6.8 \times 10^6 \text{ ohms} \\
&= 6.8 \times 1{,}000{,}000 \text{ ohms} \\
&= 6{,}800{,}000 \text{ ohms} \\
&= 68 \times 100{,}000 \text{ ohms} \\
&= 68 \times 10^5 \text{ ohms}
\end{aligned}
$$

Blue Gray Green

tech tidbit

Most resistors use a four- or five-band color coding scheme.

Some resistors use a five-band color coding scheme. The process is identical to the four-band method except that three significant digits are supplied from bands 1, 2, and 3 respectively.

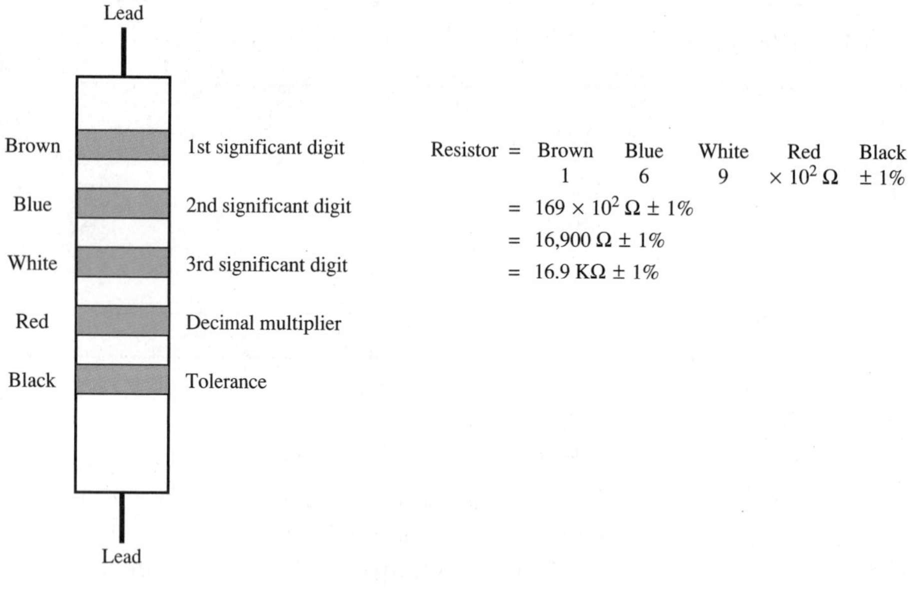

Figure 4-14 **Five-band resistor color code**

In Figure 4-14, band 1 is brown, indicating the number 1. Band 2 is blue, indicating the number 6. Band 3 is white, indicating the number 9. Thus the three significant digits are 169, respectively. The multiplier band is red indicating 10^2 or 100. The nominal value of the resistor is 16,900 ohms. The tolerance or fifth band is black, indicating a tolerance of plus or minus one percent. This means that the actual resistor value lies somewhere between 16,900 + 1% and 16,900 – 1% or 16,900 + 169 ohms and 16,900 – 169 ohms, which ranges from 17,069 to 16,731 ohms.

4.6 RESISTANCE OF WIRE

Wire is one of the most common methods of moving charge or electricity through a circuit. The resistance of a wire is determined by several factors: the type of material, the length of the material, the cross-sectional area of the material, and the temperature of the material. For a given material a given resistance will increase proportionally to the length. Thus the *longer* the material, the *more* its resistance will be. The resistance of a material will vary inversely with its cross-sectional area. That is, the *thicker* the material, the *less* the resistance. Mathematically, resistance can be expressed by the formula

The longer the wire, the greater the resistance. The thicker the wire, the less the resistance.

$$R = \rho\frac{l}{A}$$

Equation 4-3

where ρ = inherent resistivity of the material in circular mils-ohms per foot or ohm-meters

l = the length of the material in feet or meters

A = the cross-sectional area in circular mils or square meters

R = the resistance in ohms

Every material will have its own resistivity or value of ρ. For example, the resistivity, ρ, of copper is 10.37 circular mil-ohms per foot (10.37 CM-Ω/ft) or 1.72×10^{-8} ohm-meters. Table 4-3 gives the resistivity of a number of common materials.

Resistivity, ρ, is the resistance per unit length of material.

Table 4-3 Resistivity

Material	English Units ρ(CM-Ω/fr. @ 20°C)	SI Units ρ(Ω-m @ 20°C)
Silver	9.8	1.65×10^{-8}
Copper	10.37	1.72×10^{-8}
Gold	14.7	2.45×10^{-8}
Aluminum	17.0	2.83×10^{-8}
Nickel	47.0	7.81×10^{-8}
Carbon	21,000.0	$3{,}500.00 \times 10^{-8}$

SI stands for System International, the international system of units.

Circular mils is the unit used for cross-sectional area in the English system of units. One circular mil is the area of a piece of circular material with a diameter equal to 0.001 inch. For simplicity the cross-sectional area of all common copper wire sizes are listed in a table known as the American Wire Gage Standard. Table 4-4 illustrates the American Wire Gage Standard for many of the most common wire sizes.

Table 4-4 American Wire Gage Standard

AWG#	Area (CM)	Ω/1,000 ft @ 20°C (68°F)	AWG#	Area (CM)	Ω/1,000 ft @ 20°C (68°F)
0000	211,600	0.049	19	1,288	8.051
000	167,810	0.0618	20	1,022	10.15
00	133,080	0.0780	21	810	12.80
0	105,530	0.0983	22	642	16.14
1	83,694	0.1240	23	509	20.36
2	66,373	0.1563	24	404	25.67
3	52,634	0.1970	25	320	32.37
4	41,742	0.2485	26	254	40.81
5	33,102	0.3133	27	202	51.47
6	26,250	0.3951	28	160	64.90
7	20,816	0.4982	29	127	81.83
8	16,509	0.6282	30	101	103.2
9	13,094	0.7921	31	79.7	130.1
10	10,381	0.9989	32	63.2	164.1
11	8,234	1.260	33	50.1	206.9
12	6,529	1.588	34	39.8	260.9
13	5,178	2.003	35	31.5	329.0
14	4,107	2.525	36	25.0	414.8
15	3,257	3.184	37	19.8	523.1
16	2,583	4.016	38	15.7	659.6
17	2,048	5.064	39	12.47	831.8
18	1,624	6.385	40	9.9	1,049.0

The resistance of wire doubles for an increase of three wire sizes.

Note that from Table 4-4 as the AWG# or wire size *increases,* the area in circular mils *decreases* and the resistance *increases*. A simple rule of thumb to remember is that #10 copper wire is approximately 1 ohm of resistance per 1,000 feet. The resistance will *approximately double* for an increase of every three wire sizes (i.e., #13 ≈ 2 Ω/1,000 ft., #16 ≈ 4 Ω/1,000 ft., etc.). The resistance will *approximately half* for a decrease of every three wire sizes (i.e., #7 ≈ 0.5 Ω/1,000 ft., #4 ≈ 0.25 Ω/1,000 ft., etc.).

EXAMPLE 4-3

Calculate the resistance of a length of #12 copper house wire at 20°C (68°F).

(A) 2,000 feet (B) 100 feet

Solution:

From Table 4-4 #12 copper wire @20°C = 1.588 ohms/1,000 feet.

(A) $R = 2{,}000 \text{ \sout{feet}} \times \dfrac{1.588 \text{ ohms}}{1{,}000 \text{ \sout{feet}}}$

$= 3.176 \text{ ohms}$

(B) $R = 100 \text{ \sout{feet}} \times \dfrac{1.588 \text{ ohms}}{1{,}000 \text{ \sout{feet}}}$

$= 0.1588 \text{ ohms}$

Alternate Method

From Table 4-4 #12 copper wire @ 20°C

AREA = 6,529 CM

From Table 4-3 ρ for copper @ 20°C

$\rho = 10.37 \text{ CM-}\Omega\text{/ft.}$

(A) $R = \rho \dfrac{l}{A}$

$= \dfrac{(10.37 \text{ CM-}\Omega\text{/ft.})(2{,}000 \text{ ft.})}{6{,}529 \text{ CM}}$

$= 3.176 \text{ ohms}$

(B) $R = \rho \dfrac{l}{A}$

$= \dfrac{(10.37 \text{ CM-}\Omega\text{/ft.})(100 \text{ ft.})}{6{,}529 \text{ CM}}$

$= 0.1588 \text{ ohms}$

If the wire size is not known, the resistance of a wire can be obtained by measuring the diameter of the wire and then calculating the resistance.

EXAMPLE 4-4 (SI System of Units)

Calculate the resistance of 1,000 meters of circular copper wire with a diameter of 2 millimeters.

Solution:

$d = 2 \text{ millimeters} = 2 \times 10^{-3} \text{ meters}$

$A = \dfrac{\pi d^2}{4}$

$= \dfrac{(3.14)(2 \times 10^{-3} \text{ meters})^2}{4}$

$= 3.14 \times 10^{-6} \text{ meters}^2$

$R = \rho \dfrac{l}{A}$

$= \dfrac{(1.72 \times 10^{-8} \text{ ohm-meters})(1{,}000 \text{ meters})}{3.14 \times 10^{-6} \text{ meters}^2}$

$= 5.48 \text{ ohms}$

It is interesting to note that for most types of wire, the resistance will *increase* as the temperature increases. This is true for most types of conductor materials. Carbon is a nonmetallic conductor and is an example of a material whose resistance will decrease as the temperature increases. For most semiconductor materials, the resistance of a material will *decrease* as the temperature increases. The ratio of the change in resistance to the original resistance for a change in temperature is known as the **temperature coefficient** of the resistance of the material. Mathematically it is expressed as:

Equation 4-4

$$\alpha = \frac{\Delta R}{\Delta t (R_{\text{original}})}$$

✓ Self-Test 3

1. The unit of resistance is the ________________.
2. The power rating of a resistor is stated in ________________.
3. ________________ are three-terminal variable resistors.
4. What is the color code for a 12 kilohm, 1% resistor?
5. What is the resistance of 1,000 feet of #12 AWG wire?

4.7 INDUCTORS AND CAPACITORS

An **inductor** is a coil of wire that is sometimes wound around a core. Inductors store enegry in the form of a magnetic field. The coil produces a much more concentrated magnetic field than the magnetic field around a straight piece of wire. If the current through a coil is increased, the magnetic field around the coil will increase. If the current through a coil is decreased, the magnetic field around the coil will decrease. If the current through a coil is suddenly switched off, the magnetic field around the coil will collapse. The collapsing magnetic lines of flux cut across the coil and induce a voltage, *V*, into the coil. This induced voltage is known as a *counter EMF.* The induced voltage or counter EMF will be in the opposite direction of that of the applied current, as shown in Figure 4-15.

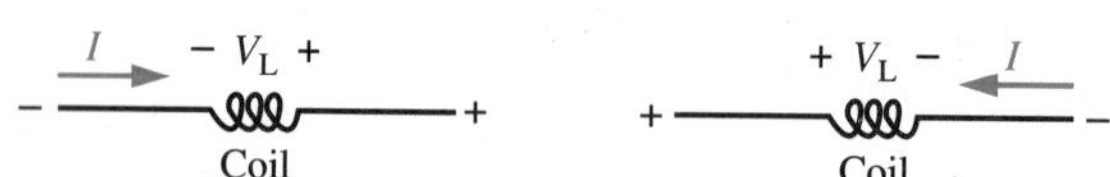

Figure 4-15 **Induced voltage in a coil**

Henry—unit of inductance.

Hence the induced voltage tends to oppose any change in current. The *faster* the lines of force collapse, the *greater* the induced voltage. Therefore it can be said that the inductor tends to oppose any change in current. Figure 4-16 shows the schematic symbol for an inductor or coil. The property of an inductor or coil to store energy is called inductance (L). The unit of inductance is the **henry** (H).

Figure 4-16 **Inductor symbols and units**

A **capacitor** consists of two parallel conductor plates separated by an insulator, as shown in Figure 4-17.

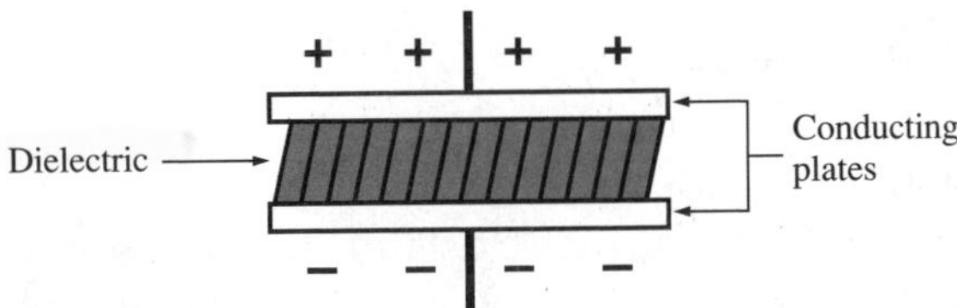

Figure 4-17 **Capacitor**

The insulator is known as a *dielectric*. When opposing charges are placed across the plates of the capacitor, energy is stored in the capacitor in the form of an electric field. The energy is stored in the dielectric material. This is similar to the way energy is stored in a rubber band. When the rubber band is stretched, energy is stored. When one end of the rubber band is let go, the energy is transferred to the other hand. In a similar manner, a capacitor stores and discharges electric energy. The *larger* the area of the plates, the *greater* the capacitor's ability to store charge. The *thinner* the dielectric, the *greater* the capacitance (ability to store energy). Figure 4-18 shows the schematic symbol for a capacitor. The property of a capacitor to store energy is called capacitance (C) and is measured in the unit **farad** (F).

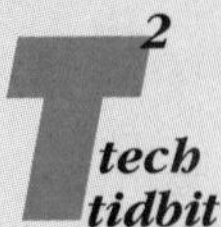

Farad—unit of capacitance.

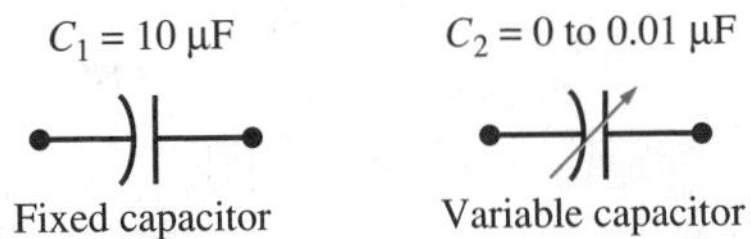

Figure 4-18 **Capacitor symbols**

Much more is said about inductors and capacitors later in the text. It should be noted, of the three electric elements, R, L, C, the resistor *dissipates* electrical energy in the form of heat. The ideal capacitor and the ideal inductor *store* energy—the inductor in the form of magnetism and the capacitor in an electric field.

4.8 SWITCHES

Switches, which are conductors, perform the basic function of opening (breaking) and closing (making) electric connections in circuits. Many types of switches can be used for different applications. Figure 4-19 shows a number of popular switch styles in use today.

Figure 4-19 **Switches**

The basic switch function is shown in Figure 4-20. When the switch is open, as in Figure 4-20A, no current exists in the switch and the circuit. When the switch is closed, as in Figure 4-20B, current exists in the switch and in the rest of the circuit. The switch shown in Figure 4-20 is sometimes referred to as a "knife" switch.

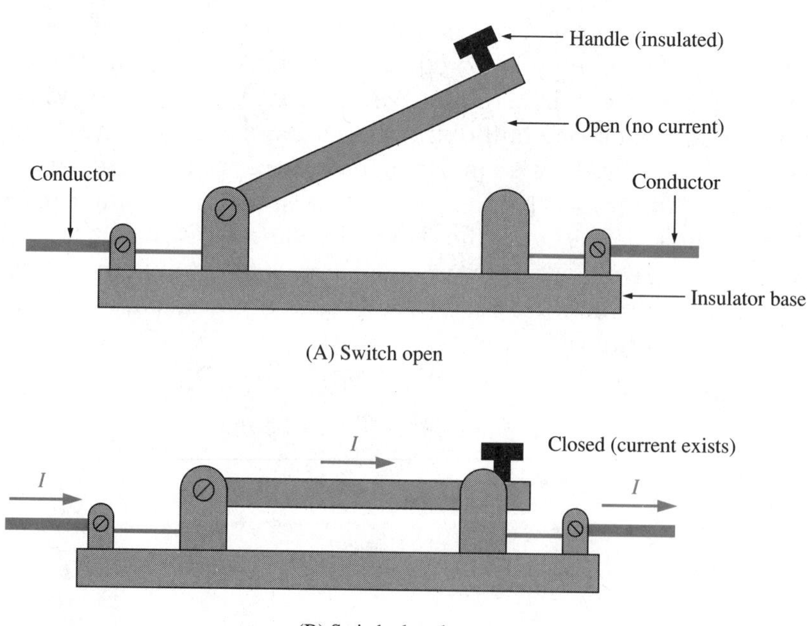

Figure 4-20 **Switch positions**

SPST—
single pole,
singe throw.

DPDT—
double pole,
double throw.

Switches are also classified by the number of circuits they can complete. Figure 4-21 shows the schematic symbols for a number of different types of switches in use today. The switch in Figure 4-21A is called a **single-pole, single-throw** switch. Single-pole refers to the number of conductors being switched. Single-throw refers to the number of places we can switch to. Figure 4-21B shows a single-pole, double-throw switch. Note that the **pole** may be switched to the left or right contact. Hence the term double throw. Figure 4-21C shows a double-pole, single-throw switch and Figure 4-21D shows a

double-pole, double-throw switch. The symbol in Figure 4-21E represents a rotary switch. It is a single-pole, five-position switch. Rotary switches are used in large switching applications where many poles and many switch positions are required.

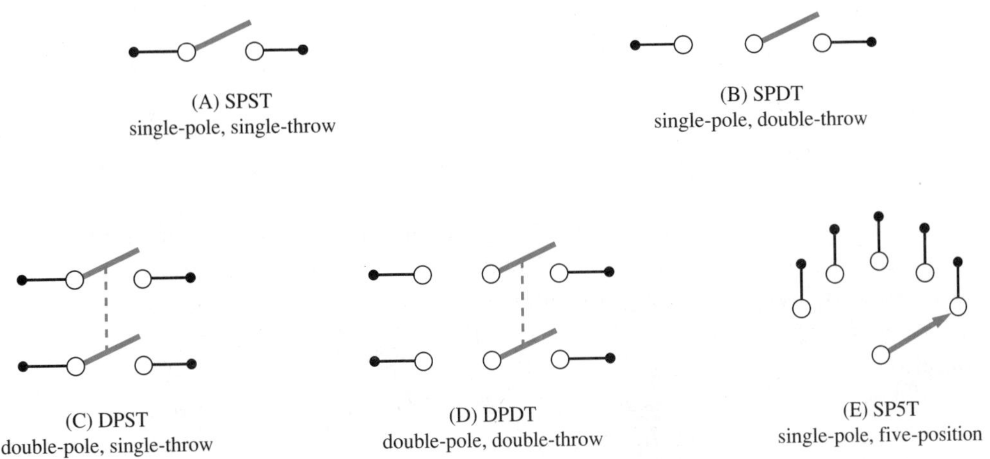

Figure 4-21 **Switch symbols**

Another popular type of switch is the push button or momentary contact switch. Push button switches come in two basic configurations, normally open and normally closed, as shown in Figure 4-22. Push button switches are also available in multipole configurations.

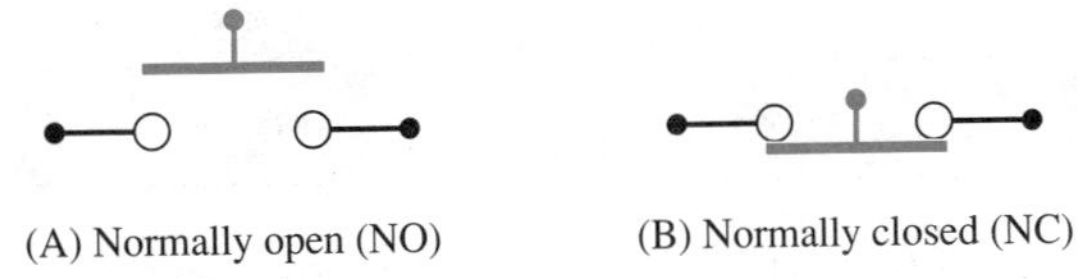

Figure 4-22 **Push button switches**

4.9 PROTECTIVE AND SAFETY DEVICES

Protective devices, as the name implies, are used to safeguard electrical devices and equipment and their operators from damage or harm. The two most common protective devices are the **fuse** and **circuit breaker.** These devices are used to protect circuits from excessive current or overloads. Fuses and circuit breakers are generally connected in series with the device they are intended to protect, as in Figure 4-23. They are intended to open before the rest of the circuit is damaged. A fuse cannot correct a faulty circuit condition. It is intended to open the circuit and stop the current (zero current) thereby preventing further damage.

Protective devices safeguard equipment and people.

Fuses open by melting or blowing out due to excessive heat energy. Circuit breakers open by either magnetism or the bimetallic principle, which are both based on excess current. The magnetic type is opened by the creation of a magnetic field. The bimetallic type opens by expansion due to heat of a bimetallic strip (a strip of metal made of two dissimilar metals bonded together,

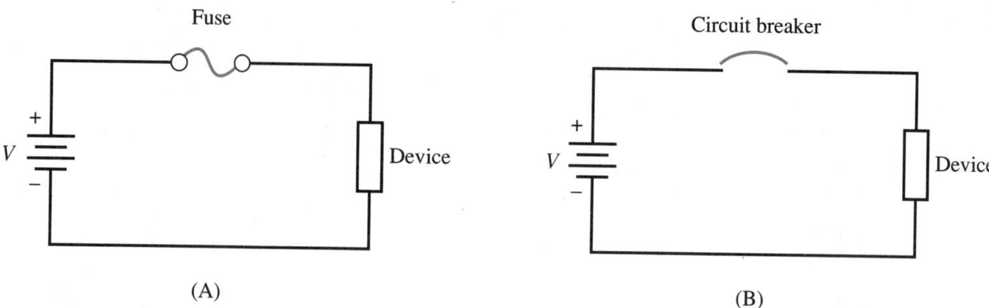

Figure 4-23 **Protection of electric circuits**

each having a different coefficient of expansion). Fuses and circuit breakers are designed to blow either quickly or slowly. Slow-blow fuses eliminate fuse blowing caused by moderate or momentary overloads. For example, a slow-blow fuse may be used to protect a motor with a high starting current and a lower operating current.

Fuses and circuit breakers can be tested by measuring continuity or resistance. A good fuse will indicate continuity or zero ohms and a defective fuse will indicate an open circuit (infinite ohms), as shown in Figure 4-24.

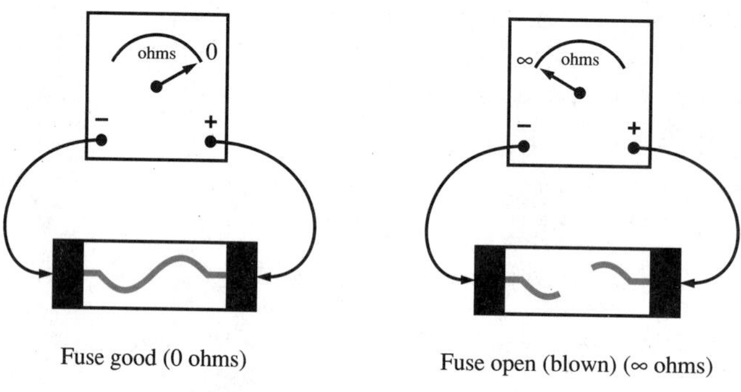

Figure 4-24 **Testing fuses**

As a safety procedure, never replace a fuse with one of greater capacity. A higher value fuse could allow excessive current and result in excessive heat and possibly a fire. Fuses must be replaced, but circuit breakers can be reset.

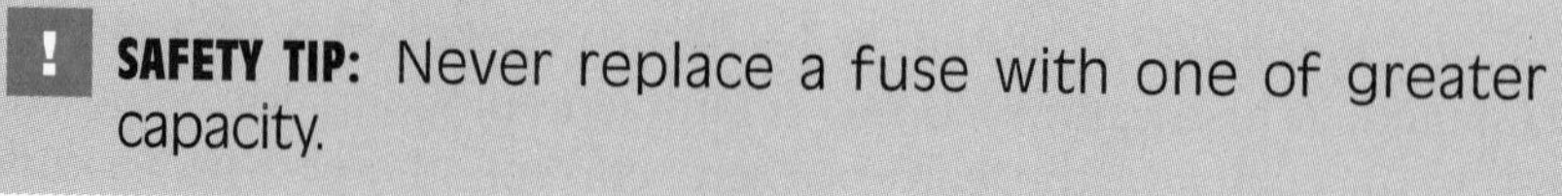

SAFETY TIP: Never replace a fuse with one of greater capacity.

tech tidbit

Fuses protect circuits, GFIs protect people.

A special type of circuit breaker is known as the **ground fault interrupter** (GFI). GFIs are actually very sensitive and fast acting circuit breakers. GFIs and many other types of protective circuit devices are discussed later in this book.

4.10 THE SCHEMATIC DIAGRAM

Electric circuits are described using symbols to represent physical devices. The use of symbols makes the drawing of an electric circuit clearer and easier to understand. Furthermore, because the symbols have become universally accepted, they are understood by people in other countries who do not speak our language. Figure 4-25 illustrates a number of common electric symbols. There are many variations of the basic symbols. However, when you understand the meaning of the basic symbols, understanding the variations becomes obvious. For example, the basic symbol for a resistor is modified with an arrow to indicate a variable resistor.

Schematic symbols are universally accepted.

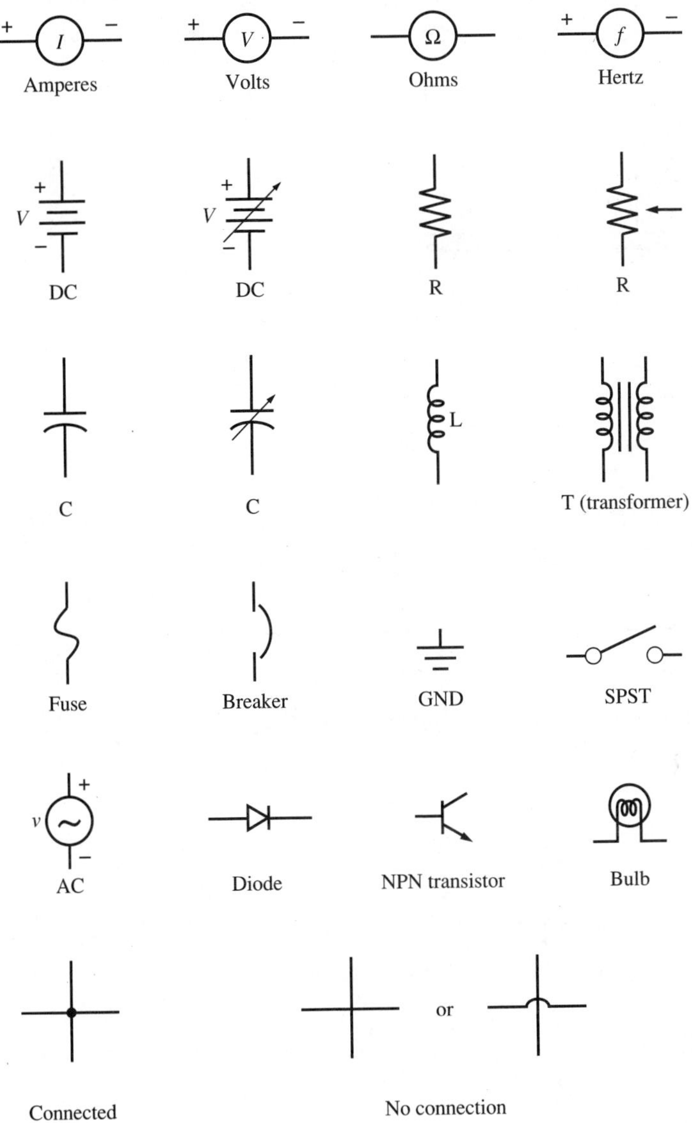

Figure 4-25 **Electric symbols**

Schematic diagrams represent electric circuits.

Combining electric symbols to represent an electric circuit is known as the schematic diagram. Figure 4-26A illustrates a pictorial of physical components in an electric circuit. Figure 4-26B illustrates the schematic diagram representation of the circuit. Note that the schematic diagram clearly identifies all circuit components and their interconnections. It is often helpful when building an electric circuit to place the physical components in the same location and orientation as they are shown in the schematic diagram. This helps to simplify component identification and troubleshooting.

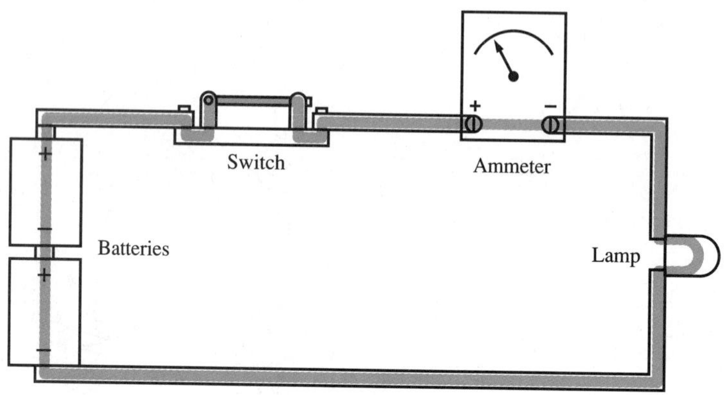

(A) Circuit drawing

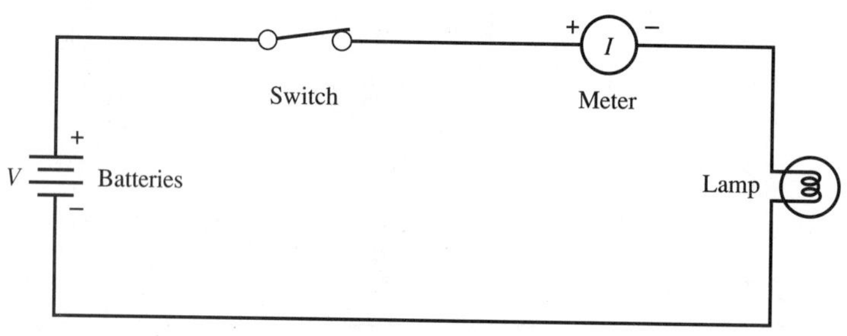

(B) Schematic diagram

Figure 4-26 **Schematic diagram**

✓ Self-Test 4

1. A ________________ produces a much more concentrated magnetic field.
2. A capacitor consists of two parallel conductors separated by an insulator or ________________.
3. The shorthand representation for a single-pole, double-throw switch is ________________.
4. As a safety procedure, never replace a fuse or circuit breaker with one of ________________ capacity.
5. Draw the schematic symbol for a voltmeter, capacitor, inductor, and a circuit breaker.

T³ MEASURING VOLTAGE, CURRENT, AND RESISTANCE

Measuring voltage, current, and resistance is a common way to determine if a circuit is operating correctly or if a circuit component is defective. The following discussion outlines the steps to be taken when performing measurements.

Voltage Measurements

Voltage measurements are the easiest and most common electrical measurements to make. Voltage measurements are made with power connected to the circuit by connecting the voltmeter across the component (in parallel), as shown in Figure 4-27.

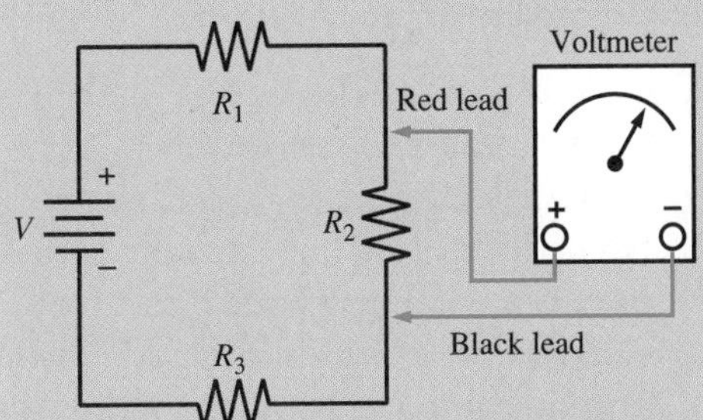

Figure 4-27 **Measuring voltage**

The following procedure is used to make voltage measurements:

1. Select the correct voltage function (AC or DC) for the type of voltage being measured in the circuit.
2. Select the range that is greater than the expected voltage. If the voltage to be measured is unknown, start on the highest range and then switch to a lower range for increased accuracy and readability.
3. Determine the polarity of the voltage to be measured by looking at the schematic diagram or at the battery terminals. This step can be omitted when measuring AC because the polarity is constantly reversing.
4. Connect the negative (black) lead of the volt-ohm-milliammeter (VOM) to the negative end of the voltage to be measured. Touch (or connect) the positive (red) lead of the meter to the positive end of the voltage. In other words, polarity must be observed when measuring DC voltage. If you do not, the meter pointer may bend when it tries to rotate in the wrong direction. On digital voltmeters, a negative voltage may be displayed.

To aid the user, many VOMs have a polarity feature built into the function switch. With the function switch selected to the +DC position, the black lead will be negative and the red lead positive. When the function switch is in the –DC position, the black lead (COM jack) becomes the positive and the red lead also reverses to become the negative. This feature can be useful when proper polarity is unknown or is incorrectly determined. Normally the polarity switch should always be left in the +DC position. In this way, the red lead will be positive. Red is the color most often used to indicate positive in electronic circuits.

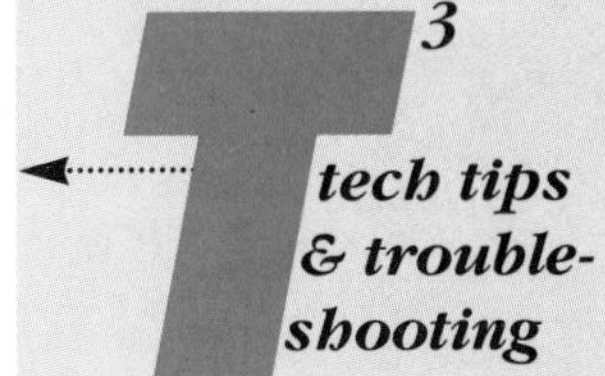

When measuring voltage, always connect the ground or black lead first.

Begin with the highest voltage range.

Always measure voltage with one hand.

Always shut the power off before opening the circuit and inserting the meter.

Never have power on when measuring resistance.

Analog meter scales are still found in many industrial and military applications.

Current Measurements

Current measurements are made much less frequently than voltage measurements. This is because the circuit has to be physically interrupted (opened) to insert the meter (in series), as shown in Figure 4-28.

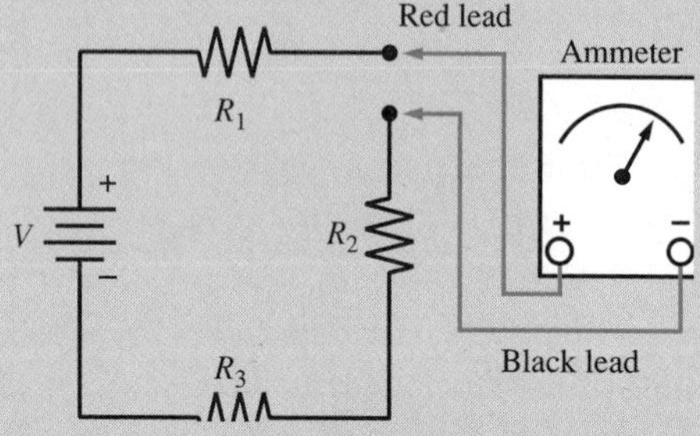

Figure 4-28 *Measuring current*

The following procedure can be used to make current measurements:

1. Select the current function.
2. Select the range that is greater than the expected current.
3. Physically interrupt (open) the circuit so that the meter can be inserted in series with the circuit. Remember, the current exists in the meter as well as the circuit.
4. Observing polarity, connect the meter between the points created by the interruption.

An ammeter can be easily damaged so it should be used with care. Reverse polarity connections may bend the pointer. However, most damage to VOMs occurs by incorrectly connecting them to a circuit. If you connect an ammeter into a circuit the same way that you would a voltmeter, you may destroy it.

Resistance Measurements

The ohmmeter function uses a battery inside the meter housing. Therefore, any other energy source must be disconnected from the circuit in which resistance is to be measured. Note that capacitors can also be a source of energy and must be discharged. The procedure used in measuring resistance is summarized as follows:

1. Remove or turn off the power from the circuit.
2. Select the appropriate range within the ohms function. The appropriate range is one that gives an indication in the right-hand position of the scale.
3. Short (touch) the test leads. Turn the ohms adjust control until the pointer reads 0 ohms.
4. Connect or touch the test leads to the terminals of the device whose resistance is to be measured. Except for some electronic components, such as diodes and transistors, the polarity of the ohmmeter leads is unimportant.

Meter Reading

Analog VOMs may look different, but all of them have functions, ranges, and scales. Functions refers to the quantity being measured (voltage, current, or resistance). For example, the DC voltage function means the meter is being set to measure DC voltage. Range refers to the *maximum* amount of the quantity that can be measured. For example, if the meter is set on the 50-volt range, the meter can measure voltage up to a maximum of 50 volts. Note that the 50 volts can be either AC or DC depending on the function to which the meter is set. Proper use of the VOM involves selection of the correct function, range, and scale. Once you understand the relationships between (1) function and scale and (2) range and scale, you should be able to operate any VOM. The following meter reading examples refer to the Simpson model 260 VOM.

EXAMPLE 4-5

The meter of Figure 4-29 is set to the +DC function. The selector switch is set to the 50-volt range. What is the value of the indicated voltage?

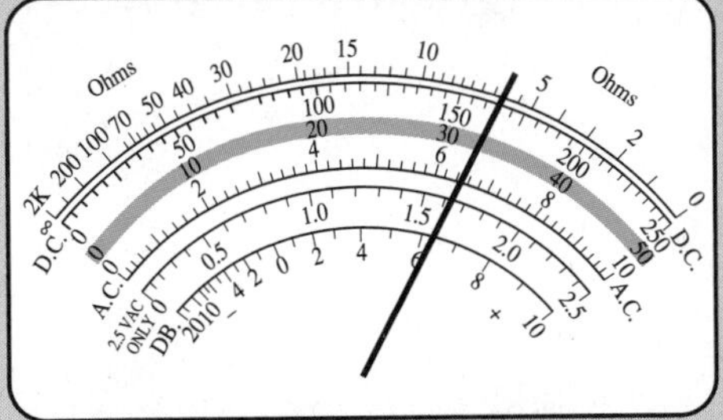

The DC meter reading using the 50-V scale is 33 V.

Figure 4-29 **Meter reading**

Solution:

All DC function measurements (voltage and current) are read on the second scale (black). There are 10 minor divisions between 30 and 40, therefore, each minor division represents 1 volt. The voltage is 33 volts.

EXAMPLE 4-6

The meter in Figure 4-30 is set to the +DC function. The selector switch is set to the 100-milliampere range. What is the value of the indicated current?

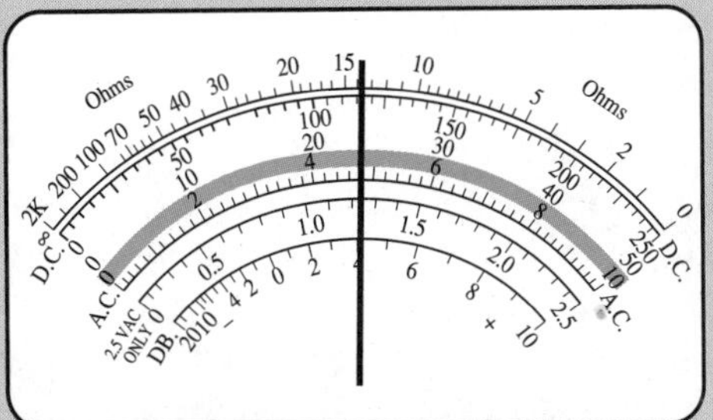

The DC meter reading using the 100-mA scale is 48 mA.

Figure 4-30 **Meter reading**

Solution:

All DC function measurements (voltage and current) are read on the second scale (black). When the scale does not equal the selected range, the scale must either be multiplied or divided by a factor of 10. In this case the 0 to 10 scale must be multiplied by 10 resulting in a 0 to 100 scale. There are 10 minor divisions between 40 and 60 (4 and 6 on the meter face). Therefore, each minor division represents 2 milliamperes. The current is 48 milliamperes.

EXAMPLE 4-7

The meter in Figure 4-31 is set to the AC function. The selector switch is set to the 250-volt range. What is the value of the indicated voltage?

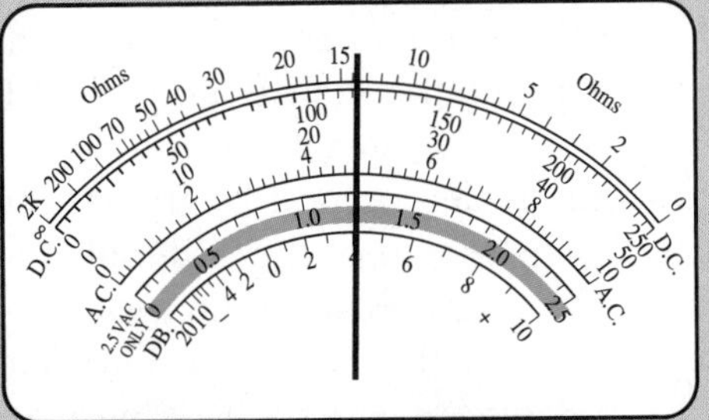

The AC meter reading using the 250-V scale is 120 V.

Figure 4-31 **Meter reading**

Solution:

All AC function measurements except for the 2.5-volt AC range are read on the third scale (red). There are 10 minor divisions between 100 and 150. Therefore, each minor division represents 5 volts. The voltage is 120 volts (AC).

EXAMPLE 4-8

The meter in Figure 4-32 is set to the +DC function. The selector switch is set to the $R \times 10$ range. What is the value of the indicated resistance?

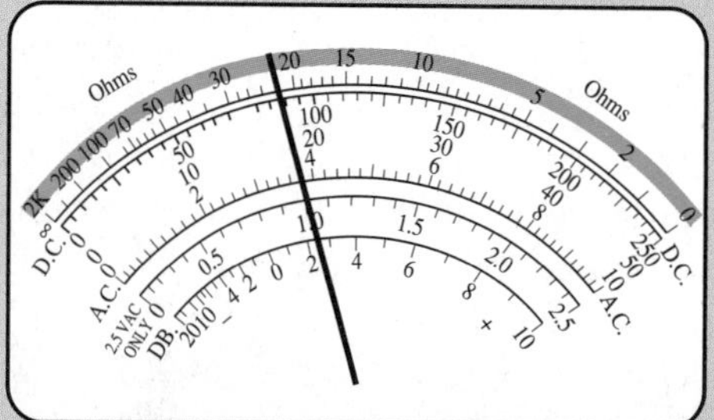

The resistance meter reading using the $R \times 10$ scale is 220 Ω.

Figure 4-32 **Meter reading**

Solution:

Resistance measurements are read on the first black scale (at the top of the meter face). The ohms scale is different from the voltage and current scales in three ways. First, the number of minor divisions between the numbered lines is not the same throughout the scale. Therefore the value of a minor division varies across the scale. The ranges within the ohms function are just multipliers. They indicate the amount by which the ohms scale reading is to be multiplied. In this case the result is multiplied by 10. There are 5 minor divisions between 20 and 30. Therefore, each minor division represents 2. The resistance is 220 ohms ($22 \times 10 = 220\ \Omega$).

SUMMARY

This chapter introduced the basic electric components and physical devices used in electric circuits. An electric circuit in its simplest form consists of an energy source, a load, and conductors that connect everything together. A schematic diagram uses symbols to represent each device in an electric circuit. The electron current standard assumes that a charge flows from minus to plus.

Batteries are sometimes given an ampere-hour rating in addition to a voltage rating. The theoretical life of a battery can be determined if the ampere-hour rating is known. Resistors are often identified using the resistor color code. Resistance is determined by the type of material, the cross-sectional area of the material, the length of the material, and the temperature of the material. For simplicity, the cross-sectional area of all common copper wire sizes are listed in a table known as the American Wire Gauge Standard.

An inductor is a coil of wire that is sometimes wound around a core. Inductors store energy in a magnetic field. The unit of inductance is the henry. A capacitor consists of two parallel conductor plates separated by an insulator known as a dielectric. Energy is stored in a capacitor in the form of an electric force field. The unit of capacitance is the farad.

Measuring voltage, current, and resistance is a common way to determine if a circuit is operating correctly. A common device used to measure voltage, current, and resistance is the VOM. VOMs come in two styles, analog and digital. The analog type uses a scale and pointer to indicate a value. The digital types display a number.

FORMULAS

$$Ah = I \cdot t \qquad LIFE = \frac{Ah}{I}$$

$$R = \frac{\rho l}{A} \qquad A = \frac{\rho l}{R} \qquad \rho = \frac{RA}{l} \qquad l = \frac{RA}{\rho}$$

$$\alpha = \frac{\Delta R}{\Delta t\,(R_{\text{original}})}$$

EXERCISES

Section 4.2

1. Name the three major parts of a simple electric circuit.
2. Name three types of energy sources found in electric circuits.
3. What is the meaning of the term *load?*
4. List three examples of electrical conductors.
5. List three examples of electrical insulators.
6. Design and draw a schematic diagram for a flashlight using two 1.5-volt cells.

Section 4.3

7. What is the major difference between a primary and a secondary cell?
8. Find the magnitude of the voltage from A to B in each of the three battery configurations of Figure 4-33.

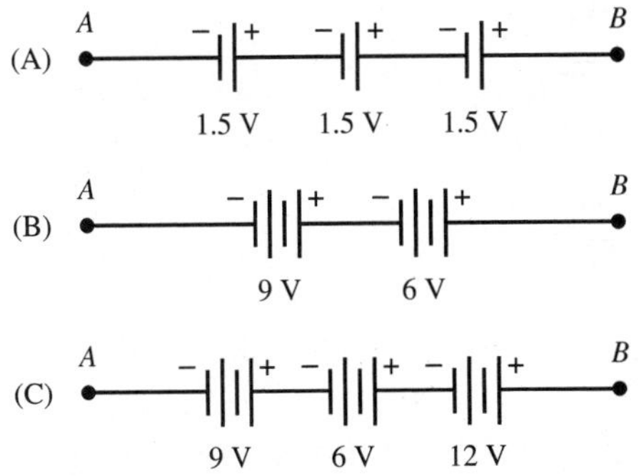

(A) A — 1.5 V — 1.5 V — 1.5 V — B

(B) A — 9 V — 6 V — B

(C) A — 9 V — 6 V — 12 V — B

Figure 4-33

9. Find the magnitude of the voltage from A to B in each of the three battery configurations of Figure 4-34.

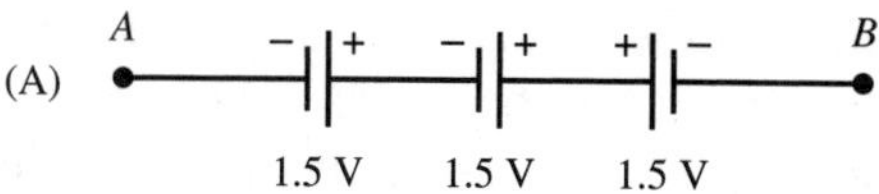

Figure 4-34

10. A 12-volt automobile battery is rated at 80 ampere-hours. What will the life of the battery be when it is supplying current to two headlights that draw 10 amperes each and two taillights that draw 5 amperes each?
11. A radio battery has a life of 20 hours. What is the ampere-hour rating of the battery if it supplies 500 milliamperes?
12. A 12 ampere-hour battery goes dead in 48 hours. How much current was the battery supplying?
13. For the 1-volt battery curve shown in Figure 4-35, calculate the life of the battery when it is supplying 0.6 ampere.

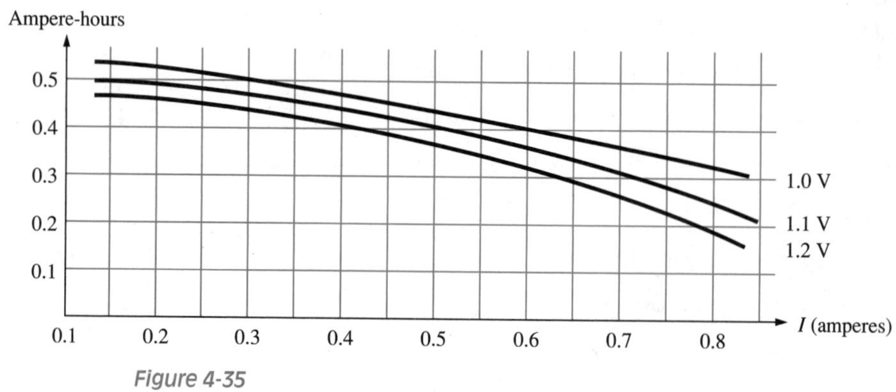

Figure 4-35

14. For the 1.2-volt battery curve shown in Figure 4-35, calculate the life of the battery when it is supplying 0.8 ampere.
15. List five types of energy sources.

Section 4.4

16. List three types of resistor constructions.
17. What factors determine the resistance of a carbon composition resistor?
18. What type of resistor is made from a length of wire?

19. What is the advantage of a film resistor?
20. Determine the value of the following resistors:

	Band 1	Band 2	Band 3
(A)	red	yellow	brown
(B)	green	blue	yellow
(C)	brown	black	black
(D)	white	brown	blue
(E)	yellow	violet	orange

21. Determine the color code for the following resistors:
    (A) 2,200 ohms
    (B) 3.9 kilohms
    (C) 39 kilohms
    (D) 150 kilohms
    (E) 1.5 megohms
22. Determine the range of the following resistors:

	Band 1	Band 2	Band 3	Band 4
(A)	brown	black	brown	gold
(B)	brown	black	red	silver
(C)	red	violet	orange	silver
(D)	blue	red	brown	gold
(E)	gray	red	red	silver

Section 4.6

23. Calculate the resistance of 500 feet of #18 AWG wire at 20°C.
24. Calculate the resistance of one mile of #0 AWG wire at 20°C.
25. How many feet of #22 AWG wire is left on a roll if the resistance of the wire measures 8.07 ohms?
26. Calculate the resistance of 300 meters of copper wire with a diameter of 3 millimeters.

Section 4.8

27. Draw the symbol and describe the following switches:
    (A) SPDT
    (B) DPST
28. Draw the symbol and describe the following switches:
    (A) SP4P Rotary Switch
    (B) NC Push Button Switch

Section 4.10

29. Give the symbol for the following electric devices:
    (A) Resistor
    (B) Inductor
    (C) Capacitor
    (D) Fuse
    (E) Voltmeter

Section T^3

30. Read the following meters:
    (A) a voltmeter using Figure 4-36 on the 10-volt DC range.
    (B) an ammeter using Figure 4-36 on the 500-milliampere DC range.
    (C) an ohmmeter using Figure 4-36 on the $R \times 100$ range.

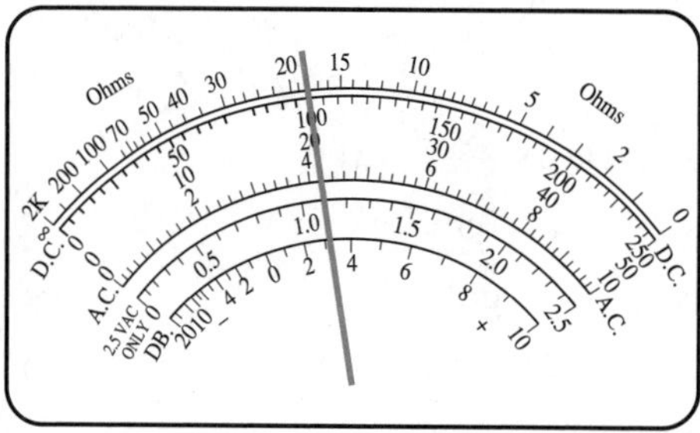

Figure 4-36

31. See Lab Manual for the crossword puzzle for Chapter 4.

CHAPTER 5
SERIES AND PARALLEL CIRCUITS

Courtesy of Cleveland Institute of Electronics, Inc.

KEY TERMS

branch
closed circuit
current divider rule
Kirchhoff's current law
Kirchhoff's voltage law
linear
node
nonlinear
Ohm's law
open circuit
parallel circuit
product over sum rule
series circuit
short circuit
ten percent rule
voltage divider rule
voltage drop
voltage rise
watt

OBJECTIVES

After completing this chapter, the student should be able to:

- state Ohm's law.
- solve problems using Ohm's law.
- define and identify series circuits.
- state Kirchhoff's voltage law.
- solve series circuit problems.
- use the voltage divider rule.
- define and identify parallel circuits.
- state Kirchhoff's current law.
- solve parallel circuit problems.
- use the current divider rule.
- identify opens and shorts in circuits.
- calculate power in series and parallel circuits.

5.1 INTRODUCTION

We have studied the basic quantities of electricity. These are described as voltage, current, and resistance. In this chapter, we will learn about the relationship between these basic electric quantities. The relationship between voltage, current, and resistance was first discovered in the early 1800s by the German scientist Georg Simon Ohm. This relationship is known as **Ohm's law.** It is the basic foundation for almost all electric calculations.

We apply Ohm's law to the analysis of simple electric circuits. This becomes the foundation for the understanding of more complex circuits.

5.2 OHM'S LAW

tech tidbit

A basic law of nature effect = cause/opposition where effect = *I*, cause = *V*, and opposition = *R*.

It can be said that the study of electricity is the study of the effects of the current and the control of the current in an electric circuit. Recall that voltage is the force or pressure that causes current to exist. Current is the rate at which electrons flow through a conductor. Resistance is defined as the opposition a material offers to the flow of electric charge (current). Ohm's law defines the relationship between the flow of current, the force of the voltage, and the opposition or resistance to the current. It states that the current in a circuit is directly proportional to the applied voltage and inversely proportional to the resistance of the circuit. Mathematically,

$$\text{Current} = \frac{\text{Voltage}}{\text{Resistance}}$$

or using symbols

Equation 5-1 ▶

$$I_{(\text{amperes})} = \frac{V_{(\text{volts})}}{R_{(\text{ohms})}}$$

This equation determines the value of the current whenever the voltage and the resistance are both known. Ohm's law can be applied to an entire circuit or to any part of a circuit such as a single resistor. Whenever any two of the three quantities are known, the third quantity can always be determined. Using algebra, we can obtain the following two relationships from Ohm's law:

Equation 5-2 ▶

$$V = I \times R$$

Equation 5-3 ▶

$$R = \frac{E}{I}$$

Equation 5-2 states that the voltage is equal to the product of the current and the resistance. Equation 5-3 states that the resistance is equal to the voltage divided by the current. The symbol *V* is sometimes interchanged with *E* to represent voltage. As an aid in remembering Ohm's law, use the Ohm's law pie chart shown in Figure 5-1. To use the Ohm's law pie chart, simply cover the quantity you wish to determine and perform the indicated operation.

tech tidbit

$I = \frac{V}{R}$

$R = \frac{V}{I}$

$V = I \bullet R$

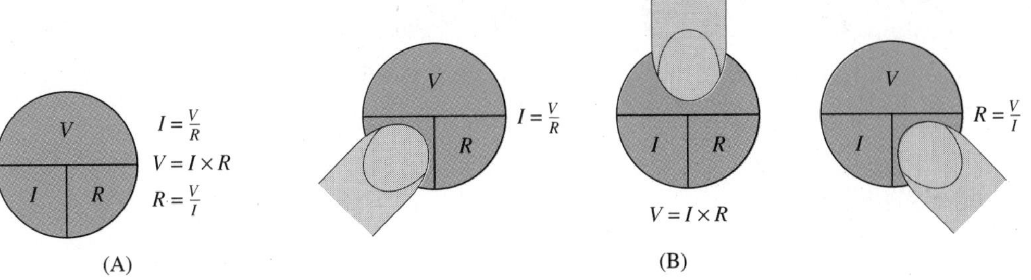

Figure 5-1 **Ohm's law pie chart**

tech tidbit

$V = E$

Now let us consider a simple electric circuit consisting of a battery, a switch, and a lamp. Figure 5-2 shows the actual circuit and the electrical schematic diagram for the circuit. As long as the switch is open, there is no complete path for the current and the lamp will not light. This is sometimes referred to as an **open circuit.** When the switch is closed, charge (current) can flow from the negative terminal of the battery to the positive terminal and the lamp will light. This is referred to as a **closed circuit.**

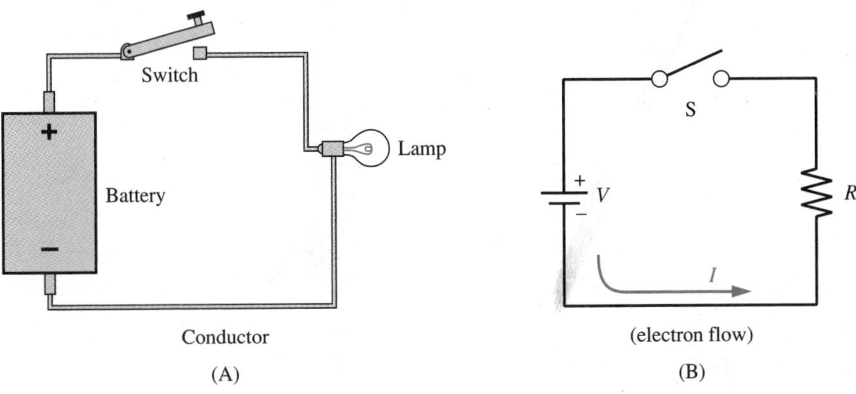

Figure 5-2 **Simple Ohm's law circuit**

If the battery has a voltage of 6 volts and the lamp has a resistance of 3 ohms, we can apply Ohm's law to calculate the current in the circuit as follows:

$$I = \frac{E}{R}$$

$$= \frac{6\ \text{V}}{3\ \Omega}$$

$$= 2\ \text{A}$$

Note that when applying Ohm's law to a load, the current, *I,* must be the current in the load. The voltage, *V,* must be the voltage across the load, and the resistance, *R,* must be the resistance of the load.

EXAMPLE 5-1

How much current is in the circuit shown in Figure 5-3?

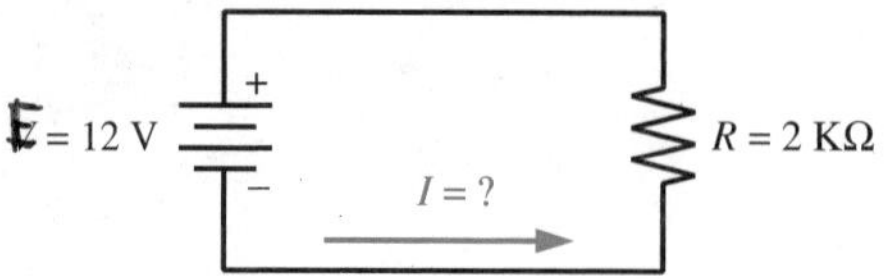

Note: In this simple circuit, the battery voltage is the voltage across *R*.

Figure 5-3

Solution:

Given $E = 12\text{ V}$

$R = 2\text{ K}\Omega = 2 \times 10^3\ \Omega = 2{,}000\ \Omega$

Find $I = ?$

$$I = \frac{E}{R} = \frac{12\text{ V}}{2{,}000\ \Omega} = 0.006\text{ A} = 6 \times 10^{-3}\text{ A} = 6\text{ mA}$$

EXAMPLE 5-2

In the circuit shown in Figure 5-4, how much voltage is required to produce 25 milliamperes of current?

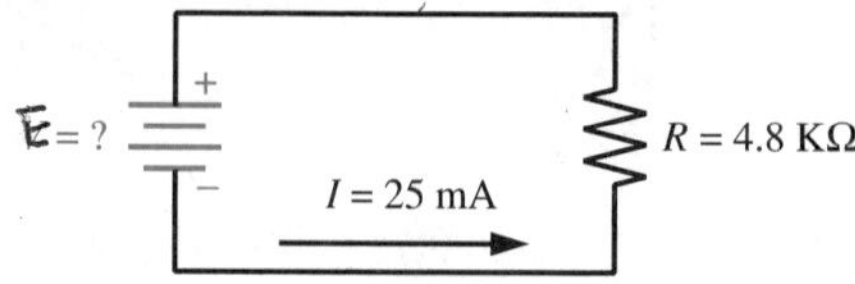

Figure 5-4

Solution:

Given $I = 25\text{ mA} = 25 \times 10^{-3}\text{ A} = 0.025\text{ A}$

$R = 4.8\text{ K}\Omega = 4.8 \times 10^3\ \Omega = 4{,}800\ \Omega$

Find $V = ?$

$$E = IR = (0.025\text{ A})(4{,}800\ \Omega) = 120\text{ V}$$

or

$$V = IR = (25\text{ mA})(4.8\text{ K}\Omega) = 120\text{ V}$$

When using prefixes:
(milli)(kilo) = 1

$\frac{1}{\text{milli}} = \text{kilo}$

$\frac{1}{\text{kilo}} = \text{milli}$

EXAMPLE 5-3

What resistance value is needed for the circuit of Figure 5-5 to draw 1.2 amperes?

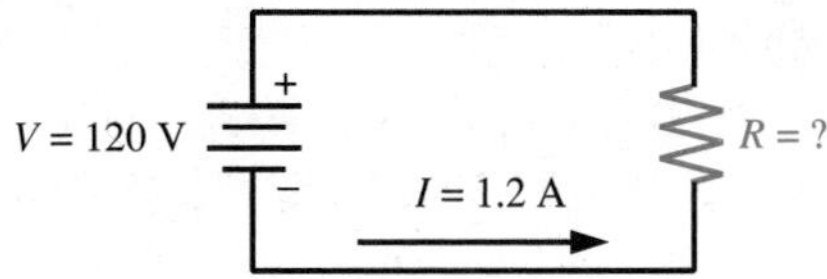

Figure 5-5

Solution:

Given $I = 1.2$ A

$E = 120$ V

Find $R = ?$

$$R = \frac{E}{I}$$

$$= \frac{120 \text{ V}}{1.2 \text{ A}}$$

$$= 100\ \Omega$$

EXAMPLE 5-4

A 120-volt dishwasher has an internal resistance of 9.6 ohms. What size fuse or circuit breaker should be used in this circuit for protection? See Figure 5-6.

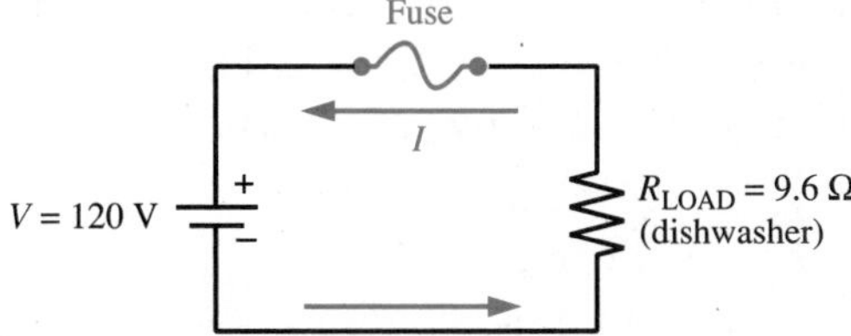

Figure 5-6

Solution:

Given $V = 120$ V

$R = 9.6\ \Omega$

Find $I = ?$

$$I = \frac{V}{R}$$

$$= \frac{120 \text{ V}}{9.6\ \Omega}$$

$$= 12.5 \text{ A}$$

NOTE: The minimum fuse value that can be used is 12.5 amperes. Practically, we would probably select a fuse with a slightly higher value, possibly 15 amperes.

> **! SAFETY TIP:** A safety factor of 25% shold be given to fuses. Example: 25% above the actual measured value of the current. The voltage rating of a fuse should be higher than the required voltage.

✓ Self-Test 1

1. The current times the resistance equals ____________.
2. When a switch is closed and a complete path for current exists, this is known as a ____________ circuit.
3. An appliance has an internal resistance of 20 ohms and operates from a 120-volt source, calculate the current.
4. Find the internal resistance of a car radio that operates on 12 volts and draws 60 milliamperes.

5.3 SERIES CIRCUITS

tech tidbit

A series circuit has only one path for current, therefore, the current is the same everywhere in the circuit.

A **series circuit** is defined as a circuit with only one continuous path for the current, as shown in Figure 5-7. You can determine if a circuit is a series arrangement by imagining yourself as an electron traveling through the circuit. If you can move through the entire circuit in a single continuous path, you have a series circuit. If you can find a way to return to the source, after you break a connection anywhere in the circuit, you do not have a series circuit. There is only one path for the current, so it follows that the current anywhere in a series circuit will always be the same value. Thus, in Figure 5-7, the value of the current at point A is the same as the value of the current at point B, point C, point D, and point E.

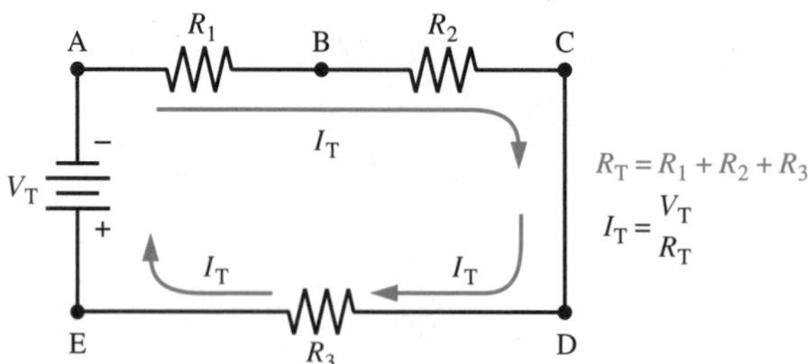

Figure 5-7 ***Series circuit***

tech tidbit

$R_T = \Sigma R$

Furthermore, referring to Figure 5-7, the total resistance of the series circuit is the sum of the individual resistors or opposition. Mathematically,

Equation 5-4 ▶

$$R_T = R_1 + R_2 + R_3 + \ldots + R_N$$

EXAMPLE 5-5

What is the total current in the circuit shown in Figure 5-8?

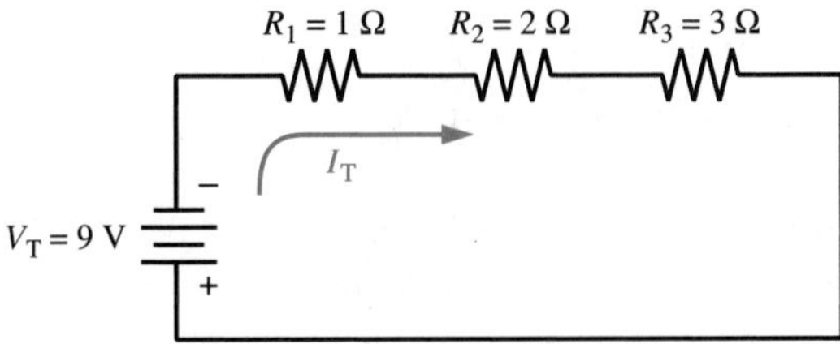

Figure 5-8

Solution:

$$R_T = R_1 + R_2 + R_3$$
$$= 1\ \Omega + 2\ \Omega + 3\ \Omega$$
$$= 6\ \Omega$$
$$I_T = \frac{V_T}{R_T}$$
$$= \frac{9\ \text{V}}{6\ \Omega}$$
$$= 1.5\ \text{A}$$

EXAMPLE 5-6

Find the value of the voltage source in the circuit of Figure 5-9.

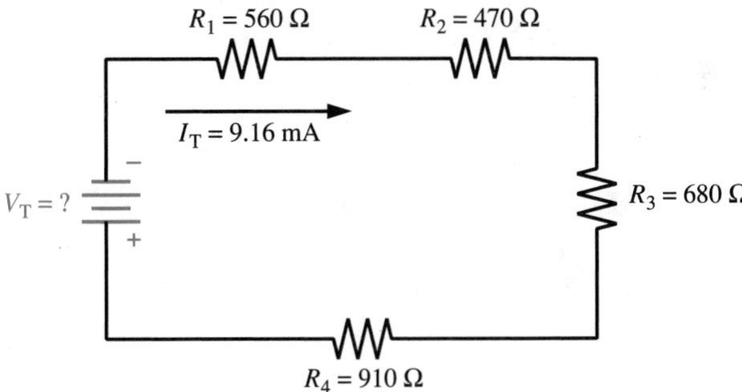

Figure 5-9

Solution:

$$R_T = R_1 + R_2 + R_3 + R_4$$
$$= 560\ \Omega + 470\ \Omega + 680\ \Omega + 910\ \Omega$$
$$= 2{,}620\ \Omega$$
$$V_T = I_T R_T$$
$$= (9.16 \times 10^{-3}\ \text{A})(2{,}620\ \Omega)$$
$$= 24\ \text{V}$$

5.4 KIRCHHOFF'S VOLTAGE LAW

German scientist Gustav Robert Kirchhoff noted that the total voltage in a series circuit is equal to the sum of the individual voltage drops across each resistor. Mathematically,

Equation 5-5

$$V_T = V_1 + V_2 + V_3 + \ldots + V_N$$

This equation is known as **Kirchhoff's voltage law** (KVL). It is an application of the conservation of energy law. Put another way, the applied voltage must equal the sum of the individual voltage drops across each element in a series circuit.

$V_T = \Sigma V$

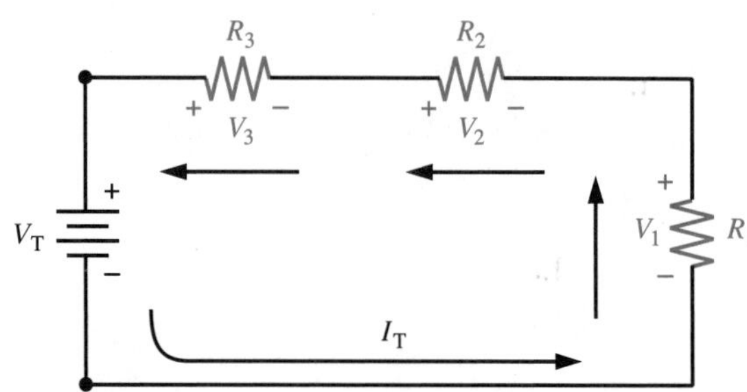

Figure 5-10 **Series circuit—Kirchhoff's voltage law**

Referring to Figure 5-10, we recall that the total resistance is equal to the sum of the individual resistors and the current in a series circuit is the same, so it follows that

$$R_T = R_1 + R_2 + R_3$$

and $$I_T = I_1 = I_2 = I_3$$

Multiplying by I_T, we obtain

$$I_T R_T = I_T R_1 + I_T R_2 + I_T R_3$$

Recalling from Ohm's law that $V = I \times R$, it follows that

$$V_T = V_1 + V_2 + V_3$$

where V_T = the applied voltage or voltage source

V_1 = the voltage across resistor R_1

V_2 = the voltage across resistor R_2

V_3 = the voltage across resistor R_3

$- \longrightarrow +$

Note that in Figure 5-10, the arrows indicate the direction of electron flow. The tail of the arrow is the minus end and the head of the arrow is the plus end. This follows our convention that negative charge (electrons) flows from minus to plus.

Another method of expressing Kirchhoff's voltage law is that the sum of the **voltage rises** and **voltage drops** around the closed path or loop is equal to zero. Algebraically

$$V_T - V_1 - V_2 - V_3 = 0$$

$\Sigma V_{rises} = \Sigma V_{drops}$

This can be derived from Figure 5-10 by following the direction of current through the circuit. As we approach the first resistor, R_1, the first polarity we see is a minus. Therefore, we assign V_1 a minus sign ($-V_1$). We continue the same approach to resistor R_2 producing $-V_2$ and resistor R_3 producing $-V_3$. As we come to the voltage source, V_T, notice that the first polarity sign is a plus, resulting in $+V_T$. Writing Kirchhoff's voltage law for the series circuit produces

$$-V_1 - V_2 - V_3 + V_T = 0$$

because the sum of the voltage rises and voltage drops around the closed loop must equal zero. Rearranging the previous equation produces the following result:

$$+V_T - V_1 - V_2 - V_3 = 0$$

Note that by convention the minus sign indicates a voltage drop and the plus sign indicates a voltage rise.

EXAMPLE 5-7

Find the total voltage required to light the five 12-volt light bulbs shown in Figure 5-11.

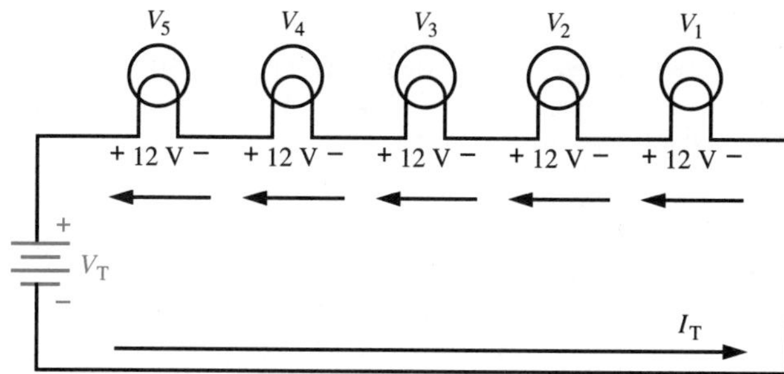

Figure 5-11

Solution:

1. Begin by drawing arrows to indicate the direction of the current, I_T, through each light bulb.
2. Next indicate the polarities.
3. Apply Kirchhoff's voltage law.

$$-V_1 - V_2 - V_3 - V_4 - V_5 + V_T = 0$$

$$-12\text{ V} - 12\text{ V} - 12\text{ V} - 12\text{ V} - 12\text{ V} + V_T = 0$$

$$-60\text{ V} + V_T = 0$$

$$V_T = 60\text{ V}$$

Note that this is a scientific method of analysis. Practically, the total voltage required is the sum of the individual voltage drops of each lamp.

EXAMPLE 5-8

Find the value of the resistor required if the five 12-volt light bulbs are connected to a 120-volt source. Assume that the current in the circuit is 300 milliamperes. See Figure 5-12.

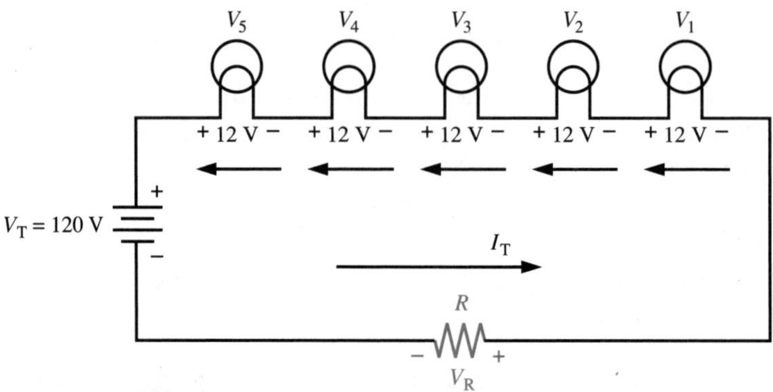

Figure 5-12

Solution:

1. Draw arrows to indicate the direction of current through each light bulb.
2. Next indicate the polarities.
3. Apply Kirchhoff's voltage law.
4. Apply Ohm's law to calculate *R*.

$$-V_R - V_1 - V_2 - V_3 - V_4 - V_5 + V_T = 0$$

$$-V_R - 12\text{ V} - 12\text{ V} - 12\text{ V} - 12\text{ V} - 12\text{ V} + 120\text{ V} = 0$$

$$-V_R - 60\text{ V} + 120\text{ V} = 0$$

$$-V_R + 60\text{ V} = 0$$

$$V_R = 60\text{ V}$$

$$R = \frac{V_R}{I}$$

$$= \frac{60\text{ V}}{}$$

$$= 200\ \Omega$$

EXAMPLE 5-9

Find (A) the total resistance, (B) current, and (C) voltage across each resistor in the circuit of Figure 5-13. (D) Check your answers using Kirchhoff's voltage law.

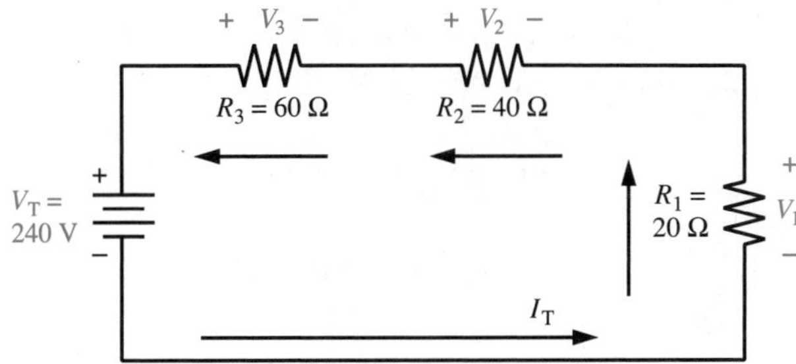

Figure 5-13

Solution:

(A)
$$R_T = R_1 + R_2 + R_3 = 20\ \Omega + 40\ \Omega + 60\ \Omega = 120\ \Omega$$

(B)
$$I_T = \frac{V_T}{R_T} = \frac{240\text{ V}}{120\ \Omega} = 2\text{ A}$$

(C)
$$V_1 = I_1R_1 = (2\text{ A})(20\ \Omega) = 40\text{ V}$$
$$V_2 = I_2R_2 = (2\text{ A})(40\ \Omega) = 80\text{ V}$$
$$V_3 = I_3R_3 = (2\text{ A})(60\ \Omega) = 120\text{ V}$$

(D) Check by Kirchhoff's voltage law (KVL)

$$-V_1 - V_2 - V_3 + V_T = 0$$
$$-40\text{ V} - 80\text{ V} - 120\text{ V} + 240\text{ V} = 0$$
$$-240\text{ V} + 240\text{ V} = 0$$
$$0\text{ V} = 0\text{ V}$$

5.5 VOLTAGE DIVIDER RULE

From Kirchhoff's voltage law, we can see that the voltage drop across individual resistors in a series circuit is proportional to the value of each resistor. We have learned that the voltage drop across any resistor in a series circuit is equal to

$$V_X = IR_X$$

but

$$I = \frac{V_T}{R_T}$$

So by substitution for I:

$$V_X = \frac{R_X \cdot V_T}{R_T}$$

◄ Equation 5-6

tech tidbit

The voltage divider rule can be used across any part of a series circuit: $V_X = \frac{R_X V_T}{R_T}$.

Equation 5-6 is commonly referred to as the **voltage divider rule.** The voltage divider rule is a useful tool for determining individual voltage values in a series circuit when the total voltage is known (without solving for the current).

EXAMPLE 5-10

Using the voltage divider rule, calculate the value of (A) V_1, (B) V_2, (C) V_3, and (D) V_{23} (voltage across resistor R_2 and R_3) in Figure 5-14.

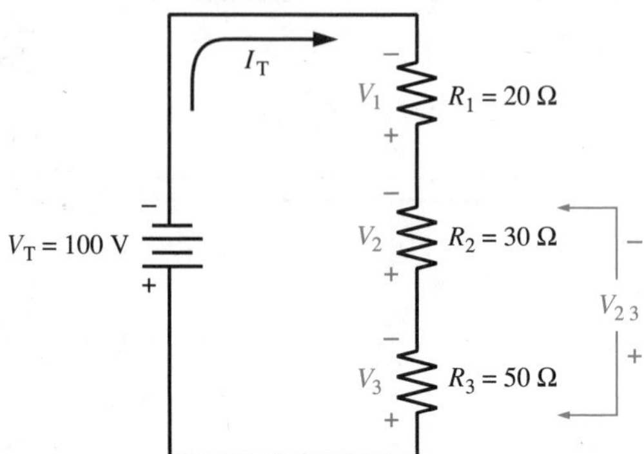

Figure 5-14

Solution:

$$R_T = R_1 + R_2 + R_3 = 20\ \Omega + 30\ \Omega + 50\ \Omega = 100\ \Omega$$

(A) $$V_1 = \frac{R_1 V_T}{R_T} = \frac{(20\ \Omega)(100\ \text{V})}{100\ \Omega} = 20\ \text{V}$$

(B) $$V_2 = \frac{R_2 V_T}{R_T} = \frac{(30\ \Omega)(100\ \text{V})}{100\ \Omega} = 30\ \text{V}$$

(C) $$V_3 = \frac{R_3 V_T}{R_T} = \frac{(50\ \Omega)(100\ \text{V})}{100\ \Omega} = 50\ \text{V}$$

(D) $$R_{23} = R_2 + R_3 = 30\ \Omega + 50\ \Omega = 80\ \Omega$$

$$V_{23} = \frac{R_{23} V_T}{R_T} = \frac{(80\ \Omega)(100\ \text{V})}{100\ \Omega} = 80\ \text{V}$$

It should be noted that in Example 5-10, the voltage across a series resistor is proportional to the value of the resistance. That is, the larger the resistor, the larger the voltage and vice versa.

In Figure 5-15, resistor R_1 is a lot greater than resistor R_2. It follows then that most of the applied voltage will be dropped across resistor R_1. This fact can be an important check when making measurements. We can tell by simply look-

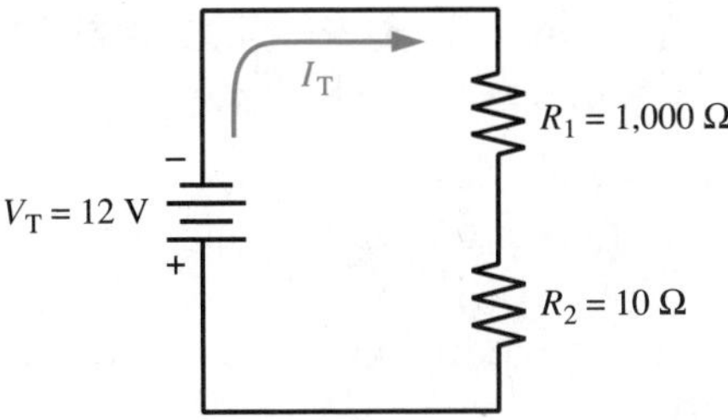

Figure 5-15 ***Ten percent rule***

ing at the circuit that because the resistance of R_1 is larger than R_2, the voltage across R_1 must be greater than the voltage across R_2. Mathematically, by the voltage divider rule, V_1 will be:

$$V_1 = \frac{R_1 V_T}{R_T}$$

$$= \frac{(1{,}000\ \Omega)(12\ \text{V})}{1{,}010\ \Omega}$$

$$= 11.88\ \text{V}$$

Notice that because resistor R_2 is so much smaller than resistor R_1 it has very little affect on the circuit. In fact, when dealing with approximations we could ignore the value of resistor R_2. For example, if we wanted to know the approximate current in the circuit, we could neglect the value of resistor R_2 and approximate the current as

$$I \approx \frac{V}{R}$$

$$\approx \frac{12\ \text{V}}{1{,}000\ \Omega}$$

$$\approx 0.012\ \text{A}$$

Note that the symbol ≈ stands for approximately.

A good rule of thumb is known as the **ten percent rule** of approximation. Whenever a resistor in a series circuit is less than 10% of another resistor, it can be ignored when making approximations.

Ten percent rule—If $R_1 = 0.1R_2$, then R_1 can be ignored.

Approximations are useful in circuit analysis to:

1. Estimate the answer before calculations are performed to avoid calculation errors.
2. Determine which range of an ammeter to use when measuring current.
3. Determine which range of a voltmeter to use when measuring voltage.
4. Determine the size of a replacement fuse.

Estimate before you calculate.

! SAFETY TIP: Never replace a fuse with one of a larger current value or lower voltage value.

5.6 APPLICATIONS OF SERIES CIRCUITS—OPENS AND SHORTS

A very common use of series circuits is seen in the Christmas tree light circuit of Figure 5-16. Each bulb is rated at 12 volts. Thus when ten bulbs are connected in series, the total applied voltage required is 120 volts. This arrangement is simple and inexpensive. It also requires only one wire to be connected from bulb to bulb.

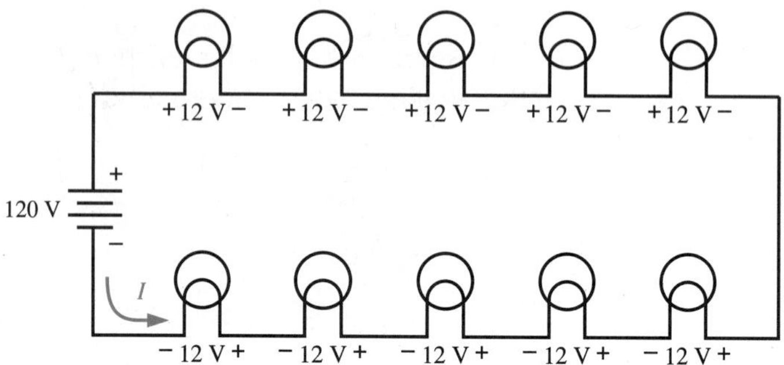

Figure 5-16 **Christmas tree lights**

In an open circuit, the current is 0 A.

However, with a series circuit, if any one element in the circuit breaks, an open circuit is said to exist. For example, if the filament in any of the bulbs in Figure 5-16 were to burn out, no (zero) current would exist in the entire circuit and none of the bulbs would light. This is illustrated in Figure 5-17. Notice that the voltage drop across any of the good bulbs is equal to 0 volts ($I \bullet R$ but $I = 0$). This is because there is no (zero) current in the circuit. Recall that $V = IR$. The voltage across the bulb with the open filament will be 120 volts. This can be verified by Kirchhoff's voltage law, which states that the applied voltage must equal the sum of the voltage drops around the circuit. All of the voltage will appear across the bad bulb because no voltage is dropped across the good bulbs.

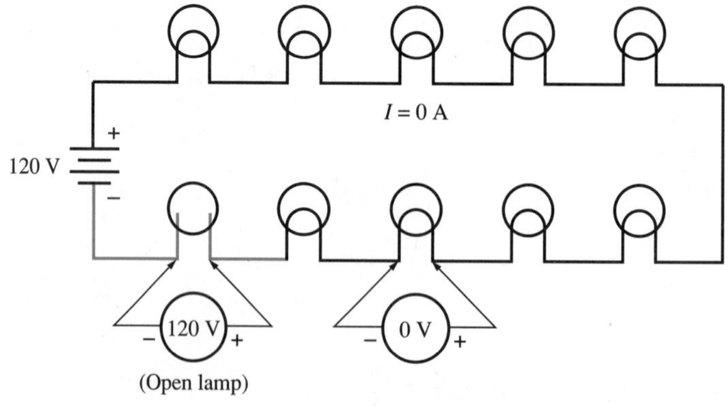

Figure 5-17 **Open circuit**

SAFETY TIP: Be careful—a voltage exists across an open circuit, in this case 120 V.

If while installing your Christmas lights, a staple that is used to hold up the wire causes a connection, as shown in Figure 5-18, a short circuit is said to exist. In this case, no current will flow through the shorted bulb and it will not light. The remaining bulbs will get brighter because the applied voltage of 120 volts will be split among the nine remaining bulbs. This will result in 120 volts ÷ 9 or 13.33 volts across each 12-volt bulb.

tech tidbit

For a short circuit, $R = 0\ \Omega$, $V_{SC} = IR = I(0) = 0\ V$.

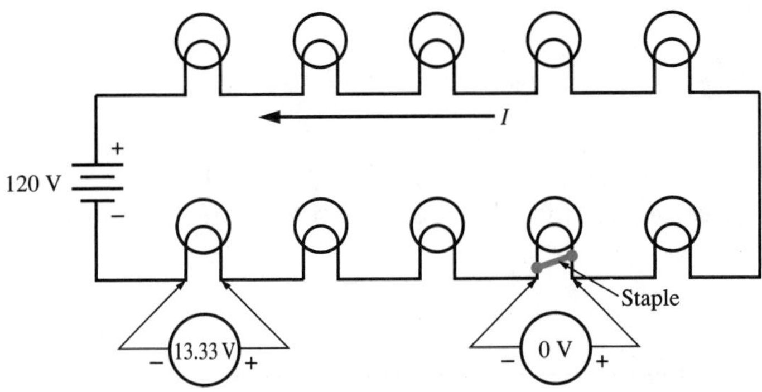

Figure 5-18 ***Short circuit***

tech tidbit

When installing wire with a staple gun, be careful not to create shorts.

Another application of a series circuit is shown in the lamp dimmer circuit of Figure 5-19. In Figure 5-19, a variable series resistor is used to control the amount of current that is allowed in the lamp. By increasing the resistance, the current is decreased and the lamp becomes dimmer (the voltage across the lamp is decreased). By decreasing the resistance, the current increases and the lamp becomes brighter (the voltage across the lamp increases). This circuit has numerous applications. For example, it can control the speed of a motor or the volume of a radio. However, it is an inefficient method of control because additional power is dissipated, in the variable resistor, in the form of heat. This point is discussed further later.

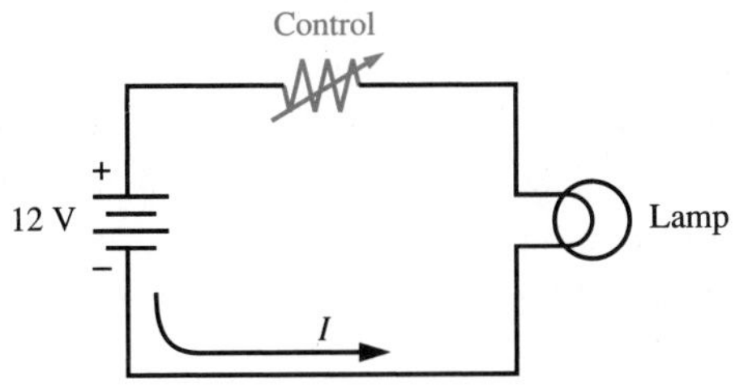

Figure 5-19 ***Lamp dimmer circuit***

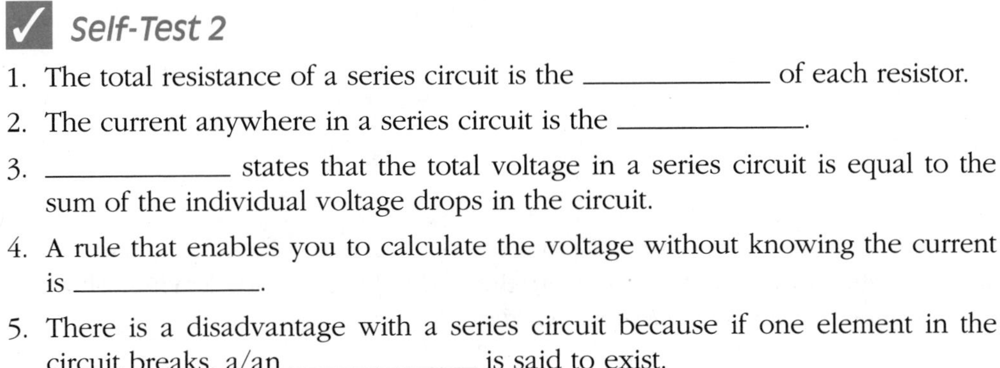

✓ Self-Test 2

1. The total resistance of a series circuit is the ____________ of each resistor.
2. The current anywhere in a series circuit is the ____________.
3. ____________ states that the total voltage in a series circuit is equal to the sum of the individual voltage drops in the circuit.
4. A rule that enables you to calculate the voltage without knowing the current is ____________.
5. There is a disadvantage with a series circuit because if one element in the circuit breaks, a/an ____________ is said to exist.

5.7 PARALLEL CIRCUITS

A **parallel circuit** is one in which the current has more than one path, as shown in Figure 5-20. Each current path is called a **branch.** Notice that for elements to be in parallel, both ends of each element must be connected together, as illustrated by points A and B. Each element or branch is similar to the rungs on a ladder. Each element is connected directly across the applied voltage, so it can be seen that

tech tidbit

Voltages in a parallel circuit are the same.

$$V_T = V_1 = V_2 = V_3$$

In other words, the voltages in a parallel circuit are the same. Note that this is the opposite of a series circuit where the current is the same throughout the circuit.

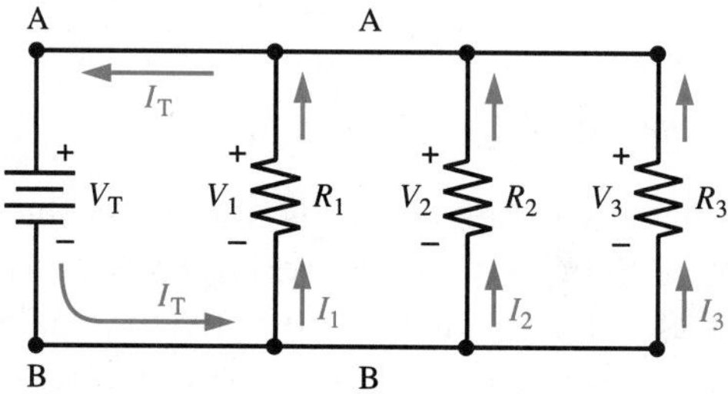

Figure 5-20 **Parallel circuit**

tech tidbit

$I_T = \Sigma I$

From Figure 5-20, we can also see that the total current in a parallel circuit is equal to the sum of the individual branch currents. Mathematically,

Equation 5-7

$$I_T = I_1 + I_2 + I_3 + \ldots + I_N$$

Equation 5-7 is sometimes referred to as **Kirchhoff's current law** (KCL). There is another way of stating Kirchhoff's current law: The currents entering a **node** minus the currents leaving a node are equal to zero. For example, in Figure 5-21, I_1, I_2, and I_4 are entering the node while I_3 is leaving the node. Therefore,

$$I_1 + I_2 - I_3 + I_4 = 0$$

$$2\text{ A} + 3\text{ A} - 9\text{ A} + 4\text{ A} = 0$$

$$9\text{ A} - 9\text{ A} = 0$$

$$0 = 0$$

tech tidbit

Goes inta-goes outta: $\Sigma_{in} = \Sigma I_{out}$

Notice that when the arrows point toward the node, the current is entering, and when the arrows point away from the node, the current is leaving. This can be thought of as the *goes inta-goes outta* rule.

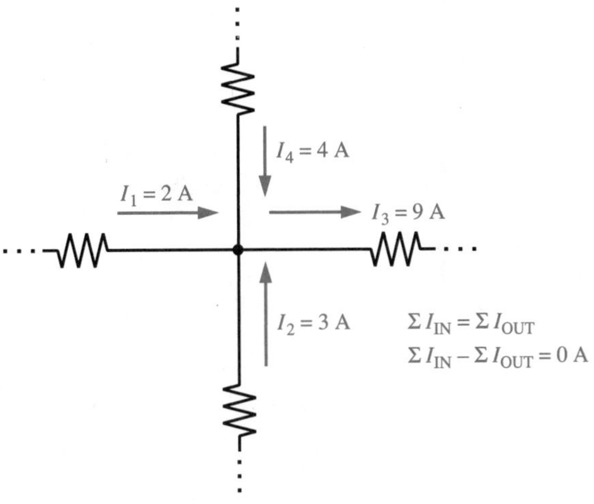

Figure 5-21 **Currents at a node (KCL)**

EXAMPLE 5-11

Find the value of I_2 in the circuit of Figure 5-22.

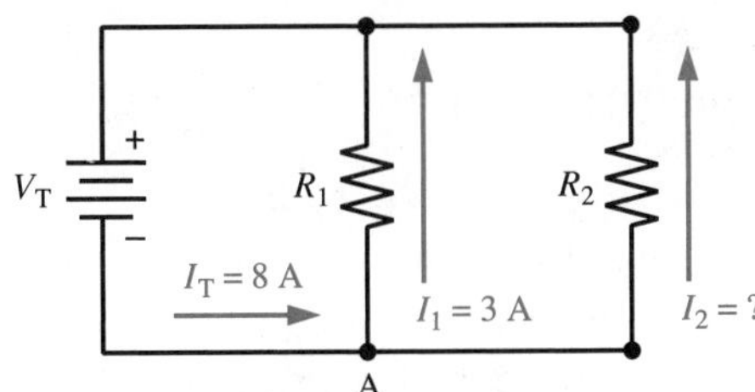

Figure 5-22

Solution:

At Node "A" (KCL)

$$I_T - I_1 - I_2 = 0$$
$$8\text{ A} - 3\text{ A} - I_2 = 0$$
$$5\text{ A} - I_2 = 0$$
$$I_2 = 5\text{ A}$$

Referring to Figure 5-23, we once again see that

at Node "A"

$$I_T - I_1 - I_2 = 0$$
$$I_T = I_1 + I_2$$

But Ohm's law states

$$I = \frac{V}{R}$$

By substitution

$$\frac{V_T}{R_T} = \frac{V_1}{R_1} + \frac{V_2}{R_2}$$

T^2 tech tidbit

KCL can also be expressed as the currents entering the node = the currents leaving the node.

But for parallel circuits

$$V_T = V_1 = V_2$$

Then

$$\frac{V_T}{R_T} = \frac{V_T}{R_1} = \frac{V_T}{R_2}$$

Dividing both sides of the equation by V_T:

Equation 5-8

$$\frac{1}{R_T} = \frac{1}{R_1} + \frac{1}{R_2}$$

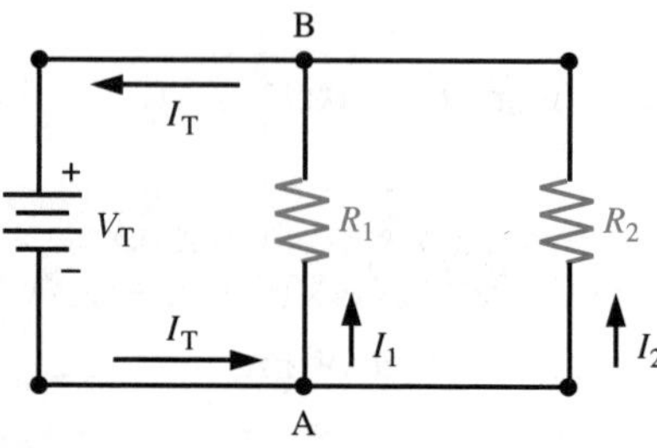

Figure 5-23 **Resistance in parallel**

$$R_T = \frac{1}{\frac{1}{R_1} + \frac{1}{R_2} + \frac{1}{R_3} + \ldots + \frac{1}{R_N}}$$

Equation 5-8 can be used to solve for the total resistance of a parallel circuit with only two resistors. In general, it can be shown that the total resistance of a parallel circuit with any number of resistors is

Equation 5-9

$$\frac{1}{R_T} = \frac{1}{R_1} + \frac{1}{R_2} + \frac{1}{R_3} + \ldots + \frac{1}{R_N}$$

Note that Equation 5-9 does not solve for R_T but its inverse. Recall that conductance is the inverse of resistance and that mathematically the conductance is

Equation 5-10

$$G_{\text{(siemens)}} = \frac{1}{R_{\text{(ohms)}}}$$

It follows then that

Equation 5-11

$$G_{T\text{(siemens)}} = G_1 + G_2 + G_3 + \ldots + G_N$$

and

Equation 5-12

$$R_T = \frac{1}{G_T}$$

EXAMPLE 5-12

For the circuit of Figure 5-24, calculate the branch currents (A) I_1 and (B) I_2, (C) the total current I_T, and (D) the total resistance R_T.

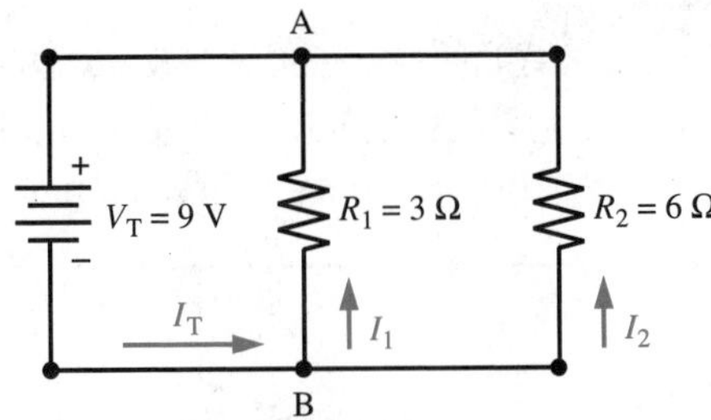

Figure 5-24

Solution:

(A) $V_T = V_1 = V_2 = 9\text{ V}$

$$I_1 = \frac{V_1}{R_1} = \frac{9\text{ V}}{3\ \Omega} = 3\text{ A}$$

(B) $$I_2 = \frac{V_2}{R_2} = \frac{9\text{ V}}{6\ \Omega} = 1.5\text{ A}$$

(C) $$I_T = I_1 + I_2 = 3\text{ A} + 1.5\text{ A} = 4.5\text{ A}$$

(D) $$R_T = \frac{V_T}{I_T} = \frac{9\text{ V}}{4.5\text{ A}} = 2\ \Omega$$

Check

$$\frac{1}{R_T} = \frac{1}{R_1} + \frac{1}{R_2}$$

$$\frac{1}{R_T} = \frac{1}{3\ \Omega} + \frac{1}{6\ \Omega}$$

$$\frac{1}{R_T} = 0.333\text{ S} + 0.167\text{ S}$$

$$\frac{1}{R_T} = 0.5\text{ S}$$

$$R_T = \frac{1}{0.5\text{ S}}$$

$$R_T = 2\ \Omega$$

$$I_T = \frac{V_T}{R_T} = \frac{9\text{ V}}{2\ \Omega} = 4.5\text{ A}$$

Notice that in Example 5-12, the total resistance R_T is less than the smallest value resistor R_1. This fact will be true for any parallel circuit because we are increasing the number of parallel paths for the current to flow. Another way to see this point is to think of a classroom with one door. If we add a second door, no matter how small the door is, the total opposition or resistance to students leaving the classroom, is decreased.

The total resistance in a parallel circuit is always less than the smallest resistor.

EXAMPLE 5-13

For the circuit of Figure 5-25, calculate the branch currents, (A) I_1, (B) I_2, (C) I_3, (D) the total current, and (E) the total resistance of the circuit.

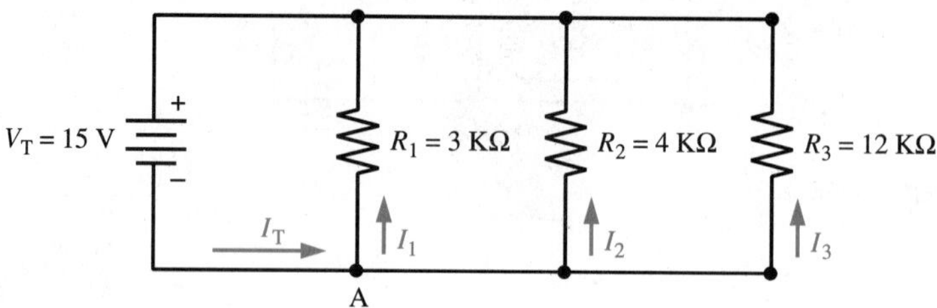

Figure 5-25

Solution:

(A) $V_T = V_1 = V_2 = V_3 = 15\text{ V}$

$$I_1 = \frac{V_1}{R_1} = \frac{15\text{ V}}{3\text{ K}\Omega} = 5\text{ mA}$$

(B) $$I_2 = \frac{V_2}{R_2} = \frac{15\text{ V}}{4\text{ K}\Omega} = 3.75\text{ mA}$$

(C) $$I_3 = \frac{V_3}{R_3} = \frac{15\text{ V}}{12\text{ K}\Omega} = 1.25\text{ mA}$$

(D) At node "A"

$$I_T = I_1 + I_2 + I_3$$

$$= 5\text{ mA} + 3.75\text{ mA} + 1.25\text{ mA}$$

$$= 10\text{ mA}$$

(E) $$R_T = \frac{V_T}{I_T}$$

$$= \frac{15\text{ V}}{10\text{ mA}}$$

$$= 1.5\text{ K}\Omega$$

Check

$$R_T = \frac{1}{\frac{1}{R_1} + \frac{1}{R_2} + \frac{1}{R_3}}$$

$$= \frac{1}{\frac{1}{3\text{ K}\Omega} + \frac{1}{4\text{ K}\Omega} + \frac{1}{12\text{ K}\Omega}}$$

$$= \frac{1}{0.333\text{ mS} + 0.250\text{ mS} + 0.0833\text{ mS}}$$

$$= \frac{1}{0.6663\text{ mS}}$$

$$= 1.5\text{ K}\Omega$$

$$I_T = \frac{V_T}{R_T}$$

$$= \frac{15\ \text{V}}{1.5\ \text{K}\Omega}$$

$$= 10\ \text{mA}$$

The mathematical solution for the total resistance in a parallel circuit is cumbersome. The use of algebra as a tool can help to simplify this process. For example, for two resistors in parallel:

$$\frac{1}{R_T} = \frac{1}{R_1} + \frac{1}{R_2}$$

$$\frac{1}{R_T} = \frac{R_2}{R_1 R_2} + \frac{R_1}{R_1 R_2}$$

$$\frac{1}{R_T} = \frac{R_1 + R_2}{R_1 R_2}$$

$$R_T = \frac{R_1 R_2}{R_1 + R_2}$$

◀ Equation 5-13

Equation 5-13 can be used to solve for the total resistance of a parallel network with only two resistors. It is known as the **product over sum rule.**

tech tidbit

Product over sum rule:
$R_T = \frac{R_1 R_2}{R_1 + R_2}$

EXAMPLE 5-14

Find the total resistance of the network in Figure 5-26.

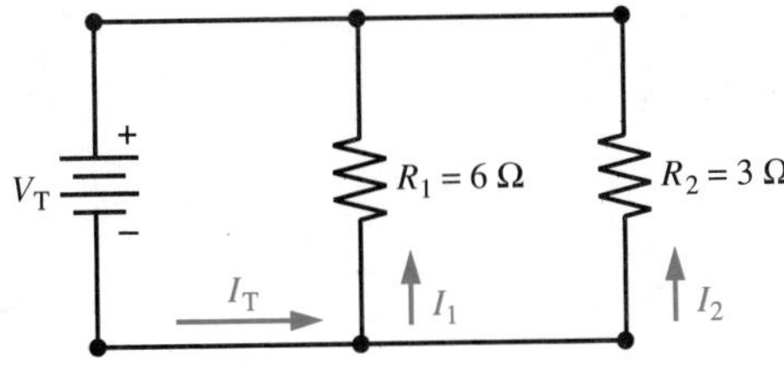

Figure 5-26

Solution:

$$R_T = \frac{R_1 R_2}{R_1 + R_2}$$

$$= \frac{(6\ \Omega)(3\ \Omega)}{6\ \Omega + 3\ \Omega}$$

$$= \frac{18\ \Omega\ \Omega}{9\ \Omega}$$

$$= 2\ \Omega$$

EXAMPLE 5-15

Find the total resistance of the network in Figure 5-27.

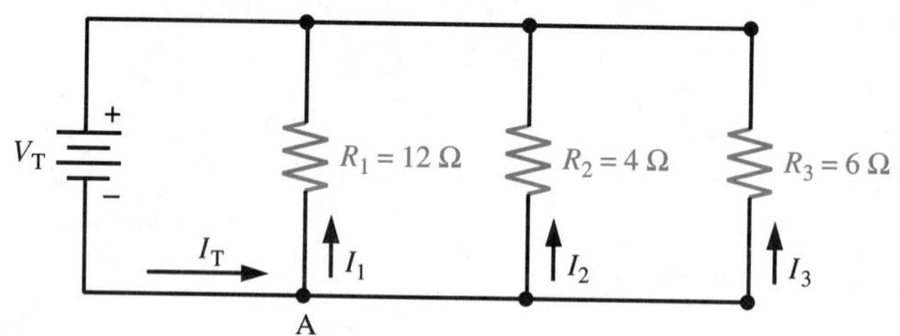

Figure 5-27

Solution:

$$R_{12} = \frac{R_1 R_2}{R_1 + R_2} = \frac{(12\ \Omega)(4\ \Omega)}{12\ \Omega + 4\ \Omega} = \frac{48\ \Omega\ \Omega}{16\ \Omega} = 3\ \Omega$$

$$R_T = \frac{R_{12} R_3}{R_{12} + R_3} = \frac{(3\ \Omega)(6\ \Omega)}{3\ \Omega + 6\ \Omega} = \frac{18\ \Omega\ \Omega}{9\ \Omega} = 2\ \Omega$$

Check by Fractions

$$\frac{1}{R_T} = \frac{1}{R_1} + \frac{1}{R_2} + \frac{1}{R_3}$$

$$\frac{1}{R_T} = \frac{1}{12\ \Omega} + \frac{1}{4\ \Omega} + \frac{1}{6\ \Omega}$$

$$\frac{1}{R_T} = \frac{4}{48\ \Omega} + \frac{12}{48\ \Omega} + \frac{8}{48\ \Omega}$$

$$\frac{1}{R_T} = \frac{24}{48\ \Omega}$$

$$R_T = \frac{48\ \Omega}{24}$$

$$R_T = 2\ \Omega$$

Check by Calculator

$$R_T = \frac{1}{\frac{1}{R_1} + \frac{1}{R_2} + \frac{1}{R_3}} = \frac{1}{\frac{1}{12\ \Omega} + \frac{1}{4\ \Omega} + \frac{1}{6\ \Omega}} = \frac{1}{0.0833\ \text{S} + 0.25\ \text{S} + 0.1667\ \text{S}} = \frac{1}{0.5\ \text{S}} = 2\ \Omega$$

T² tech tidbit

If all resistors in parallel are the same value, $R_T = \frac{R}{N}$

For parallel resistors, if all the resistors are the same value, the total resistance can be found using equation 5-14:

Equation 5-14 ▶

$$R_T = \frac{R}{N}$$

where R = the individual resistor value and

N = the number of resistors that are in parallel.

EXAMPLE 5-16

Calculate the total resistance of Figure 5-28.

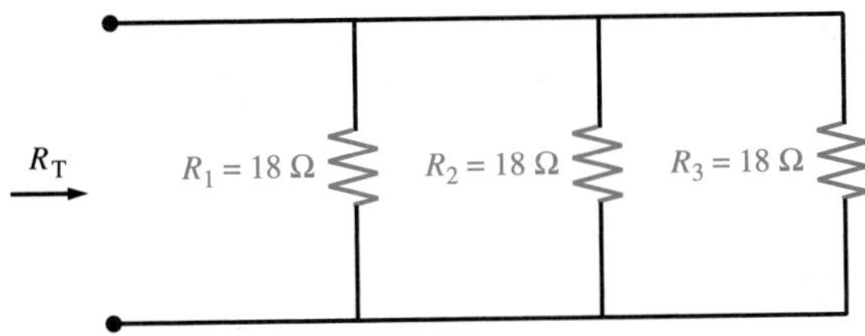

Figure 5-28

Solution:

$$R_T = \frac{R}{N}$$

$$= \frac{18\ \Omega}{3}$$

$$= 6\ \Omega$$

EXAMPLE 5-17

Find the resistance, R_1, of the lamp that is connected in parallel with the two 8-ohm resistive loads in Figure 5-29.

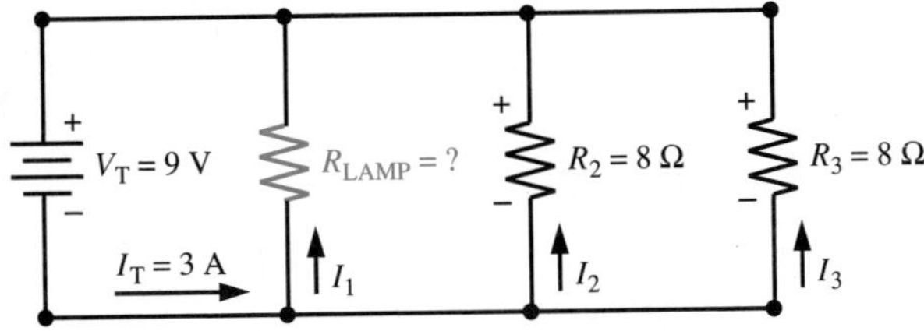

Figure 5-29

Solution:

Method 1

$$R_T = \frac{V_T}{I_T}$$

$$= \frac{9\text{ V}}{3\text{ A}}$$

$$= 3\ \Omega$$

$$R_{LOADS} = \frac{R}{N}$$

$$= \frac{8\ \Omega}{2}$$

$$= 4\ \Omega$$

$$\frac{1}{R_T} = \frac{1}{R_{LOADS}} + \frac{1}{R_{LAMP}}$$

$$\frac{1}{3\ \Omega} = \frac{1}{4\ \Omega} + \frac{1}{R_{LAMP}}$$

$$\frac{1}{R_{LAMP}} = \frac{1}{3\ \Omega} - \frac{1}{4\ \Omega}$$

$$\frac{1}{R_{LAMP}} = 0.333\ \text{S} - 0.250\ \text{S}$$

$$\frac{1}{R_{LAMP}} = 0.083\ \text{S}$$

$$R_{LAMP} = \frac{1}{0.083\ \text{S}}$$

$$R_{LAMP} = 12\ \Omega$$

Method 2

$$V_T = V_{LAMP} = V_2 = V_3 = 9\ \text{V}$$

$$I_2 = \frac{V_2}{R_2} = \frac{9\ \text{V}}{8\ \Omega} = 1.125\ \text{A}$$

$$I_3 = \frac{V_3}{R_3} = \frac{9\ \text{V}}{8\ \Omega} = 1.125\ \text{A}$$

$$I_T = I_1 + I_2 + I_3 \quad \text{(KCL)}$$

$$3\ \text{A} = I_1 + 1.125\ \text{A} + 1.125\ \text{A}$$

$$3\ \text{A} = I_1 + 2.25\ \text{A}$$

$$I_1 = 3\ \text{A} - 2.25\ \text{A}$$

$$I_1 = 0.75\ \text{A}$$

$$R_{LAMP} = \frac{V_{LAMP}}{I_1} = \frac{9\ \text{V}}{0.75\ \text{A}} = 12\ \Omega$$

5.8 CURRENT DIVIDER RULE

The **current divider rule** can be used for determining branch currents when the total current is known. It can be thought of as similar to the voltage divider rule in a series circuit. The general form of the current divider rule can be developed referring to Figure 5-30 as follows.

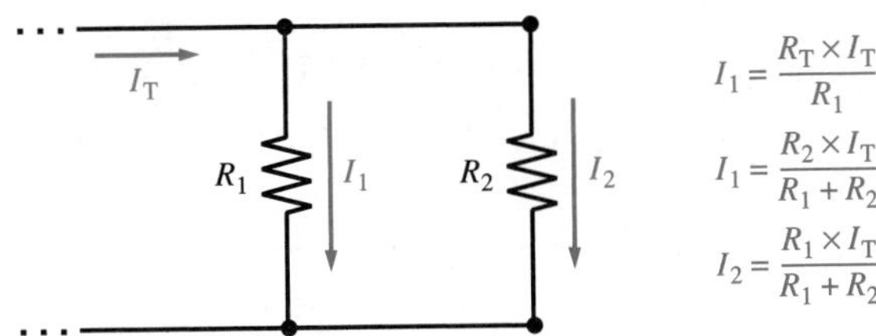

Figure 5-30 **Current divider rule**

From Ohm's law

$$I_X = \frac{V_X}{R_X}$$

But in a parallel circuit

$$V_X = V_T$$

Therefore

$$I_X = \frac{V_T}{R_X} \qquad (1)$$

But

$$V_T = I_T R_T \qquad (2)$$

So by substitution of 2 into 1

$$I_X = \frac{I_T R_T}{R_X}$$

◀ *Equation 5-15*

where the subscript X represents branch 1 or branch 2.

Notice that the general form of the current divider rule requires the user to calculate the total resistance, R_T. When only two resistors are in parallel, a simpler form of the current divider rule can be derived as follows:

$$I_1 = \frac{R_T I_T}{R_1} \qquad \text{General Form}$$

but

$$R_T = \frac{R_1 R_2}{R_1 + R_2} \qquad \text{when two resistors are in parallel}$$

T² *tech tidbit*

Note, in the current divider rule, the opposite resistor is in the numerator.

By substitution

$$I_1 = \frac{\left(\frac{R_1 R_2}{R_1 + R_2}\right) I_T}{R_1}$$

Simplifying

$$I_1 = \frac{R_2 I_T}{R_1 + R_2}$$

◀ *Equation 5-16*

Similarly it can be shown that

$$I_2 = \frac{R_1 I_T}{R_1 + R_2}$$

◀ *Equation 5-17*

Note that the opposite resistor in both equations, 5-16 and 5-17, appears in the numerator of the rule. **This is opposite to that of the voltage divider rule.**

EXAMPLE 5-18

Calculate the branch currents (A) I_1 and (B) I_2 for the circuit of Figure 5-31.

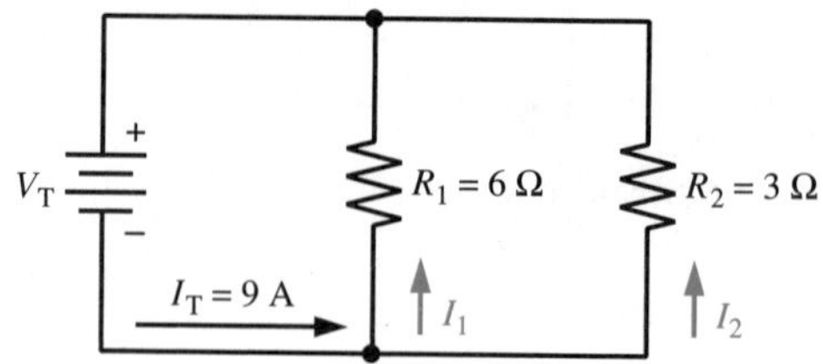

Figure 5-31

Solution:

As a check by KCL:
$I_T = I_1 + I_2$
$9\text{ A} = 3\text{ A} + 6\text{ A}$
$9\text{ A} = 9\text{ A}$

Method 1

(A)
$$I_1 = \frac{R_2 I_T}{R_1 + R_2} = \frac{(3\ \Omega)(9\text{ A})}{6\ \Omega + 3\ \Omega} = 3\text{ A}$$

(B)
$$I_2 = \frac{R_1 I_T}{R_1 + R_2} = \frac{(6\ \Omega)(9\text{ A})}{6\ \Omega + 3\ \Omega} = 6\text{ A}$$

Method 2

(A)
$$R_T = \frac{R_1 R_2}{R_1 + R_2} = \frac{(6\ \Omega)(3\ \Omega)}{6\ \Omega + 3\ \Omega} = 2\ \Omega$$

$$I_1 = \frac{R_T I_T}{R_1} = \frac{(2\ \Omega)(9\text{ A})}{6\ \Omega} = 3\text{ A}$$

(B)
$$I_2 = \frac{R_T I_T}{R_2} = \frac{(2\ \Omega)(9\text{ A})}{3\ \Omega} = 6\text{ A}$$

5.9 APPLICATIONS OF PARALLEL CIRCUITS—OPENS AND SHORTS

A parallel circuit has the advantage that if one branch fails, the other branches can continue to operate. In Figure 5-32, we see a simple parallel circuit consisting of two light bulbs labeled R_1 and R_2. If bulb R_1 burns out or opens, no (zero) current will be in R_1. However, bulb R_2 will continue to operate. The total current in the circuit will now be equal to I_2 only. If light bulb R_2 were to be shorted, then no current would flow through R_2 and R_2 would not light. All of the current would take the path of least resistance and be equal to the short circuit current (I_S). Note also that no current would be in bulb R_1 and that bulb R_1 would not light.

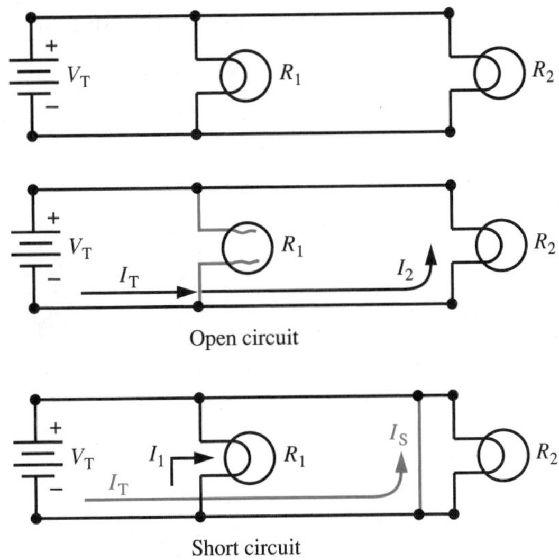

Figure 5-32 **Opens and shorts in parallel circuits**

Short circuits can be very dangerous because the resistance of a short circuit is close to zero ohms. Recall from Ohm's law that $I = V/R$. Therefore, $I = V/0$ will approach infinity amperes. As a practical note, the short circuit may not last long. A safety device such as a fuse or circuit breaker will open the circuit, stopping current. If no safety device is present, the heat caused by the excess current will cause the conductor or short to burn up and open the circuit. However, a serious fire hazard could exist.

> **! SAFETY TIP:** Shorts in parallel circuits can be very dangerous because high currents will exist.

✓ Self-Test 3

1. The voltages in a parallel circuit are the ____________.
2. ____________ states that the current entering a node minus the current leaving the node is equal to zero.
3. State the product over sum rule.
4. The resistance of a short circuit is ____________ ohms.

5.10 POWER IN SERIES AND PARALLEL CIRCUITS

Power, whether electrical or mechanical, is always the rate of doing work. The total power supplied by a source is equal to the sum of the powers dissipated by each individual load. It does not matter whether the circuit is series or parallel, the total power will always equal:

$$P_T = P_1 + P_2 + P_3 + \ldots + P_N$$

◀ *Equation 5-18*

If power is defined as the rate of doing work, then mathematically,

$$P = \frac{W}{t}$$

Recall that

$$V = \frac{W}{Q} \text{ or } W = QV$$

Substitute, as in the following:

$$P = \frac{W}{t} = \frac{QV}{t} = \frac{Q}{t} \times V$$

But recall that

$$I = \frac{Q}{t}$$

Therefore by substitution:

$$P = \frac{Q}{t} \times V = IV$$

Equation 5-19 ▶

Or $P = IV$

tech tidbit

This is sometimes referred to as the *PIE* equation because $E = V$. $P = IV$ or $P = IE$

where P = the power in watts

I = the current in amperes

V = the voltage in volts

Equation 5-19 is the basic power formula, which states that power is equal to current times voltage and is measured in the unit of **watts.** By substituting Ohm's law into the power equation (Equation 5-19), we can obtain the following additional relationships:

tech tidbit

$P = IV$

$P = \frac{V_2}{R}$

$P = I^2R$

$$P = IV$$

but $I = \frac{V}{R}$

therefore $P = \frac{V}{R} \cdot V$

Equation 5-20 ▶

$$P = \frac{V^2}{R}$$

Also if $P = IV$

and $V = IR$

then $P = I \cdot I \cdot R$

Equation 5-21 ▶

$$P = I^2R$$

Figure 5-33 summarizes the Ohm's law and power relationships.

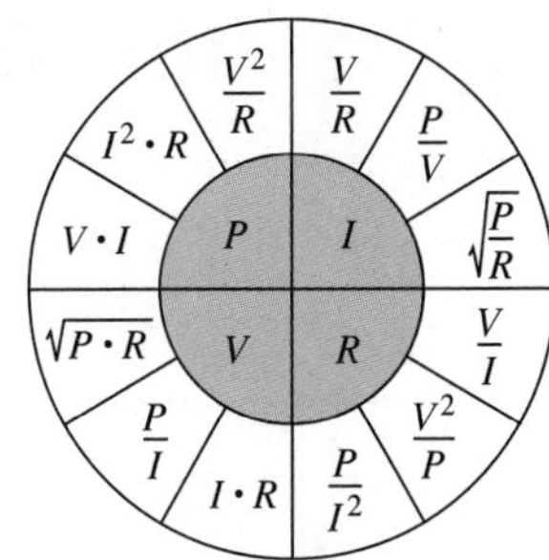

Figure 5-33 **Ohm's law and power wheel**

EXAMPLE 5-19

Calculate (A) the voltage, (B) power for each resistor, and (C) the total power in the series circuit of Figure 5-34.

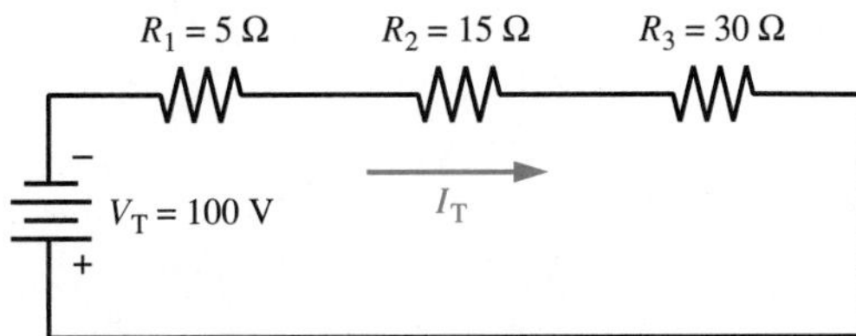

Figure 5-34

Solution:

(A)
$$R_T = R_1 + R_2 + R_3 = 5\ \Omega + 15\ \Omega + 30\ \Omega = 50\ \Omega$$

$$I_T = \frac{V_T}{R_T} = \frac{100\ \text{V}}{50\ \Omega} = 2\ \text{A}$$

$$V_1 = I_1R_1 = (2\ \text{A})(5\ \Omega) = 10\ \text{V}$$

$$V_2 = I_2R_2 = (2\ \text{A})(15\ \Omega) = 30\ \text{V}$$

$$V_3 = I_3R_3 = (2\ \text{A})(30\ \Omega) = 60\ \text{V}$$

(B)
$$P_1 = I_1V_1 = (2\ \text{A})(10\ \text{V}) = 20\ \text{W}$$

$$P_2 = \frac{V_2^2}{R_2} = \frac{(30\ \text{V})^2}{15} = 60\ \text{W}$$

$$P_3 = I_3^2R_3 = (2\ \text{A})^2(30\ \Omega) = 120\ \text{W}$$

$$P_T = P_1 + P_2 + P_3 = 20\ \text{W} + 60\ \text{W} + 120\ \text{W} = 200\ \text{W}$$

Check

$$P_T = I_TV_T = (2\ \text{A})(100\ \text{V}) = 200\ \text{W}$$

It is good design practice to double the calculated wattage value. This is due to the actual physical heat dissipation of the resistor.

EXAMPLE 5-20

Calculate the power dissipated by each resistor (A) P_1, (B) P_2, and (C) the total power delivered for the circuit of Figure 5-35.

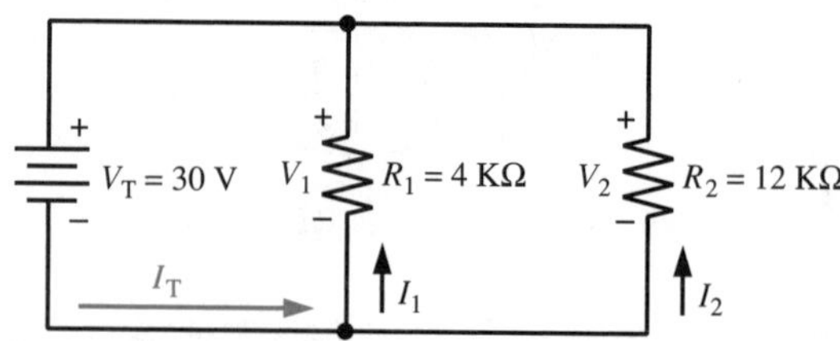

Figure 5-35

Solution:

(A) $V_T = V_1 = V_2 = 30$ V

$$P_1 = \frac{V_1^2}{R_1} = \frac{(30 \text{ V})^2}{4{,}000\ \Omega} = 0.225 \text{ W} = 225 \times 10^{-3} \text{ W} = 225 \text{ mW}$$

(B)

$$P_2 = \frac{V_2^2}{R_2} = \frac{(30 \text{ V})^2}{12{,}000\ \Omega} = 0.075 \text{ W} = 75 \times 10^{-3} \text{ W} = 75 \text{ mW}$$

(C)

$$P_T = P_1 + P_2 = 225 \text{ mW} + 75 \text{ mW} = 300 \text{ mW}$$

✓ Self-Test 4

1. The total power supplied by a source is equal to the ____________ of the power used by each individual load.
2. The unit of power is ____________.
3. State the three forms of the power equation.
4. A 100-ohm resistor has 0.1 amperes passing through it. What should the wattage rating of the resistor be?

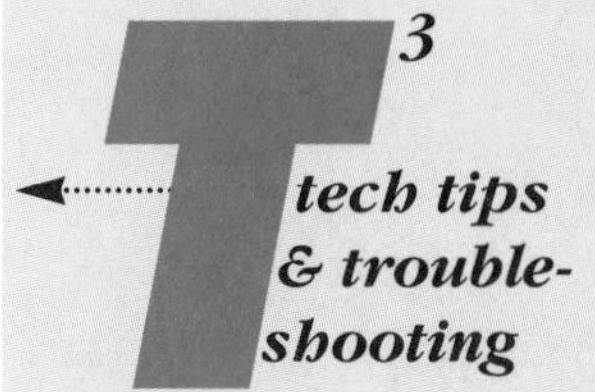

T3 GRAPHS

It has been said that a picture is worth a thousand words. A graph is a pictorial analysis of a device or circuit. Engineers and technicians use graphs to define the operation and characteristics of electric devices and circuits. It is important to be able to read, analyze, and create data in graphical form. Many manufacturers publish specifications and data sheets in graphical form. These specification sheets explain the operation of a device. Consider the typical graphical specification for a thermistor, as illustrated in Figure 5-36.

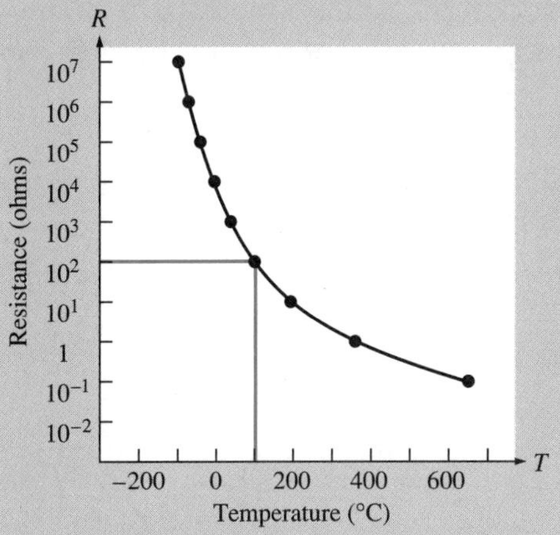

R (ohms)	T (°C)
10^7	−100
10^6	−75
10^5	−50
10^4	0
10^3	50
10^2	100
10^1	200
$10^0 = 1$	350
10^{-1}	650

Figure 5-36 **Thermistor characteristics**

A thermistor is a device that changes its resistance with temperature. From this one graph, we can quickly understand a lot about this device. For example, if we look at the graph, we can see that as the temperature increases, the resistance of the thermistor decreases. The slope of the curve tells us how fast the resistance changes with temperature. Notice that the slope for this device is different at different parts of the curve. We call this a **nonlinear** curve. A **linear** curve would have a constant slope. A straight-line graph is an example of a linear curve.

To determine the resistance of the device at 100°C, draw a vertical line at the 100°C point. At the point where this vertical line intersects the curve, draw a horizontal line. We can see that the resistance is 100 ohms. The table of Figure 5-36 describes the resistance at many fixed points. If we start with the table of data, we could plot the individual points on a graph. By connecting the points in a smooth curve, we would create the graph. This can be done by beginning at the top data point and sketching in a smooth curve connecting the first three data points (points 1, 2, and 3) together. Move to the second highest data point and again connect with a smooth line the three data points (points 2, 3, and 4). Continue the process until all the data points are connected. Finally, make a final darker curve of all of the points. Occasionally, the points will be radically different and unsmoothable. You may wish to circle this point and exclude it from the curve. Tools such as French curves are very useful when plotting graphs. You may wish to purchase a set of French curves if you do a lot of graphing.

Linear—constant slope.

SUMMARY

Ohm's law describes the fundamental relationship between voltage, current, and resistance. Mathematically it states that the current is equal to the voltage divided by the resistance. A series circuit is defined as a circuit with only one continuous path for current. There is only one path for current, so it follows that the current in a series circuit will always be the same value. The total resistance in a series circuit is the sum of the individual resistors. Kirchhoff's voltage law states that the sum of the voltage rises and the voltage drops around any closed path or loop is equal to zero. The voltage divider rule states that the voltage drop across individual resistors in a series circuit is proportional to the value of each resistor.

A parallel circuit is one in which the current has more than one path. Each current path is called a branch. For elements to be in parallel, both ends of each element must be connected together. Because each element in a parallel circuit is connected directly across the applied voltage, the voltages across each element are the same as the applied voltage. Kirchhoff's current law states that the currents entering a node minus the currents leaving a node are equal to zero. The total resistance of a parallel circuit is always less than the smallest resistor in the circuit. Thus, as more resistors are added in parallel, the total resistance decreases. The current divider rule is a handy tool for determining branch currents when the total current is known.

Power is the rate of doing work. The total power supplied by a source is equal to the sum of the power dissipated by each individual load. Electrical Power is equal to current times voltage. It is measured in the unit of watts.

FORMULAS

$$I = \frac{V}{R} \qquad V = IR \qquad R = \frac{V}{I}$$

$$G = \frac{1}{R} \qquad R = \frac{1}{G}$$

$$R_T = R_1 + R_2 + R_3 + \ldots + R_N \quad \text{(series)}$$

$$V_T = V_1 + V_2 + V_3 + \ldots + V_N \quad \text{(series)}$$

$$V_X = \frac{R_X V_T}{R_T} \quad \text{(VDR-series)}$$

$$I_T = I_1 + I_2 + I_3 + \ldots + I_N \quad \text{(parallel)}$$

$$\frac{1}{R_T} = \frac{1}{R_1} + \frac{1}{R_2} + \frac{1}{R_3} + \ldots + \frac{1}{R_N} \quad \text{(parallel)}$$

$$R_T = \frac{R_1 R_2}{R_1 + R_2} \quad \text{(2 resistors in parallel)}$$

$$R_T = \frac{R}{N} \quad \text{(equal resistors in parallel)}$$

$$I_X = \frac{I_T R_T}{R_X} \quad \text{(CDR, general—parallel)}$$

$$I_1 = \frac{R_2 I_T}{R_1 + R_2} \qquad \text{(CDR, 2 resistors—parallel)}$$

$$I_2 = \frac{R_1 I_T}{R_1 + R_2} \qquad \text{(CDR, 2 resistors—parallel)}$$

$$P = \frac{W}{t} \qquad W = P \cdot t \qquad t = \frac{W}{P}$$

$$P = IV \qquad I = \frac{P}{V} \qquad V = \frac{P}{I}$$

$$P = I^2R \qquad I = \sqrt{\frac{P}{R}} \qquad R = \frac{P}{I^2}$$

$$P = \frac{V^2}{R} \qquad V = \sqrt{P \cdot R} \qquad R = \frac{V^2}{P}$$

$$P_T = P_1 + P_2 + P_3 + \ldots + P_N \qquad \text{(series or parallel)}$$

EXERCISES

Section 5-2

1. What is the current in a 6-ohm resistor that has a voltage drop of 24 volts across it?
2. What is the current in a 2-kilohm resistor that has a voltage drop of 12 volts across it?
3. What is the current in a 4.7-kilohm resistor that has a voltage drop of 125 volts across it?
4. Calculate the voltage that must be applied across a motor to draw 2 amperes if the resistance of the motor is 60 ohms.
5. What is the voltage drop across a lamp that draws 21 milliamperes and has a resistance of 300 ohms?
6. The current in a 3-megohm resistor is 4 microamperes. What is the voltage?
7. Calculate the resistance of a soldering iron that has a potential of 120 volts across it and a current of 1.5 amperes.
8. A voltmeter reads 60 volts and draws 3 milliamperes. Calculate the internal resistance of the meter.
9. A heating element connected to a 120-volt source draws 5.45 amperes. What is the resistance of the heating element?

Section 5-3

10. Three resistors are in series. Find the total resistance if their values are 7 ohms, 14 ohms, and 9 ohms.
11. A 2-kilohm, 3.3-kilohm, and a 4.7-kilohm resistor are connected in series. Calculate the total resistance.
12. A 2-ohm and a 3-ohm resistor are connected in series across a 100-volt source. What is the total current in the circuit?

13. Three resistors connected in series have values of 220 ohms, 330 ohms, and 470 ohms respectively. If they are connected to a 24-volt source, what will be the current in the circuit?
14. Calculate the current for the circuit of Figure 5-37A.
15. Calculate the current for the circuit of Figure 5-37B.
16. Calculate the current for the circuit of Figure 5-37C.

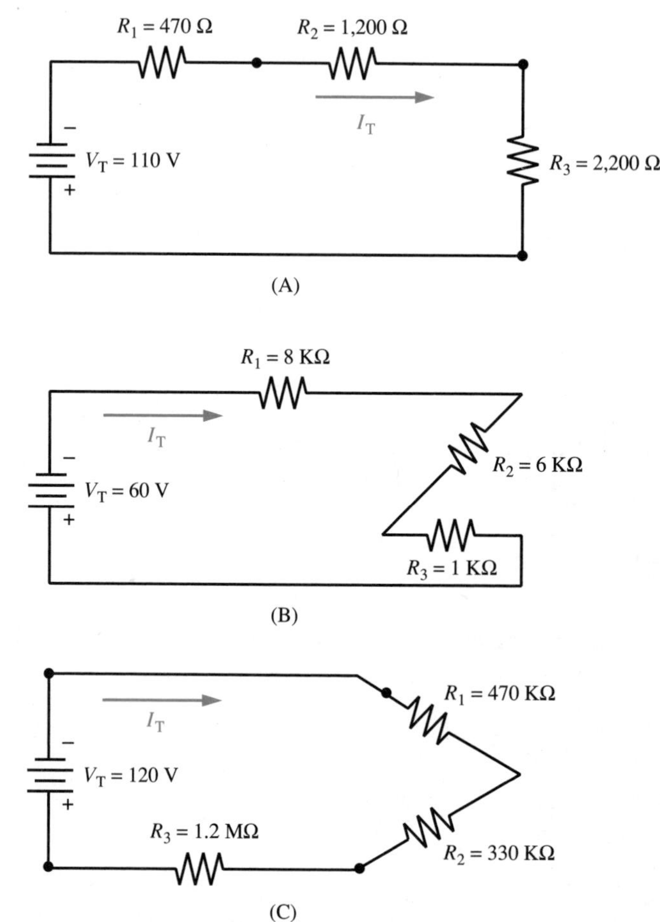

Figure 5-37

17. Calculate the resistance of the lamp in Figure 5-38A.
18. Calculate the resistance of the lamp in Figure 5-38B.

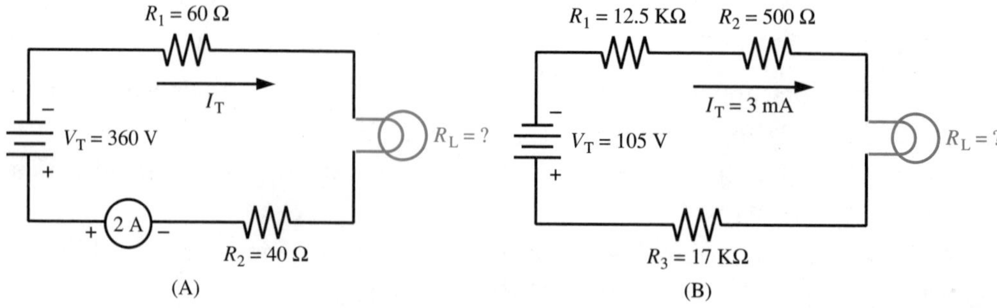

Figure 5-38

Section 5-4

19. Find the total voltage required to light three 9-volt lamps and four 6-volt lamps connected in series.
20. What size resistor connected in series would allow you to operate the lamps in problem 19 from a 100-volt source. Assume a current of 100 milliamperes.
21. Find the total resistance, current, and voltage across each resistor in the circuit of Figure 5-39A. Check your answers using Kirchhoff's voltage law.
22. Find the total resistance, current, and voltage across each resistor in the circuit of Figure 5-39B. Check your answers using Kirchhoff's voltage law.

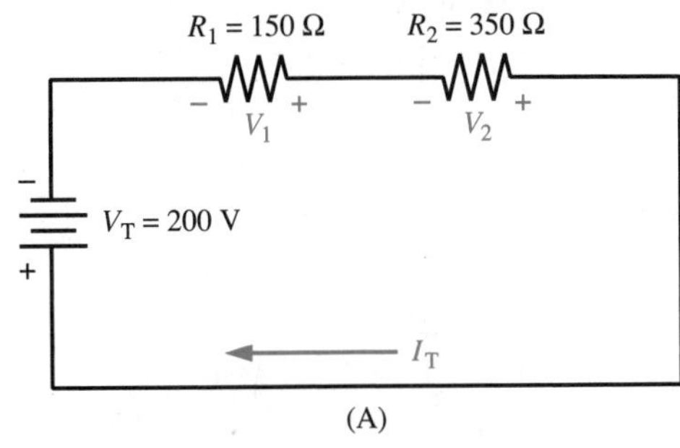

(A)

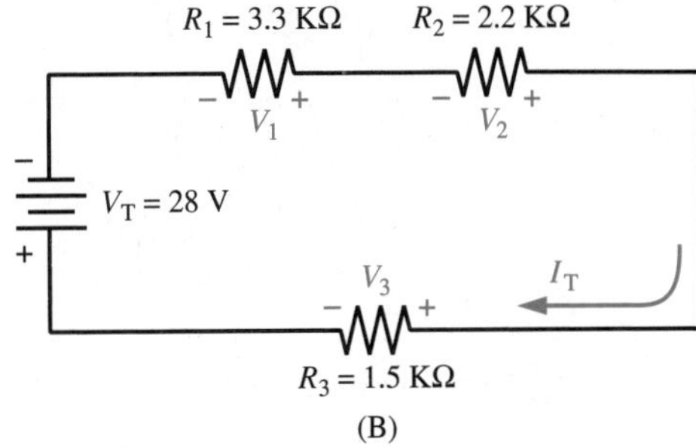

(B)

Figure 5-39

23. Find the voltage V_{AB} for each configuration of Figure 5-40.

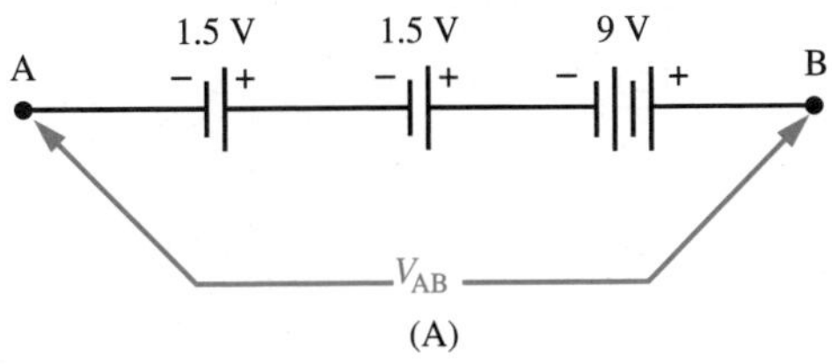

(A)

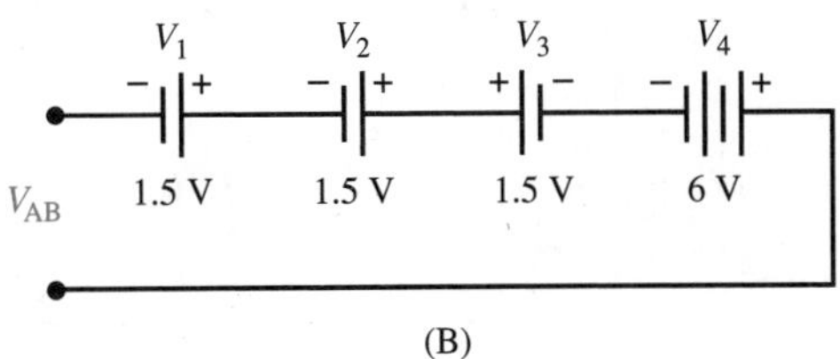

(B)

Figure 5-40

24. For the configuration of Figure 5-41, find V_{AB}, V_{AC}, V_{BC}, V_{BA}, V_{CA}, and V_{CB}.

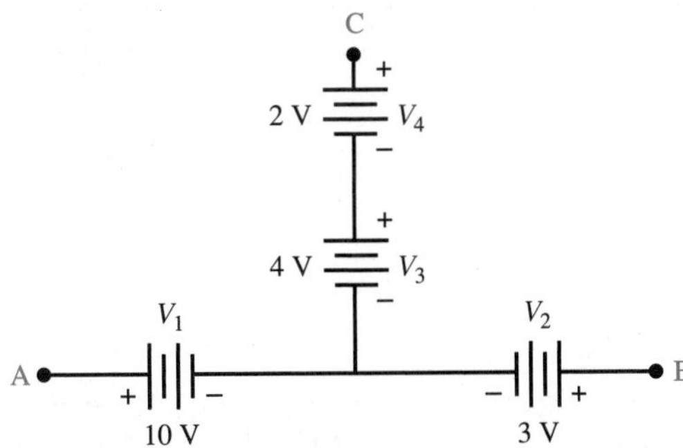

Figure 5-41

25. Calculate the current in the circuit of Figure 5-42.
26. For the circuit of Figure 5-42, calculate V_{AB}, V_{BC}, V_{CD}, V_{AC}, and V_{BD}.

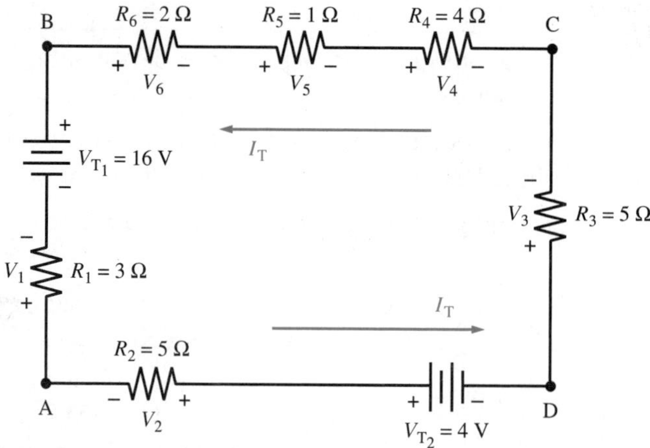

Figure 5-42

Section 5.5

27. Using the voltage divider rule, calculate V_1 and V_2 in Figure 5-43A.
28. Using the voltage divider rule, calculate V_1, V_2, V_3, V_{12}, and V_{23} in Figure 5-43B.

No Divider Rule

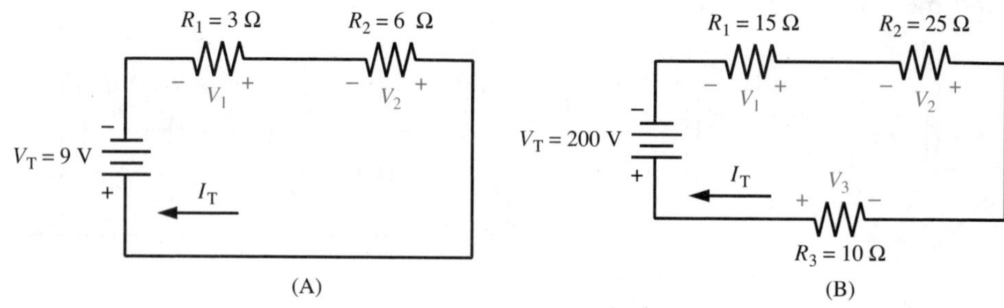

Figure 5-43

Section 5.6

29. Calculate the current and voltage across each lamp in the circuit of Figure 5-44A.
30. Calculate the current and voltage across each lamp in the circuit of Figure 5-44B.

Section 5.7

31. Find the value of I_4 in the circuit of Figure 5-45A if $I_1 = 3$ A, $I_2 = 5$ A, and $I_3 = 2$ A.
32. Find the value of I_2 in the circuit of Figure 5-45A if $I_1 = 1$ A, $I_3 = 3$ A, and $I_4 = 9$ A.

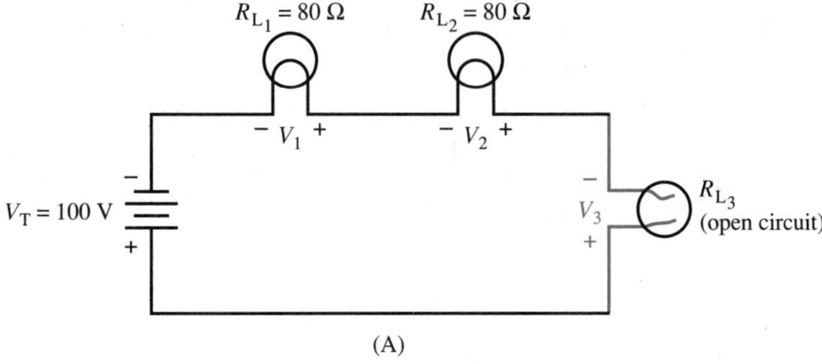

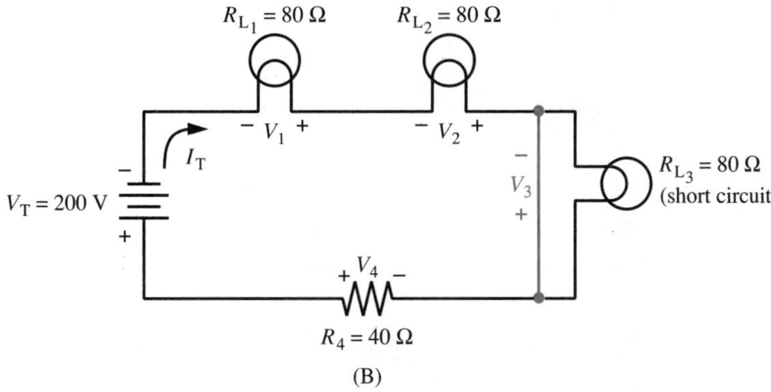

Figure 5-44

33. Find the value of I_4 in the circuit of Figure 5-45A if $I_1 = 6$ A, $I_2 = 10$ A, and $I_3 = -4$ A.

34. Find the value of I_7 in the circuit of Figure 5-45B if $I_1 = 2$ A, $I_2 = 3$ A, $I_3 = 4$ A, $I_5 = 2.5$ A, $I_6 = 3.5$ A, and $I_8 = 5$ A.

35. Find the value of I_8 in the circuit of Figure 5-45B if $I_1 = 1$ A, $I_2 = 2$ A, $I_3 = 3$ A, $I_5 = 2$ A, $I_6 = 4$ A, and $I_7 = 2$ A.

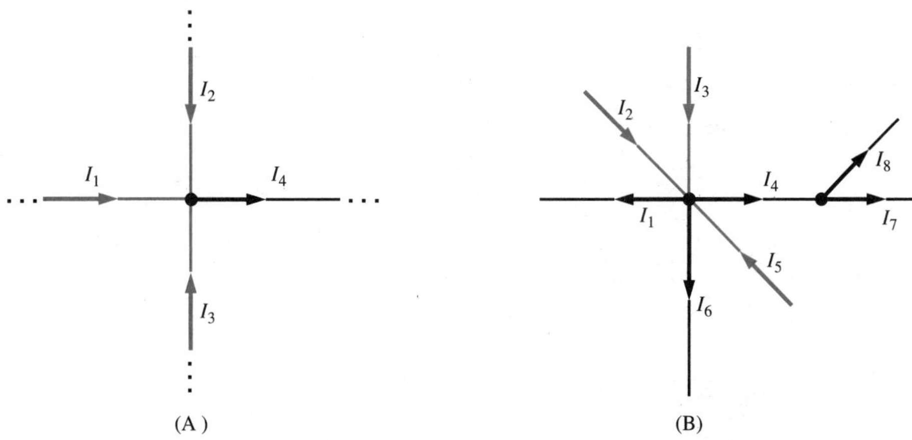

Figure 5-45

36. For the circuit of Figure 5-46A, calculate the branch currents I_1 and I_2, the total current I_T, and the total resistance, R_T.

37. For the circuit of Figure 5-46B, calculate the branch currents I_1, I_2, and I_3, the total current I_T, and the total resistance, R_T.

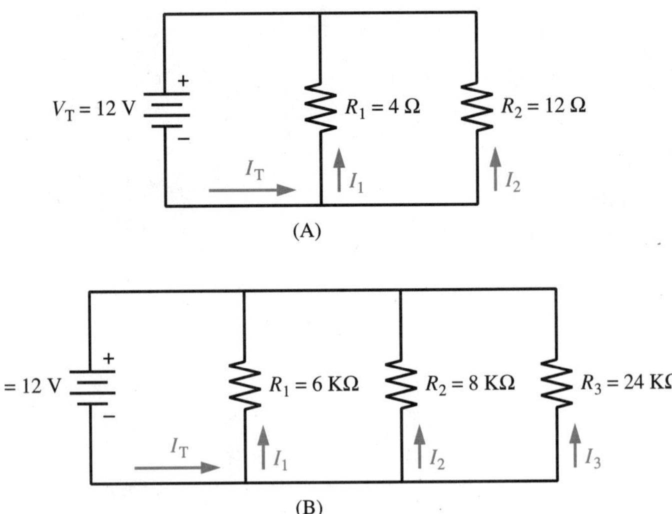

Figure 5-46

38. Find the total conductance and the total resistance of the circuit of Figure 5-47A.
39. Find the total conductance and the total resistance of the circuit of Figure 5-47B.
40. Find the total conductance and the total resistance of the circuit of Figure 5-47C.

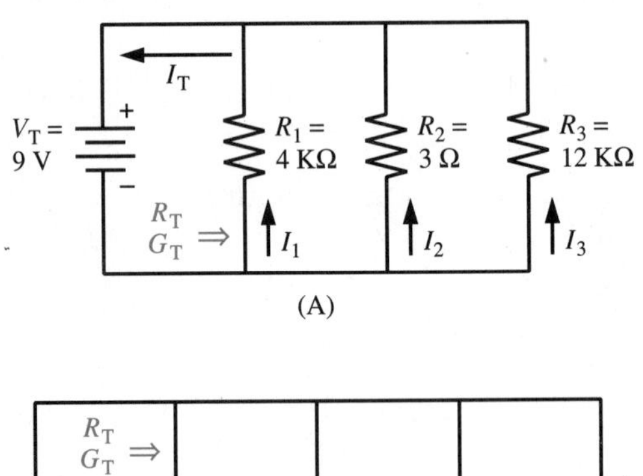

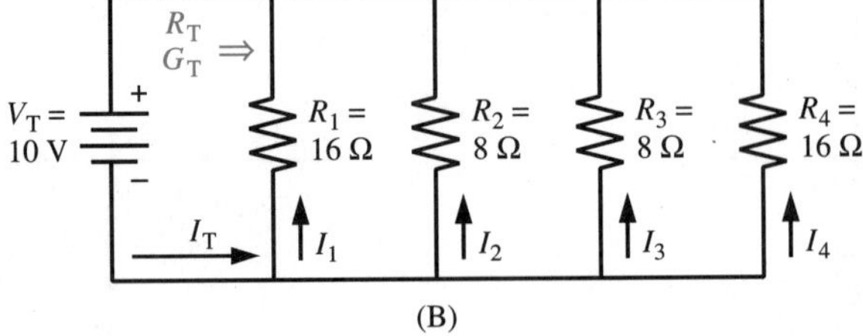

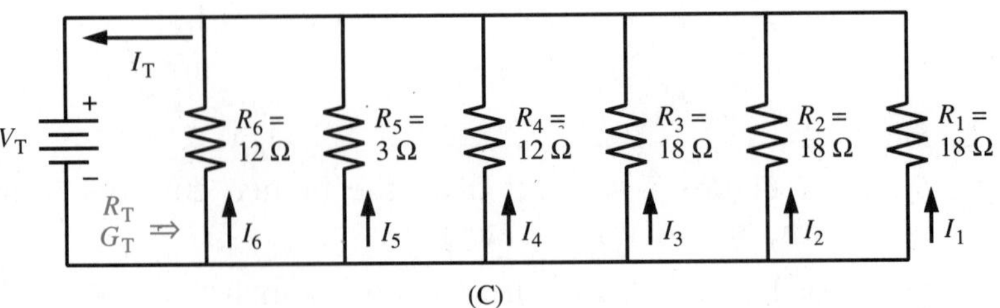

Figure 5-47

41. Find the value of resistor R_2 in the circuit of Figure 5-48.

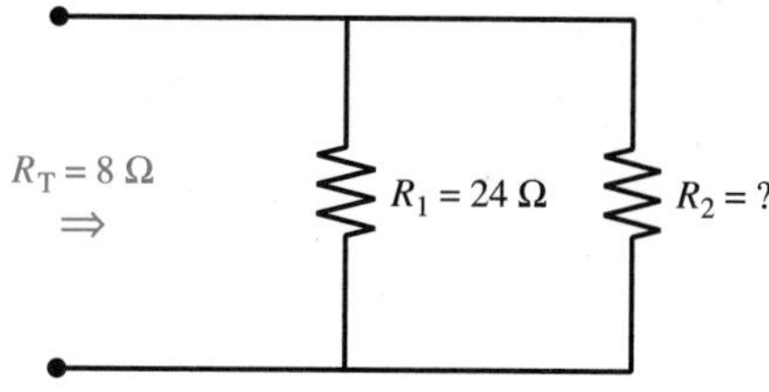

Figure 5-48

Section 5.8

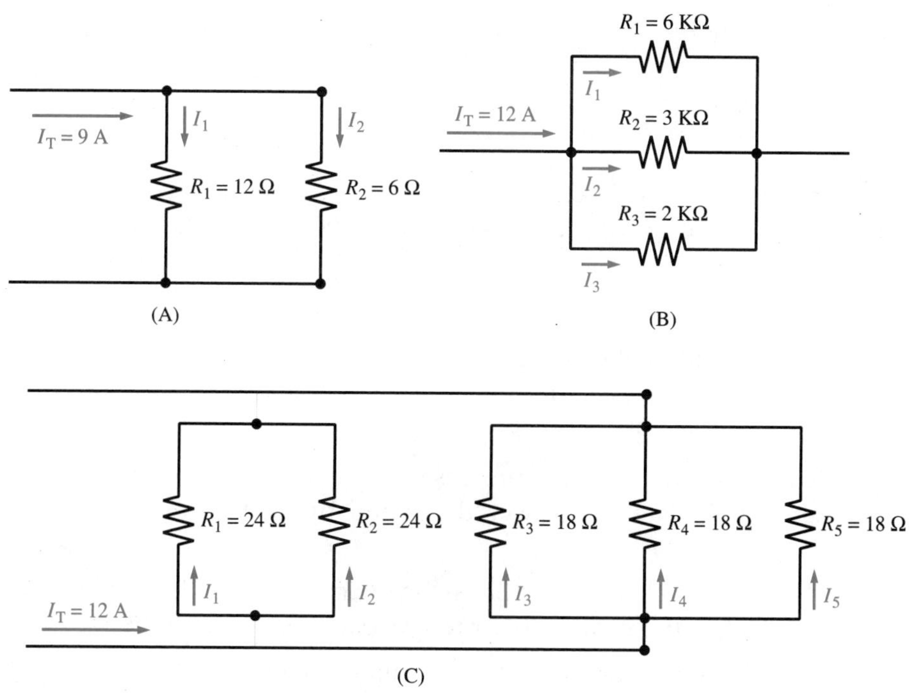

Figure 5-49

42. Using the current divider rule, calculate the value of I_1 and I_2 in Figure 5-49A.

43. Using the current divider rule, calculate the value of I_1, I_2, and I_3 in Figure 5-49B.

44. Using the current divider rule, calculate the value of I_1, I_2, I_3, I_4, and I_5 in Figure 5-49C.

Section 5-9

45. Find the value of I_T in Figure 5-50A with and without the short circuit at lamp 3.

46. Find the value of I_T in Figure 5-50B with and without the open circuit (broken lamp) at lamp 3.

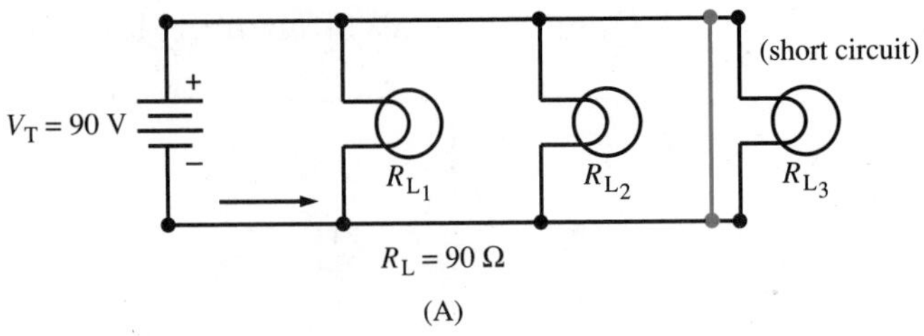

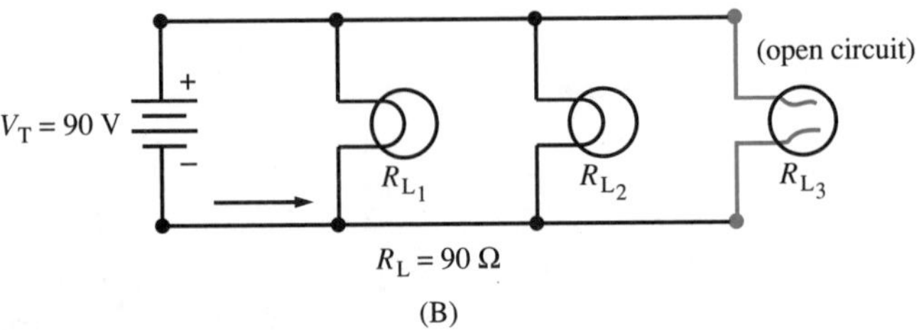

Figure 5-50

Section 5.10

47. Calculate the total power dissipated by five 100-watt light bulbs operating at rated power when:

    (A) connected in series

    (B) connected in parallel

48. Calculate the total power dissipated by a circuit when the total voltage is 120 volts and the total current is 5 amperes.

49. A 10-ohm, 20-ohm, and 30-ohm resistor are connected in series drawing 2 amperes. Calculate the power dissipated by each resistor and the total power.

50. A 10-ohm and a 20-ohm resistor are connected in parallel across a 10-volt source. Calculate the power dissipated by each resistor and the total power.

51. Find the unknown quantities for the circuit of Figure 5-51.

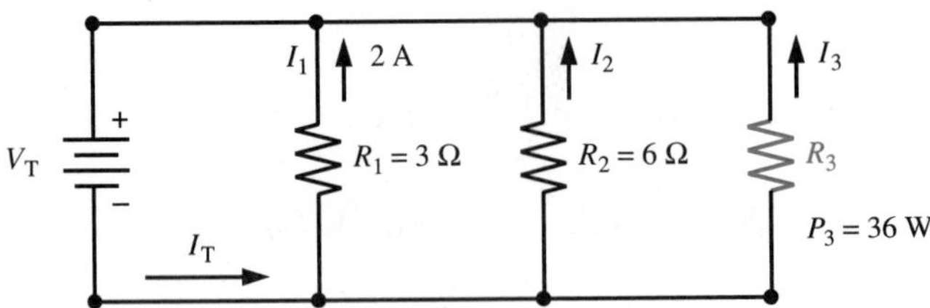

Figure 5-51

Section T^3

52. A technician records the following data:

Voltage	Current
5 V	1 A
10 V	2 A
15 V	3 A
20 V	4 A
25 V	5 A

Draw a graph for the data shown. From the graph determine the current at 12.5 volts.

53. See Lab Manual for the crossword puzzle for Chapter 5.

CHAPTER 6

NETWORK ANALYSIS

Courtesy of Eastman Kodak Company

OBJECTIVES

After completing this chapter, the student should be able to:

- identify series and parallel branches of series-parallel circuits.
- label nodes in series-parallel circuits.
- analyze and solve series-parallel circuits.
- define ideal voltage sources.
- define ideal current sources.
- perform source conversions.
- state Thevenin's theorem.
- solve circuits using Thevenin's theorem.
- solve circuits using Norton's theorem.
- state the maximum power transfer theorem.
- calculate the maximum power delivered to a load.

KEY TERMS

ideal current source
ideal voltage source
internal resistance
maximum power
network
Norton's theorem
series-parallel
source conversion
Thevenin's theorem

6.1 INTRODUCTION

Series-parallel circuits are made up of series branches and parallel branches.

We have looked at the fundamental series circuit and the fundamental parallel circuit. Many electric circuits are formed by a combination of series and parallel circuits. These circuit configurations or **networks** are called **series-parallel** circuits. Series-parallel circuits are analyzed using Ohm's law and Kirchhoff's laws. We first divide the series-parallel network into its series and its parallel branches. Next we apply the rules for series circuits to the series part of the circuit and the rules for parallel branches to the parallel part of the circuit.

Ohm's law can be very inefficient when used to solve complex series-parallel circuits. Therefore, more advanced techniques have been developed to make it easier to solve complex series-parallel circuits. Among them are, Thevenin's theorem, Norton's theorem, superposition, mesh (loop) analysis, and nodal analysis, to name just a few.

In this chapter, we first study series-parallel circuits using conventional Ohm's law techniques. We then build on this knowledge and learn how to apply some of the advanced theorems and methods of analysis to the solution of series-parallel circuits.

6.2 IDENTIFYING SERIES AND PARALLEL CIRCUITS

When applying Ohm's law to series-parallel circuits, the general method is to replace groups of series or parallel resistances with equivalent single resistances. The first step is to learn to recognize which components are in series and which components are in parallel. Figure 6-1 illustrates a series-parallel circuit.

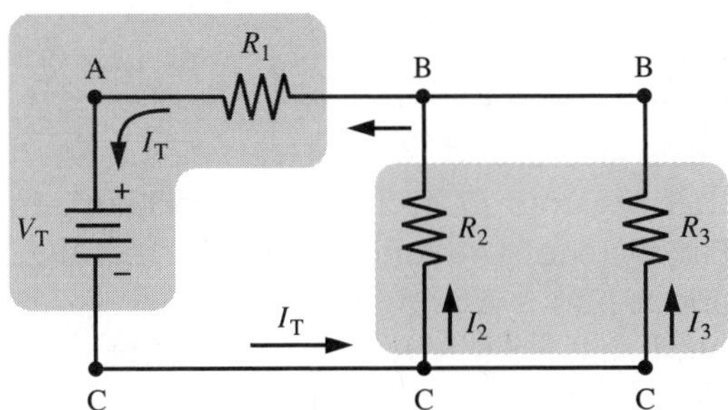

Figure 6-1 **Series parallel circuit**

In a series circuit, the current is the same.

In Figure 6-1, we have a series-parallel circuit consisting of three resistors and a voltage source. In this circuit, the voltage source, V_T, and resistor R_1, form a series circuit branch. Recall that in a series circuit, there is only one path for the current. From node B to A to C we can see that there is only one path for the current. Resistors R_2 and R_3 form a parallel branch. Recall that in a parallel circuit, both ends of the components must be connected *directly* to each other. This allows for more than one path for the current. Also recall that the voltage across parallel components is the same. Resistors R_2 and R_3 are connected directly across nodes B and C. The voltage across B and C will be the voltage across R_2 and R_3. Note that at node C, the total current entering the node divides into currents I_2 and I_3. This is also indicative of a parallel circuit.

In a parallel circuit, the voltage is the same.

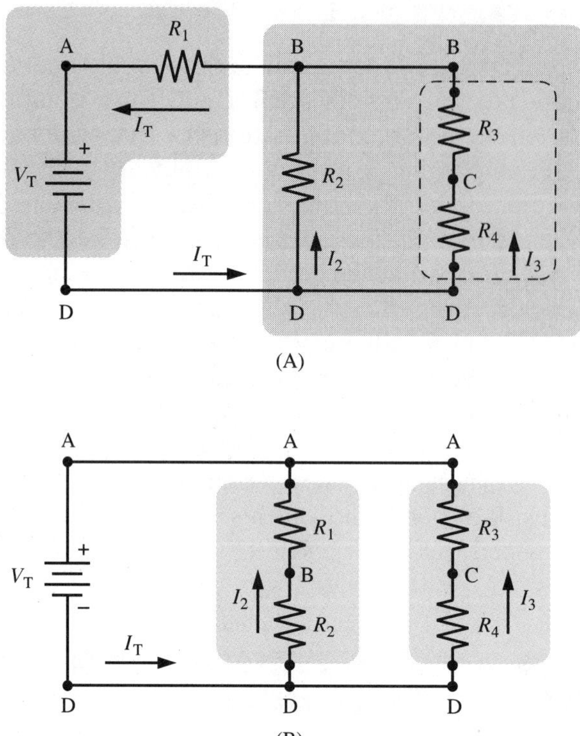

Figure 6-2 ***Identifying series and parallel branches***

In Figure 6-2, we have two more examples of series-parallel circuits. In Figure 6-2A, voltage source V_T is in series with resistor R_1. We also note that resistor R_3 is in series with resistor R_4. A parallel branch is formed between nodes B and D by resistor R_2 in parallel with the series branch of R_3 and R_4.

In Figure 6-2B, resistors R_1 and R_2 form a series branch between nodes A, B, and D. Also note that resistors R_3 and R_4 form another series branch between nodes A, C, and D. Finally, the two series branches and the voltage source, V_T, form a parallel network at nodes A and D.

It should be noted that in a series branch there must be at least three nodes connected. In a parallel branch all of the elements must be connected across two nodes.

Self-Test 1

1. ____________ electric circuits are formed by a combination of series and parallel branches.

2. Name three advanced technique theorems developed to solve complex circuits.

3. All of the elements must be connected across two nodes in a ____________.

6.3 SERIES-PARALLEL CIRCUIT ANALYSIS

As previously stated, series-parallel circuits are combinations of both series and parallel connected circuits. We begin our study of series-parallel circuits by considering the circuit of Figure 6-3A. First we identify the series and parallel branches of the circuit. Voltage source, V_T, and resistor, R_1, are in series. Resistors, R_2 and R_3, are in parallel. Figure 6-3B illustrates both the series branch and the parallel branch of the circuit.

When analyzing series-parallel circuits, begin by identifying the series and parallel branches.

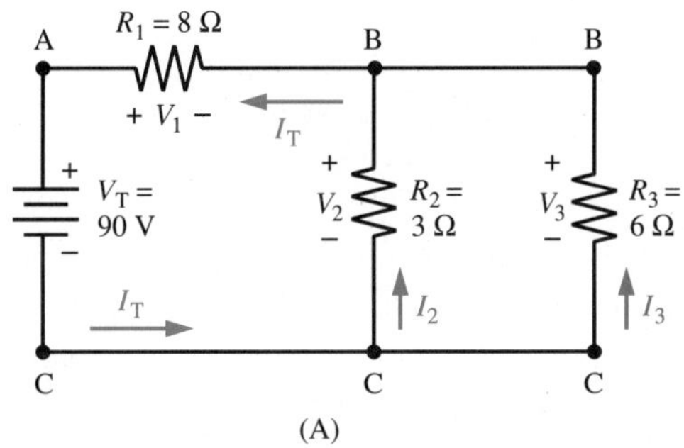

(A)

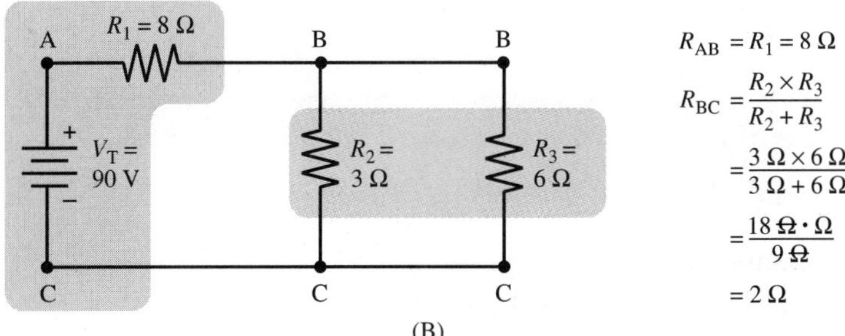

$$R_{AB} = R_1 = 8\ \Omega$$

$$R_{BC} = \frac{R_2 \times R_3}{R_2 + R_3} = \frac{3\ \Omega \times 6\ \Omega}{3\ \Omega + 6\ \Omega} = \frac{18\ \cancel{\Omega} \cdot \Omega}{9\ \cancel{\Omega}} = 2\ \Omega$$

(B)

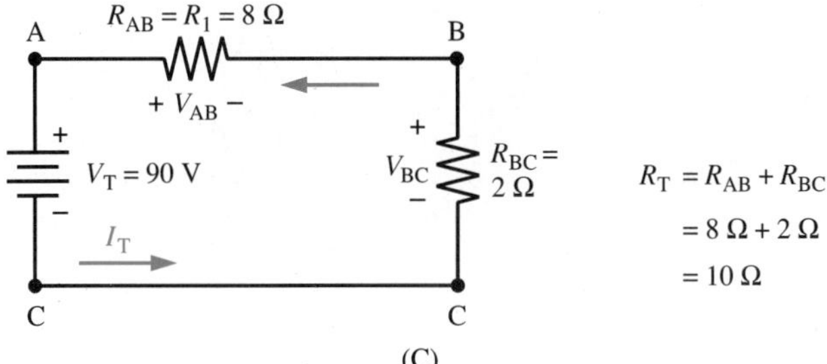

$$R_T = R_{AB} + R_{BC} = 8\ \Omega + 2\ \Omega = 10\ \Omega$$

(C)

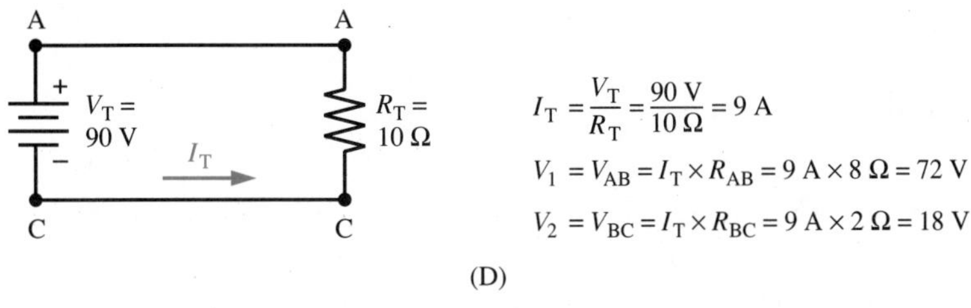

$$I_T = \frac{V_T}{R_T} = \frac{90\ \text{V}}{10\ \Omega} = 9\ \text{A}$$

$$V_1 = V_{AB} = I_T \times R_{AB} = 9\ \text{A} \times 8\ \Omega = 72\ \text{V}$$

$$V_2 = V_{BC} = I_T \times R_{BC} = 9\ \text{A} \times 2\ \Omega = 18\ \text{V}$$

(D)

Figure 6-3 *Series parallel circuit analysis* *(continued on page 126)*

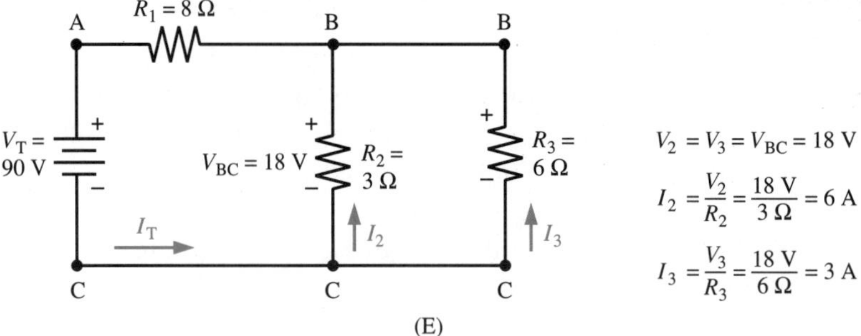

Figure 6-3 (cont.)

Applying the rules for parallel resistors, we can obtain an equivalent resistance for the parallel branch. This equates to 2 ohms, as shown in Figure 6-3B. From Figure 6-3C, we can see that the equivalent parallel resistance, R_{BC}, is in series with R_1. By applying the rules of series resistance, we can obtain that R_T = 10 ohms. In Figure 6-3D, knowing V_T and R_T we can compute I_T by applying Ohm's law. Next we calculate V_1, the voltage across resistor R_1, and V_2, the voltage across the parallel equivalent resistance. Referring to Figure 6-3E, we can see that $V_2 = V_3 = V_{BC}$ because resistor R_2 and R_3 are in parallel. Finally we calculate I_2, the current through resistor R_2 and I_3, the current through resistor R_3 using Ohm's law.

EXAMPLE 6-1

Calculate I_4, the current through resistor R_4 in Figure 6-4.

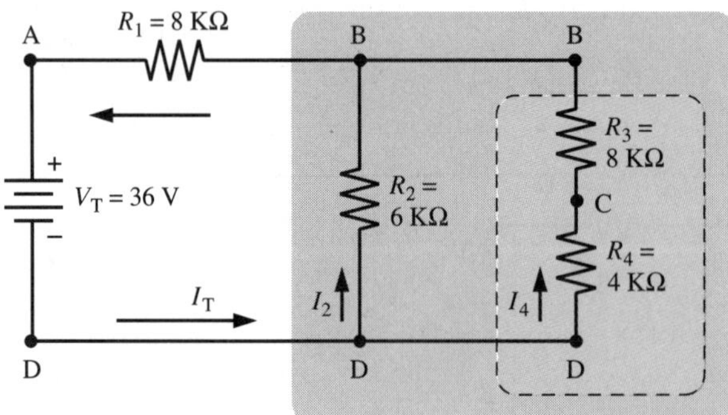

Figure 6-4

Solution:

(See Figure 6-5)

R_1 = 8 KΩ

R_2 = 6 KΩ

R_{34} = $R_3 + R_4$

= 8 KΩ + 4 KΩ

= 12 KΩ (See Figure 6-5A)

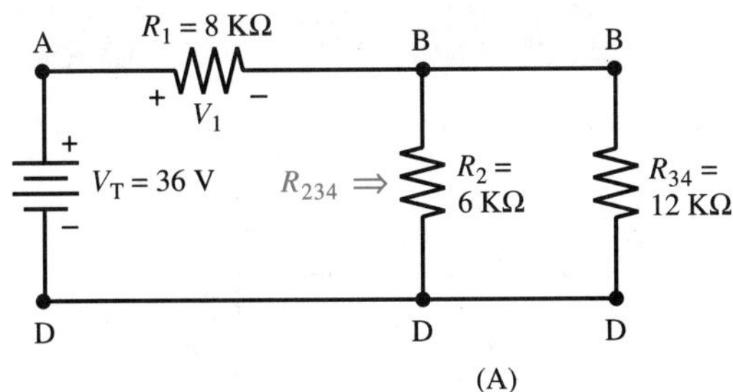

(A)

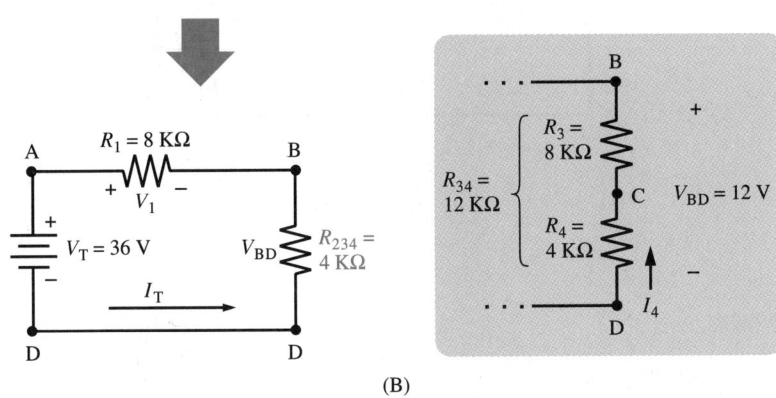

(B)

Figure 6-5

R_2 in parallel with R_{34}

$$R_{234} = R_2 \parallel R_{34}$$

$$= \frac{R_2 R_{34}}{R_2 + R_{34}}$$

$$= \frac{(6\text{ K}\Omega)(12\text{ K}\Omega)}{6\text{ K}\Omega + 12\text{ K}\Omega}$$

$$= \frac{72\text{ K}\Omega\Omega}{18\text{ K}\Omega}$$

$$= 4\text{ K}\Omega \text{ (See Figure 6-5B)}$$

$$R_T = R_1 + R_{234}$$

$$= 8\text{ K}\Omega + 4\text{ K}\Omega$$

$$= 12\text{ K}\Omega$$

$$I_T = \frac{V_T}{R_T}$$

$$= \frac{36\text{ V}}{12\text{ K}\Omega}$$

$$= 3\text{ mA}$$

$$V_{BD} = I_T R_{234}$$

$$= (3\text{ mA})(4\text{ K}\Omega)$$

$$= 12\text{ V}$$

$$I_4 = \frac{V_{BD}}{R_{34}}$$

$$= \frac{12\text{ V}}{12\text{ K}\Omega}$$

$$= 1\text{ mA}$$

T² tech tidbit

The mathematical symbol for parallel is ||.

EXAMPLE 6-2

Find (A) V_1 and (B) V_4 for the network of Figure 6-6.

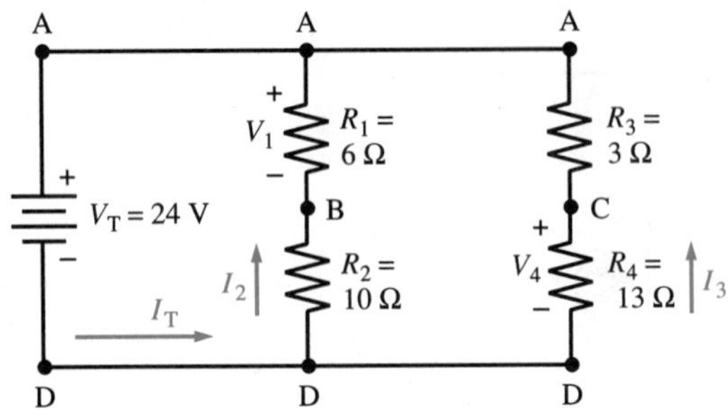

Figure 6-6

Solution:

By the voltage divider rule:

(A) For branch ABD

$$V_1 = \frac{R_1 V_T}{R_1 + R_2} = \frac{(6\ \Omega)(24\ \text{V})}{6\ \Omega + 10\ \Omega} = 9\ \text{V}$$

(B) For branch ACD

$$V_4 = \frac{R_4 V_T}{R_3 + R_4} = \frac{(13\ \Omega)(24\ \text{V})}{3\ \Omega + 13\ \Omega} = 19.5\ \text{V}$$

T^2 tech tidbit

The voltage divider rule can be used for series branches when the total series branch voltage is known.

EXAMPLE 6-3

Find the value of R_{AB} in the network of Figure 6-7.

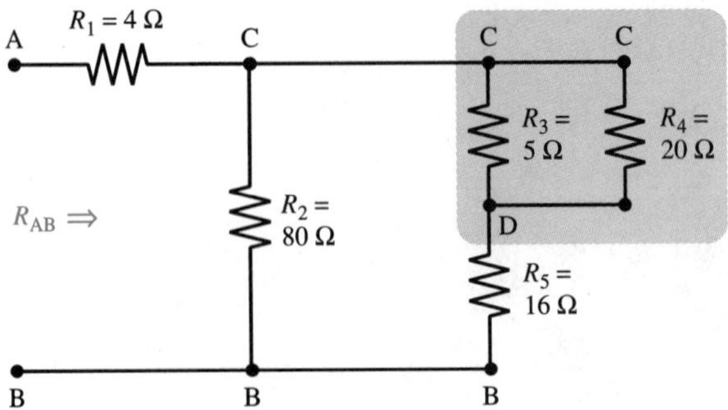

Figure 6-7

Solution:

(See Figure 6-8)

$$R_{34} = \frac{R_3 R_4}{R_3 + R_4}$$

$$= \frac{(5\ \Omega)(20\ \Omega)}{5\ \Omega + 20\ \Omega}$$

$$= \frac{100\ \Omega\Omega}{25\ \Omega}$$

$= 4\ \Omega$ (See Figure 6-8A)

$$R_{345} = R_{34} + R_5$$

$$= 4\ \Omega + 16\ \Omega$$

$= 20\ \Omega$ (See Figure 6-8B)

$$R_{2345} = \frac{R_2 R_{345}}{R_2 + R_{345}}$$

$$= \frac{(80\ \Omega)(20\ \Omega)}{80\ \Omega + 20\ \Omega}$$

$$= \frac{1{,}600\ \Omega\Omega}{100\ \Omega}$$

$= 16\ \Omega$ (See Figure 6-8C)

$$R_{AB} = R_1 + R_{2345}$$

$$= 4\ \Omega + 16\ \Omega$$

$$= 20\ \Omega$$

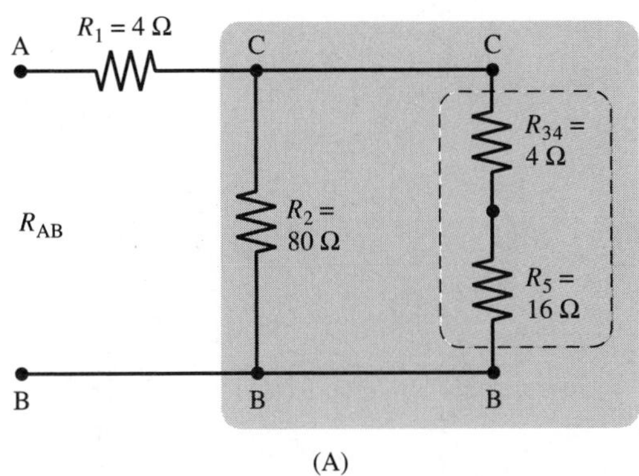

(A)

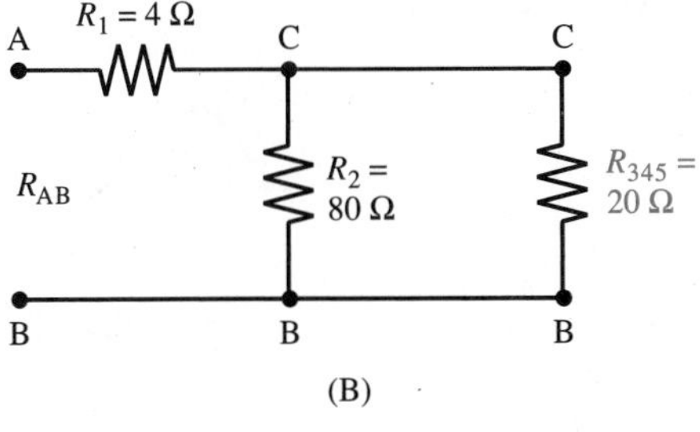

(B)

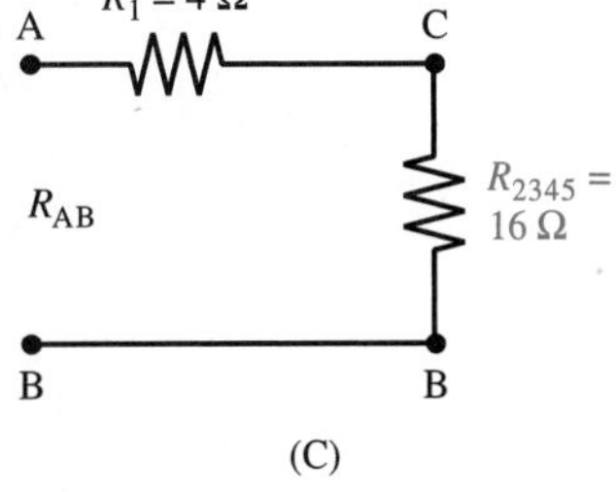

(C)

Figure 6-8

EXAMPLE 6-4

For the circuit of Figure 6-9, find (A) R_T, (B) I_A, (C) I_B, and (D) I_C.

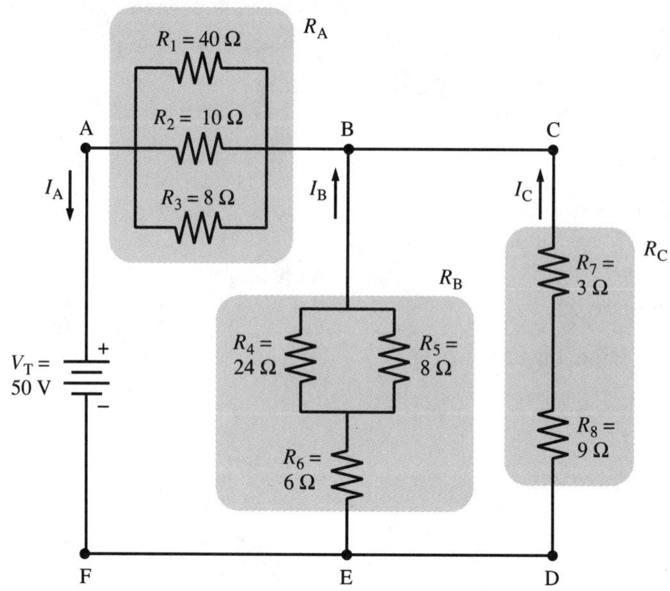

Figure 6-9

Solution:

(See Figure 6-10)

(A)
$$R_A = \cfrac{1}{\cfrac{1}{R_1} + \cfrac{1}{R_2} + \cfrac{1}{R_3}}$$

$$= \cfrac{1}{\cfrac{1}{40\ \Omega} + \cfrac{1}{10\ \Omega} + \cfrac{1}{8\ \Omega}}$$

$$= \cfrac{1}{\cfrac{1}{40\ \Omega} + \cfrac{4}{40\ \Omega} + \cfrac{5}{40\ \Omega}}$$

$$= \cfrac{1}{\cfrac{10}{40\ \Omega}}$$

$$= \frac{1}{0.25\ \text{S}}$$

$$= 4\ \Omega$$

$$R_{45} = \frac{R_4 R_5}{R_4 + R_5}$$

$$= \frac{(24\ \Omega)(8\ \Omega)}{24\ \Omega + 8\ \Omega}$$

$$= 6\ \Omega$$

$$R_B = R_{45} + R_6$$

$$= 6\ \Omega + 6\ \Omega$$

$$= 12\ \Omega$$

$$R_C = R_7 + R_8$$

$$= 3\ \Omega + 9\ \Omega$$

$$= 12\ \Omega$$

$$R_T = R_A + R_B \parallel R_C$$

$$= 4\ \Omega + 12\ \Omega \parallel 12\ \Omega$$

$$= 4\ \Omega + \frac{12\ \Omega}{2}$$

$$= 4\ \Omega + 6\ \Omega$$

$$= 10\ \Omega$$

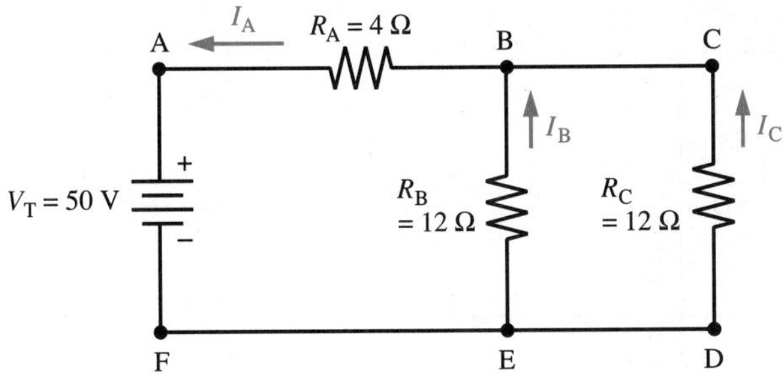

Figure 6-10

(B) $I_A = \dfrac{V_T}{R_T}$

$= \dfrac{50 \text{ V}}{10\ \Omega}$

$= 5$ A

(C) Because $R_B = R_C$,

$$I_B = I_C = \frac{I_A}{2} = \frac{5 \text{ A}}{2} = 2.5 \text{ A}$$

EXAMPLE 6-5

Find (A) I_1, (B) I_2, (C) I_3, (D) I_4, (E) I_5, and (F) I_6 in the circuit of Figure 6-11.

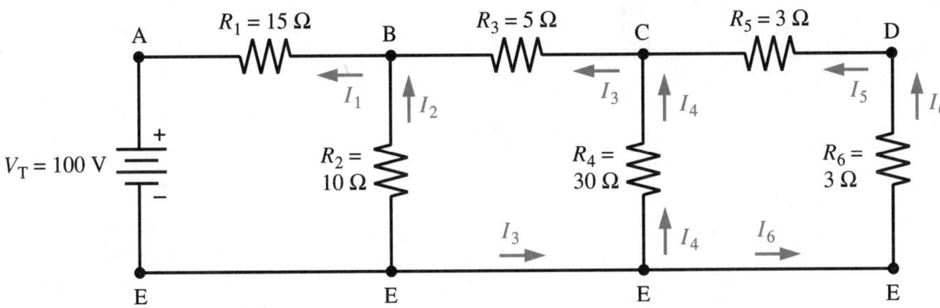

Figure 6-11

Solution:

(See Figure 6-12)

(A) (See Figure 6-12D)

$I_1 = \dfrac{V_T}{R_T}$

$= \dfrac{100 \text{ V}}{20\ \Omega}$

$= 5$A

(B) (See Figure 6-12C)

$I_2 = \dfrac{(R_{3456})(I_1)}{R_{3456} + R_2}$

$= \dfrac{(10\ \Omega)(5 \text{ A})}{10\ \Omega + 10\ \Omega}$

$= \dfrac{50\ \Omega \cdot \text{A}}{20\ \Omega}$

$= 2.5$ A

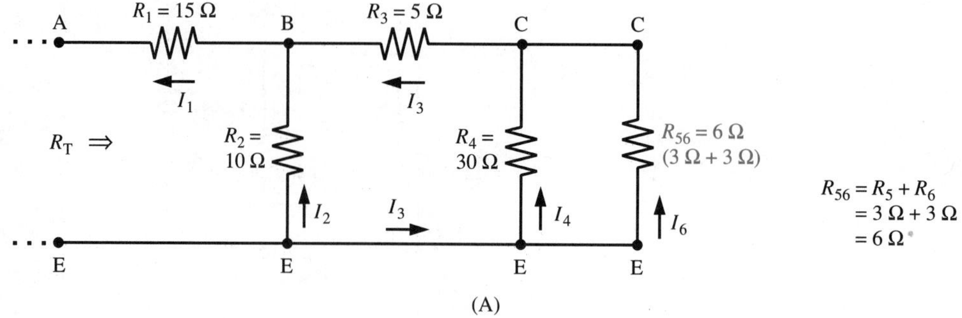

(A)

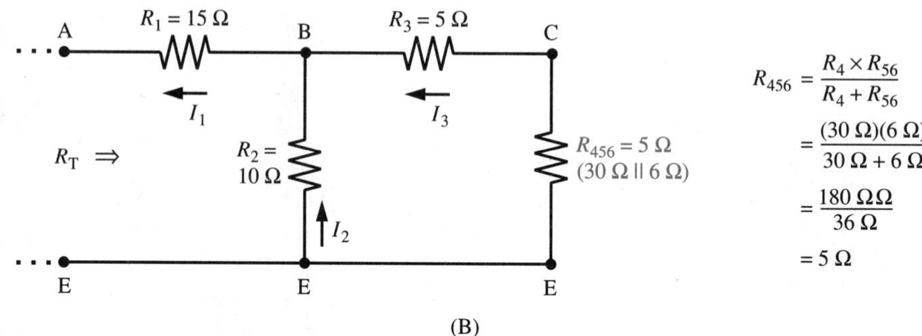

(B)

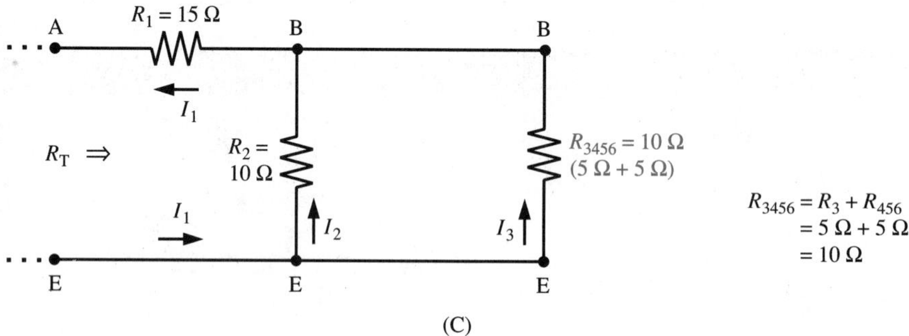

(C)

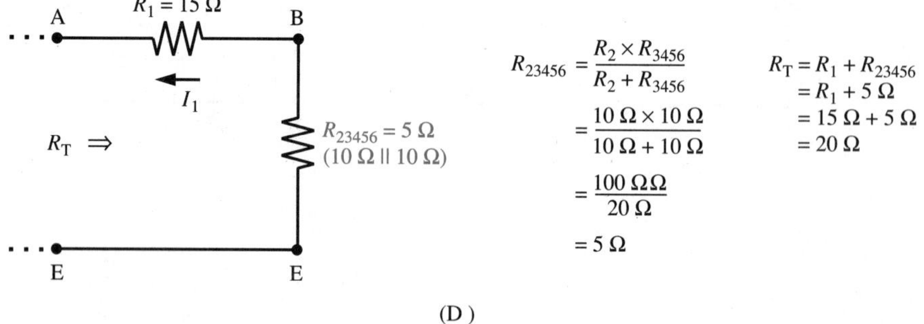

(D)

Figure 6-12

(C) $I_3 = I_2 = 2.5\ \text{A}$

(D) (See Figure 6-12A)

$$I_4 = \frac{(R_{56})(I_3)}{R_{56} + R_4} = \frac{(6\ \Omega)(2.5\ \text{A})}{6\ \Omega + 30\ \Omega} = \frac{15\ \Omega \cdot \text{A}}{36\ \Omega} = 0.416\ \text{A}$$

(E) (See Figure 6-11)

$$I_5 = I_3 - I_4$$
$$= 2.5\text{ A} - 0.416\text{ A}$$
$$= 2.084\text{ A}$$

(F) (See Figure 6-11)

$$I_6 = I_5$$
$$= 2.084\text{ A}$$

✓ Self-Test 2

1. In Figure 6-2A, if V_T = 100 V, R_1 = 2 KΩ, R_2 = 6 KΩ, R_3 = 2 KΩ, and R_4 = 4 KΩ, find R_T, I_T, I_2, and I_3.
2. In Figure 6-2B, if V_T = 12 V, R_1 = 6 KΩ, R_2 = 6 KΩ, R_3 = 1 KΩ, and R_4 = 3 KΩ, find R_T, I_T, I_2, and I_3.

6.4 VOLTAGE AND CURRENT SOURCES

An **ideal voltage source** is a device that supplies a *constant voltage* to a load. It adjusts its current according to the resistance of the load. An **ideal current source** is a device that supplies a *constant current* to a load. It adjusts its voltage according to the resistance of the load. Figure 6-13 shows the symbols for an ideal voltage source and an ideal current source. It is important to understand that the voltage value of an ideal voltage source does not change even if the resistance of the load does change. Similarly, the output current of an ideal current source does not change even if the resistance of the load does change.

tech tidbit

An ideal voltage source supplies a constant voltage to a load.

tech tidbit

An ideal current source supplies a constant current to a load.

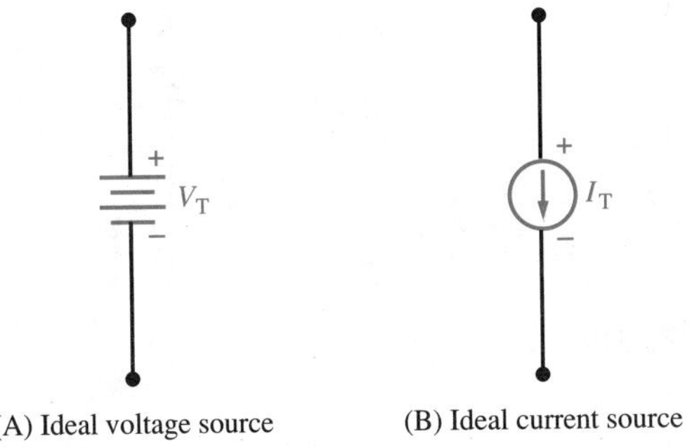

Figure 6-13 **Ideal sources**

Note that in Figure 6-13B, the arrow in the ideal current source indicates the direction of the current.

Obviously, ideal sources do not exist. For example, no battery can supply any amount of current required by a load without its voltage dropping because every source has some **internal resistance.** Figure 6-14 illustrates the practical configuration of a voltage and current source.

In Figure 6-14A, we see that the terminal (output) voltage of a battery is actually the internal cell voltage minus the *IR* drop across the internal resistance of the battery, R_S. If a load were to draw more current the terminal (output) voltage of the battery will drop. Similarly, in Figure 6-14B, we see that the terminal (output) current of a current source is the ideal current minus the current through the internal parallel resistance, R_S. Note that it is customary to consider voltage sources with series resistors and current sources with parallel resistors. However, either source can be expressed in either way.

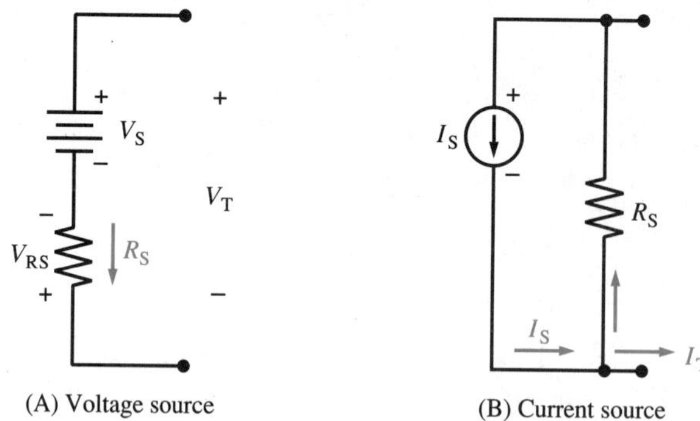

(A) Voltage source (B) Current source

Figure 6-14 **Practical sources**

tech tidbit

In source conversion, the magnitude of R_S remains unchanged.

Mathematically, a voltage source can be converted into a current source and a current source can be converted into a voltage source. Figure 6-15 shows the equivalent sources with the conversion equations. Note that in either **source conversion** the magnitude of R_S remains unchanged and its affect on the load will be the same. It is often more convenient to solve certain circuits with one type of source instead of another. Source conversion allows us to convert from a voltage source to a current source or vice versa. For example, transistor circuits are often modeled as current sources.

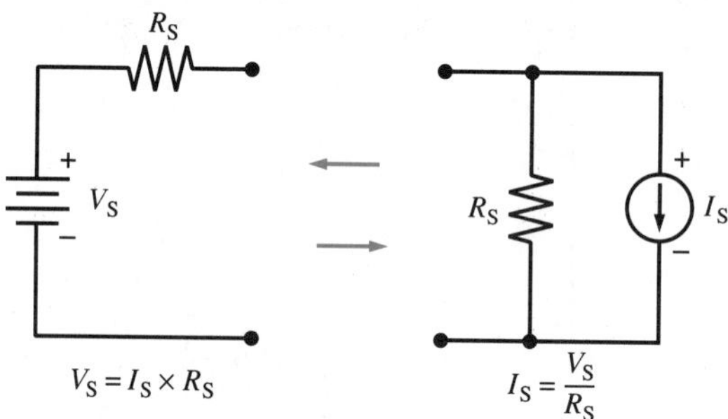

Figure 6-15 **Source conversion**

EXAMPLE 6-6

(A) Convert the voltage source of Figure 6-16 to a current source and (B) calculate the current through the load for each source.

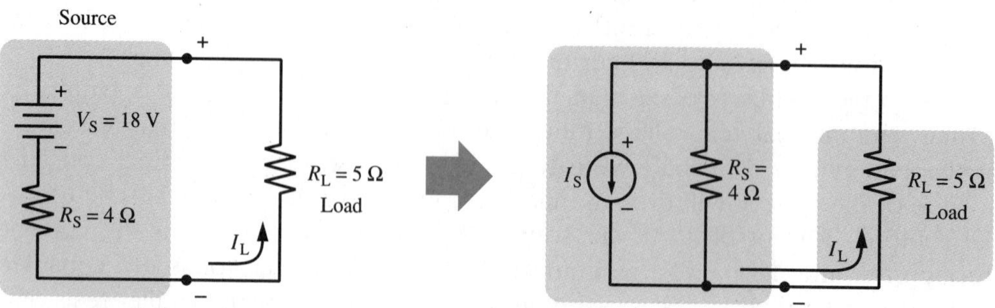

Figure 6-16

Solution:

(A) $I_S = \dfrac{V_S}{R_S}$

$= \dfrac{18\text{ V}}{4\ \Omega}$

$= 4.5\text{ A}$

(B) Method 1

$I_L = \dfrac{V_S}{R_S + R_L}$

$= \dfrac{18\text{ V}}{4\ \Omega + 5\ \Omega}$

$= \dfrac{18\text{ V}}{9\ \Omega}$

$= 2\text{ A}$

Method 2 (Current Divider Rule)

$I_L = \dfrac{R_S \times I_S}{R_S + R_L}$

$= \dfrac{4\ \Omega \times 4.5\text{ A}}{4\ \Omega + 5\ \Omega}$

$= 2\text{ A}$

EXAMPLE 6-7

(A) Convert the current source of Figure 6-17 to a voltage source and (B) find the current through the load for each source.

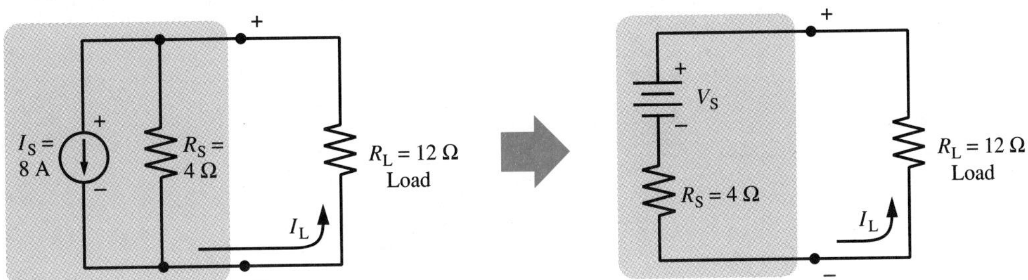

Figure 6-17

Solution:

(A) $V_S = I_S \times R_S$

$= 8\text{ A} \times 4\ \Omega$

$= 32\text{ V}$

(B) Method 1 (Current Divider Rule)

$I_L = \dfrac{R_S \times I_S}{R_S + R_L}$

$= \dfrac{4\ \Omega \times 8\text{ A}}{4\ \Omega + 12\ \Omega}$

$= 2\text{ A}$

Method 2

$I_L = \dfrac{V_S}{R_S + R_L}$

$= \dfrac{32\text{ V}}{4\ \Omega + 12\ \Omega}$

$= 2\text{ A}$

✓ Self-Test 3

1. An ideal voltage source is a device that supplies a __________ to a load.
2. An ideal current source is a device that supplies a __________ to a load.
3. Referring to Figure 6-16, if V_S = 12 V and R_S = 0.5 Ω, calculate the equivalent current source.
4. Referring to Figure 6-16, if I_S = 5 μA and R_S = 10 MΩ, calculate the equivalent voltage source.

6.5 THEVENIN'S THEOREM

Thevenin's theorem is a very useful tool for analyzing complex electric circuits. It allows you to solve for the effects of any part of a circuit. For example, if you want to solve for the current through a light bulb that is part of a complex circuit, Thevenin's theorem allows you to do so without solving for anything else in the circuit. Furthermore, the Thevenin equivalent circuit, with the attached light bulb load, becomes a simple series circuit. This can be very useful and efficient when considering "what if?" situations. For example, what will happen to the current if we change to a different light bulb?

In a simplified sense, Thevenin's theorem states: Any resistive circuit can be converted into an equivalent circuit that consists of a voltage source, V_{TH}, and a series resistor, R_{TH}, as shown in Figure 6-18.

2 tech tidbit

Thevenin's theorem—any resistive circuit can be converted into an equivalent circuit that consists of a voltage source and a series resistor.

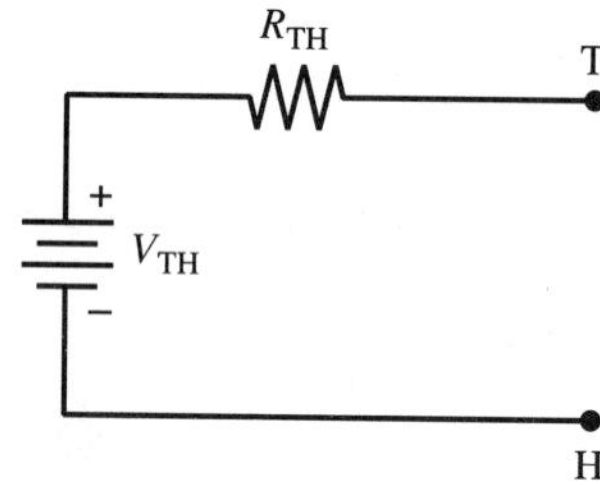

Figure 6-18 **Thevenin's circuit**

To "Theveninize" or convert a circuit into its Thevenin equivalent circuit, we perform the following steps:

1. Identify and remove the portion or component you wish to analyze from the circuit. This becomes the Thevenin load and will be added to the Thevenin circuit later.
2. Mark the open circuit terminals created by removing the Thevenin load with the letters T and H.
3. Calculate V_{TH}, which is equal to the open circuit voltage across terminals T and H.
4. Calculate R_{TH} by first setting all sources to zero (voltage sources are replaced by short circuits and current sources are replaced by open circuits) and find the resultant resistance between terminals T and H.
5. Draw the Thevenin equivalent circuit with the Thevenin load replaced across terminals T and H and calculate the original required information.

EXAMPLE 6-8

Refer to the circuit shown in Figure 6-19 and calculate the current through R_3 using Thevenin's theorem.

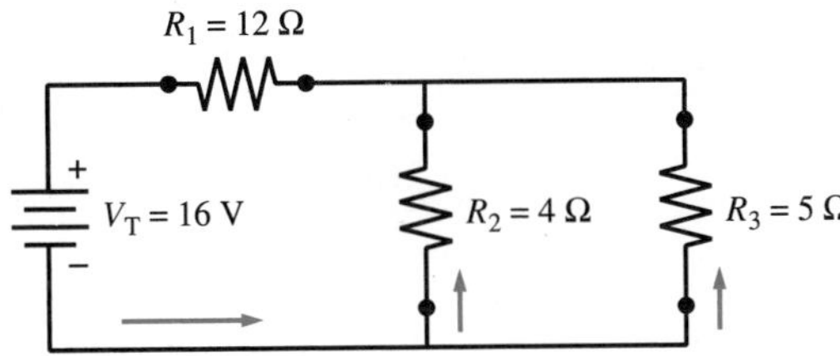

Figure 6-19

Solution:

(See Figure 6-20)

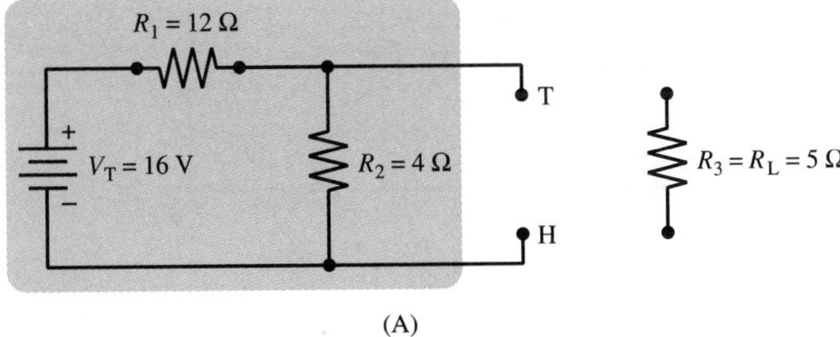

(A)

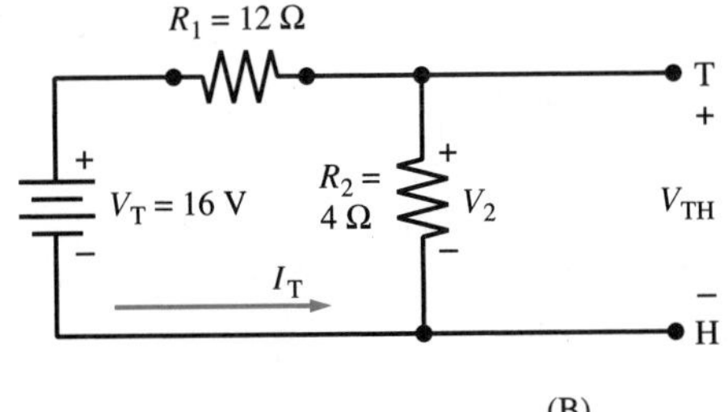

Voltage Divider Rule

$$V_{TH} = V_2 = \frac{R_2 \times V_T}{R_1 + R_2}$$

$$= \frac{4\ \Omega \times 16\ \text{V}}{12\ \Omega + 4\ \Omega}$$

$$= \frac{64\ \cancel{\Omega} \cdot \text{V}}{16\ \cancel{\Omega}}$$

$$= 4\ \text{V}$$

(B)

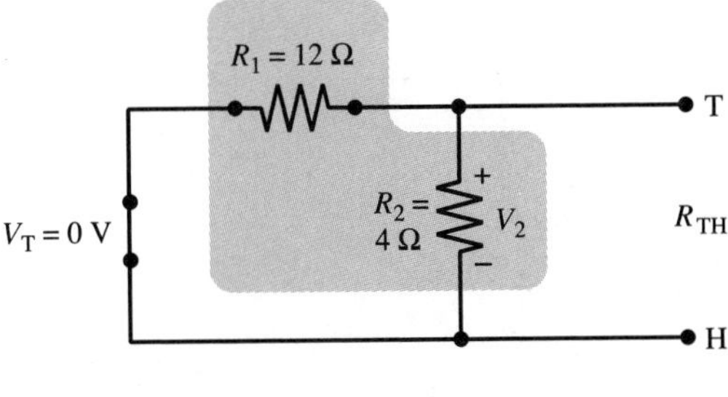

R_1 in parallel with R_2

$$R_{TH} = R_1 \parallel R_2$$

$$= \frac{R_1 \times R_2}{R_1 + R_2}$$

$$= \frac{12\ \Omega \times 4\ \Omega}{12\ \Omega + 4\ \Omega}$$

$$= \frac{48\ \cancel{\Omega} \cdot \Omega}{16\ \cancel{\Omega}}$$

$$= 3\ \Omega$$

Note: R_1 is now in parallel with R_2.

(C)

Figure 6-20 (continued on page 138)

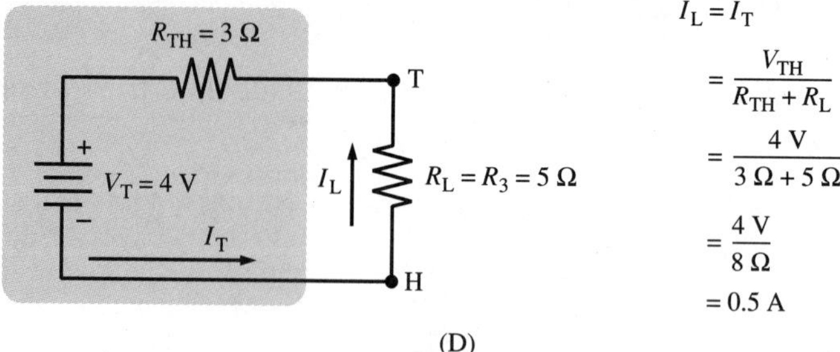

Figure 6-20 (cont.)

EXAMPLE 6-9

Refer to the circuit shown in Figure 6-21 and calculate the current through R_3 using Thevenin's theorem.

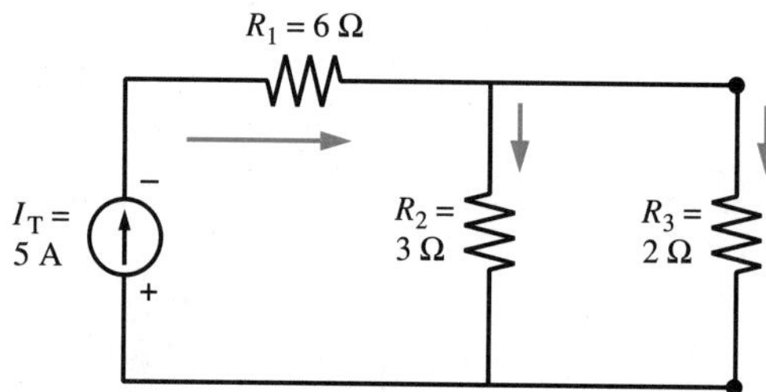

Figure 6-21

Solution:

(See Figure 6-22)

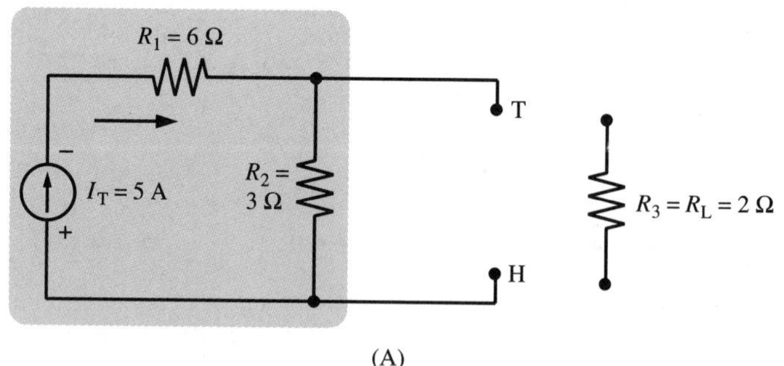

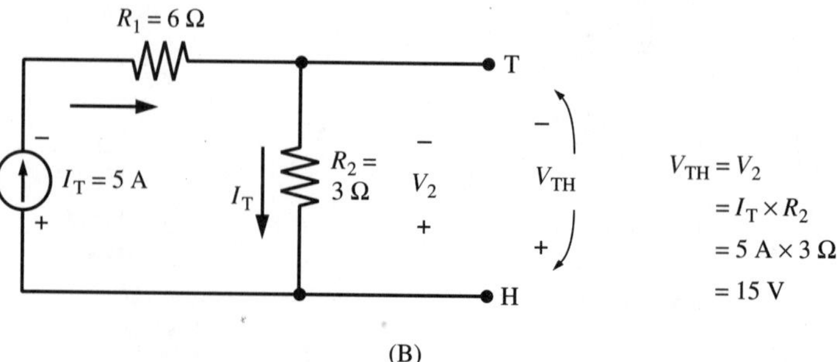

Figure 6-22 (continued on page 139)

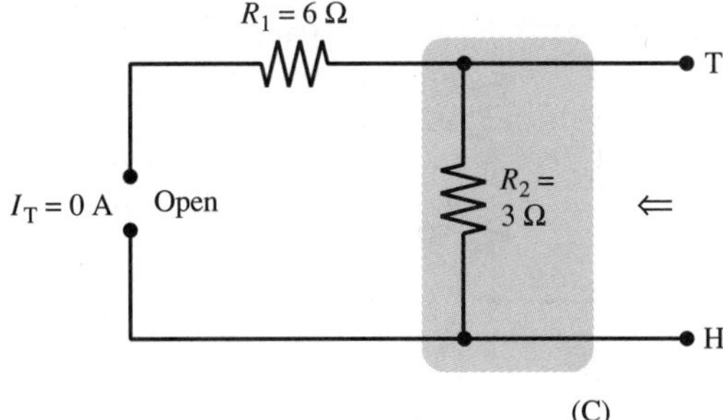

(C)

$R_T = R_2 = 3\ \Omega$

Note: With $I_T = 0$ A, the branch with R_1 in it is an open circuit.

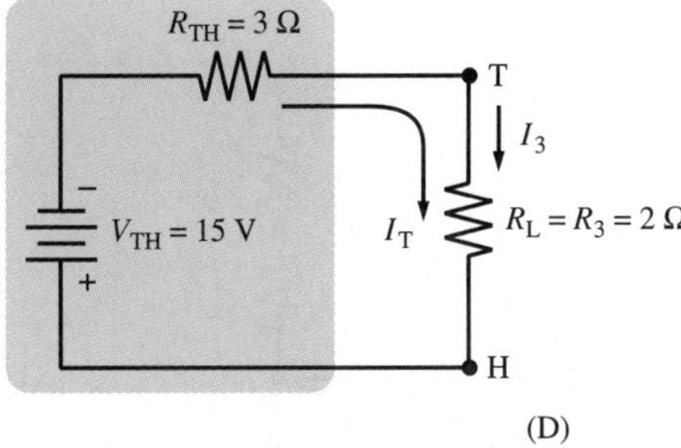

(D)

$$I_3 = I_T = \frac{V_{TH}}{R_{TH} + R_L} = \frac{15\text{ V}}{(3\ \Omega + 2\ \Omega)} = \frac{15\text{ V}}{5\ \Omega} = 3\text{ A}$$

Figure 6-22 (cont.)

EXAMPLE 6-10

Refer to the circuit shown in Figure 6-23 and calculate the voltage across R_2 using Thevenin's theorem.

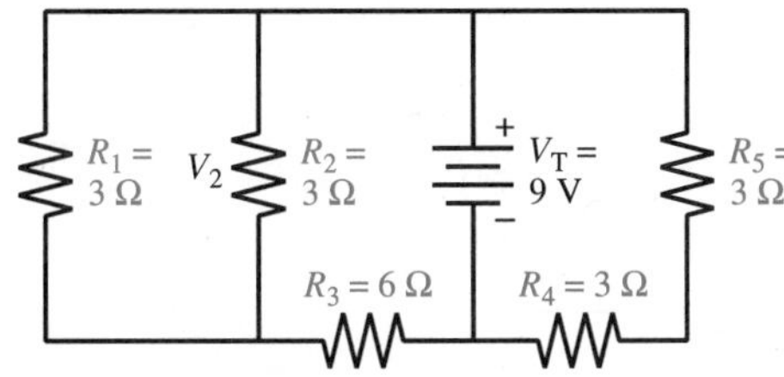

Figure 6-23

Solution:

(See Figure 6-24)

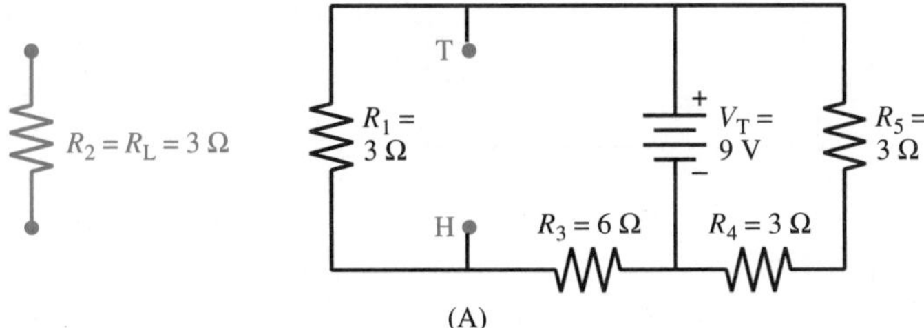

(A)

Figure 6-24 (continued on page 140)

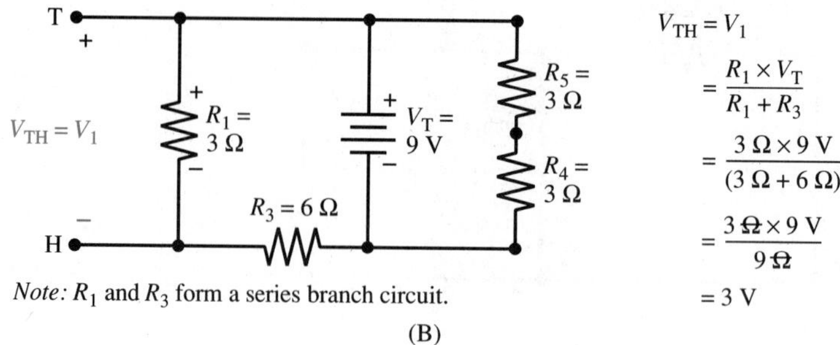

$$V_{TH} = V_1 = \frac{R_1 \times V_T}{R_1 + R_3} = \frac{3\ \Omega \times 9\ \text{V}}{(3\ \Omega + 6\ \Omega)} = \frac{3\ \not{\Omega} \times 9\ \text{V}}{9\ \not{\Omega}} = 3\ \text{V}$$

Note: R_1 and R_3 form a series branch circuit.

(B)

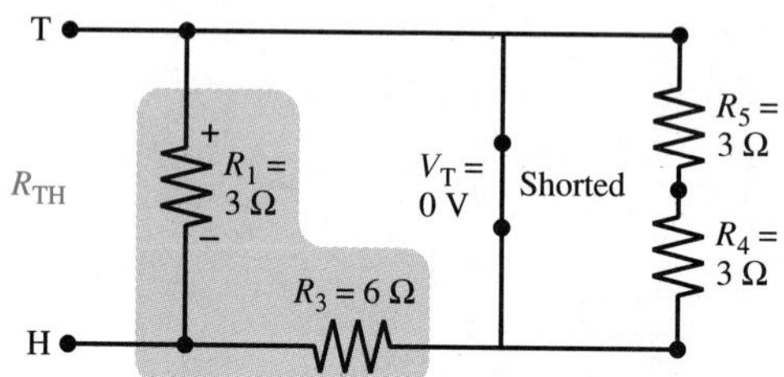

Note: R_4 and R_5 are shorted out of circuit (0 Ω).
R_1 and R_3 are in parallel.

(C)

T
R_{TH}
H
$R_1 =$ 3 Ω
$R_3 =$ 6 Ω

Therefore the equivalent-circuit is:
R_1 in parallel with R_3

$$R_{TH} = R_1 \parallel R_3 = \frac{R_1 \times R_3}{R_1 + R_3} = \frac{3\ \Omega \times 6\ \Omega}{3\ \Omega + 6\ \Omega} = \frac{18\ \not{\Omega} \cdot \Omega}{9\ \not{\Omega}} = 2\ \Omega$$

(D)

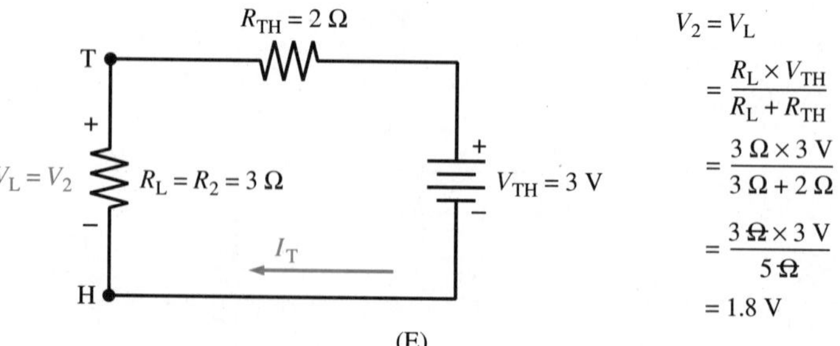

$$V_2 = V_L = \frac{R_L \times V_{TH}}{R_L + R_{TH}} = \frac{3\ \Omega \times 3\ \text{V}}{3\ \Omega + 2\ \Omega} = \frac{3\ \not{\Omega} \times 3\ \text{V}}{5\ \not{\Omega}} = 1.8\ \text{V}$$

(E)

Figure 6-24 (cont.)

EXAMPLE 6-11

Refer to the circuit shown in Figure 6-25 and calculate the current through R_5 using Thevenin's theorem.

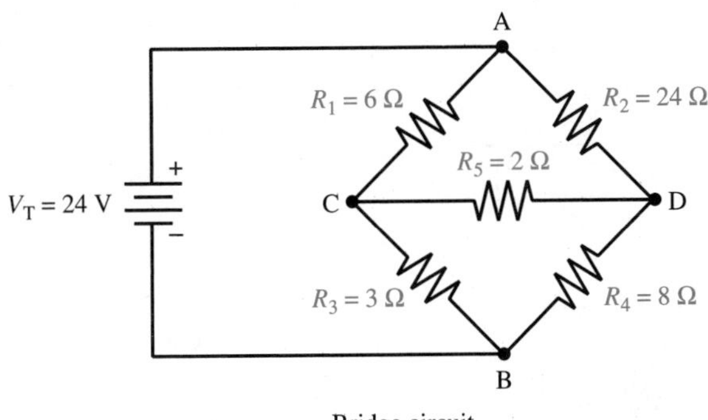

Bridge circuit

Figure 6-25

Solution:

(See Figure 6-26)

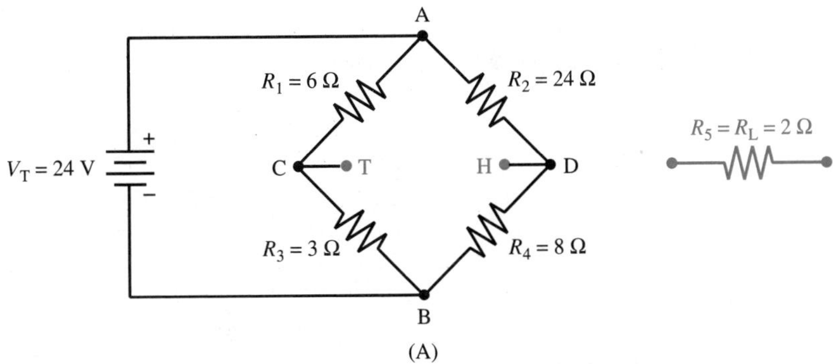

(A)

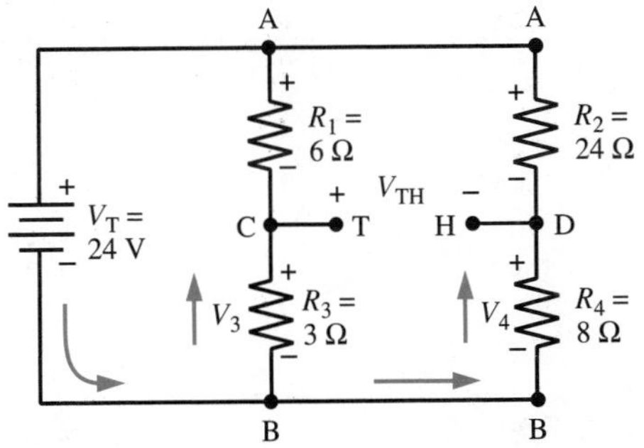

Redrawing the circuit and applying Kirchhoff's voltage law
Note: R_1 and R_2 are in series, R_3 and R_4 are in series.

By voltage divider rule:

$$V_3 = \frac{R_3 V_T}{R_1 + R_3} = \frac{(3\ \Omega)(24\ \text{V})}{6\ \Omega + 3\ \Omega} = 8\ \text{V}$$

$$V_4 = \frac{R_4 V_T}{R_2 + R_4} = \frac{(8\ \Omega)(24\ \text{V})}{24\ \Omega + 8\ \Omega} = 6\ \text{V}$$

By Kirchhoff's voltage law in the counter-clockwise direction for loop BDCB:

$$V_4 + V_{TH} - V_3 = 0$$

$$6\ \text{V} + V_{TH} - 8\ \text{V} = 0$$

$$V_{TH} = 2\ \text{V}$$

(B)

Figure 6-26 (continued on page 142)

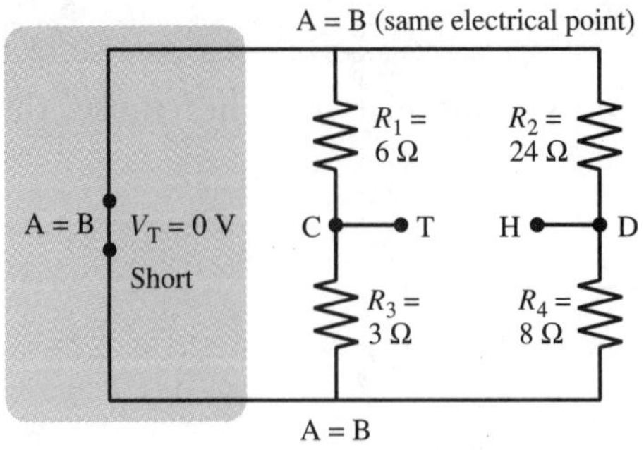

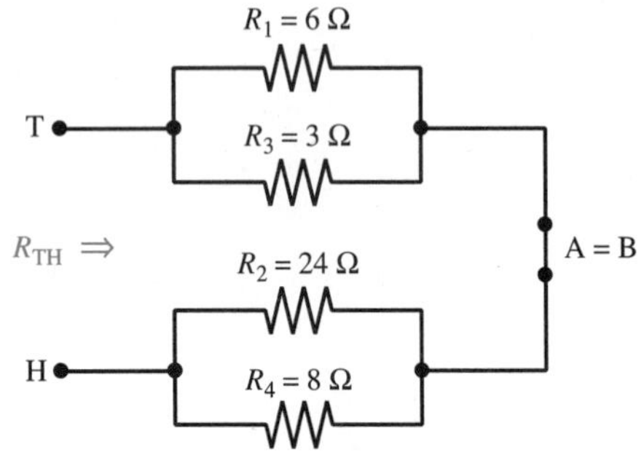

Note: Node A and B are connected.

$$R_{13} = R_1 \parallel R_3 = \frac{R_1 \times R_3}{R_1 + R_3} = \frac{6\ \Omega \times 3\ \Omega}{6\ \Omega + 3\ \Omega} = 2\ \Omega$$

$$R_{24} = R_2 \parallel R_4 = \frac{R_2 \times R_4}{R_2 + R_4} = \frac{24\ \Omega \times 8\ \Omega}{24\ \Omega + 8\ \Omega} = 6\ \Omega$$

(C)

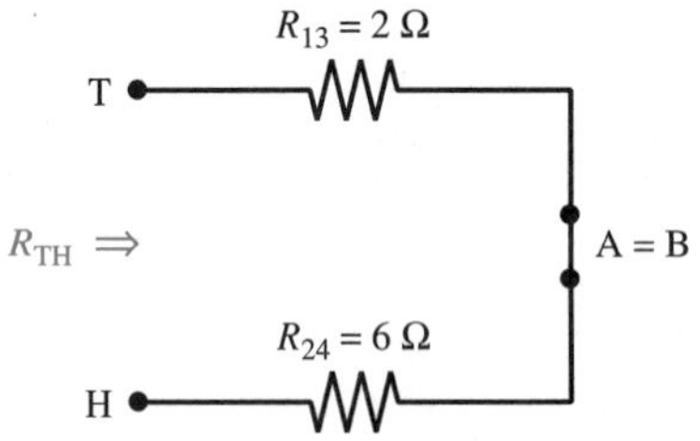

Therefore:

$R_{13} = 2\ \Omega$,
$R_{24} = 6\ \Omega$

$$R_{TH} = R_{13} + R_{24} = 2\ \Omega + 6\ \Omega = 8\ \Omega$$

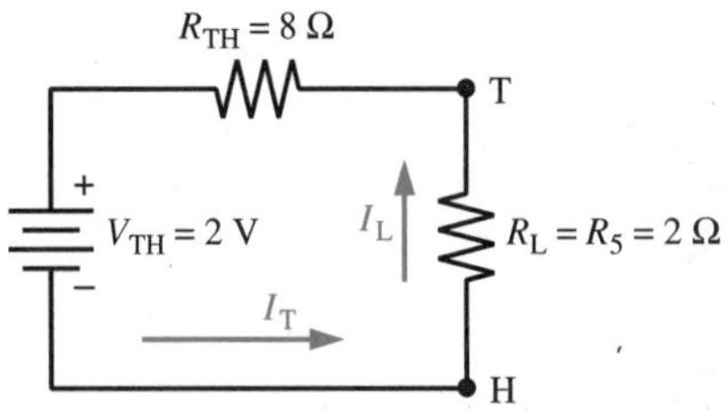

$$I_L = I_T = \frac{V_{TH}}{R_{TH} + R_L} = \frac{2\ \text{V}}{8\ \Omega + 2\ \Omega} = \frac{2\ \text{V}}{10\ \Omega} = 0.2\ \text{A}$$

(D)

Figure 6-26 (cont.)

Norton's theorem—any resistive circuit can be converted into an equivalent circuit that consists of a current source and a parallel resistor.

6.6 NORTON'S THEOREM

In a simplified sense, **Norton's theorem** states: Any resistive circuit can be converted into an equivalent circuit that consists of a current source and a parallel resistor, as shown in Figure 6-27.

The conversion process is similar to the Thevenin conversion process. The current source I_N is equal to the short circuit current between the removed components. The parallel resistance R_N is equal to the resistance between the termi-

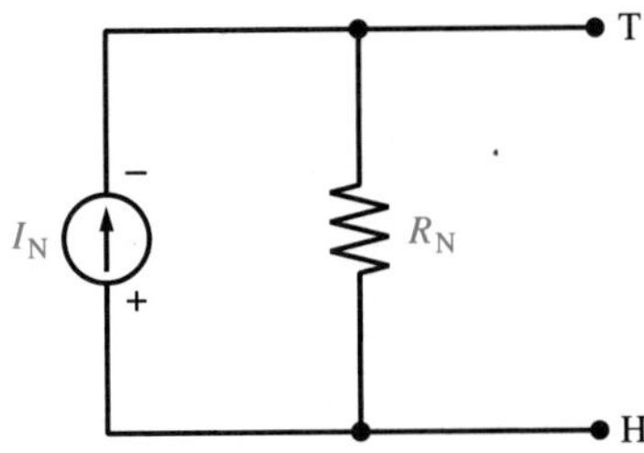

Figure 6-27 **Norton's circuit**

nals of the removed components with all sources set to zero. Note that R_N is calculated exactly the same way as R_{TH} and in fact $R_N = R_{TH}$.

Instead of having to learn and memorize two processes, one for Thevenin and one for Norton, we can obtain a Norton equivalent circuit by first obtaining the Thevenin equivalent circuit and applying the rules of source conversion as shown in Figure 6-28.

2 **tech tidbit**

Thevenin circuits can be converted to Norton circuits by source conversion. Norton circuits can be converted to Thevenin circuits by source conversion.

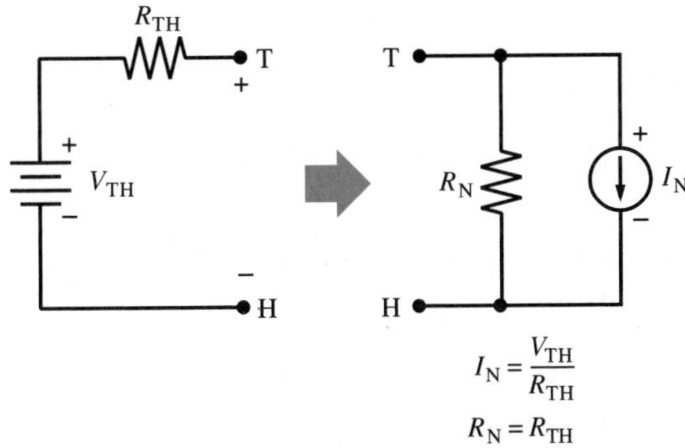

Figure 6-28 **Thevenin to Norton conversion**

EXAMPLE 6-12

Repeat Example 6-8 using Norton's theorem.

Solution:

Refer to Figure 6-29.

From Example 6-8 (Figure 6-19), we have:

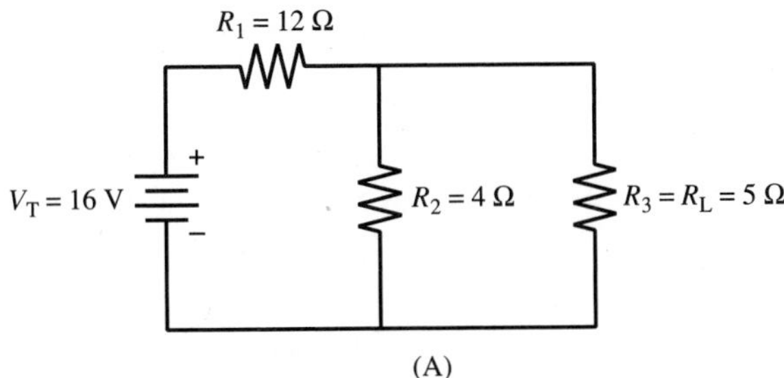

Figure 6-29 **(continued on page 144)**

From Figure 6-20D, we have:

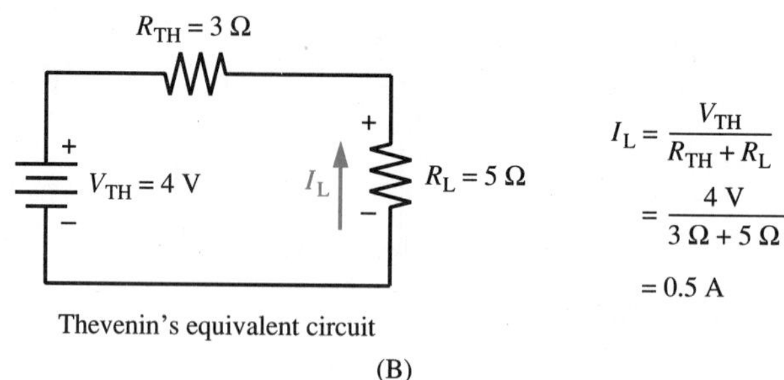

$$I_L = \frac{V_{TH}}{R_{TH} + R_L} = \frac{4\text{ V}}{3\ \Omega + 5\ \Omega} = 0.5\text{ A}$$

(B)

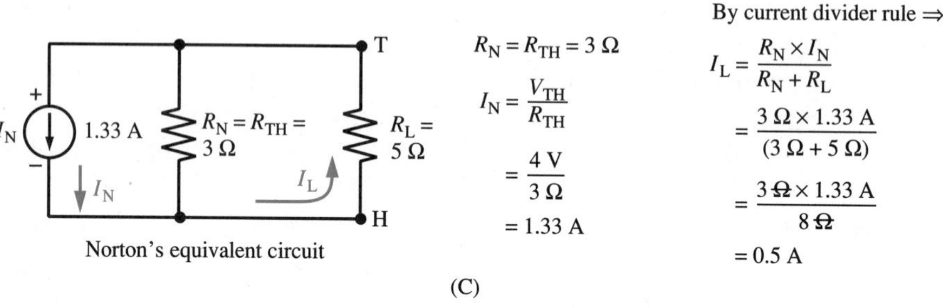

$$R_N = R_{TH} = 3\ \Omega$$

$$I_N = \frac{V_{TH}}{R_{TH}} = \frac{4\text{ V}}{3\ \Omega} = 1.33\text{ A}$$

By current divider rule ⇒

$$I_L = \frac{R_N \times I_N}{R_N + R_L} = \frac{3\ \Omega \times 1.33\text{ A}}{(3\ \Omega + 5\ \Omega)} = \frac{3\ \Omega \times 1.33\text{ A}}{8\ \Omega} = 0.5\text{ A}$$

(C)

Figure 6-29 (cont.)

Tech tidbit 2

Maximum power transfer theorem—a load will receive maximum power when it equals the Thevenin resistance of the circuit.

6.7 MAXIMUM POWER TRANSFER THEOREM

A load will receive maximum power (watts) from a circuit when its resistive value is equal to the Thevenin resistance, R_{TH}, of the circuit. The amount of maximum power transferred to the load will equal

Equation 6-1 ▶

$$P_{MAX} = \frac{V_{TH}^2}{4\,R_{TH}}$$

Tech tidbit 2

$P_{MAX} = \frac{V_{TH}^2}{4R_{TH}}$

or

$P_{MAX} = I_L^2 R_{TH}$

The **maximum power** transfer theorem can be verified by considering the Thevenin circuit of Figure 6-30.

In Figure 6-30, we plot a graph of power at the load versus the resistance of the load for the Thevenin circuit shown. Recall that $P_L = I_L^2 \cdot R_L$. To obtain the current we divide the total voltage, V_{TH}, by the total resistance ($R_{TH} + R_L$). The table of Figure 6-30 shows the results of all of the calculations. The calculations are then plotted on the graph. Note that maximum power is obtained when $R_L = R_{TH} = 5$ ohms. Maximum power could have also been calculated using Equation 6-1 as follows:

$$P_{MAX} = \frac{V_{TH}^2}{4\,R_{TH}} = \frac{(10\text{ V})^2}{(4)(5\ \Omega)} = 5\text{ W}$$

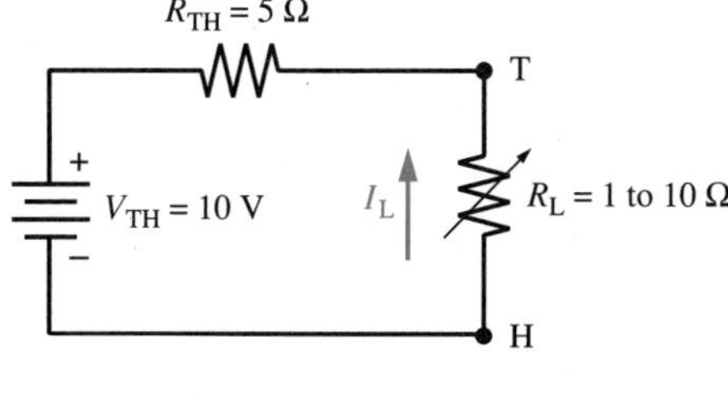

Calculations

If $R_L = 1\ \Omega$

then $I_L = \dfrac{V_{TH}}{R_{TH} + R_L}$

$$I_L = \frac{10\text{ V}}{(5\ \Omega + 1\ \Omega)}$$

$$= \frac{10\text{ V}}{6\ \Omega}$$

$$= 1.67\text{ A}$$

$$P_L = (I_L)^2 R_L$$

$$= (1.67\text{ A})^2 \times 1\ \Omega$$

$$= 2.78\text{ W}$$

If $R_L = R_{TH} = 5\ \Omega$

then $I_L = \dfrac{10\text{ V}}{(5\ \Omega + 5\ \Omega)}$

$$= \frac{10\text{ V}}{10\ \Omega}$$

$$= 1\text{ A}$$

$$P_L = (1\text{ A})^2 \times 5\ \Omega$$

$$= 5\text{ W}$$

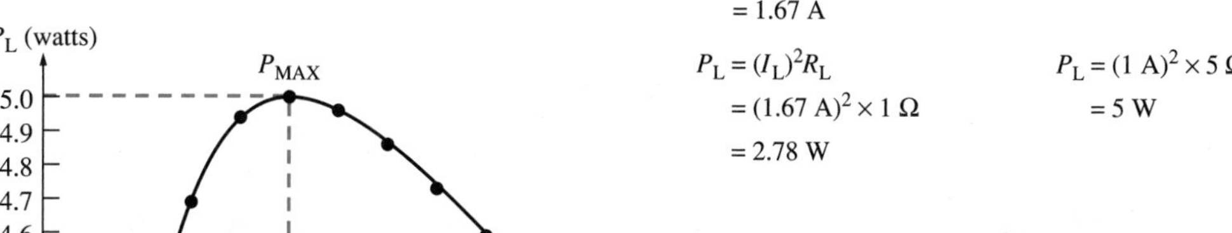

R_L (Ω)	I_L (A)	P_L (W)
1	1.67	2.78
2	1.43	4.08
3	1.25	4.69
4	1.11	4.94
5	1	5
6	0.91	4.96
7	0.83	4.86
8	0.77	4.73
9	0.71	4.59
10	0.67	4.44

$R_L = R_{TH}$ → (row 5) ← Maximum power

Figure 6-30 **Maximum power transfer**

EXAMPLE 6-13

(A) Find the value of R_L in Figure 6-31 for maximum power transfer to the load.

(B) Calculate the maximum power transferred to the load.

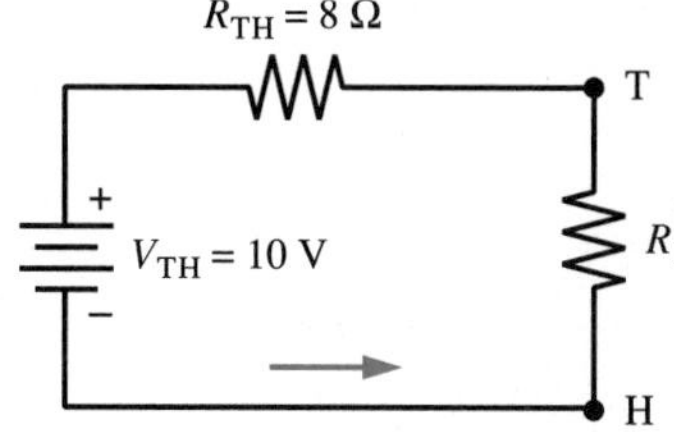

Figure 6-31

Solution:

(A) For maximum power transfer to the load

$$R_L = R_{TH} = 8\ \Omega$$

(B) Therefore

$$P_{MAX} = \frac{V_{TH}^2}{4R_{TH}} = \frac{(10\ \text{V})^2}{4 \times 8\ \Omega} = 3.125\ \text{W}$$

EXAMPLE 6-14

Find the value of R_L in Figure 6-32A for maximum power transfer to the load. Calculate the maximum power transferred to the load.

Solution:

Refer to Figure 6-32.

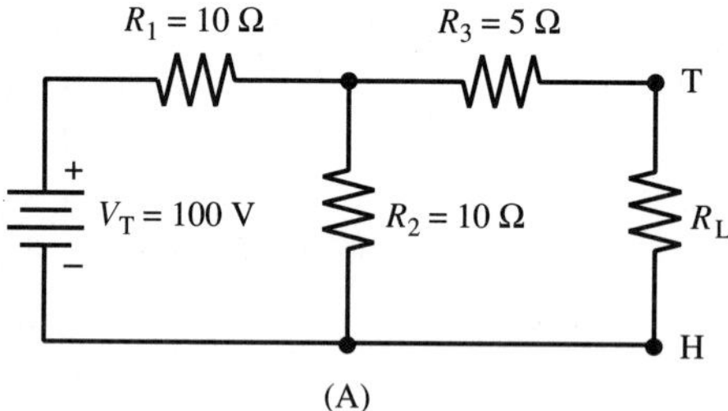

For maximum power transfer, select $R_L = R_{TH}$.
Remove load and mark nodes.

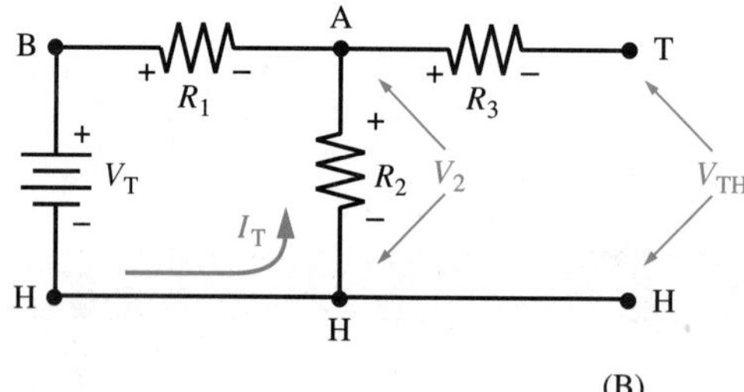

Find V_{TH}

$$V_{TH} = V_2 = \frac{R_2 \times V_T}{R_1 + R_2} = \frac{10\ \Omega \times 100\ \text{V}}{10\ \Omega + 10\ \Omega} = 50\ \text{V}$$

Find R_{TH}; set $V_T = 0$ V.

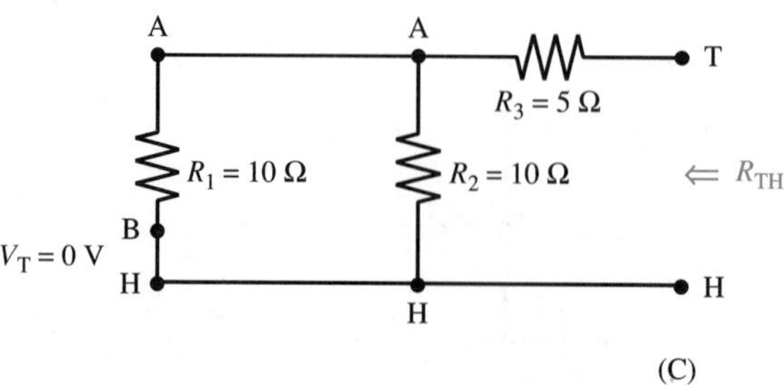

For two equal resistors, $N = 2$

$$R_{12} = \frac{R}{N} = \frac{10\ \Omega}{2} = 5\ \Omega$$

$$R_{TH} = R_3 + (R_1 \parallel R_2) = R_3 + R_{12} = 5\ \Omega + 5\ \Omega = 10\ \Omega$$

Figure 6-32 (continued on page 147)

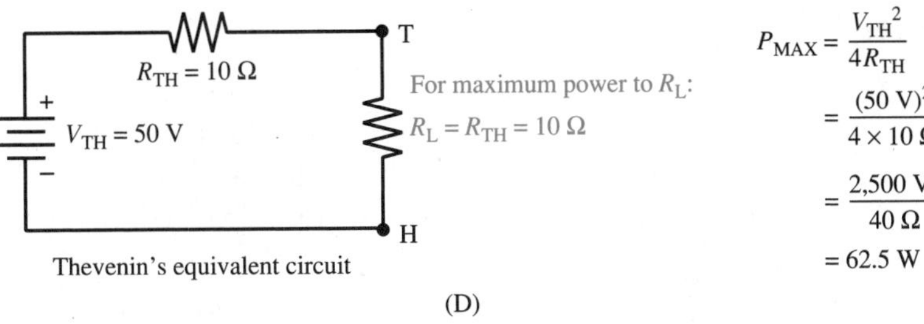

Figure 6-32 (cont.)

✓ Self-Test 4

1. Which theorem allows us to convert a circuit into an equivalent circuit that consists of a voltage source and a series resistor?
2. Which theorem allows us to convert a circuit into an equivalent circuit that consists of a current source and a parallel resistor?
3. Referring to Figure 6-18, if V_{TH} = 24 V and R_{TH} = 6 KΩ, what value load resistor should be added for maximum power transfer?
4. What is the maximum power delivered to the load resistor in question 3?

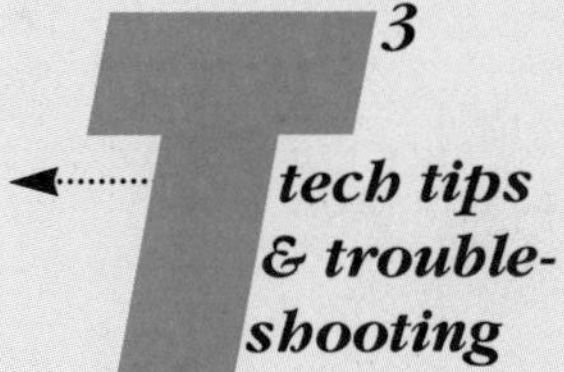

T3 CIRCUIT MEASUREMENTS

Circuit analysis enables us to determine what the voltage, current, or resistance should be in a circuit. However, circuit measurements are also required in the design, testing, repair, and maintenance of electric circuits to fully understand what is going on. Remember that high voltages and currents can cause serious injury or extensive equipment damage. Use only well-designed and well-maintained equipment to test, repair, and maintain electric circuits and equipment. Use appropriate safety equipment such as safety glasses and insulated gloves. Make sure that meters contain adequate protection on all inputs. For example, fuse protection should be provided on all current measurement jacks. Make certain that all power has been turned off when connecting a meter in any situation where one might actually come in contact with the circuit. Make sure that the person making the measurement is the only one who can turn the power back on.

Figure 6-33 shows the schematic symbol for a voltmeter and an ammeter in an electric circuit. The voltage that is being measured must appear across the input terminals of the voltmeter. Voltmeters are always connected in parallel with the load.

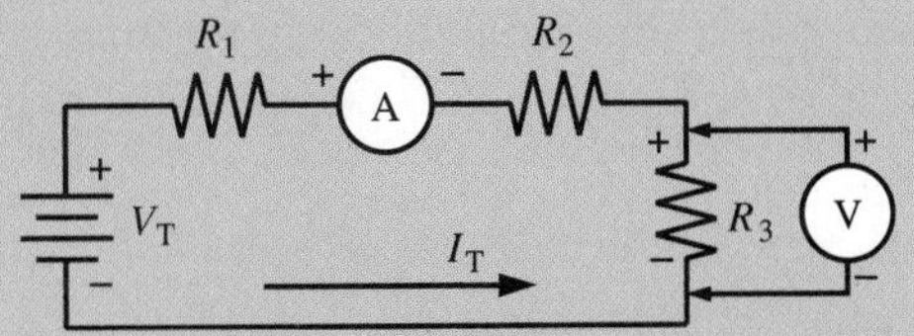

Figure 6-33 **Schematic symbol for voltmeter and ammeter**

An ammeter measures current. For this reason, the current being measured must pass through the ammeter. Thus, ammeters must always be connected in series. Always remember when connecting something in series, the circuit must be broken open. Opening a circuit is often difficult. For this reason, we often measure the voltage across a resistor and calculate the current using Ohm's law.

Remember, when making measurements, be careful to select the proper scale or range before the power is turned on. If the exact range is not known, always start at the highest range and work down.

Remember: High voltages and currents can cause serious injury or extensive damage to equipment.

Voltmeters are connected in parallel and ammeters are connected in series.

It is often easier to measure the voltage and calculate the current.

SUMMARY

Series-parallel circuits are combinations of series and parallel circuits. Series-parallel circuits are analyzed using Ohm's law and Kirchhoff's laws. We first divide the series-parallel network into its series and its parallel branches. Next we apply the rules for series circuits to the series part of the circuit and the rules for parallel branches to the parallel part of the circuit. Circuit reduction is the key to the analysis and solution of series-parallel circuits. Be certain to redraw the circuit at each stage of circuit reduction.

As with voltage sources in series, current sources in parallel can be combined into equivalent sources. Furthermore, ideal voltage sources can be converted into ideal current sources and vice versa. Source conversion can be a powerful tool when analyzing complex electric circuits. Thevenin's theorem reduces a complex circuit into a simple circuit that consists of a voltage source and a series resistor. Norton's theorem reduces a complex circuit into a simple circuit that consists of a current source and a parallel resistor. The maximum power transfer theorem is an extension of Thevenin's and Norton's theorems. It states that a load will receive maximum power from a circuit when its resistive value is equal to the Thevenin resistance, R_{TH}, or Norton resistance, R_N, of the circuit.

It is evident by the formula list that there are very few new formulas to learn in this chapter. The content of this chapter is based on the application of previously learned facts. You may want to review Chapter 5 on Series and Parallel Circuits as you apply these principles to network analysis.

FORMULAS

$I_S = \frac{V_S}{R_S}$ $V_S = I_S R_S$ (source conversion)

$P_{MAX} = \frac{V_{TH}^2}{4R_{TH}}$ $P_{MAX} = I_L^2 R_{TH}$ (maximum power transfer)

$R_L = R_{TH} = R_N$ (maximum power transfer)

EXERCISES

Section 6.2

1. For the diagrams of Figure 6-34, identify which components are in series and which components are in parallel.

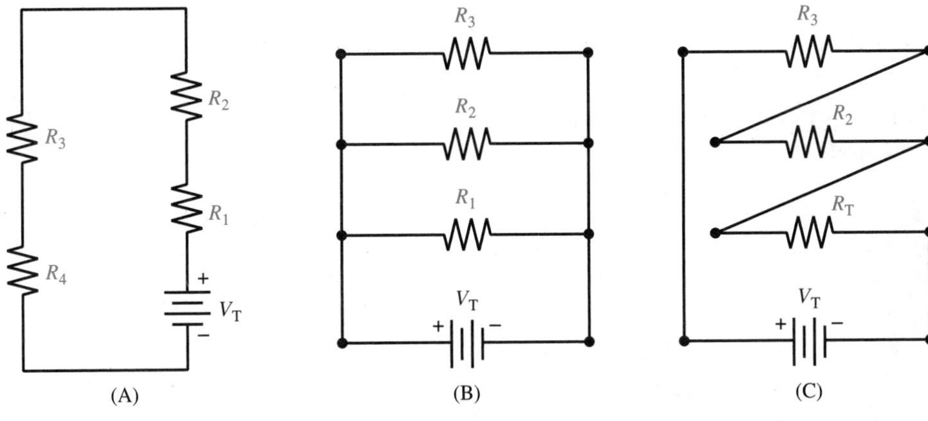

Figure 6-34

2. For the diagrams of Figure 6-35, identify which components are in series and which components are in parallel.

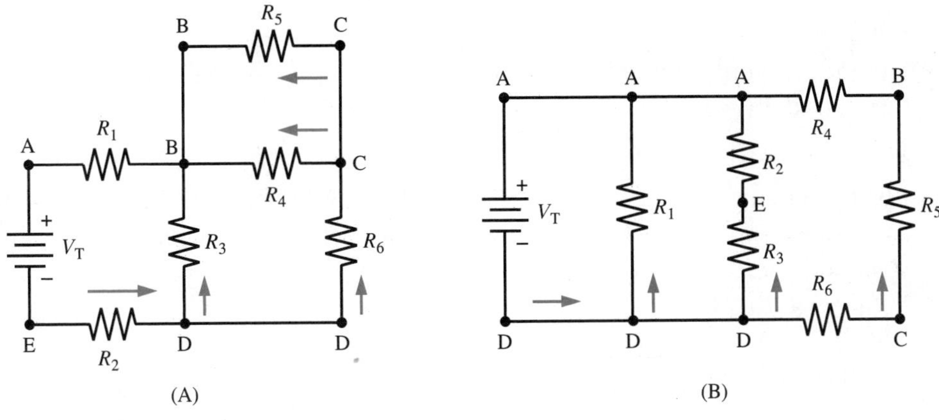

Figure 6-35

Section 6.3

3. For the circuit diagram of Figure 6-36, V_T = 10 V, R_1 = 4 Ω, R_2 = 24 Ω, and R_3 = 8 Ω. Calculate R_T, I_T, V_1, V_2, and V_3.

4. Repeat problem 3 when V_T = 44 V, R_1 = 100 Ω, R_2 = 200 Ω, and R_3 = 300 Ω.

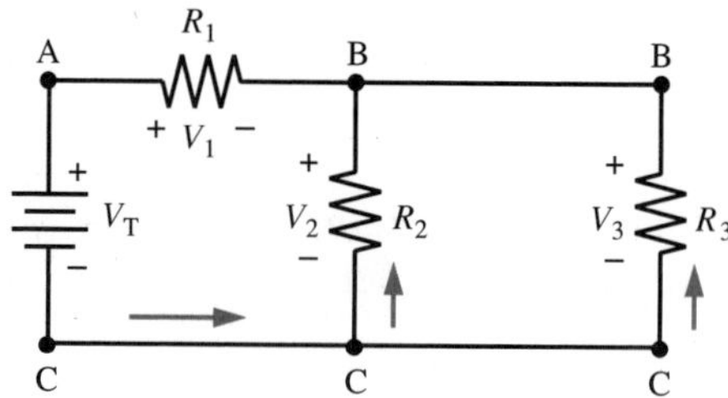

Figure 6-36

5. For the circuit diagram of Figure 6-37, V_T = 10 V, R_1 = 3 Ω, R_2 = 2 Ω, R_3 = 4 Ω, R_4 = 1 Ω, and R_5 = 2 Ω. Calculate R_T, I_T, I_2, and I_4.

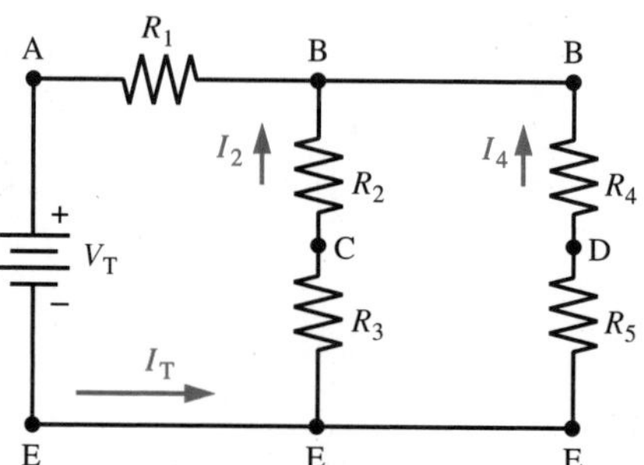

Figure 6-37

6. For problem 5, calculate V_{CD}.

7. Repeat problem 5 when V_T = 120 V, R_1 = 2 KΩ, R_2 = 5 KΩ, R_3 = 3 KΩ, R_4 = 7 KΩ, and R_5 = 1 KΩ.

8. For problem 7, calculate V_{CD}.

9. For the circuit diagram of Figure 6-38, V_T = 12 V, R_1 = 2 KΩ, R_2 = 4 KΩ, R_3 = 12 KΩ, R_4 = 3 KΩ, R_5 = 18 KΩ, and R_6 = 18 KΩ. Calculate R_T, I_T, I_1, I_4, V_{BD}, and V_{CD}.

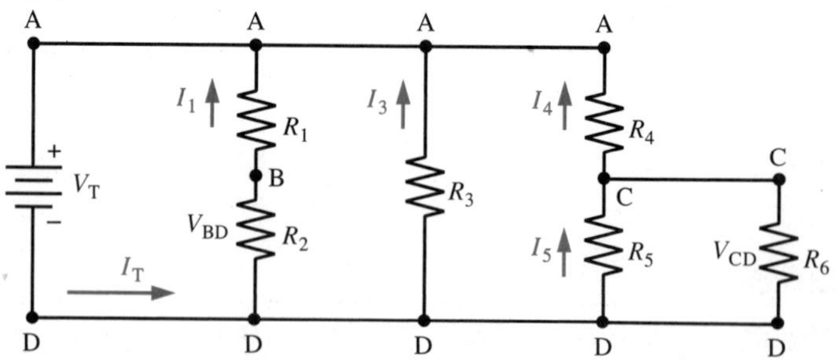

Figure 6-38

10. For the circuit diagram of Figure 6-39, V_T = 100 V, R_1 = 100 Ω, R_2 = 1 KΩ, R_3 = 1.1 KΩ, R_4 = 900 Ω, R_5 = 1 KΩ, and R_6 = 9 KΩ. Calculate R_T, I_T, V_{AB}, and V_{BC}.

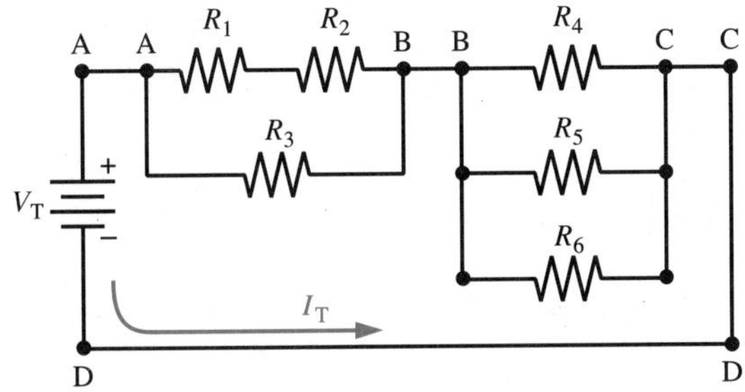

Figure 6-39

11. Find R_T for the circuits of Figure 6-40.

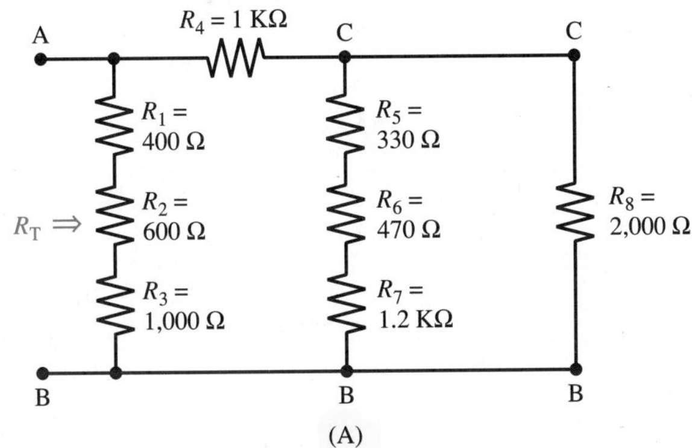

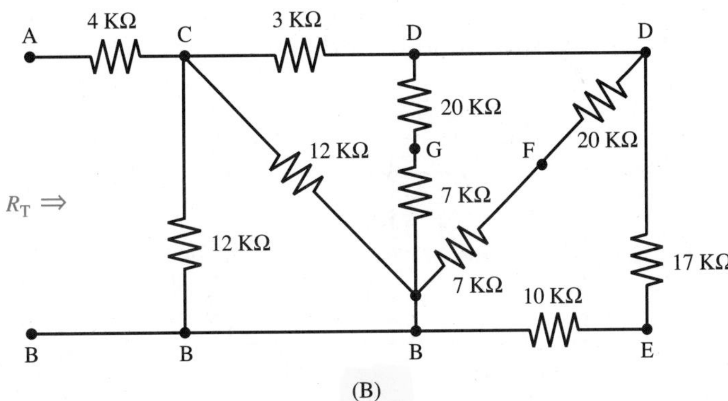

Figure 6-40

Section 6.4

12. Convert the voltage sources of Figure 6-41 to equivalent current sources.

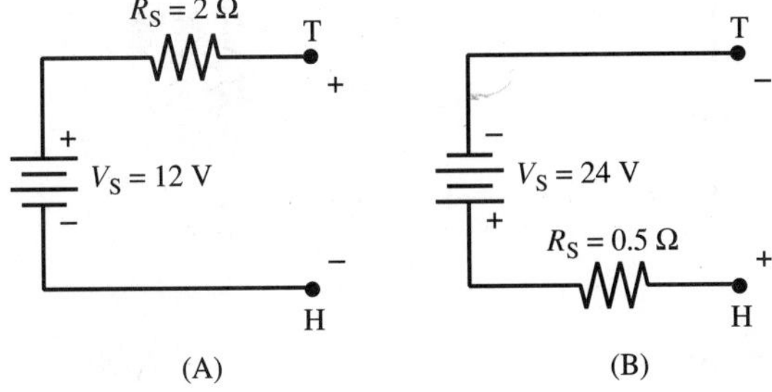

Figure 6-41

13. Convert the current sources of Figure 6-42 to equivalent voltage sources.

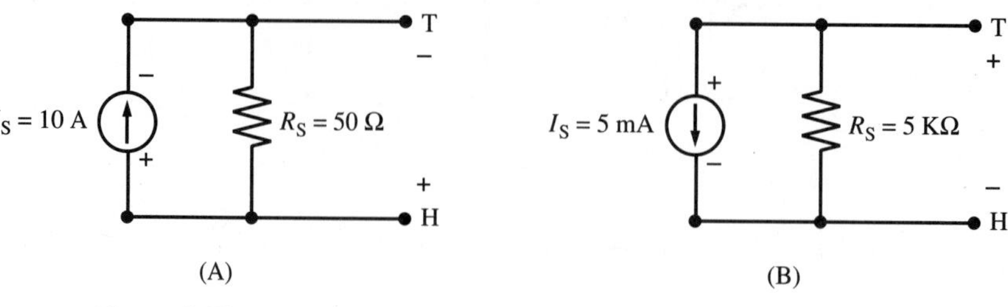

Figure 6-42

14. In Figure 6-43, find the voltage across the 6-ohm resistor. Convert the voltage source to a current source and again solve for the voltage across the 6-ohm resistor. Compare your results.

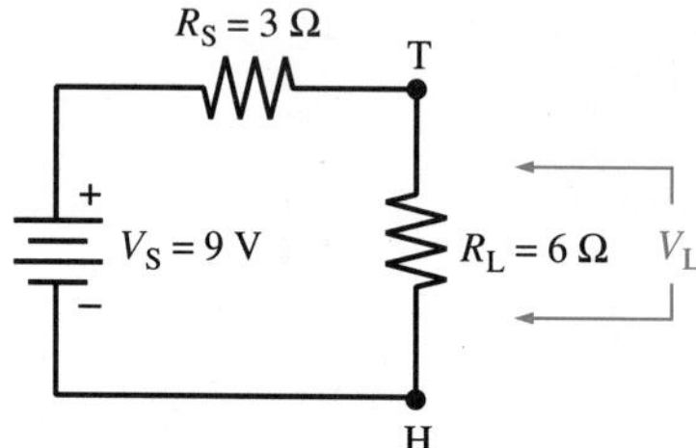

Figure 6-43

Section 6.5

15. Find the Thevenin equivalent circuit for the network external to R_L in Figure 6-44. Using the Thevenin circuit, find the voltage across R_L when R_L = 8 ohms.

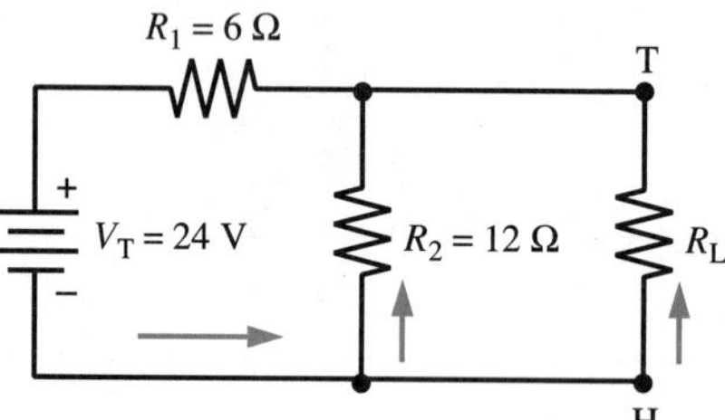

Figure 6-44

16. Find the Thevenin equivalent circuit for the network external to R_L in Figure 6-45. Find the current through R_L when R_L = 13 ohms.

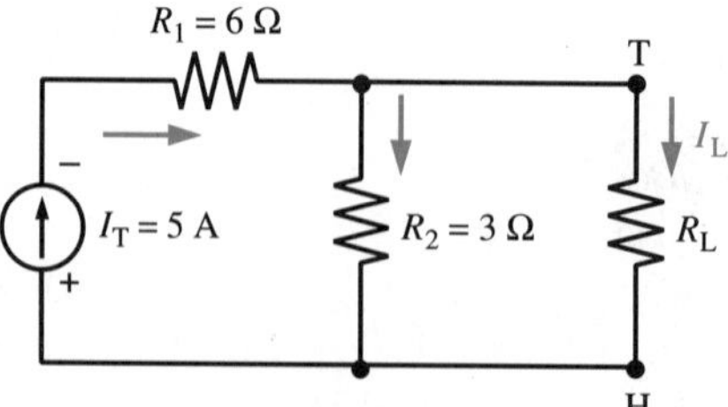

Figure 6-45

17. For the circuit diagram of Figure 6-46, solve for the current I_4 using Thevenin's theorem.

18. Find the Thevenin equivalent circuit for the network external to R_L in Figure 6-46.

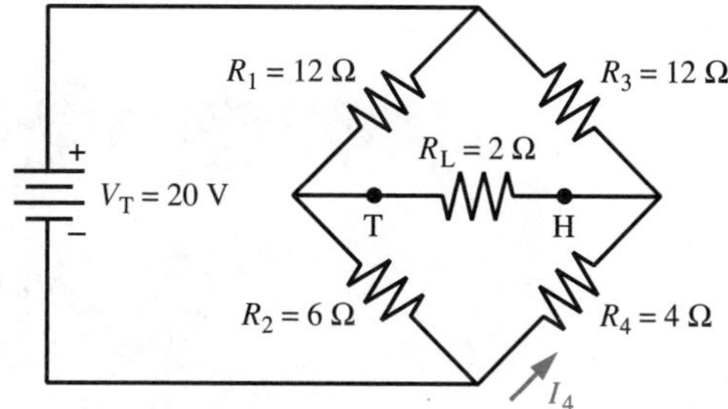

Figure 6-46

Section 6.6

19. Find the Norton equivalent circuit for the network external to R_L in Figure 6-44.
20. Find the Norton equivalent circuit for the network external to R_L in Figure 6-45.
21. Find the Norton equivalent circuit for the network external to R_4 in Figure 6-46.
22. Find the Norton equivalent circuit for the network external to R_L in Figure 6-46.

Section 6.7

23. Find the power delivered to the load R_L in Figure 6-44.
24. In Figure 6-44, what should R_L be changed to for maximum power to be delivered to R_L? Calculate the value of the maximum power that can be delivered to R_L.
25. See Lab Manual for the crossword puzzle for Chapter 6.

CHAPTER 7

MULTISOURCE CIRCUIT ANALYSIS

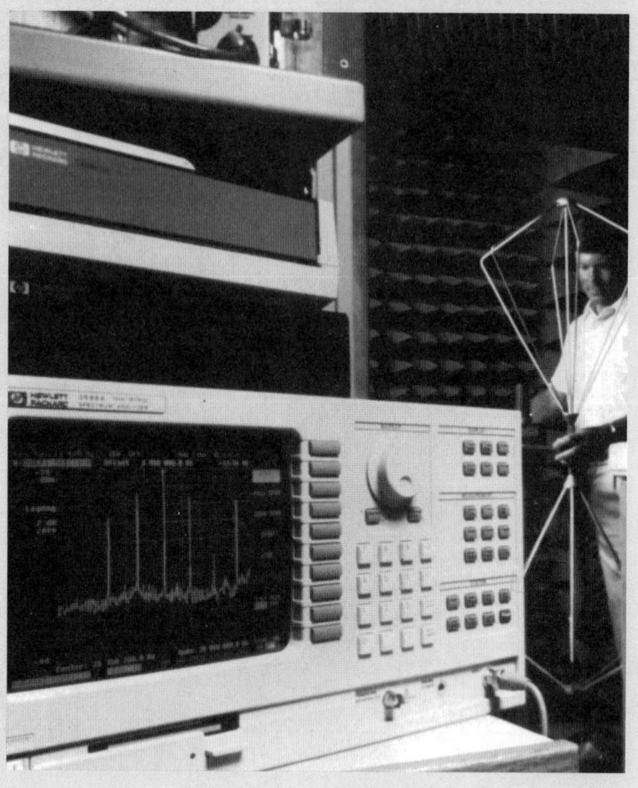

Courtesy of Hewlett Packard

KEY TERMS

linear
loop analysis
multisource
nodal analysis
node
nonlinear
simultaneous equations
superposition theorem

OBJECTIVES

After completing this chapter, the student should be able to:

- simplify and analyze multisource circuits.
- determine the effects of voltage sources in series.
- determine the effects of current sources in parallel.
- perform source conversions.
- apply the Superposition theorem.
- use Kirchhoff's voltage law to solve multisource circuits (loop analysis).
- use Kirchhoff's current law to solve multisource circuits (nodal analysis).

7.1 INTRODUCTION

Up to this point we have studied circuits and networks having a single voltage or current source. In this chapter, we investigate electric circuits containing more than one energy source. Many electric circuits require multiple energy sources or voltage levels to operate. Techniques have been developed to simplify the analysis of **multisource** circuits. Among these are source conversion, superposition, loop analysis, and nodal analysis. Furthermore, several computer programs are available that are capable of analyzing multisource circuits. These computer programs are based on the techniques presented in this chapter.

T^2 *tech tidbit*

Multisource circuits—more than one energy source.

7.2 VOLTAGE SOURCES IN SERIES

Kirchhoff's voltage law gives us a mathematical technique for analyzing multiple voltage sources in series. This technique was discussed in Chapter 5. However, voltage sources in series can be analyzed logically using common sense.

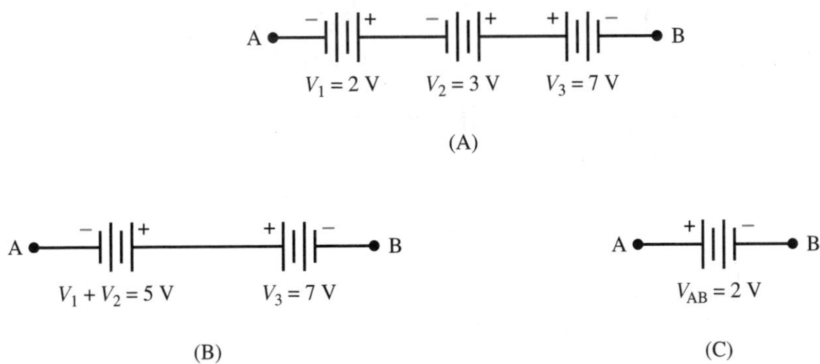

Figure 7-1 **Voltage sources in series**

In Figure 7-1A, we see three voltage sources that are connected in series. Voltage sources V_1 and V_2 are both pushing electrons in the same direction and therefore can be added together. Voltage source V_3 is pushing electrons in the opposite direction from V_1 and V_2. Figure 7-1B illustrates the effect of combining V_1 and V_2 together. Source V_3 is pushing electrons in the opposite direction from V_1 and V_2, so the net effect is a subtraction. Figure 7-1C illustrates the overall effect of the three voltage sources connected in series. There are two general rules or steps to follow:

Add all sources pushing electrons in the same direction.

1. Combine by adding all sources pushing electrons in the *same* direction.
2. Combine by subtracting all sources pushing electrons in the opposite direction.

Subtract all sources pushing electrons in opposite directions.

Remember that with series circuits the components may be rearranged in any order. This is true because the same current exists in each component in the circuit. This principle can be very helpful when analyzing series circuits with multiple voltage sources.

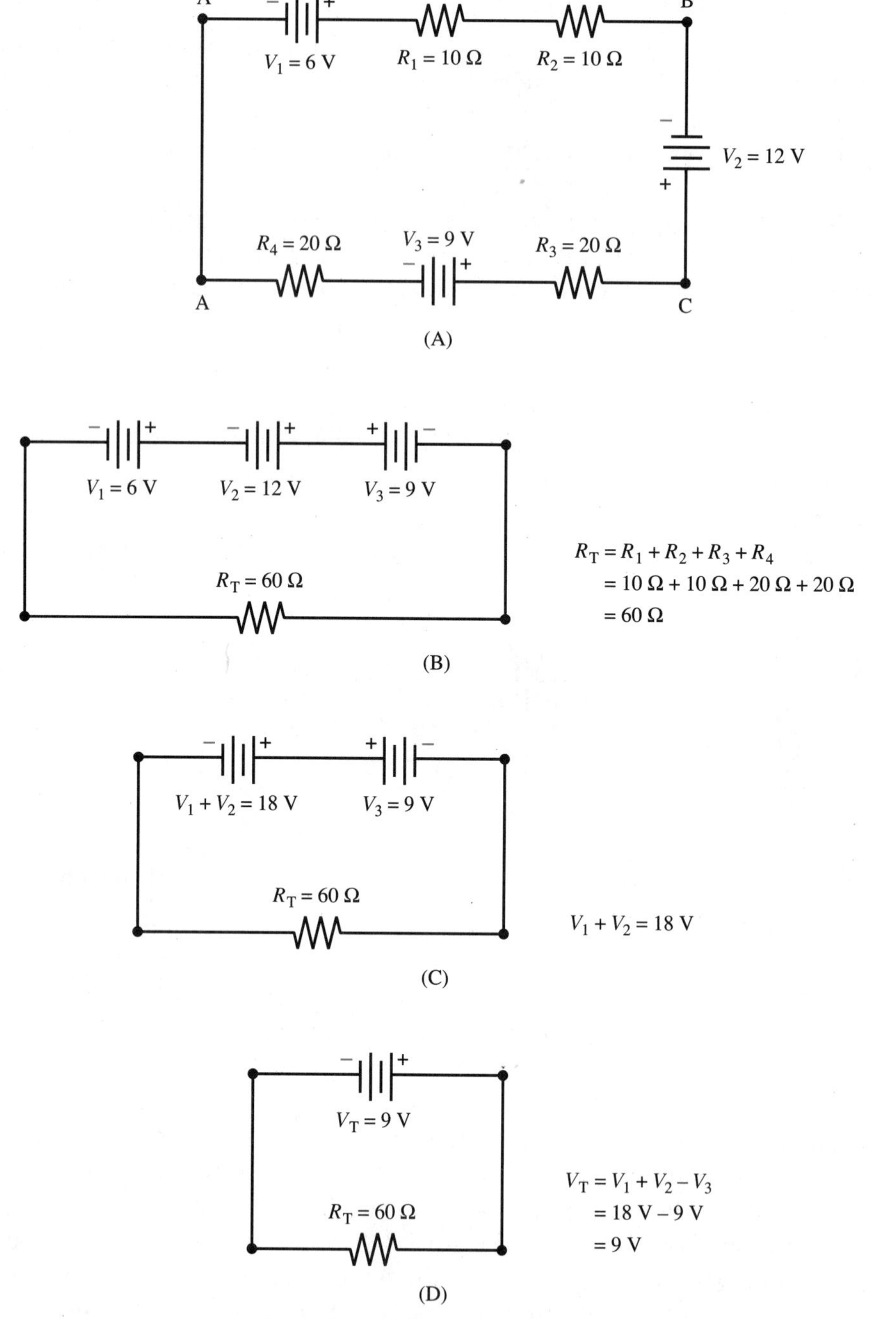

Figure 7-2 **Voltage sources in a series circuit**

In Figure 7-2A, we see a series circuit with three voltage sources. By rearranging the components and combining the series resistors, we can create an equivalent circuit to analyze. Figure 7-2B illustrates this rearrangement of components. Notice that the voltage sources can be analyzed in the same way as shown in Figure 7-1. Voltage sources V_1 and V_2 are pushing electrons in the same direction and are combined in Figure 7-2C. The overall effect is illustrated in Figure 7-2D.

EXAMPLE 7-1

Find the current in the circuit of Figure 7-3.

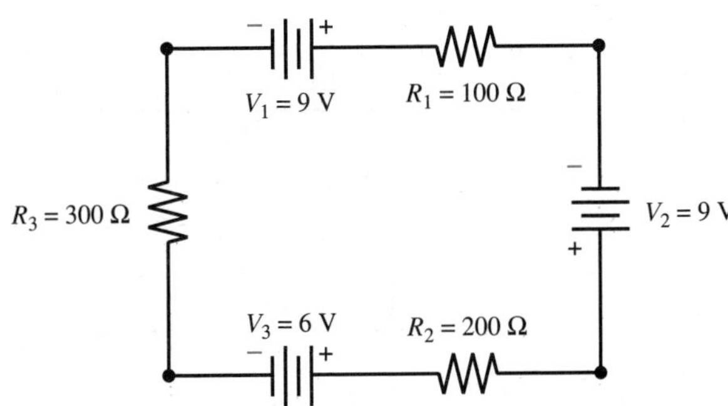

Figure 7-3

Solution:

See Figure 7-4.

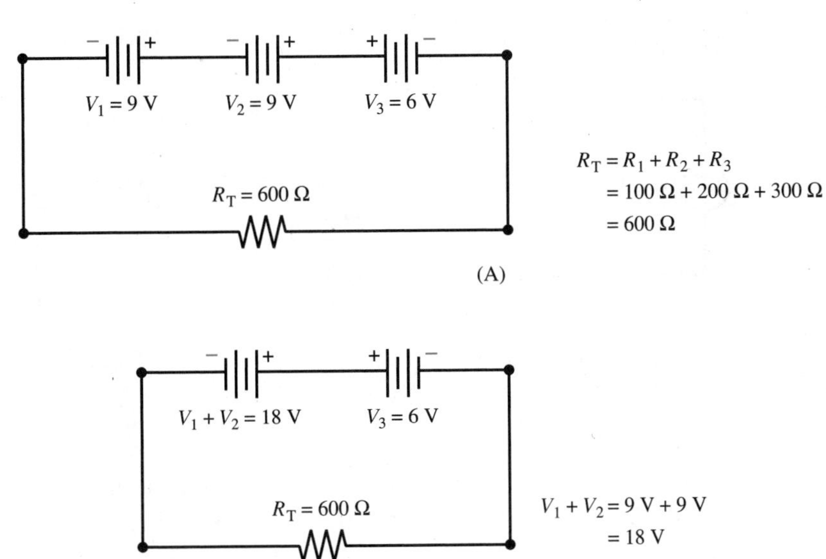

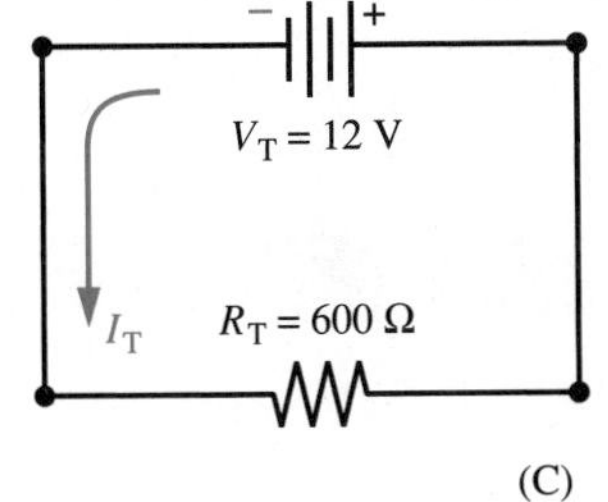

$$
\begin{aligned}
V_T &= V_1 + V_2 - V_3 \\
&= 18\text{ V} - 6\text{ V} \\
&= 12\text{ V}
\end{aligned}
\qquad
\begin{aligned}
I_T &= \frac{V_T}{R_T} \\
&= \frac{12\text{ V}}{600\ \Omega} \\
&= 0.02\text{ A or } 20\text{ mA}
\end{aligned}
$$

(C)

Figure 7-4

7.3 CURRENT SOURCES IN PARALLEL

2 **tech tidbit**

Current sources pushing electrons in the same direction are added. Current sources pushing electrons in opposite directions are subtracted.

Current sources that push electrons in the same direction can be combined by addition in the same manner as voltage sources in series. Current sources that push electrons in opposite directions are combined by subtraction.

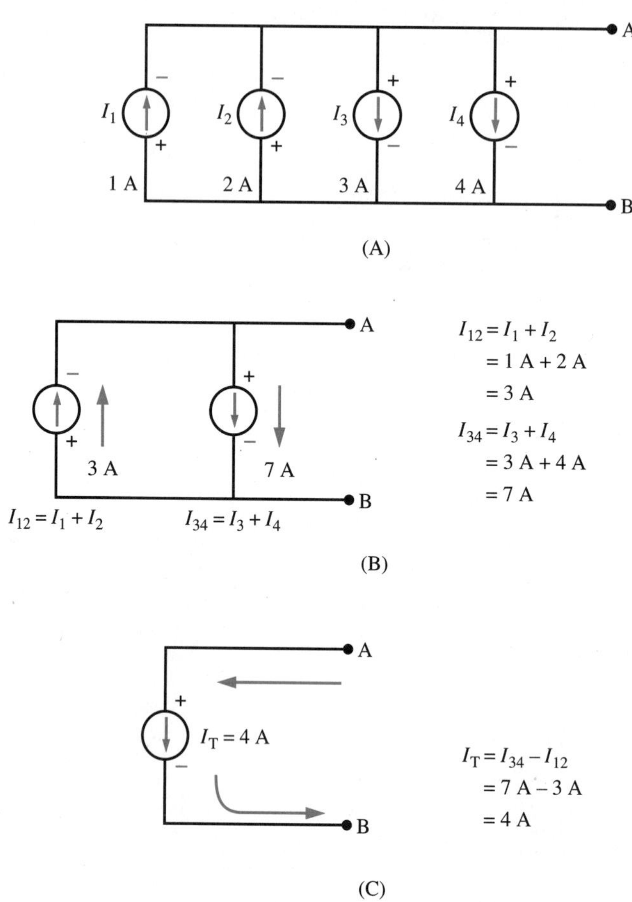

Figure 7-5 **Current sources in parallel**

In Figure 7-5A, current sources I_1 and I_2 are pushing electrons in the upward direction. Current sources I_3 and I_4 are pushing electrons in the downward direction. Figure 7-5B illustrates the effect of combining current sources that are pushing electrons in the same direction. Figure 7-5C illustrates the overall effect by combining current sources that are pushing electrons in opposite directions. The overall direction is determined by the larger source.

EXAMPLE 7-2

Find the voltage, V_{AB}, in the circuit of Figure 7-6.

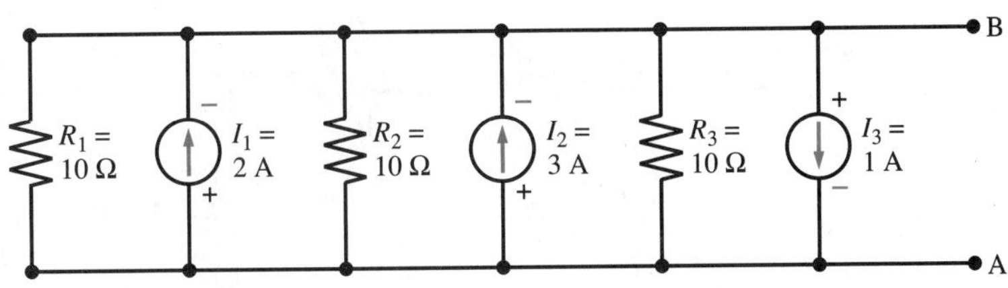

Figure 7-6

Solution:

See Figure 7-7.

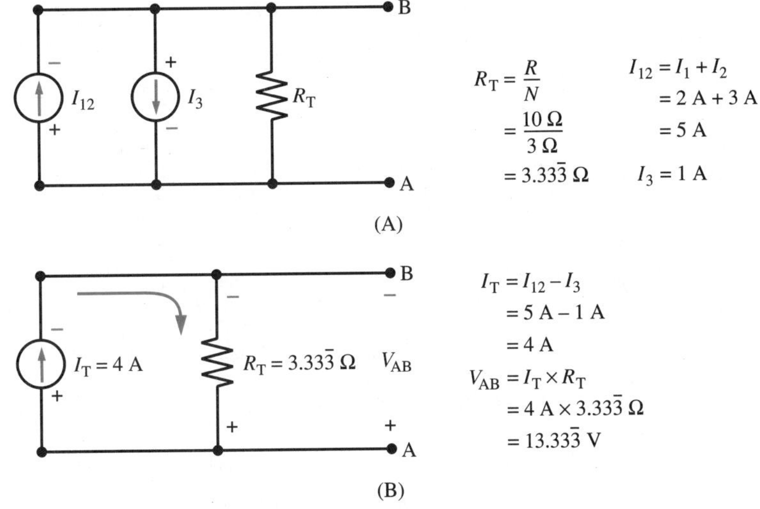

Figure 7-7

✓ Self-Test 1

1. List four techniques to solve multisource circuits.
2. If two voltage sources are pushing current in the same direction, the voltage sources may be _______________.
3. If two voltage sources are pushing current in opposite directions, the voltage sources may be _______________.
4. If a 3-ampere current source and a 4-ampere current source are both supplying current in the same direction, what is the total current?

Equivalent sources have the same affect on the load.

7.4 USE OF SOURCE CONVERSION

Recall that a voltage source with a series resistor can be converted into a current source with a parallel resistor and vice versa. Source conversion can be a very useful technique for solving multisource circuits. Consider the circuit of Figure 7-8. The circuit looks difficult to solve because of the two different energy sources. However the solution is not difficult at all.

Current sources have a resistor in parallel. Voltage sources have a resistor in series.

Suppose we wish to solve for the voltage, V_{AB}, across resistor, R_2. We begin by converting the current source and parallel resistor into a voltage source with a series resistor. This results in a voltage source of V_A = 100 volts and R_2 = 10 ohms, as shown in Figure 7-9A. Next we rearrange the components, as

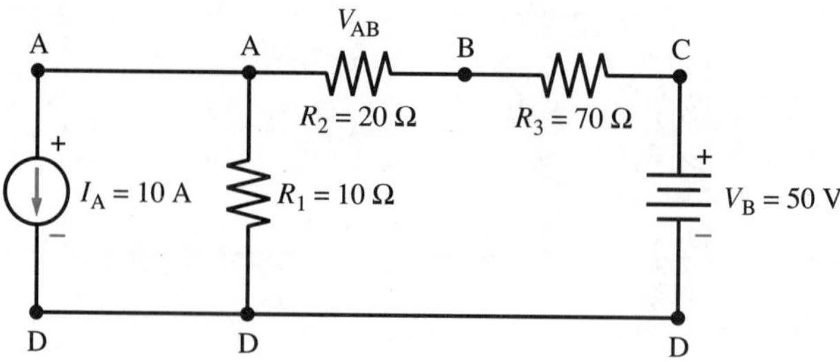

Figure 7-8 **Multisource circuit**

shown in Figure 7-9B. Combining voltage sources V_A and V_B results in a single 50-volt source. Combining R_1 and R_3 results in a single equivalent resistor, R_{13}, which equals 80 ohms, as shown in Figure 7-9C. Applying the voltage divider rule results in V_{AB} = 10 volts.

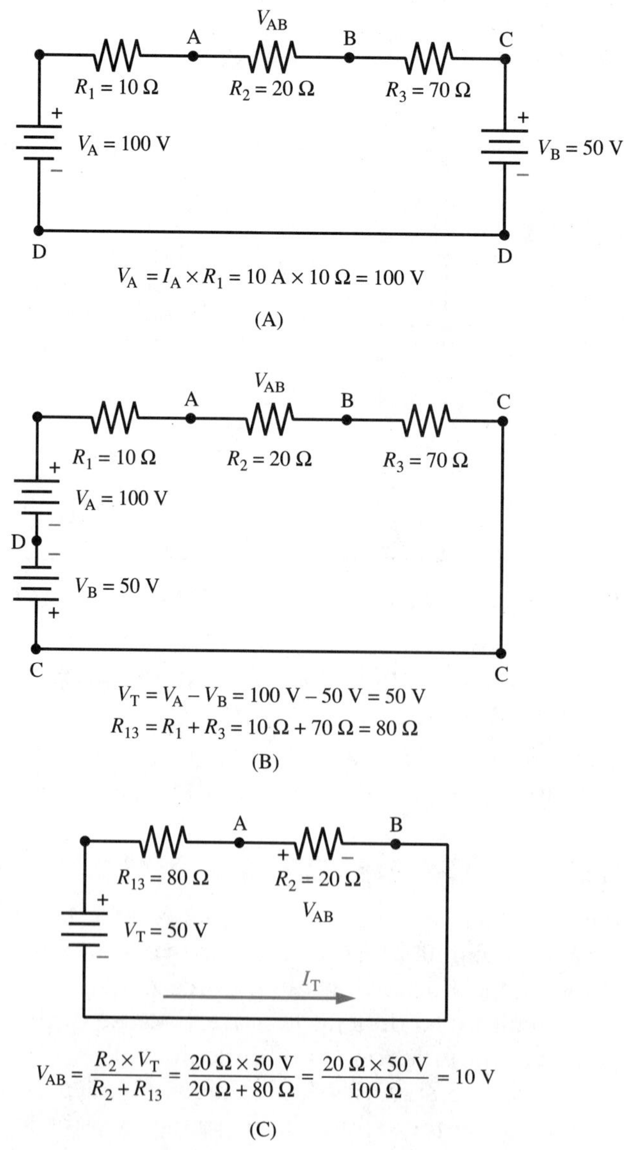

Figure 7-9 **Use of source conversion**

EXAMPLE 7-3

Find the voltage, V_{AB}, in the circuit of Figure 7-10.

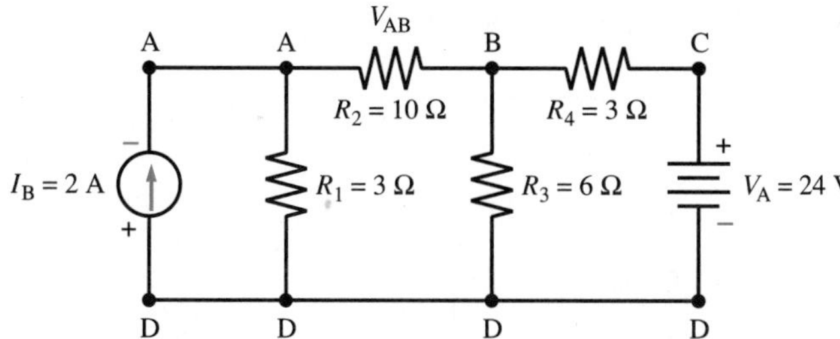

Figure 7-10

Solution:

See Figure 7-11.

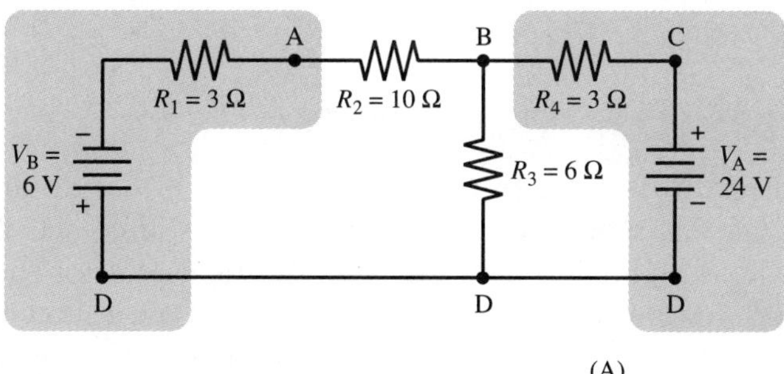

(A)

Convert I_B and R_1 to a voltage sourc

$V_B = I_B \times R_1 = 2\ \text{A} \times 3\ \Omega = 6\ \text{V}$

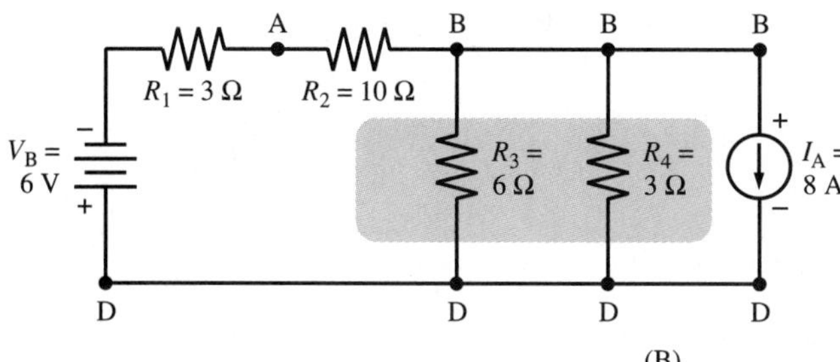

(B)

Convert V_A to a current source. This will put R_3 and R_4 in parallel with th current source.

$$I_A = \frac{V_A}{R_4} = \frac{24\ \text{V}}{3\ \Omega} = 8\ \text{A}$$

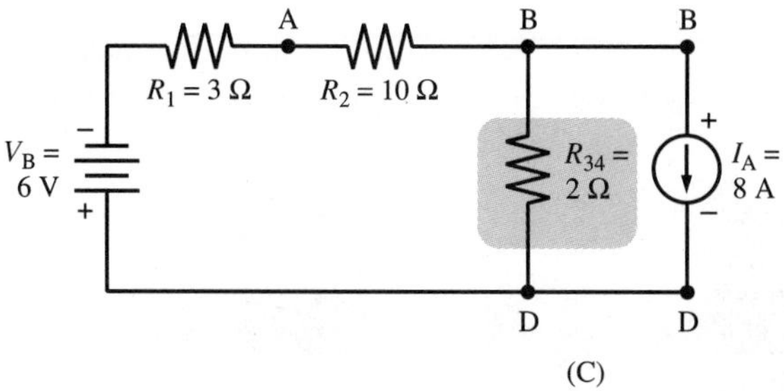

(C)

Now $R_{34} = R_3 \parallel R_4 = 2\ \Omega$.

Figure 7-11 (continued on page 162)

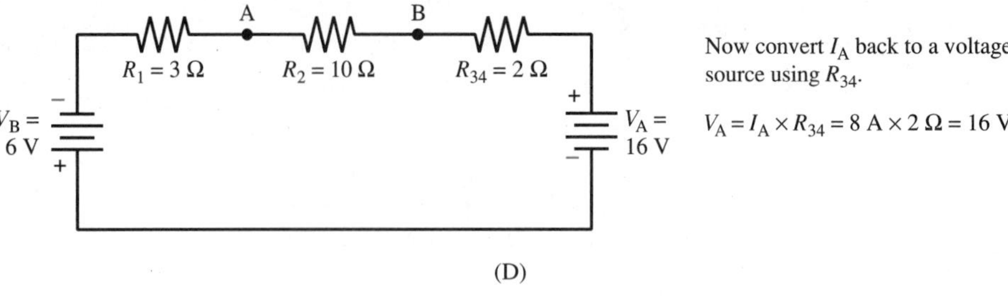

(D)

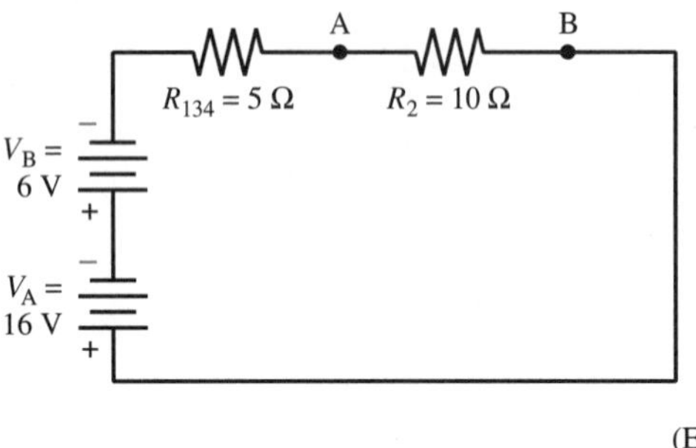

$$R_T = R_1 + R_2 + R_{34} = 3\ \Omega + 10\ \Omega + 2\ \Omega = 15\ \Omega$$

and $V_T = V_A + V_B$ (series adding)

$$= 6\ V + 16\ V = 22\ V$$

$$V_{AB} = \frac{R_2 \times V_T}{R_T} = \frac{10\ \Omega \times 22\ V}{15\ \Omega} = 14.667\ V$$

(E)

Figure 7-11 (cont.)

tech tidbit

To set sources to zero: Voltage sources are replaced with shorts, current sources are replaced with opens.

7.5 SUPERPOSITION THEOREM

The **superposition theorem** is actually a theorem of nature that applies to multisource circuits. It states that when more than one energy source acts on a body or circuit, the results can be analyzed by considering the effects of each energy source separately and summing the results algebraically. The following are four general rules for superposition as applied to resistive circuits:

1. Set all energy sources except for one equal to zero. To set energy sources to zero, voltage sources are replaced by shorts and current sources are replaced with opens.
2. Find the resulting current and voltage for the desired component or components.
3. Repeat steps one and two for other energy sources.
4. Find the algebraic (total) sum of the separate currents or voltages solved for relative to the desired components. The result is the answer for all energy sources in the circuit acting simultaneously.

tech tidbit

Find the effect due to each source separately and add the results algebraically.

It should be noted that the superposition theorem does not apply to power calculations. This is because power varies as the square of the current and is therefore **nonlinear.**

EXAMPLE 7-4

Using superposition, find the voltage, V_{AB}, across resistor R_1 in Figure 7-12.

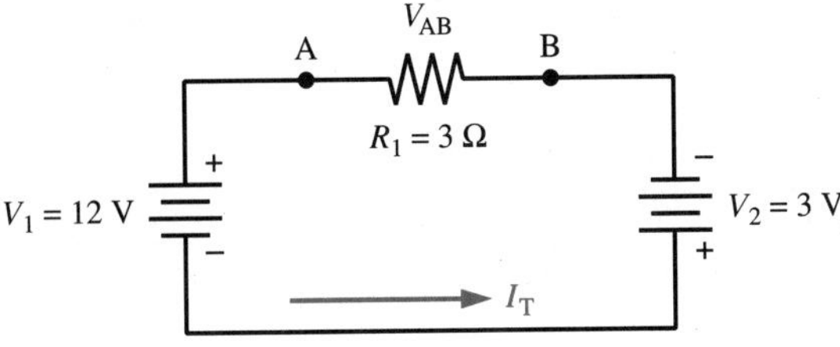

Figure 7-12

Solution:

See Figure 7-13.

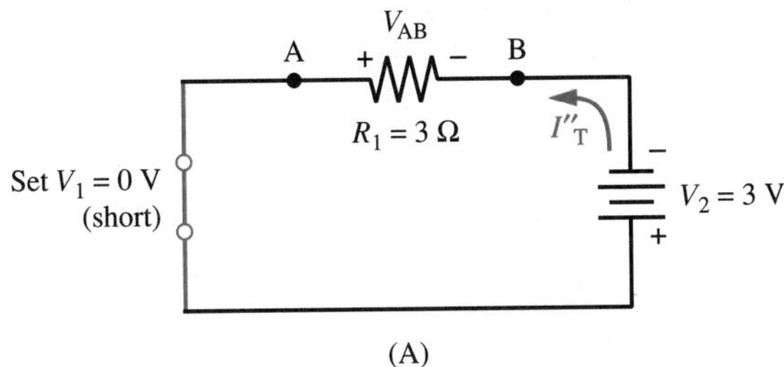

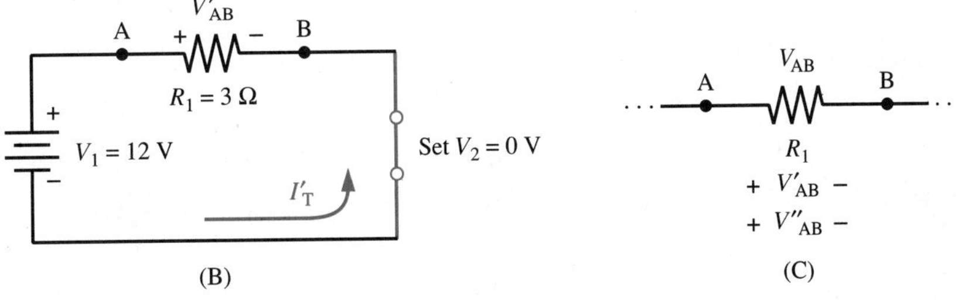

Figure 7-13

Find voltage V_{AB} from source V_2. See Figure 7-13A.

$$V''_{AB} = V_2 = 3\text{ V}$$

or

$$I''_T = \frac{V_2}{R_1} = \frac{3\text{ V}}{3\ \Omega} = 1\text{ A}$$

and

$$V''_{AB} = I''_T \times R_1 = 1\text{ A} \times 3\ \Omega = 3\text{ V}$$

Note: Node A (+) and node B (−).

Find voltage V_{AB} from source V_1 by setting $V_2 = 0$ V and $V_1 = 12$ V. See Figure 7-13B.

$$V'_{AB} = V_1 = 12 \text{ V}$$

or

$$I'_T = \frac{V_1}{R_1} = \frac{12 \text{ V}}{3\ \Omega} = 4 \text{ A}$$

and

$$V'_{AB} = I'_T \times R_1 = 4 \text{ A} \times 3\ \Omega = 12 \text{ V}$$

Find the total sum of V_{AB}. See Figure 7-13C.

$$V_{AB} = V'_{AB} + V''_{AB} = 12 \text{ V} + 3 \text{ V} = 15 \text{ V}$$

Note: V''_{AB} refers to the solution of V_{AB} due to the second energy source. V'_{AB} refers to the solution of V_{AB} due to the first energy source.

EXAMPLE 7-5

Using the superposition theorem, find the voltage, V_{AB}, across resistor R_3 in Figure 7-14.

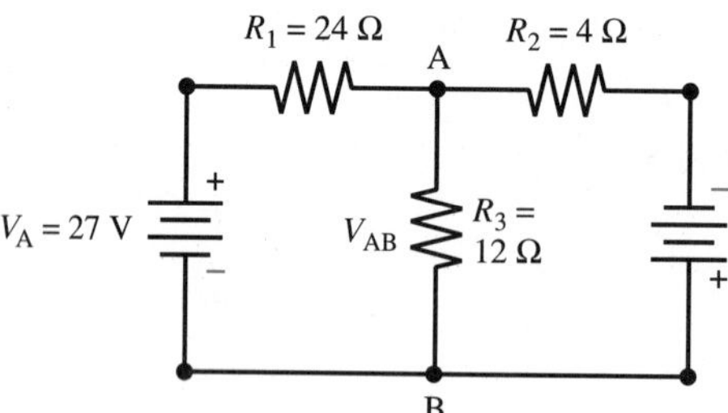

Figure 7-14

Solution:

See Figure 7-15.

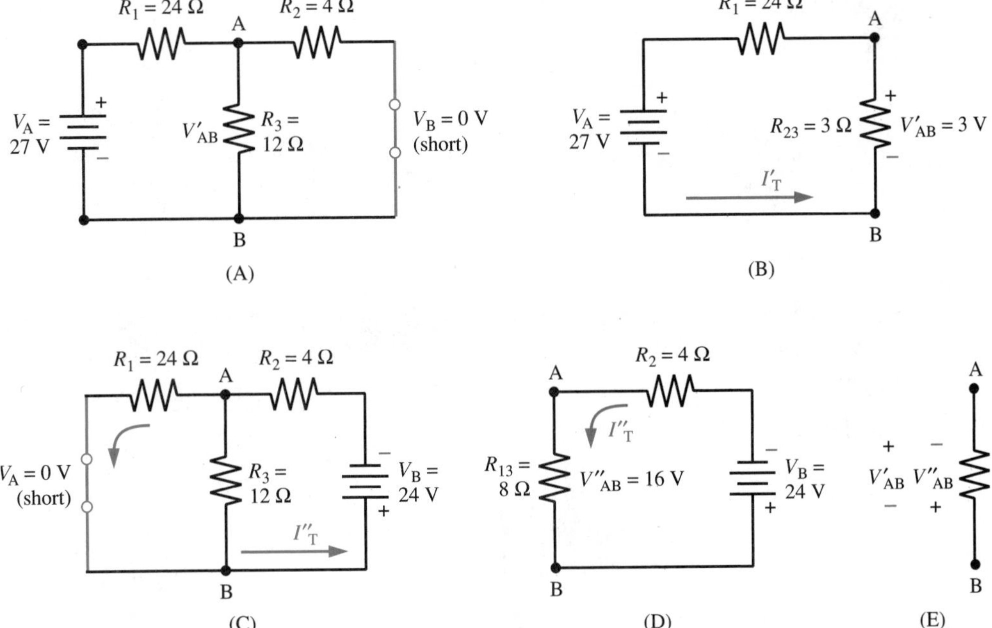

Figure 7-15

Set V_B = 0 V and solve for effects of V_A. See Figure 7-15A.

Find V'_{AB}. See Figure 7-15B.

$$R_{23} = R_2 \parallel R_3 = \frac{R_2 \times R_3}{R_2 + R_3} = \frac{4\ \Omega \times 12\ \Omega}{4\ \Omega + 12\ \Omega} = 3\ \Omega$$

and

$$V'_{AB} = \frac{R_{23} \times V_A}{R_{23} + R_1} = \frac{3\ \Omega \times 27\ \text{V}}{3\ \Omega + 24\ \Omega} = \frac{3\ \Omega \times 27\ \text{V}}{27\ \Omega} = 3\ \text{V}$$

Set V_A = 0 V and V_B = 24 V. Solve for effects of V_B. See Figure 7-15C.

Find V''_{AB}. See Figure 7-15D.

$$R_{13} = R_1 \parallel R_3 = \frac{R_1 \times R_3}{R_1 + R_3} = \frac{24\ \Omega \times 12\ \Omega}{24\ \Omega + 12\ \Omega} = 8\ \Omega$$

and

$$V''_{AB} = \frac{R_{13} \times V_B}{R_{13} + R_2} = \frac{8\ \Omega \times 24\ \text{V}}{8\ \Omega + 4\ \Omega} = 16\ \text{V}$$

Find V_{AB}. See Figure 7-15E.

$$V_{AB} = V''_{AB} - V'_{AB}$$
$$= 16\text{ V} - 3\text{ V}$$
$$= 13\text{ V}$$

EXAMPLE 7-6

Using the superposition theorem, find the current, I_3, through resistor R_3 in Figure 7-16.

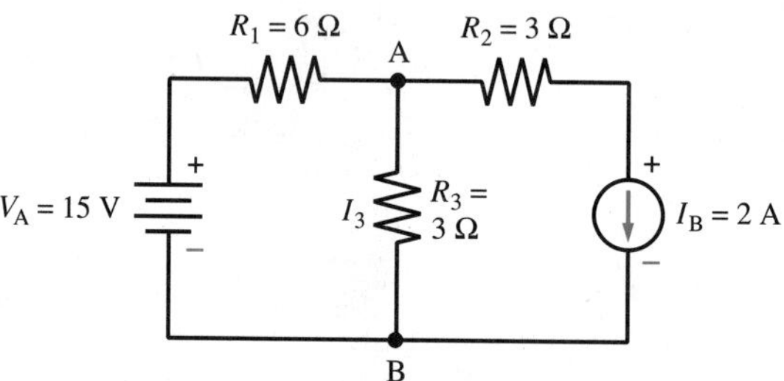

Figure 7-16

Solution:

See Figure 7-17.

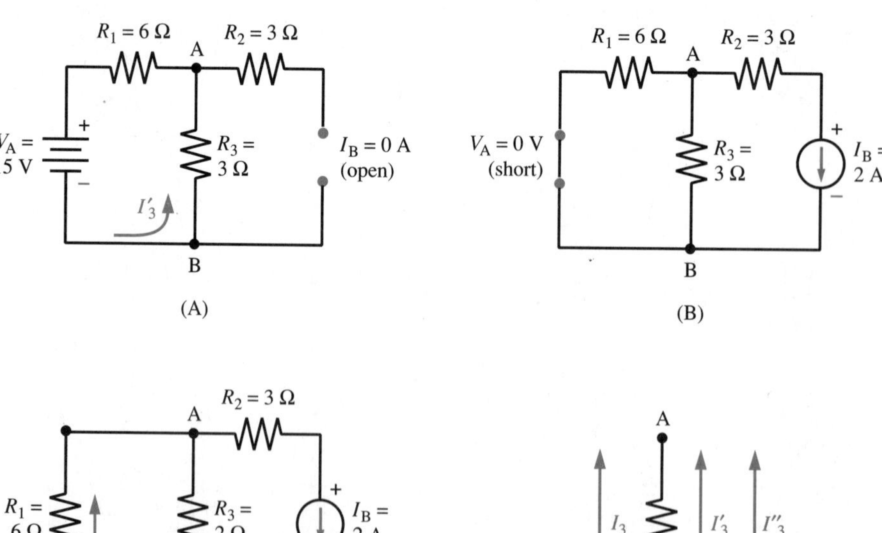

Figure 7-17

Set $I_B = 0$ A (open circuit). See Figure 7-17A.

Note: When $I_B = 0$ A, it is replaced by an open circuit.

Solve for effects of V_A.

$$I'_3 = \frac{V_A}{R_1 + R_3}$$

$$= \frac{15\text{ V}}{6\ \Omega + 3\ \Omega}$$

$$= \frac{15\text{ V}}{9\ \Omega}$$

$$= 1.667\text{ A}$$

Set $V_A = 0$ V (short circuit). See Figure 7-17B.

See Figure 7-17C. By the current divider rule:

$$I''_3 = \frac{I_B \times R_1}{R_1 + R_2}$$

$$= \frac{2\text{ A} \times 6\ \Omega}{6\ \Omega + 3\ \Omega}$$

$$= \frac{2\text{ A} \times 6\ \Omega}{9\ \Omega}$$

$$= 1.333\text{ A}$$

Find I_3. See Figure 7-17D.

$$I_3 = I'_3 + I''_3$$

$$= 1.667\text{ A} + 1.333\text{ A}$$

$$= 3\text{ A}$$

7.6 LOOP ANALYSIS

Loop analysis can be applied to any **linear** (load components that do not change value) electric circuit or network. Some circuits cannot be readily solved by applying the principles of Ohm's law, series circuits, and parallel circuits. For this reason, we need to apply some other method of analysis. Loop analysis is actually another application of Kirchhoff's voltage law. The following general steps should be followed when applying loop analysis to a circuit:

Loop analysis is also known as mesh analysis.

1. Define a complete loop, current and current direction to each complete loop in the circuit. The direction of current can be arbitrarily assumed.
2. Indicate the polarities for each component in the circuit as determined from the arbitrarily assumed direction of current.
3. Develop voltage equations by applying Kirchhoff's voltage law around each loop.
4. Solve the equations for the desired quantities.

Loop analysis is an application of Kirchhoff's voltage law.

EXAMPLE 7-7

Find the loop currents in the circuit of Figure 7-18.

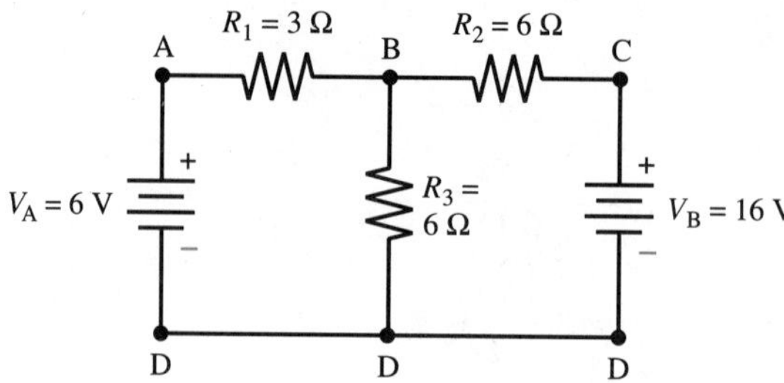

Figure 7-18

Solution:

See Figure 7-19.

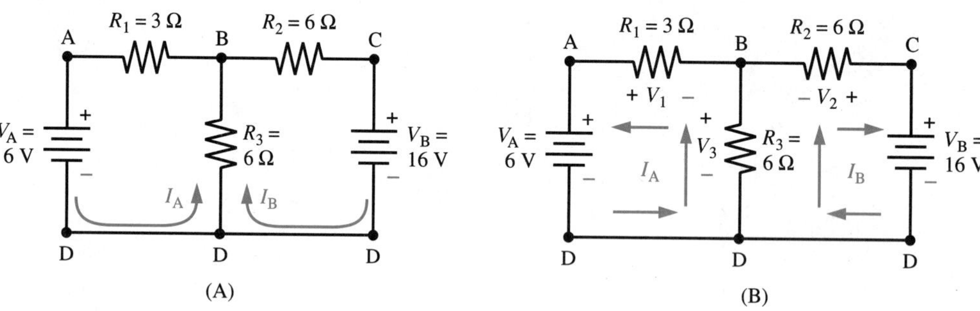

Figure 7-19

Assign loops and current directions. See Figure 7-19A.

Find voltage drops. See Figure 7-19B. From Ohm's law:

$$V_1 = (3\ \Omega)I_A$$

$$V_2 = (6\ \Omega)I_B$$

$$V_3 = (6\ \Omega)I_A + (6\ \Omega)I_B$$

Apply Kirchhoff's voltage law to each loop.

Loop A DDBA: $$V_A - V_3 - V_1 = 0\text{ V}$$

or $$V_1 + V_3 = V_A$$

$$(3\ \Omega)I_A + (6\ \Omega)I_A + (6\ \Omega)I_B = 6\text{ V}$$

$$(9\ \Omega)I_A + (6\ \Omega)I_B = 6\text{ V}$$

Note: There are two currents I_A and I_B through R_3.

Loop C DDBC: $V_B - V_3 - V_2 = 0\text{ V}$

or $V_2 + V_3 = V_B$

$$(6\ \Omega)I_B + (6\ \Omega)I_A + (6\ \Omega)I_B = 16\text{ V}$$

$$(6\ \Omega)I_A + (12\ \Omega)I_B = 16\text{ V}$$

Note: There are two currents I_A and I_B through R_3.

Solve equations.

Eq. 1: $(9\ \Omega)I_A + (6\ \Omega)I_B = 6\text{ V}$

Eq. 2: $(6\ \Omega)I_A + (12\ \Omega)I_B = 16\text{ V}$

To solve loop equations, we need one unknown current (I_B) to drop out of the final equation.

Multiply Eq. 1 by –2.

$$-2[(9\ \Omega)I_A + (6\ \Omega)I_B = 6\text{ V}]$$

New Eq. 1: $(-18\ \Omega)I_A - (12\ \Omega)I_B = -12\text{ V}$

Next add Eq. 2 to new Eq. 1.

New Eq. 1: $(-18\ \Omega)I_A - (12\ \Omega)I_B = -12\text{ V}$

+ Eq. 2: $(6\ \Omega)I_A + (12\ \Omega)I_B = 16\text{ V}$

Final Eq.: $(-12\ \Omega)I_A + 0 = 4\text{ V}$

$$(-12\ \Omega)I_A = 4\text{ V}$$

$$I_A = \frac{4\text{ V}}{-12\ \Omega}$$

$$I_A = -0.333\text{ A}$$

Note: The minus sign indicates the current is really going in the opposite direction.

Find I_B. Replace I_A with –0.333 A.

$$(6\ \Omega)I_A + (12\ \Omega)I_B = 16\text{ V}$$

$$(6\ \Omega)(-0.333\text{ A}) + (12\ \Omega)I_B = 16\text{ V}$$

$$-2\text{ V} + (12\ \Omega)I_B = 16\text{ V}$$

$$(12\ \Omega)I_B = 16\text{ V} + 2\text{ V}$$

$$(12\ \Omega)I_B = 18\text{ V}$$

$$I_B = \frac{18\text{ V}}{12\ \Omega}$$

$$I_B = 1.5\text{ A}$$

EXAMPLE 7-8

Find the loop currents and the branch currents for each resistor in the circuit of Figure 7-20.

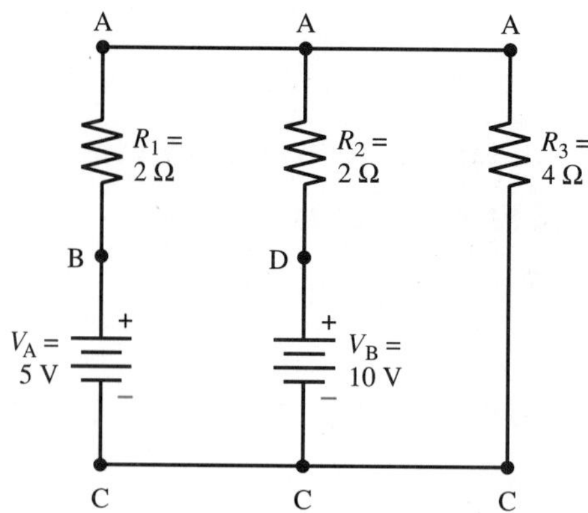

Figure 7-20

Solution:

See Figure 7-21.

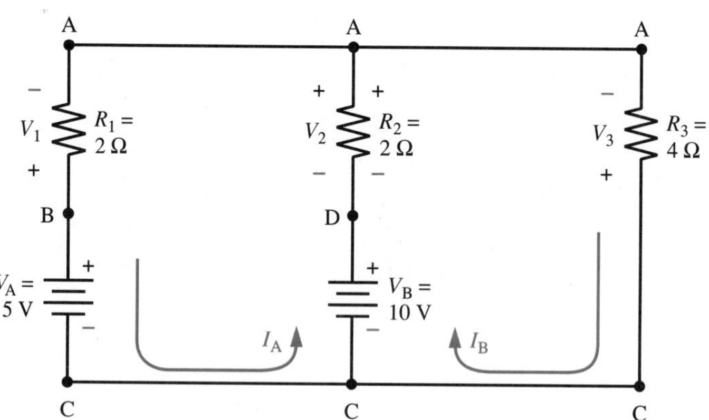

Figure 7-21

Note: I_A and I_B can be assigned any direction.

From Ohm's law:

$$V_1 = (2\ \Omega)I_A$$

$$V_2 = (2\ \Omega)I_A + (2\ \Omega)I_B$$

$$V_3 = (4\ \Omega)I_B$$

Loop A CCDAAB:

$$V_A - V_B - V_2 - V_1 = 0\text{ V}$$

or

$$V_1 + V_2 = V_A - V_B$$

$$(2\ \Omega)I_A + (2\ \Omega)I_A + (2\Omega)I_B = 5\text{ V} - 10\text{ V}$$

Eq. 1:

$$(4\ \Omega)I_A + (2\Omega)I_B = -5\text{ V}$$

Loop B CCDAA:

$$-V_3 - V_B - V_2 = 0$$

or

$$V_2 + V_3 = -V_B$$

$$(2\ \Omega)I_A + (2\ \Omega)I_B + (4\Omega)I_B = -10\text{ V}$$

Eq. 2: $$(2\ \Omega)I_A + (6\Omega)I_B = -10\text{ V}$$

Eq. 1: $$(4\ \Omega)I_A + (2\ \Omega)I_B = -5\text{ V}$$

Eq. 2: $$(2\ \Omega)I_A + (6\ \Omega)I_B = -10\text{ V}$$

Multiply Eq. 2 by −2.

$$-2[(2\ \Omega)I_A + (6\ \Omega)I_B = -10\text{ V}]$$

New Eq. 2: $$(-4\ \Omega)I_A - (12\ \Omega)I_B = +20\text{ V}$$

Solve for I_B.

Eq. 1: $$(4\ \Omega)I_A + (2\ \Omega)I_B = -\ 5\text{ V}$$

New Eq. 2: $$\underline{(-4\ \Omega)I_A - (12\ \Omega)I_B = +20\text{ V}}$$

$$(-10\ \Omega)I_B = 15\text{ V}$$

$$I_B = \frac{15\text{ V}}{-10\ \Omega}$$

$$I_B = -1.5\text{ A}$$

Solve for I_A.

Eq. 1: $$(4\ \Omega)I_A + (2\ \Omega)I_B = -5\text{ V}$$

$$(4\ \Omega)I_A + (2\ \Omega)(-1.5\text{A}) = -5\text{ V}$$

$$(4\ \Omega)I_A - 3\text{ V} = -5\text{ V}$$

$$\underline{+3\text{ V} = +3\text{ V}}$$

$$(4\ \Omega)I_A = -2\text{ V}$$

$$I_A = \frac{-2\text{ V}}{4\ \Omega}$$

$$I_A = -0.5\text{ A}$$

∴ for R_1, $I_1 = I_A = -0.5$ A (opposite direction)

for R_3, $I_3 = I_B = -1.5$ A (opposite direction)

for R_2, $I_2 = I_A + I_B = (-0.5\text{ A}) + (-1.5\text{ A}) = -2.0$ A (opposite direction0

7.7 NODAL ANALYSIS

Recall that loop analysis is an application of Kirchhoff's voltage law. **Nodal analysis** is an application of Kirchhoff's current law. A **node** is a junction of two or more branches. When applying nodal analysis, one node is considered the reference node. Often the reference node is the ground or zero potential node. The general procedure for applying nodal analysis to a circuit is as follows:

Nodal analysis is an application of Kirchhoff's current law.

1. Convert all voltage sources to current sources.
2. Select the reference node and label the remaining nodes.
3. Develop the nodal equations for each node (except the reference node) by applying Kirchhoff's current law.
4. Solve the equations for the desired quantities.

EXAMPLE 7-9

Find the voltage across resistor R_1 in Figure 7-22 using nodal analysis.

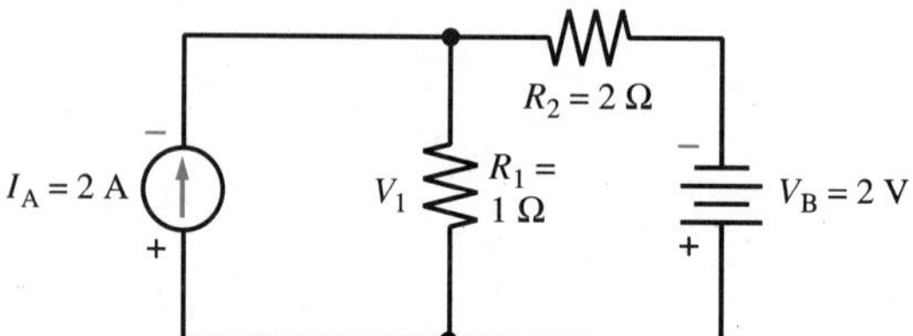

Figure 7-22

Solution:

See Figure 7-23.

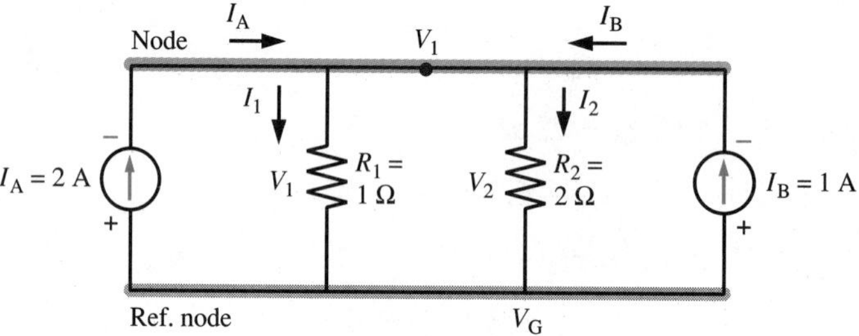

Figure 7-23

Convert $V_B = 2$ V to a current source using R_2.

$$I_B = \frac{V_B}{R_2} = \frac{2\text{ V}}{2\ \Omega} = 1\text{ A}$$

Assign reference node V_G and mark node V_1. I_1 and I_2 are assigned as shown. Apply Kirchhoff's current law to node V_1.

$$\Sigma I_{IN} = \Sigma I_{OUT}$$

$$I_A + I_B = I_1 + I_2$$

Note: I_A and I_B are entering node V_1 and I_1 and I_2 where assigned leaving node V_1.

$$I_A + I_B = I_1 + I_2$$

$$2\text{ A} + 1\text{ A} = I_1 + I_2$$

$$3\text{ A} = I_1 + I_2$$

Using Ohm's law:

$$I_1 = \frac{V_1}{R_1} = \frac{V_1}{1\ \Omega}$$

$$I_2 = \frac{V_2}{R_2} = \frac{V_2}{2\ \Omega}$$

$$3\text{ A} = I_1 + I_2$$

$$3\text{ A} = \frac{V_1}{1\ \Omega} + \frac{V_2}{2\ \Omega}$$

Solve for V_1.

Note: $V_1 = V_2$.

$$3\text{ A} = \frac{V_1}{1\ \Omega} + \frac{V_1}{2\ \Omega}$$

$$3\text{ A} = (1\text{ S})V_1 + (0.5\text{ S})V_1$$

$$3\text{ A} = (1.5\text{ S})V_1$$

$$V_1 = \frac{3\text{ A}}{1.5\text{ S}}$$

$$V_1 = 2\text{ V}$$

Note: $G = \frac{1}{R}$; conductance is expressed in siemens (S).

$$G_T = 1.0\text{ S} + 0.5\text{ S} = 1.5\text{ S}$$

Knowing V_1, we can solve for the current. $I_1 = I_{R_1}$.

$$I_{R_1} = \frac{V_1}{R_1} = \frac{2\text{ V}}{1\ \Omega} = 2\text{ A}$$

EXAMPLE 7-10

Find V_3, the voltage across R_3, in Figure 7-24 using nodal analysis.

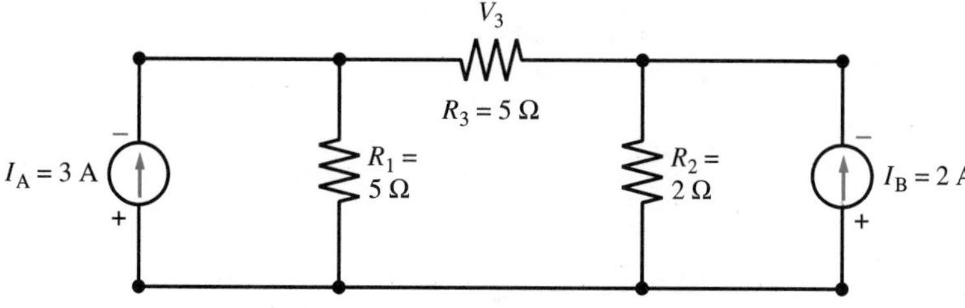

Figure 7-24

Solution:

See Figure 7-25.

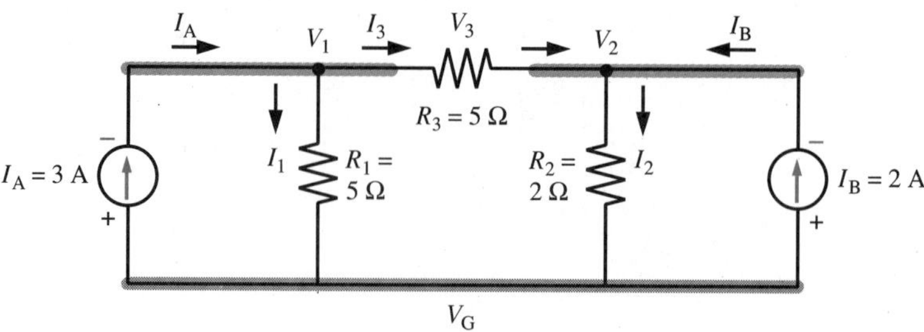

Figure 7-25

Assume V_1 is greater than V_2.

Using Kirchhoff's voltage law:

for node V_1:

$$\Sigma I_{IN} = \Sigma I_{OUT}$$

$$I_A = I_1 + I_3$$

for node V_2:

$$I_B + I_3 = I_2$$

Using Ohm's law:

$$I_1 = \frac{V_1}{R_1} = \frac{V_1}{5\ \Omega} = (0.2\ \text{S})V_1$$

$$I_2 = \frac{V_2}{R_2} = \frac{V_2}{2\ \Omega} = (0.5\ \text{S})V_2$$

$$I_3 = \frac{V_1 - V_2}{R_3} = \frac{V_1 - V_2}{5\ \Omega} = (0.2\ \text{S})V_1 - (0.2\ \text{S})V_2$$

for node V_1:

$$I_1 + I_3 = I_A$$

$$(0.2\ \text{S})V_1 + (0.2\ \text{S})V_1 - (0.2\ \text{S})V_2 = 3\ \text{A}$$

or

$$(0.4\ \text{S})V_1 - (0.2\ \text{S})V_2 = 3\ \text{A}$$

for node V_2:

$$I_B + I_3 = I_2$$

$$I_2 - I_3 = I_B$$

$$(0.5\ \text{S})V_2 - [(0.2\ \text{S})V_1 - (0.2\ \text{S})V_2] = 2\ \text{A}$$

$$(0.5\ \text{S})V_2 - (0.2\ \text{S})V_1 + (0.2\ \text{S})V_2 = 2\ \text{A}$$

$$(-0.2\ \text{S})V_1 + (0.7\ \text{S})V_2 = 2\ \text{A}$$

Eq. 1: $(0.4\text{ S})V_1 - (0.2\text{ S})V_2 = 3\text{ A}$

Eq. 2: $(-0.2\text{ S})V_1 + (0.7\text{ S})V_2 = 2\text{ A}$

Multiply Eq. 2 by 2.

New Eq. 2: $2[(-0.2\text{ S})V_1 + (0.7\text{ S})V_2 = 2\text{ A}]$

Solve for V_2.

Eq. 1: $(0.4\text{ S})V_1 - (0.2\text{ S})V_2 = 3\text{ A}$

New Eq. 2: $\underline{(-0.4\text{ S})V_1 + (1.4\text{ S})V_2 = 4\text{ A}}$

Add Eq. 1 and Eq. 2: $0 + (1.2\text{ S})V_2 = 7\text{ A}$

$$(1.2\text{ S})V_2 = 7\text{ A}$$

$$V_2 = \frac{7\text{ A}}{1.2\text{ S}}$$

$$V_2 = 5.8\overline{3}\text{ V}$$

Solve for V_1.

Eq. 1: $(0.4\text{ S})V_1 - (0.2\text{ S})V_2 = 3\text{ A}$

$$(0.4\text{ S})V_1 - (0.2\text{ S})(5.8\overline{3}\text{ V}) = 3\text{ A}$$

$$(0.4\text{ S})V_1 - 1.16\overline{6}\text{ A} = 3\text{ A}$$

$$(0.4\text{ S})V_1 = 4.16\overline{6}\text{ A}$$

$$V_1 = \frac{4.16\overline{6}\text{ A}}{0.4\text{ S}}$$

$$V_1 = 10.415\text{ V}$$

Find V_3. V_3 is the difference between nodes V_1 and V_2.

Because V_1 is greater than V_2, V_3 is:

$$V_3 = V_1 - V_2$$

$$V_3 = 10.415 - 5.8\overline{3}\text{ V}$$

$$V_3 = 4.585\text{ V}$$

✓ Self-Test 2

1. A voltage source and a series resistor can be converted to a current source and a _______________ resistor.
2. Which theorem considers the effect of each energy source separately?
3. Which method of circuit analysis uses Kirchhoff's voltage law?
4. Which method of circuit analysis uses Kirchhoff's current law?

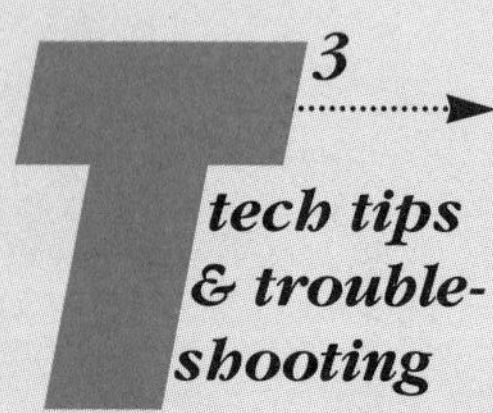

T3 SOLVING EQUATIONS

Solving equations with more than one unknown quantity or variable requires more than one equation. The number of equations required is equal to the number of unknown quantities or variables. These sets of equations are known as **simultaneous equations.** For example in the equation $7I_1 + 4I_2 = 12$ we have two unknowns, I_1 and I_2. Therefore, to solve this equation, we require a second equation that involves I_1 and I_2. Thus two unknowns require two equations, three unknowns require three equations, and so forth.

The general method of solution involves the adding of the equations to eliminate one variable. The general procedure is as follows:

1. Multiply the equations by a constant that will cause elimination of a variable.
2. Add the equations to eliminate the variable.
3. Solve the resultant one variable equation.
4. Solve for the second variable by substitution.

EXAMPLE 7-11

Solve the following set of simultaneous equations for I_1 and I_2.

Equation 1 $\quad 2I_1 - 3I_2 = 15$

Equation 2 $\quad 5I_1 - 4I_2 = -1$

Solution:

Multiply Equation 1 by 4 and Equation 2 by −3 to cause elimination of I_2.

$$4 \times (2I_1 - 3I_2 = 15)$$
$$-3 \times (5I_1 - 4I_2 = -1)$$
$$8I_1 - 12I_2 = 60$$
$$-15I_1 + 12I_2 = 3$$

Add equations

$$8I_1 - 12I_2 = 60$$
$$\underline{-15I_1 + 12I_2 = 3}$$
$$-7I_1 = 63$$
$$I_1 = -9$$

Substitute $I_1 = -9$ into any of the previous equations

Equation 1 $\quad 2I_1 - 3I_2 = 15$

$$2(-9) - 3I_2 = 15$$
$$-18 - 3I_2 = 15$$
$$-3I_2 = 33$$
$$I_2 = -11$$

Be careful to multiply each term in the parentheses.

SUMMARY

In this chapter we investigated electric circuits that require more than one energy source. Source conversion can be a very useful technique for analyzing and solving multisource circuits. It can lead to multiple voltage sources in series that can be combined or multiple current sources in parallel that can also be combined.

The superposition theorem is actually a theorem of nature. It states that when more than one energy source acts on a circuit, the results can be analyzed by considering the effects of each energy source separately and summing the results algebraically. Usually, the network analysis for each source separately is simpler than with all the sources present at the same time.

Loop and nodal analysis are closely linked to Kirchhoff's voltage law and Kirchhoff's current law respectively. Although they are more mathematical, they can induce a higher level of understanding and confidence in circuit analysis. Furthermore, the actual mathematical calculation can be left to a computer or scientific calculator. For example, many of the new scientific calculators can solve up to five level determinants automatically.

FORMULAS

$$I_S = \frac{V_S}{R_S} \qquad V_S = I_S R_S \qquad \text{(source conversion)}$$

$$\left.\begin{aligned} V_T &= V_1 + V_2 + V_3 + \ldots + V_N \\ V_T - V_1 - V_2 - V_3 - &\ldots - V_N = 0 \end{aligned}\right\} \quad \text{(Kirchhoff's voltage law)}$$

$$\left.\begin{aligned} I_T &= I_1 + I_2 + I_3 + \ldots + I_N \\ I_T - I_1 - I_2 - I_3 - &\ldots - I_N = 0 \end{aligned}\right\} \quad \text{(Kirchhoff's current law)}$$

EXERCISES

Section 7.2

1. Find the voltage, V_{AB}, in each of the networks of Figure 7-26.

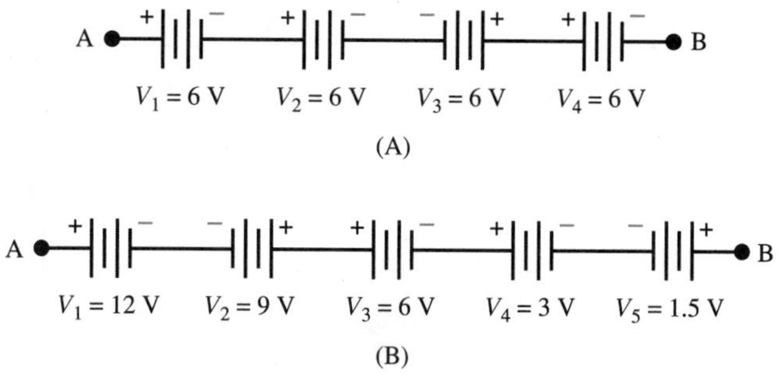

A
B
$V_1 = 6$ V
$V_2 = 6$ V
$V_3 = 6$ V
$V_4 = 6$ V
(A)
A
B
$V_1 = 12$ V
$V_2 = 9$ V
$V_3 = 6$ V
$V_4 = 3$ V
$V_5 = 1.5$ V
(B)

Figure 7-26

2. In the network of Figure 7-27, find V_{AD}, V_{AC}, V_C, and V_D.

A
B
C
D
$V_1 = 9$ V
$V_2 = 6$ V
$V_3 = 3$ V
$V_4 = 10$ V

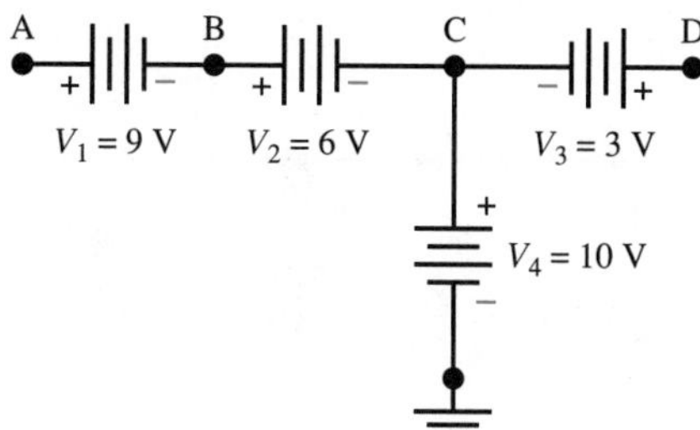

Figure 7-27

3. Find the current in the circuit of Figure 7-28.

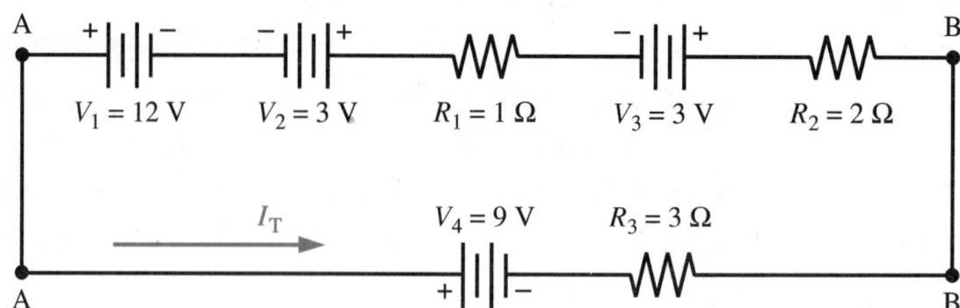

Figure 7-28

Section 7.3

4. Find the total current represented by the current sources in each of the networks of Figure 7-29.

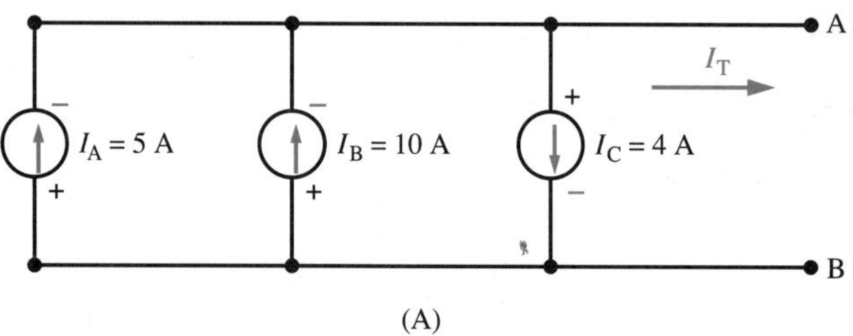

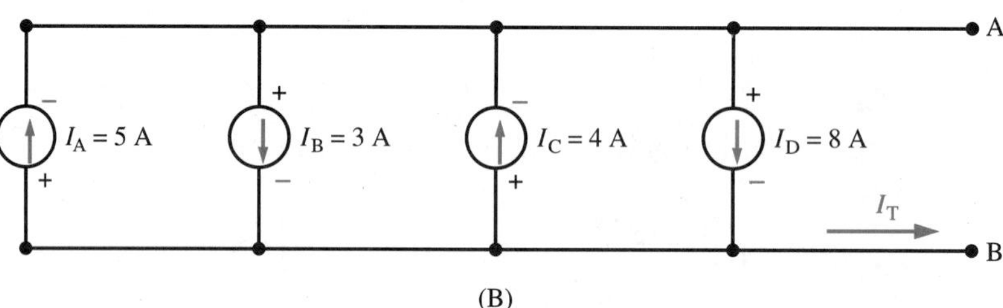

Figure 7-29

5. Find the voltage, V_{AB}, in the circuit of Figure 7-30.

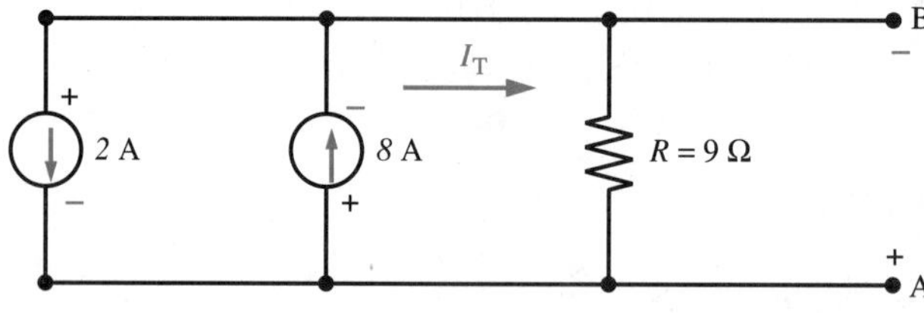

Figure 7-30

6. Find the voltage, V_{AB}, in the circuit of Figure 7-31.

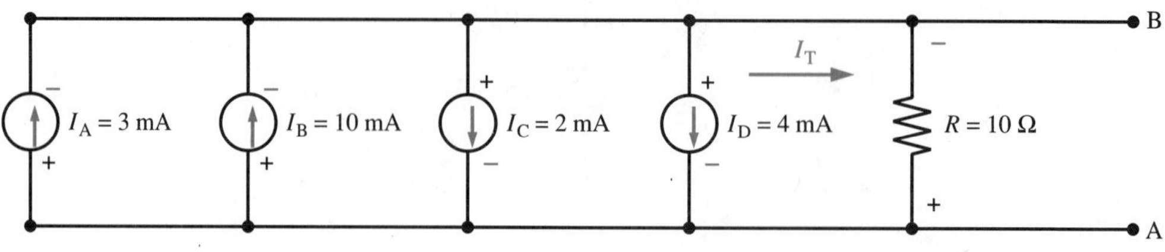

Figure 7-31

Section 7.4

7. Find the voltage, V_{AB}, in the circuit of Figure 7-32.

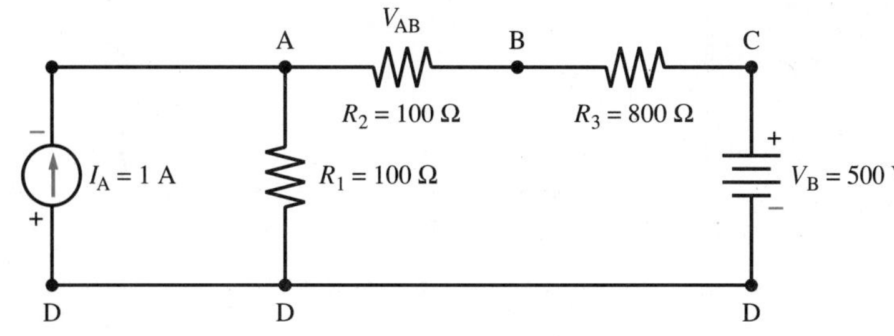

Figure 7-32

8. Find I_2, the current through resistor R_2, in the circuit of Figure 7-33.

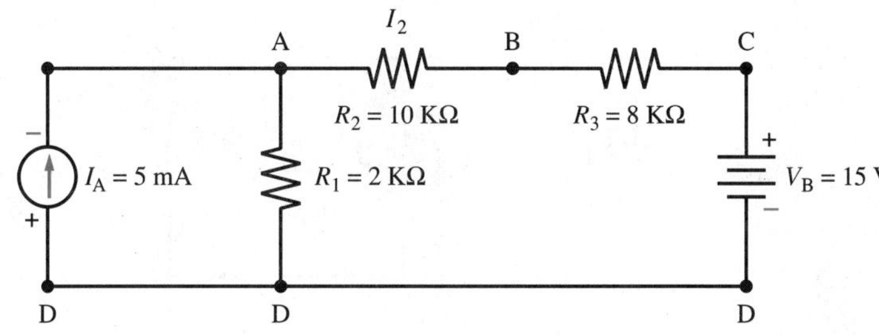

Figure 7-33

9. Find I_3, the current through resistor R_3, in the circuit of Figure 7-34.

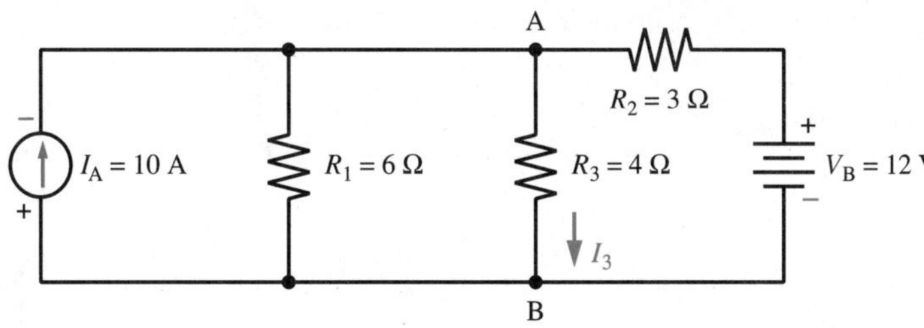

Figure 7-34

Section 7.5

10. Using superposition, find the voltage, V_{AB}, across resistor R_1 in Figure 7-35.

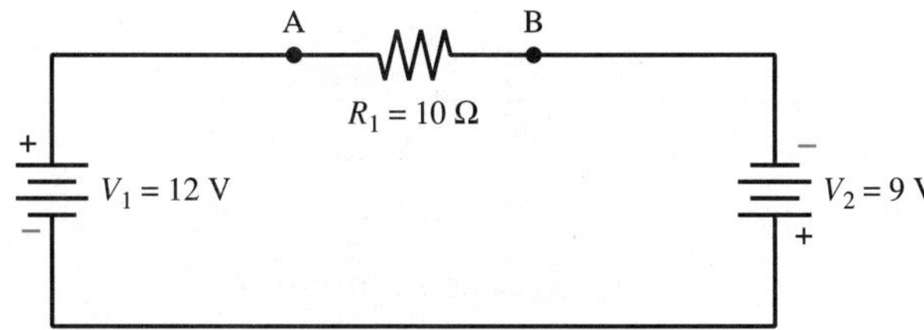

Figure 7-35

11. Repeat problem 10 for the network of Figure 7-36.

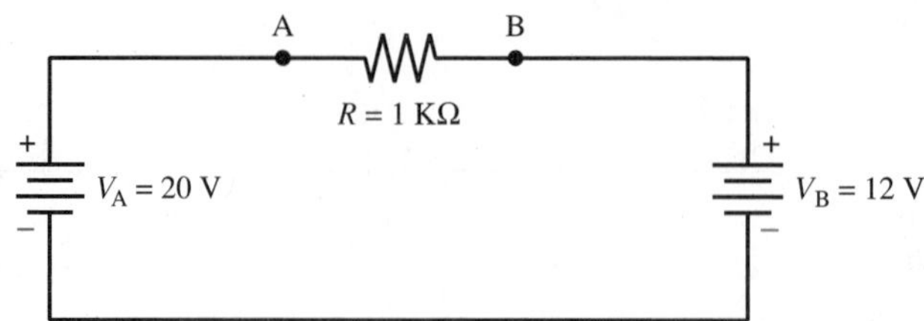

Figure 7-36

12. Using the superposition theorem, find the voltage, V_{AB}, across resistor R_3 in Figure 7-37.

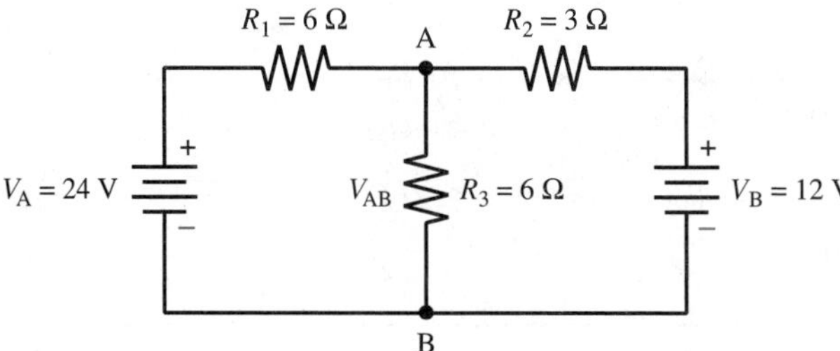

Figure 7-37

13. Repeat problem 12 for the circuit of Figure 7-38.

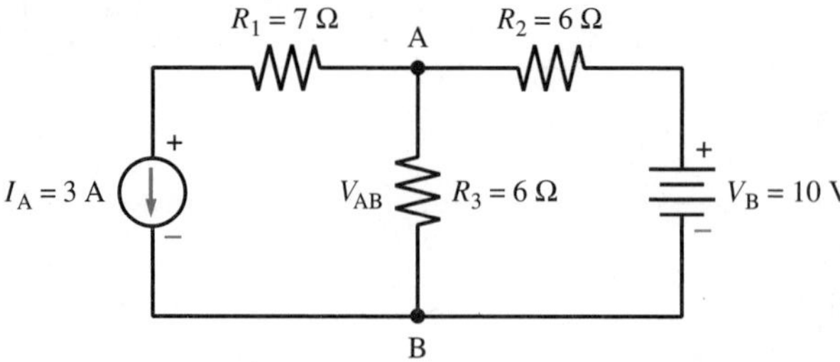

Figure 7-38

Section 7.6

14. Write the loop equations for the interior current paths of Figure 7-37.
15. Solve the loop equations of problem 14.
16. Write the loop equations for I_A and I_B in Figure 7-39.
17. Write the loop equations for I_A and I_B in Figure 7-40.

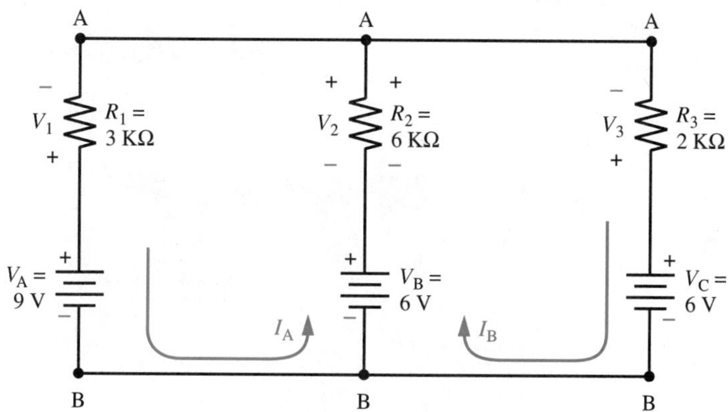

Figure 7-39

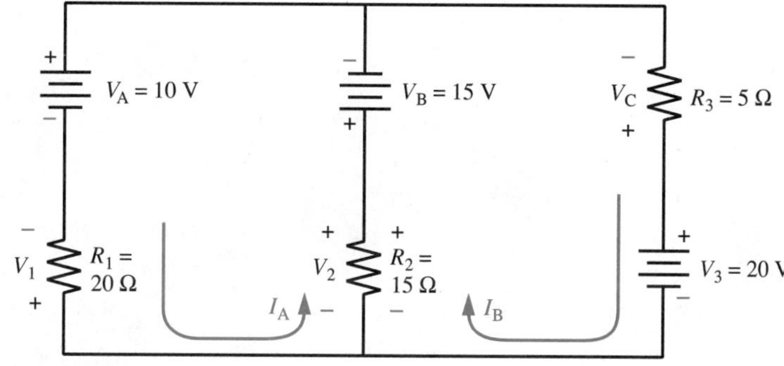

Figure 7-40

Section 7.7

18. Using nodal analysis, find the voltage across resistor R_1 in Figure 7-41.

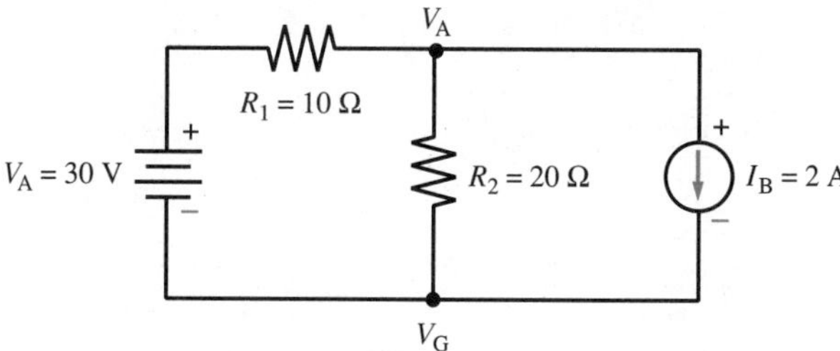

Figure 7-41

19. Write the nodal equations for nodes V_A and V_B in Figure 7-42.

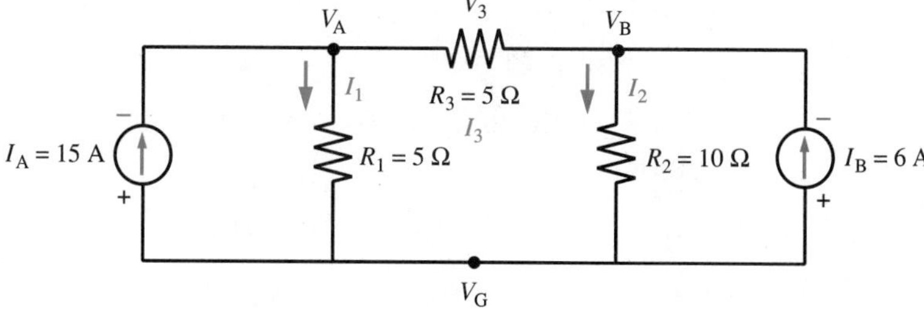

Figure 7-42

20. See Lab Manual for the crossword puzzle for Chapter 7.

CHAPTER 8

MAGNETISM

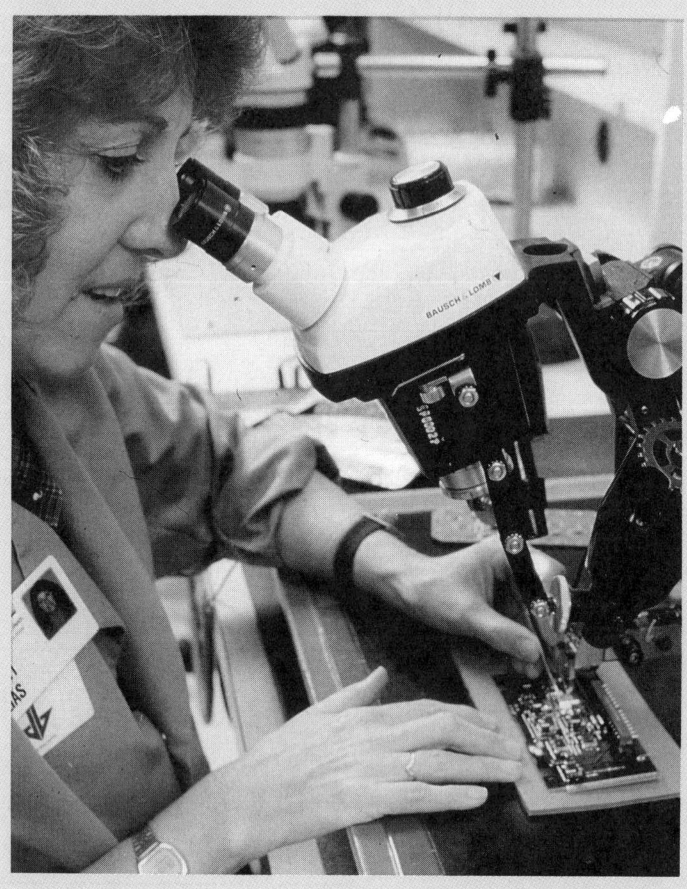

Courtesy of Motorola, Inc.

KEY TERMS

artificial magnet
electromagnet
electromotive force (EMF)
Faraday's law
flux
magnetic field
magnetic induction
natural magnet
permanent magnet
permeability
poles
relay
reluctance
retentivity
temporary magnet

OBJECTIVES

After completing this chapter, the student should be able to:

- classify magnets by type.
- define magnetic field.
- explain how the poles of a magnet are created.
- identify like and unlike magnetic poles.
- state the relationship between electricity and magnetism.
- state the principle of electromagnetic induction.
- describe Oersted's and Faraday's experiments.
- list and explain applications of electromagnetic principles.
- define terminology related to magnetism.

8.1 Introduction

Magnetism and electricity are interrelated. The ancient Greeks were the first to observe that pieces of a black mineral, magnetite, also known as lodestone, were able to pick up small pieces of iron. The Chinese were the first to observe that small pieces of magnetite when suspended by a string would always point in a north–south direction. This led to the development of the compass. In 1820, Hans Christian Oersted discovered the relationship between magnetism and electricity. This led to the creation of powerful magnets that are made by using electricity. These electromagnets are the basis for the operation of many electric devices such as relays, transformers, motors, and loudspeakers.

Magnetism was first observed in magnetite or lodestone.

8.2 Magnetic Fields

Magnets can be classified into three general types. A **natural magnet** is produced from the mineral magnetite. An **artificial magnet** is a magnet created by rubbing a piece of soft iron with a natural magnet. The rubbing transfers some of the magnetic properties of the natural magnet into the soft iron creating the artificial magnet. This principle allows a technician to magnetize a screwdriver, which can aid in the insertion of screws. An **electromagnet** is a magnet produced by allowing electric current through a coil of wire. Whenever an electric current exists in a conductor, a magnetic field will be produced.

Three types of magnets: natural magnets, artificial magnets, electromagnets.

Magnets can be further classified as permanent magnets and temporary magnets. A **permanent magnet** retains its magnetic properties over a long period of time. A **temporary magnet** loses most of its magnetic strength in a relatively short period of time.

Magnets can be permanent or temporary.

Recall that magnets tend to align themselves in a north–south direction. Magnets obey laws similar to those of positive and negative charges. The ends of a magnet are known as **poles.** These poles are referred to as north poles and south poles. The force or strength of a magnet is known as its **magnetic field.** The magnetic field is strongest at the poles or ends of the magnets because the density of the magnetic lines called **flux** is greatest at the poles. Density refers to the number of lines in a given area of space. The lines of flux are said to travel from the north pole of a magnet to the south pole, as shown in Figure 8-1.

The ends of a magnet are the poles.

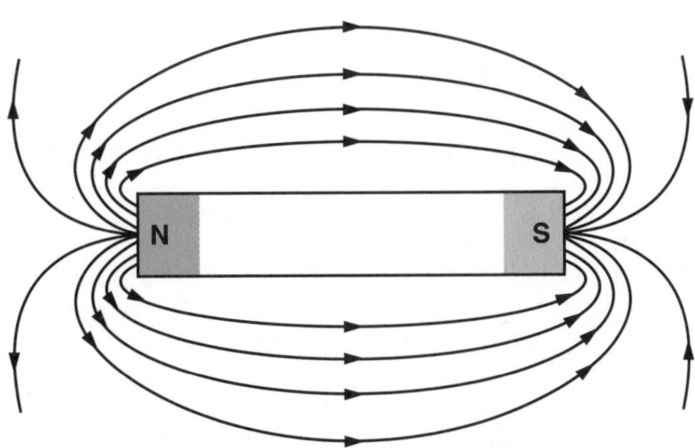

Figure 8-1 Magnetic field

The magnetic lines of force are called flux.

The charge of an electron that is spinning in an atom, produces a magnetic field just like an electron passing through a wire produces a magnetic field. The direction the electron spins determines the pole of the magnet. When the majority of the electrons in a material are spinning in the same direction, they are said to be aligned. This alignment creates the north and south poles of the magnet. In a nonmagnetic material, the electrons spin randomly, thus there is no alignment of the poles. Figure 8-2 illustrates the alignment and nonalignment of magnetic poles in a magnet.

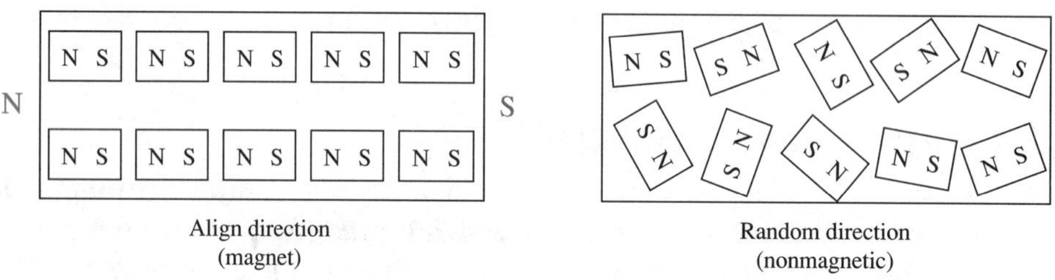

Figure 8-2 **Material alignment**

Like magnetic poles repel; unlike magnetic poles attract.

Like magnetic poles repel each other, and *unlike* magnetic poles attract each other. This is similar to Coulomb's law of electric charge. Figure 8-3 illustrates the laws of attraction and repulsion of magnets.

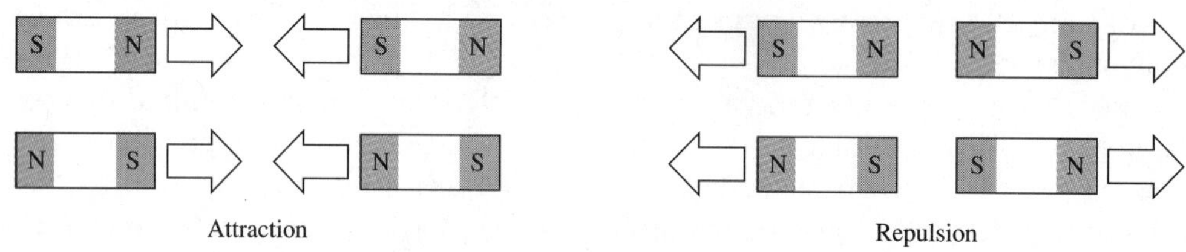

Figure 8-3 **Magnetic attraction and repulsion**

According to Coulomb's law of electric charge, unlike charges attract and like charges repel. The force of attraction (or repulsion) between two charges (Q_1 and Q_2) can be expressed mathematically as

Equation 8-1

$$F = \frac{kQ_1Q_2}{d^2}$$

where F = the force of attraction or repulsion in newtons

Q_1 and Q_2 = the charge in coulombs

d = the distance between the charges in meters

$k = 9 \times 10^9$ (a constant)

Note that the d^2 term in the denominator produces a very rapid decrease in the force of attraction (or repulsion) as the distance between the charges increases.

As with Coulomb's law, the closer the magnetic poles are to each other, the stronger the force of attraction or repulsion. Magnetic force varies inversely as the square of the distance. Thus, if the distance is doubled, the force will become one-fourth as great.

Magnets are made of metallic or ceramic (a ferromagnetic material in a ceramic body) materials. Iron, nickel, and cobalt are used to produce some of the strongest magnets. The ease in which a material becomes magnetized is known as **permeability.** Thus, permeability is the ability of a material to accept magnetic lines of force. **Retentivity** is the ability of a magnetic material to hold its magnetism after a magnetizing force has been removed. The opposition to a magnetic field is called **reluctance** (similar to resistance in an electric circuit).

Permeability—ease in which a material becomes magnetized.

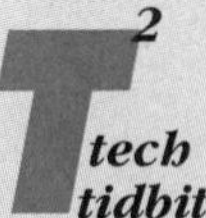

Retentivity—ability of a material to hold magnetism.

✓ Self-Test 1

1. A ______________ magnet is produced by the material magnetite.
2. Rubbing a piece of soft iron with a magnet produces a ______________.
3. An ______________ is a magnet produced by electric current in a coil of wire.
4. The force or strength of a magnet is known as its ______________.
5. The magnetic lines are called ______________.

8.3 Electricity and Magnetism

Reluctance—opposition to a magnetic field.

In 1820, Hans Christian Oersted accidentally discovered that a compass placed near a wire with an electric charge flowing (current) through it caused the compass needle to deflect and align itself perpendicular to the electric wire. He then changed the direction of the current and noted that the compass needle deflected in the opposite direction. No matter where he moved the wire, the needle of the compass would always be perpendicular (at right angles) to the wire. When the current was turned off, the needle of the compass returned to its normal north–south direction. Hence, Oersted's discovery that electric current is always surrounded by a magnetic field. Furthermore, the direction of the magnetic field is perpendicular to the direction of the current in the wire. Oersted's experiment is illustrated in Figure 8-4.

Magnetic fields are perpendicular to direction of current.

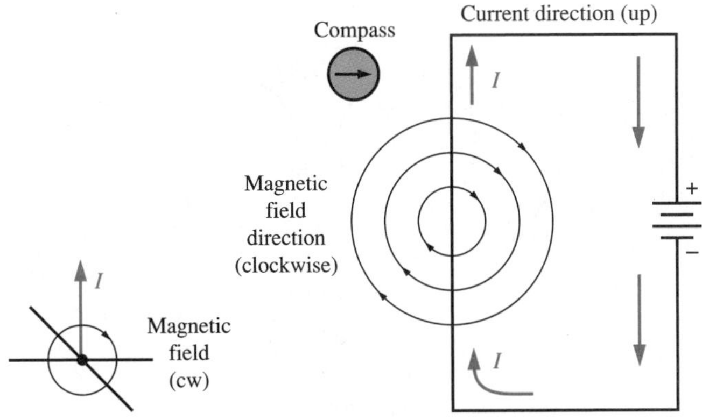

Figure 8-4 **Oersted's experiment**

Coiling a wire around an iron core increases the strength of the magnetic field.

Lefthand rule—thumb points north.

Whenever an electric current exists in a wire a magnetic field is produced around the wire. Wrapping the wire in a coil allows the magnetic fields of each turn of the wire to accumulate. Wrapping the coil around an iron core also increases the strength of a magnetic field. The iron core focuses the lines of force like a lens will focus light and thus increases the strength of the field. The strength of a magnetic field is a function of the number of turns of wire and the amount of current that exists in a wire. The greater the number of turns, the greater the strength of the magnetic field. The greater the amount of current, the greater the strength of the magnetic field. The polarity of an electromagnet can be determined by grasping the coil with the left hand with the fingers pointing in the direction of current, as illustrated in Figure 8-5. The thumb then points in the direction of the north pole.

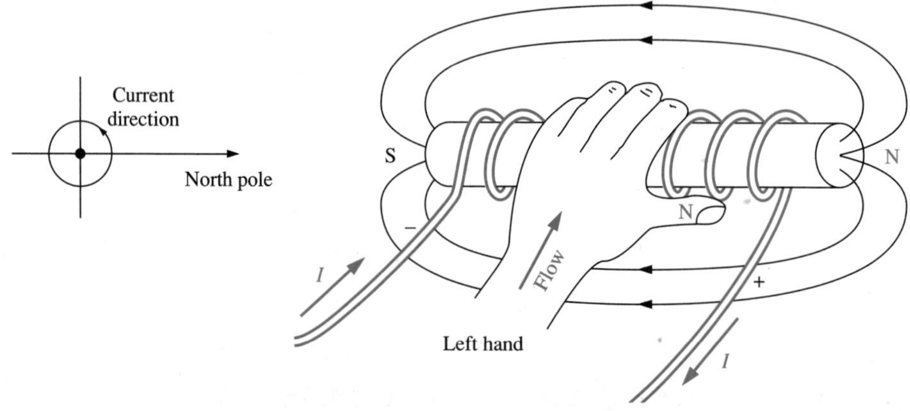

Figure 8-5 **Lefthand rule**

8.4 Magnetic Induction

After Oersted discovered the relationship between electricity and magnetism, many scientists started to work on producing electricity from magnetism. Michael Faraday, through a series of experiments, discovered the principle of electromagnetic induction. Faraday connected a coil of wire directly to a galvanometer (current meter). He found that when he moved a bar magnet in and out of the coil a momentary current was read on the meter. Furthermore, the faster he moved the magnet, the higher the current reading on the meter. If the magnet was held stationary, no current was read on the meter. Figure 8-6 illustrates this experiment.

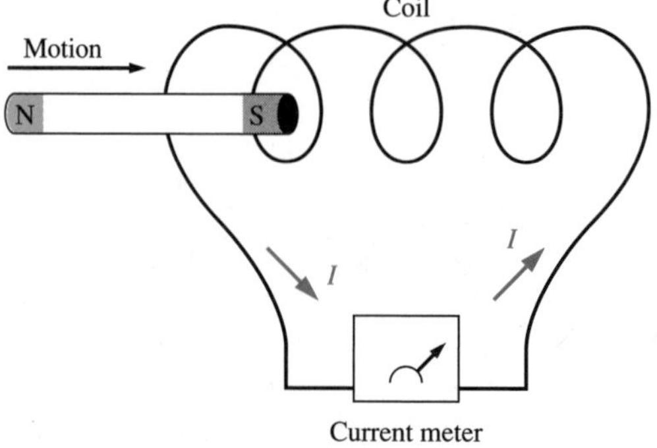

Figure 8-6 **Magnetic induction**

Faraday's law states that the induced voltage in a conductor is directly proportional to the rate at which the conductor cuts the magnetic lines of force.

In another experiment, Faraday arranged two coils of wire side by side. He connected one coil through a switch to a battery and another coil to a galvanometer. When the switch was closed, a momentary current was seen on the galvanometer. If the switch was opened, another momentary current was noted on the galvanometer but in the opposite direction. Whenever the switch was left open or closed for any length of time, no current was noted on the galvanometer. Faraday concluded that only the change in current flow produced the effect on the meter and that nothing happened when the current remained unchanged. Figure 8-7 illustrates this experiment.

Faraday's law—induced voltage is proportional to the rate at which a conductor cuts the magnetic lines of force.

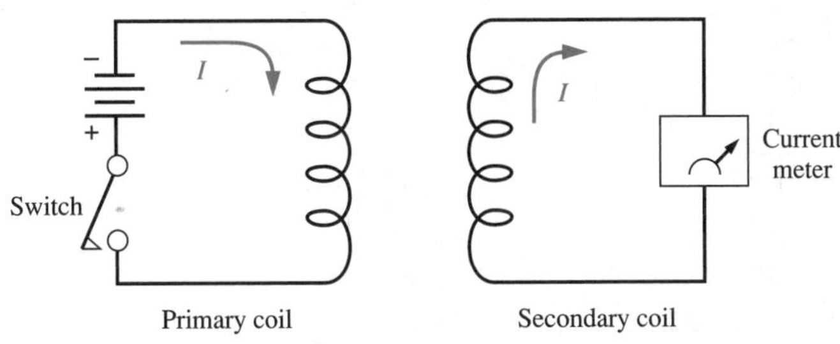

Figure 8-7 ***Induction between two coils***

Magnetic induction is the effect a magnet has on a conductor without physically touching it. Magnetic induction is the principle behind the generation of electricity. When a conductor passes through a magnetic field, a voltage is induced in that conductor. This voltage only exists while the conductor or magnetic field is moving. The voltage produced is called the induced voltage or **electromotive force (EMF).**

EMF—electromotive force or voltage.

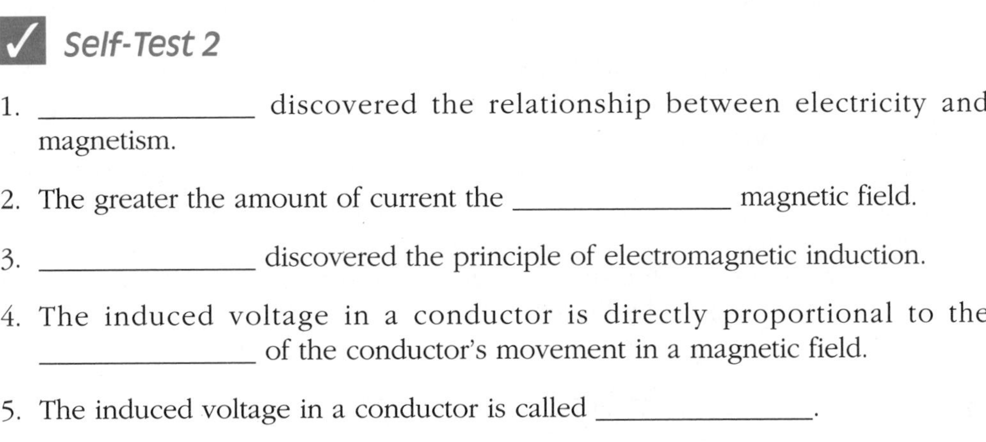

✓ Self-Test 2

1. _______________ discovered the relationship between electricity and magnetism.
2. The greater the amount of current the _______________ magnetic field.
3. _______________ discovered the principle of electromagnetic induction.
4. The induced voltage in a conductor is directly proportional to the _______________ of the conductor's movement in a magnetic field.
5. The induced voltage in a conductor is called _______________.

8.5 Electromagnetic Applications

One of the primary uses of electromagnetic principles is the generation of electricity. The electric generator converts mechanical energy of rotation into electric energy. The rotating mechanical energy is used to produce the motion

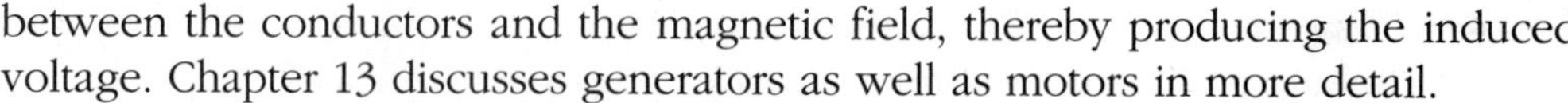

Relay—remote control electromagnetic switch.

between the conductors and the magnetic field, thereby producing the induced voltage. Chapter 13 discusses generators as well as motors in more detail.

A **relay** is a device that utilizes the electromagnetic principle to produce a remote-controlled switch. The relay, or electromechanical switch, opens or closes when a coil produces a magnetic field as current passes through it. The magnetic field activates or attracts the movable element of the relay. Figure 8-8 illustrates the operation of a relay. When the coil is energized, a magnetic field is produced that pulls the armature down to point B (against the action of the spring). When the circuit is de-energized (no magnetic field), the spring returns the armature back to point A.

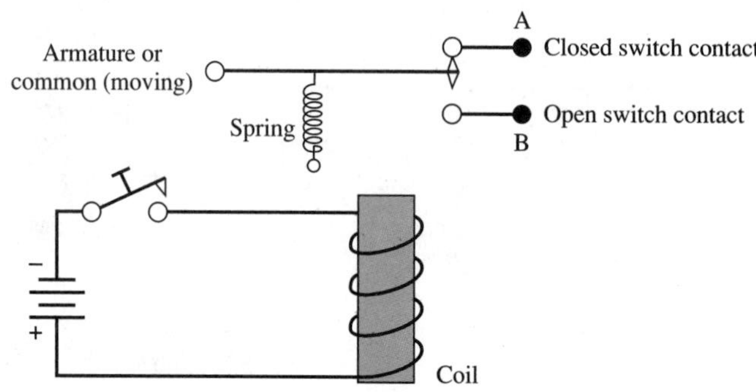

Figure 8-8 **Relay**

Another application of the relay principle is the doorbell. In this case, a small hammer is attached to the movable element of the relay. When a current exists in the relay coil a magnetic field is produced that causes the hammer to strike the bell. As the hammer moves, the circuit that energizes the coil is broken. The magnetic field collapses and the hammer is pulled back by a spring. This in turn allows charge to flow (current) through the coil and the process is repeated.

Relays are often used where low current circuits are used to control high current circuits. This may be for safety or for convenience. Relays are being replaced in many applications by devices called solid state relays. Solid state relays have no moving parts and do not operate magnetically.

Magnetic recording is used in cassette recorders, VCRs, and disk drives.

Magnetic recording is another application of the electromagnetic principle. It is used in cassette recorders, VCRs, and computer disk drives. Magnetic recording involves the storing of information onto a magnetic media like a magnetic tape or a magnetic disk. A recording or playback head, which is basically a coil of wire, is used to transfer the information to and from the media via magnetic induction. Figure 8-9 illustrates the magnetic recording process.

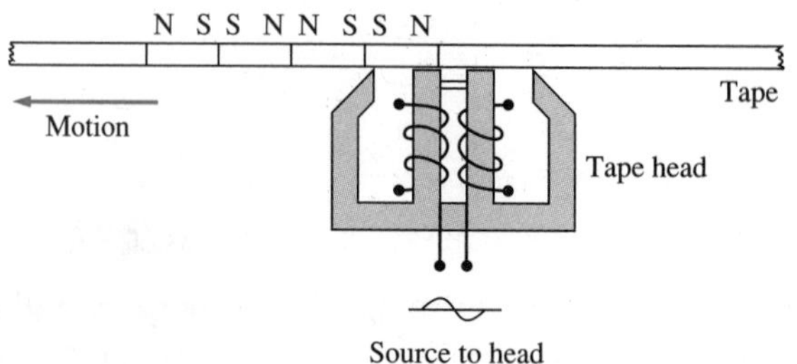

Figure 8-9 **Magnetic tape recording**

Speakers are still another example of electromagnetic induction. Speakers are constructed with a movable coil of wire around a permanent magnet. As current is passed through the coil, it produces a magnetic field that reacts with the magnetic field of the permanent magnet. The coil is attached to the paper cone of the speaker. As the coil moves back and forth, so does the cone, which in turn pushes the air to produce the sound. This is another example of a transducer, the conversion of electric energy to sound or acoustic energy.

T3 MAGNETIC MATERIALS

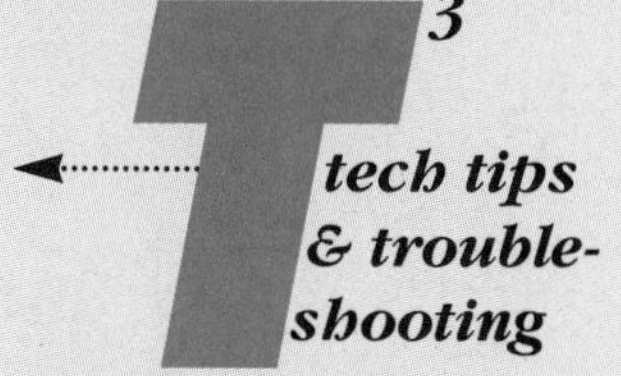

Magnetic materials can be magnetized or demagnetized using permanent magnets or electromagnets. Magnetizing a screwdriver is helpful when inserting screws into tight spaces. Figure 8-10 illustrates how to magnetize a screwdriver using a permanent magnet. While holding the magnet in one hand, hit the end of the screwdriver blade against one pole of the magnet. Repeat this several times on the same pole. The screwdriver will become magnetized. The end of the blade will become the opposite pole of the magnet you were using. The blade of the screwdriver will then hold screws that contain ferrous (iron) materials.

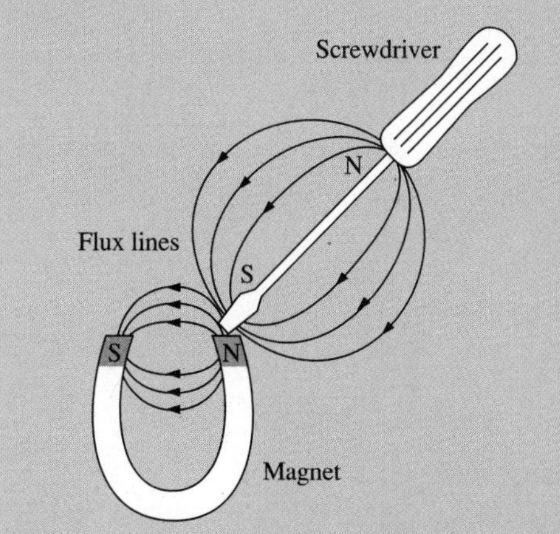

Figure 8-10 **Magnetizing a screwdriver with a magnet**

Figure 8-11 illustrates how to magnetize a screwdriver with an electromagnet. When current exists in the coil, the magnetic field of the coil will flow through the blade of the screwdriver and magnetize the screwdriver.

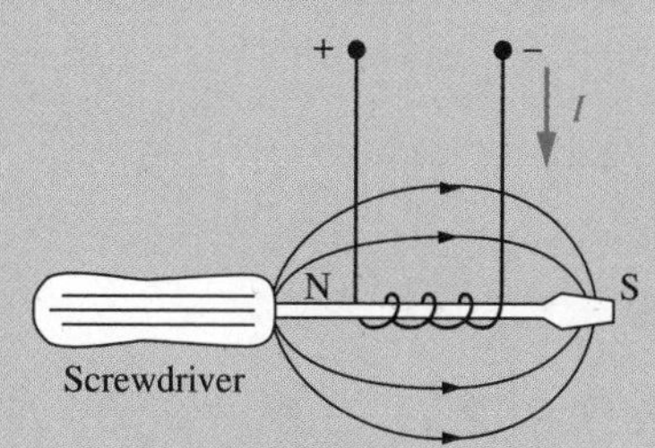

Figure 8-11 **Magnetizing a screwdriver with electric current**

The screwdriver can be demagnetized or neutralized by placing the blade in an alternating magnetic field. A standard tape head demagnetizer can be used for this purpose, as shown in Figure 8-12.

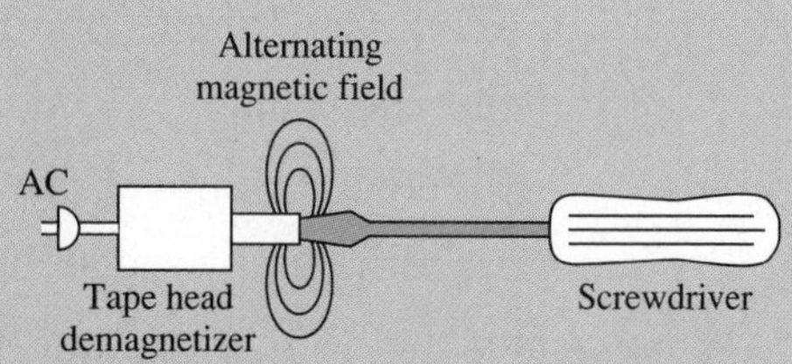

Figure 8-12 **Demagnetizing a screwdriver**

SUMMARY

Hans Christian Oersted discovered the relationship between electricity and magnetism in 1820. This led to the creation of powerful electromagnets using electricity. Generally, magnets are classified as natural, artificial, or electromagnets. They are further classified as permanent or temporary magnets. The ends of the magnet are known as the poles. *Like* magnetic poles repel each other, and *unlike* magnetic poles attract each other. The force or strength of a magnet is known as its magnetic field. Retentivity is the ability of a magnet to hold its magnetism. Permeability is the ease in which a material becomes magnetized. The opposition to a magnetic field is called reluctance.

Oersted discovered that whenever an electric current exists in a wire a magnetic field is produced around the wire. This led to a series of experiments by Michael Faraday, who discovered the principle of electromagnetic induction. Electromagnetic induction is the effect that a magnet has on a conductor without physically touching it. Electromagnetic induction is the principle behind the generation of electricity. Faraday's law states that the induced voltage in a conductor is directly proportional to the rate at which the conductor cuts the magnetic lines of force.

FORMULA

$$F = \frac{kQ_1Q_2}{d^2} \qquad \text{(Coulomb's law)}$$

EXERCISES

Section 8.1

1. What natural material exhibits magnetic qualities?
2. Who discovered the relationship between electricity and magnetism?
3. Name three devices that use magnetism as the basis for operation.

Section 8.2

4. What are the three general types of magnets?
5. What is the difference between a permanent magnet and a temporary magnet?
6. How are the ends of a magnet identified?
7. What is the force or strength of a magnet called?
8. What is the term for magnetic lines of force?
9. What is the direction of travel of magnetic lines of force?
10. Explain what is meant by magnetic density.
11. What creates the north and south poles of a magnet?
12. State the laws of magnetic attraction and repulsion.

13. What is known as the ability of a magnetic material to accept magnetic lines of force?
14. Explain the term *retentivity*.
15. What is the opposition to a magnetic field called?

Section 8.3

16. When making an electromagnet, why do we coil the wire?
17. When making an electromagnet, why do we use iron cores?
18. What is the effect of increasing the current through an electromagnet?
19. The direction of a magnetic field is determined by what rule?

Section 8.4

20. Who discovered the principle of electromagnetic induction?
21. State the law of electromagnetic induction.

Section 8.5

22. List three devices in your home that use the electromagnetic principle.
23. What electromagnetic device converts:
    (A) mechanical energy into electrical energy?
    (B) electrical energy into mechanical energy?
    (C) electrical energy into (sound) acoustic energy?

Section T^3

24. Explain two methods of magnetizing a screwdriver.
25. Explain how to demagnetize a screwdriver.
26. See Lab Manual for the crossword puzzle for Chapter 8.

CHAPTER 9

ALTERNATING CURRENT FUNDAMENTALS

Courtesy of Hewlett Packard

KEY TERMS

alternating current
amplitude
average value
cycle
duplex receptacle
effective value
frequency
hertz
lags
leads
oscilloscope
peak
peak-to-peak
period
phase angle
power lost
pulsating DC
pure DC
radians
rectangular wave
RMS
saw tooth
sine wave
square wave
terminal
transformer

OBJECTIVES

After completing this chapter, the student should be able to:

- define alternating current.
- draw DC and AC waveforms.
- describe the generation of a sine wave.
- define AC terminology.
- calculate period and frequency.
- calculate average, effective, and peak-to-peak values.
- determine phase relationships.
- state the advantages of alternating current.
- perform basic duplex receptacle wiring.

9.1 Introduction

Direct current sources cause electric charge to flow (current) in one direction only. In other words, the polarity of the voltage source never changes. An example of a direct current energy source is a battery. Figure 9-1 shows a graphical representation of some DC voltage sources. The magnitude of the voltage source is plotted on the ordinate (*Y* or vertical axis) and time is plotted on the abscissa (*X* or horizontal axis).

DC—polarity never changes.

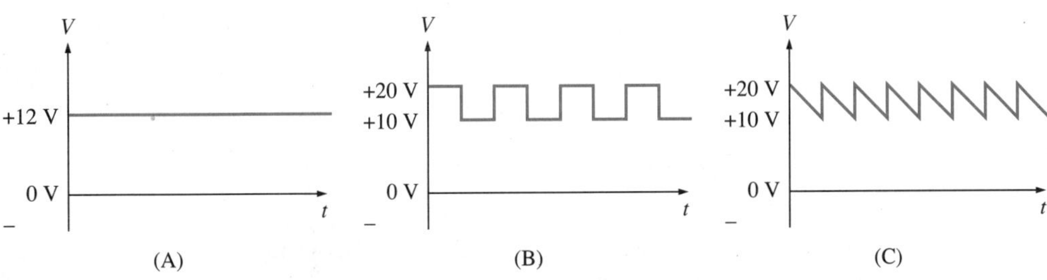

Figure 9-1 **DC waveforms**

The DC voltage, which is represented as a straight line in Figure 9-1A, is sometimes referred to as **pure DC,** because the magnitude of the voltage remains constant such as the voltage produced by a battery. When the magnitude of the voltage changes, as in Figures 9-1B and C, this is known as **pulsating DC.** Notice that the polarity (direction) of the voltage force does not change, however, the magnitude of the pulsating DC voltage does vary over time. Let's explore the behavior of the waveforms of Figure 9-1 when applied to the simple electric circuit of Figure 9-2 that lights a lamp.

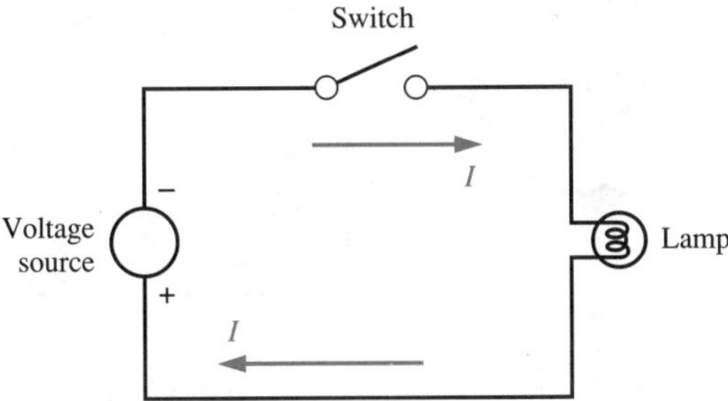

Figure 9-2 **Simple circuit**

First consider the waveform in Figure 9-1A. When the switch is closed, the voltage applied to the light will instantaneously rise from 0 to 12 volts. Current will exist in the wires in a clockwise direction to light the lamp. If the voltage changes in magnitude, as shown in Figure 9-1B and C, current will still exist in the same direction because the polarity does not change; however, the amount of current that will exist will vary with the amount of voltage available at any instant in time. The greater the voltage, the greater the current in the circuit; the smaller the voltage, the smaller the current. Thus the amount of current in an electric circuit is said to vary directly with the amount of voltage available.

AC—polarity is always changing.

Alternating current sources are those where the polarity of the energy source changes periodically. That is, electrons are first pushed in one direction and then in the opposite direction. In Figure 9-3, we see three examples of AC voltages represented graphically.

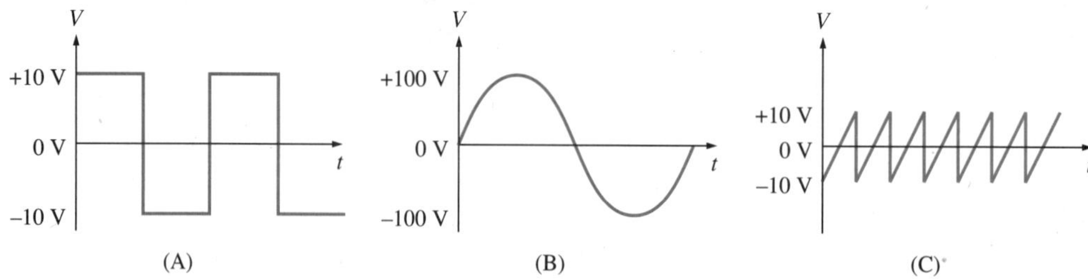

Figure 9-3 **AC waveforms**

Notice that the polarity and the magnitude of the voltage are both changing. By comparing the waveforms in Figure 9-1 and 9-3, we can see that DC can be considered a special case of AC in which the polarity never changes.

One complete sine wave is produced for each 360° revolution of a generator.

9.2 Generating a Sine Wave

The AC waveform in Figure 9-3B is called a **sine wave.** It can be produced when an AC generator shaft rotates in a magnetic field, as shown in Figure 9-4. One AC sine wave is produced for each 360-degree revolution of the generator shaft. The generation of the sine wave can be developed by plotting the rotation of the radius of the armature of an AC generator.

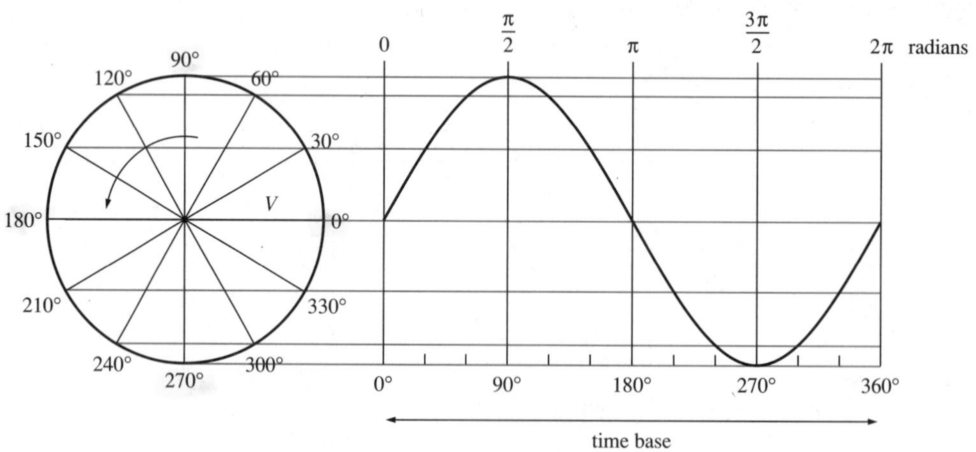

Figure 9-4 **Generating an AC sine wave voltage**

Note that the sine wave has a positive maximum value at 90 degrees and decreases to 0 at 180 degrees. It then increases to a negative maximum value at 270 degrees and returns to 0 at 360 degrees. Electrons flow in one direction during the positive polarity duration and in the opposite direction during the negative polarity duration. As was the case in DC, the amount of current that exists in either direction will vary directly with the amount or magnitude of the voltage at any given instant in time.

✓ Self-Test 1

1. When only the magnitude of a waveform changes, this is known as ________________.
2. When the magnitude and polarity of a waveform both change, this is known as ________________.
3. One AC sine wave is produced for each ________________ revolution of an AC generator shaft.

9.3 AC Definitions

The **amplitude** of a sine wave refers to the maximum or **peak** value of the waveform. In Figure 9-5 the amplitude or peak value is said to be 10 volts ($V_{peak} = V_P = 10$ V).

Amplitude is the peak value.

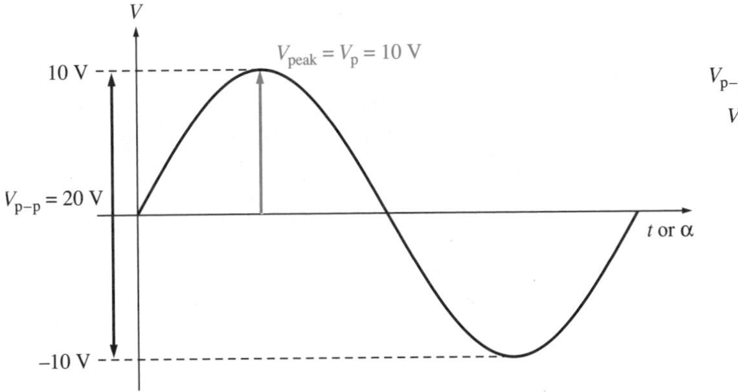

$$V_{p-p} = 2 \times V_p = 2 \times 10 \text{ V} = 20 \; V_{p-p}$$
$$V_p = \tfrac{1}{2} \times V_{p-p} = \frac{20 \text{ V}}{2} = 10 \; V_p$$

Figure 9-5 **Amplitude of a sine wave**

Note: $v = V_P \sin \alpha$ (instantaneous voltage of a general sine wave)
The number that multiplies the sine wave is its peak or amplitude.
∴ for the sine wave shown: $v = 10 \sin \alpha$

The **peak-to-peak** value of the waveform is the peak amplitude value in the positive direction plus the absolute value of the peak amplitude value in the negative direction. In Figure 9-5 the peak-to-peak voltage is said to be 20 volts ($V_{P-P} = 20$ V).

The peak-to-peak value equals two times the peak value.

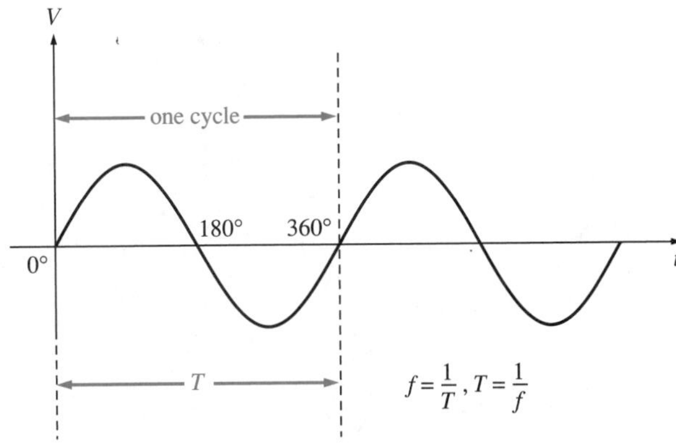

Figure 9-6 **Frequency and period of a sine wave**

One cycle equals 360°.

During the time that the armature rotates through 360 degrees, the voltage developed is said to have gone through one complete **cycle.** After this time, the waveform repeats itself. The time required to complete one full cycle is called the **period** (T), as shown in Figure 9-6. The number of complete cycles in one second is known as the **frequency** (f), which is measured in units called **hertz** (Hz), or cycles per second (cps). Mathematically,

$$1 \text{ hertz} = 1 \text{ cycle/second}$$

Equation 9-1

$$f = \frac{1}{T}$$

where f = the frequency in hertz or cycles per second

T = the period in seconds

The period is the time for one cycle.

The standard frequency in the United States is 60 hertz. In Europe, the Far East, and South America, the standard frequency is 50 hertz.

EXAMPLE 9-1

A technician observes a 340-volt peak-to-peak sine wave on an **oscilloscope** (an electronic device that observes waveforms). Determine the peak value of the voltage.

tech tidbit

Frequency is the number of cycles per second:

$f = \frac{1}{T}$

$T = \frac{1}{f}$

Solution:

$$V_{\text{p}} = \frac{V_{\text{p-p}}}{2}$$

$$= \frac{340 \text{ V}}{2}$$

$$= 170 \; V_{\text{(peak)}}$$

EXAMPLE 9-2

What is the period of a 60-hertz sine wave?

Solution:

$$f = \frac{1}{T}$$

$$T = \frac{1}{f}$$

$$= \frac{1}{60 \text{ Hz}}$$

$$= 0.016\overline{6} \text{ s}$$

$$= 16.\overline{6} \times 10^{-3} \text{ s}$$

$$= 16.\overline{6} \text{ ms}$$

EXAMPLE 9-3

A technician observes the waveform of Figure 9-7 on an oscilloscope. Determine V_{P-P}, V_P, the period of the waveform, and the frequency.

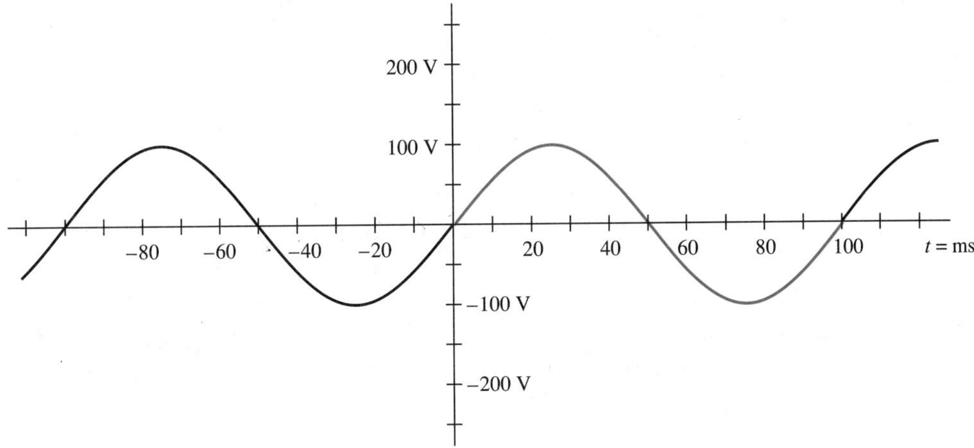

Figure 9-7

Solution:

$$V_{P-P} = 200 \text{ V, from } +100 \text{ V to } -100 \text{ V}$$

$$V_P = \frac{V_{P-P}}{2} = \frac{200 \text{ V}}{2} = 100 \text{ V}$$

The time for one cycle is:

$$T = 100 \text{ ms}$$

The frequency is:

$$f = \frac{1}{T} = \frac{1}{100 \text{ ms}} = \frac{1}{\phantom{100 \text{ s}}} = \frac{10^3}{100 \text{ s}} = \frac{1{,}000}{100 \text{ s}} = \frac{10 \text{ cycles}}{\text{s}} = 10 \text{ Hz}$$

✓ Self-Test 2

1. The maximum amplitude of a sine wave is called the ______________ value.
2. The ______________ value of an AC waveform is measured from the positive maximum amplitude to the negative maximum amplitude.
3. The time for one complete cycle is called the ______________.
4. The unit of frequency is ______________.

9.4 The Average Value of an AC Waveform

Referring to the sine wave in Figure 9-8, it is clear that the **average value** of the waveform over one complete cycle is zero, because the wave shape is symmetrical about the X axis. For this reason, the average value is always considered over one-half of a cycle (or one alternation).

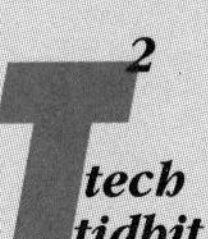

The average value of a complete sine wave is zero.

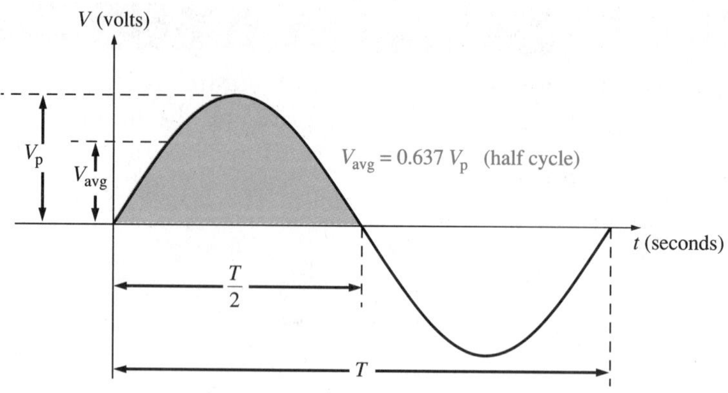

Figure 9-8 **Average value of a sine wave**

tech tidbit

The average value of a sine wave pulse is 0.637 V_{peak}.

For a sine wave, it can be shown mathematically that the average value of one alternation is equal to 0.637 of the peak value. This is arrived at by dividing the area of the positive half cycle by the period of the positive half cycle. Mathematically, for sine waves,

Equation 9-2

$$V_{avg} = 0.637\ V_{peak}$$

and

Equation 9-3

$$I_{avg} = 0.637\ I_{peak}$$

EXAMPLE 9-4

Find the average value for a half cycle of a 30-volt peak-to-peak sine wave.

tech tidbit

$$\text{average} = \frac{\Sigma\ \text{area}}{\text{period}}$$

average value of any waveform

Solution:

$$V_{peak} = \frac{V_{p-p}}{2} = \frac{30\ \text{V}}{2} = 15\ \text{V}$$

$$V_{avg} = 0.637\ V_{peak} = (0.637)(15\ \text{V}) = 9.55\ \text{V}$$

The average value over one cycle can be computed for any waveform using Equation 9-4:

Equation 9-4

$$\text{Average} = \frac{\Sigma\ \text{area}}{\text{period}}$$

where Σ area = the sum of the areas above and below the horizontal axis. Areas below the horizontal axis are to be considered negative areas.

period = the time for one cycle

Consider the waveform of Figure 9-9.

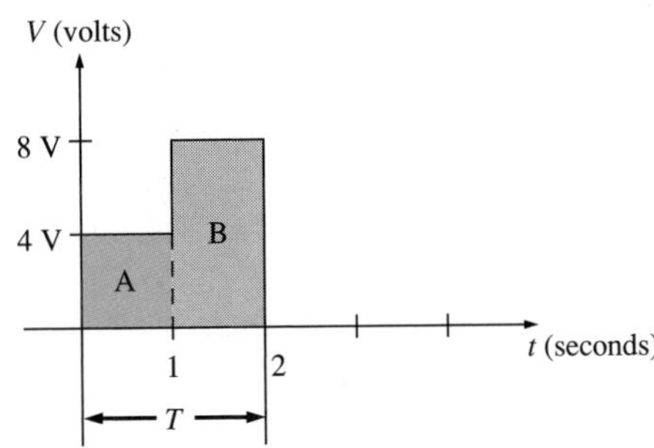

Figure 9-9 **Average value for a nonsinusoidal waveform**

The area of each section of the waveform can be computed and added together to obtain the total areas as follows:

section	altitude	base	area
A	4 V	1 s	4 Vs (rectangular area 1 s × 4 V)
B	8 V	1 s	8 Vs (rectangular area 1 s × 8 V)
Total		period = 2 s	Σ area = 12 Vs

The average value of the waveform is then:

$$\text{Average} = \frac{\sum \text{area}}{\text{period}}$$

$$\text{Average} = \frac{12 \text{ Vs}}{2 \text{ s}} = 6 \text{ V}$$

EXAMPLE 9-5

Find the average value of the waveform shown in Figure 9-10.

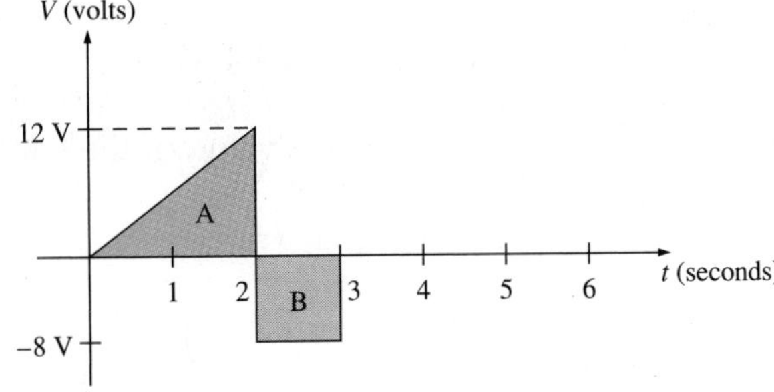

Figure 9-10

Solution:

section	altitude	base	area
A	12 V	2 s	12 Vs triangular area
B	−8 V	1 s	− 8 Vs rectangular area
Total		period = 3 s	Σ area = 4 Vs

$$\text{Average} = \frac{\Sigma \text{ area}}{\text{period}}$$

$$\text{Average} = \frac{4 \text{ Vs}}{3 \text{ s}} = 1.33 \text{ V}$$

Note:

$$\text{Area of a triangle} = \frac{1}{2}(\text{base} \times \text{altitude})$$

$$= \frac{1}{2}(2 \text{ s} \times 12 \text{ V})$$

$$= 12 \text{ Vs}$$

$$\text{Area of a rectangle} = \text{base} \times \text{altitude}$$

$$= 1 \text{ s} \times -8 \text{ V}$$

$$= -8 \text{ Vs}$$

T² tech tidbit

Effective value is the work value of an AC waveform.

9.5 The Effective Value of AC Waveforms

T² tech tidbit

$V_{eff} = 0.707\ V_p$ for sine waves only.
$V_p = 1.414\ V_{eff}$ for sine waves only.

Although the magnitude of an alternating current or voltage sine wave is constantly changing, there must be some value of alternating current or voltage that will perform the same "rate of work" as an equivalent amount of direct current. One way to compare the work performed is by measuring the heating effect of the current through the same value of resistance. Therefore, an effective value of 1 volt of AC will produce the same heat as 1 volt of DC in the same amount of time. Using this definition, it can be shown that the **effective value** of an alternating current sine wave is equal to the peak value divided by the square root of two. Mathematically for sine-wave voltages only,

Equation 9-5

$$V_{eff} = \frac{V_p}{\sqrt{2}} = 0.707\ V_p$$

and

Equation 9-6

$$V_p = \sqrt{2}\ V_{eff} = 1.414\ V_{eff}$$

Similarly, it can be shown that the effective value of the current is

Equation 9-7

$$I_{eff} = \frac{I_p}{\sqrt{2}} = 0.707\ I_p$$

and

Equation 9-8

$$I_p = \sqrt{2}\ I_{eff} = 1.414\ I_{eff}$$

Because of the mathematical process for determining the effective value of an AC waveform, the effective value is also known as the **RMS** value (root-mean-square), as shown in Figure 9-11.

RMS value equals the effective value.

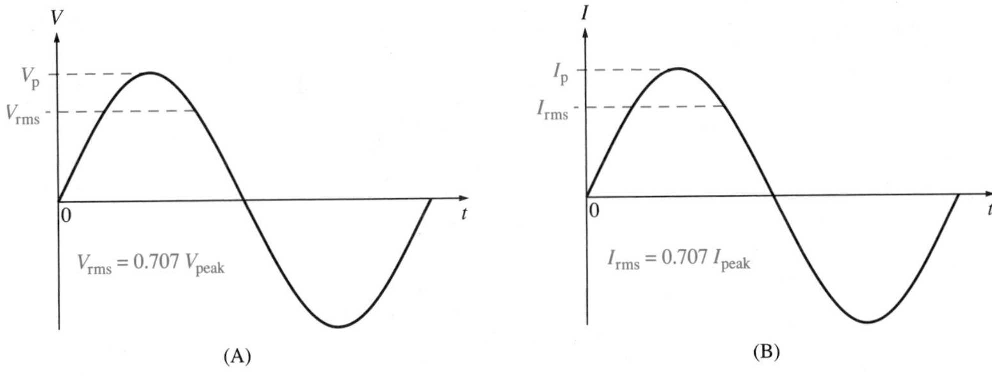

Figure 9-11 ***Effective value of a sine wave***

It is important to note that all AC voltage and current values are always considered as equivalent DC (effective or RMS) values unless otherwise stated. Furthermore, AC electric meters also indicate RMS or effective values unless otherwise stated. In addition, AC will always represent a sine wave unless otherwise stated.

EXAMPLE 9-6

An AC sine wave voltage source has a peak value of 100 volts. Find the RMS value of the voltage.

Solution:

$$\begin{aligned} V_{eff} &= 0.707\ V_p \\ &= 0.707\ (100\ \text{V}) \\ &= 70.7\ \text{V (RMS)} \end{aligned}$$

EXAMPLE 9-7

An AC ammeter measures an effective current of 50 milliamperes. Find the peak value of the current.

Solution:

$$\begin{aligned} I_{peak} &= 1.414\ I_{eff} \\ &= (1.414)(50 \times 10^{-3}\ \text{A}) \\ &= 70.7 \times 10^{-3}\ \text{A} \\ &= 70.7\ \text{mA (peak)} \end{aligned}$$

EXAMPLE 9-8

An AC voltage of 120 volts RMS is applied to a circuit. Draw the peak-to-peak waveform.

Solution:

See Figure 9-12.

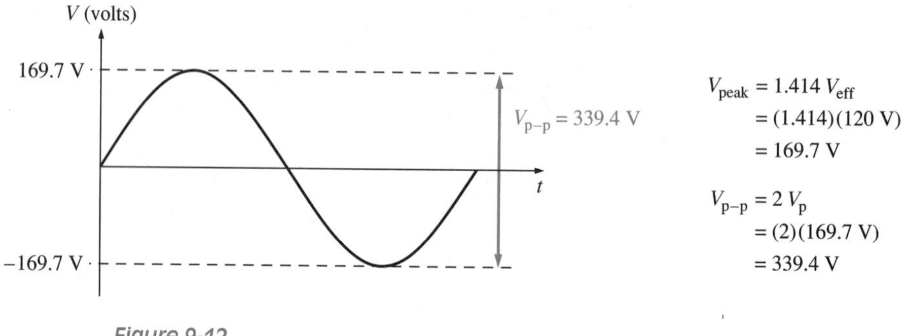

Figure 9-12

✓ Self-Test 3

1. State the formula for the average value of a sine wave pulse.
2. State the formula for the average value of any waveform.
3. What is the peak value of a 120-volt RMS AC sine wave?
4. What is the effective value of a 440-volt peak-to-peak sine wave?

9.6 Phase Relationships for Sine Waves

tech tidbit

The phase angle is the difference in time between two waveforms of the same frequency.

The phase relationship is a measurement of the difference in time between two AC waveforms. This is usually measured in degrees and is referred to as the **phase angle.** To determine the phase angle, both sine waves must have the same frequency. Generally, it is necessary to select a common reference point on both waveforms. Phase relationship also indicates which waveform leads or lags another sine wave. Phase relationships are independent of the peak values.

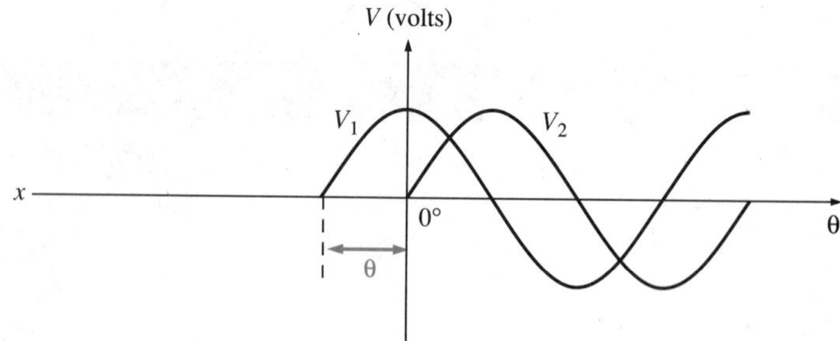

Figure 9-13 **Phase relationships between sine waves of the same frequency**

For example, in Figure 9-13, the reference is taken as the *X* axis. V_1 crosses the reference before V_2, so we say that V_1 **leads** V_2 by θ degrees or V_1 starts out earlier in time than V_2. We could also say that V_2 **lags** V_1 by θ degrees, or V_2 starts out later in time than V_1.

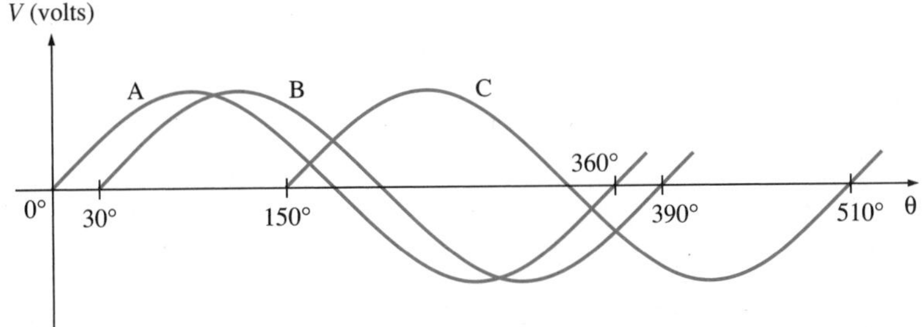

Figure 9-14 **Phase relationships between sine waves of the same frequency**

In Figure 9-14 it can be stated that:

A leads B by 30 degrees.

A leads C by 150 degrees.

B leads C by 120 degrees.

or

B lags A by 30 degrees.

C lags A by 150 degrees.

C lags B by 120 degrees.

Angles can also be measured in units called **radians** (rad). One complete cycle contains 2π radians (where π = 3.14). In other words, 360 degrees is equal to 2π radians, as shown in Figure 9-15.

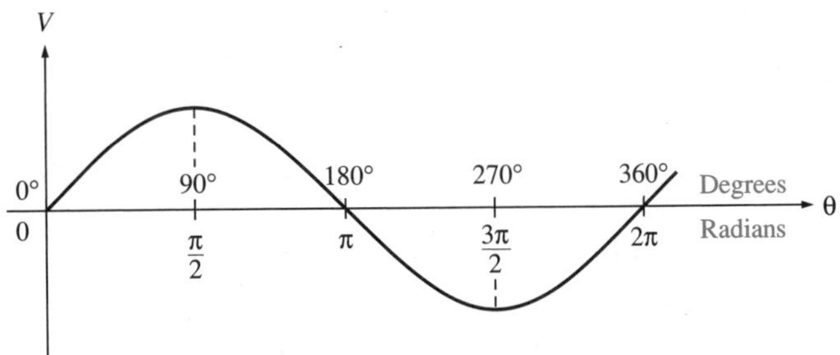

Figure 9-15 **Degrees versus radians**

tech tidbit

To convert from degrees to radians multiply by π/180.

Because 2π radians is equal to 360 degrees, it follows that to convert from degrees to radians we must multiply the degrees by π/180.

EXAMPLE 9-9

Convert the following degrees to radians:

(A) 30° (B) 60° (C) 90° (D) 180°

Solution:

(A) $30° \times \dfrac{\pi \text{ rad}}{180°} = \dfrac{\pi}{6}\text{rad} = 0.524 \text{ rad}$

(B) $60° \times \frac{\pi \text{ rad}}{180°} = \frac{\pi}{3}\text{rad} = 1.047 \text{ rad}$

(C) $90° \times \frac{\pi \text{ rad}}{180°} = \frac{\pi}{2}\text{rad} = 1.571 \text{ rad}$

(D) $180° \times \frac{\pi \text{ rad}}{180°} = \pi \text{rad} = 3.141 \text{ rad}$

$\therefore \quad 180° = \pi \text{rad}$

T² tech tidbit

To convert from radians to degrees multiply by 180/π.

To convert from radians to degrees we must multiply radians by 180/π because

$$\cancel{\text{radians}} \times \frac{360 \text{ degrees}}{2\pi \cancel{\text{ radians}}} = \frac{180}{\pi} \text{ degrees.}$$

EXAMPLE 9-10

Convert the following radian angles to degrees:

(A) 0.8 rad (B) 1.6 rad (C) −2.78 rad (D) −0.4 rad (E) 1 rad

Solution:

(A) $0.8 \text{ rad} \times \frac{180°}{\pi \text{ rad}} = 45.8°$

(B) $1.6 \text{ rad} \times \frac{180°}{\pi \text{ rad}} = 91.7°$

(C) $-2.78 \text{ rad} \times \frac{180°}{\pi \text{ rad}} = -159.3°$

(D) $-0.4 \text{ rad} \times \frac{180°}{\pi \text{ rad}} = -22.9°$

(E) $1 \text{ rad} \times \frac{180°}{\pi \text{ rad}} = 57.3°$

$\therefore \quad 1 \text{ rad} = 57.3°$

9.7 Other Waveforms

Another very common type of waveform is the **square wave.** In a square wave, the voltage or current increases from zero to a maximum value instantaneously. It stays at the maximum value for a period of time and then decreases to zero, reverses direction and increases to a maximum value in the opposite direction, all instantaneously. It then stays at this negative maximum value for the same period of time and returns to zero. Figure 9-16A illustrates the shape of a square waveform. Figure 9-16B illustrates a variation of the square wave called the **rectangular wave.** Square waves and rectangular waves are frequently used in digital computer circuits.

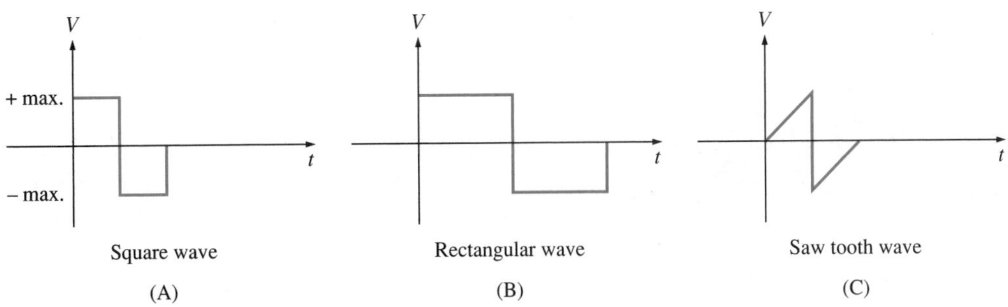

Figure 9-16 ***Other waveforms***

The waveform in Figure 9-16C is called a **saw tooth.** It looks very much like the tooth of a saw blade. The saw tooth waveform starts at zero and increases linearly to a maximum value. It then drops to zero, reverses direction, and increases to a negative maximum value instantaneously. It then decreases linearly and returns to zero. Saw tooth waveforms are frequently used in the horizontal section of TV sets and monitors.

9.8 The Reasons for Alternating Current

Direct current was used in the first large-scale power-distribution systems. These DC systems were developed and built by Thomas A. Edison. However, it was George Westinghouse who demonstrated the advantages of AC over DC for commercial power distribution. The long-time dispute was finally won by the advantages of the Westinghouse AC system. Hence, the majority of the electric power used today is AC.

There are two main reasons why AC is preferred over DC in the power distribution system. First, AC can be produced easier and more efficiently than DC. Second, AC can be distributed over long distances far more economically than DC. This means that less power is wasted when transmitting power from the power station to your home.

AC can be distributed over long distances more economically than DC.

In Figure 9-17, we see a simplified power transmission system. Power is generated at the power station and transmitted long distances using wires to the end user. In an ideal system, all of the power generated would get to the user. However, there are no ideal systems. Practically speaking, some of the power generated is lost in the form of heat (I^2R) due to the resistance of the wires. Thus the power (watts) supplied to the end user equals the power generated minus the power lost. Mathematically,

$$P_{\text{supplied}} = P_{\text{generated}} - P_{\text{lost}}$$

Equation 9-9

Furthermore, the **power lost** as heat is a function of the current in the wire and the resistance of the wire. Recall that mathematically,

$$P_{\text{lost}} = I^2R \text{ (watts)}$$

where I = effective (RMS) value for AC

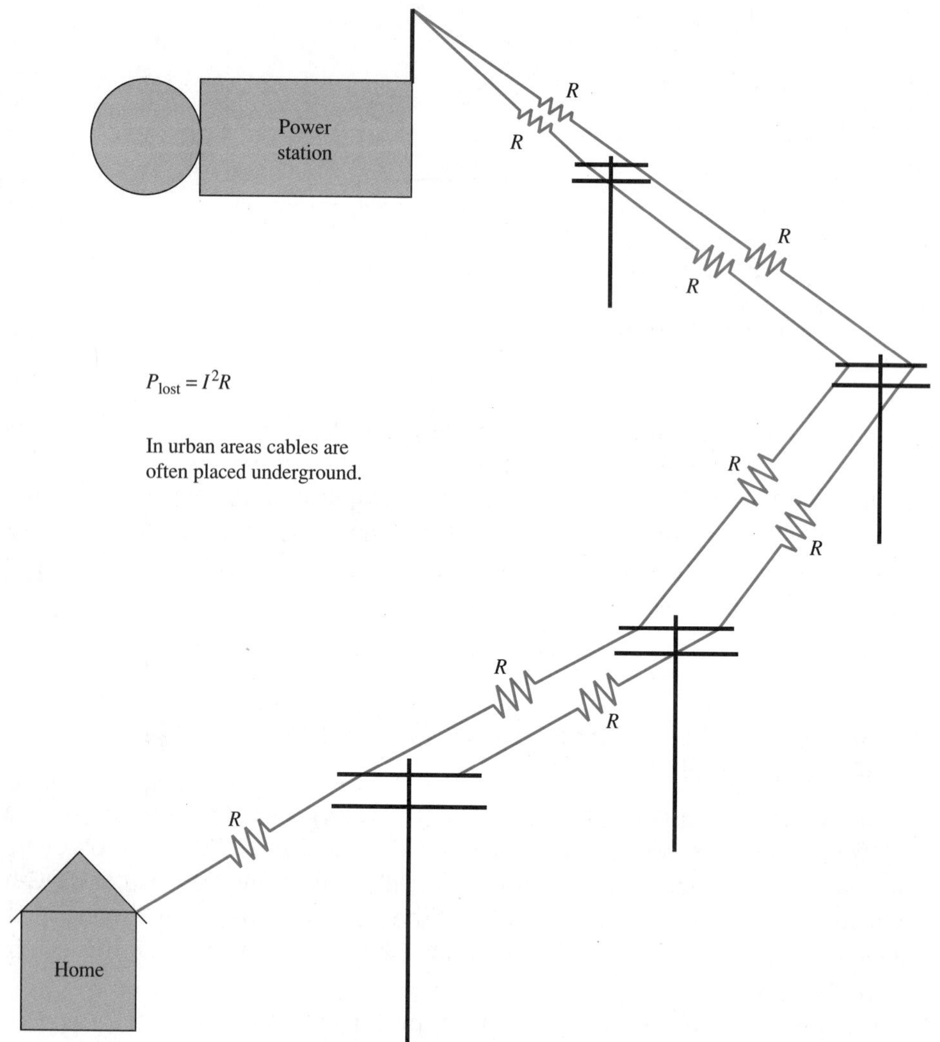

Figure 9-17 **Power distribution**

Transformers are devices that convert voltage and current levels.

AC has the advantage of being able to change voltage and current values while keeping the power constant through the use of a device called the transformer. **Transformers** are devices that very efficiently convert voltage and current levels and are discussed in Chapter 11. When sending power long distances, the voltage is made as high as possible in order to minimize the current for a fixed amount of power. This minimizes the power losses because for a fixed amount of power the high voltage must result in a low current, as seen by the fundamental power equation

$$P = IV$$

For example, if we transmit 1,000 watts of power at 100 volts, the current must be 10 amperes. If we transmit 1,000 watts of power at 1,000 volts, the current is only 1 ampere.

Generating power at high voltage has three disadvantages. First, it is extremely dangerous. Second, the high voltage must be transformed back to a lower voltage before it actually reaches the end user. Finally, high voltage emits radiation.

It would not be practical to build electric power stations near every home and factory. Furthermore, power generating stations are usually located near natural sources of energy like large rivers. The electric generating stations then must send their electric power over long distances to the end users. Large power losses would result with DC because it cannot be easily and efficiently converted to a high or low voltage.

✓ Self-Test 4

1. Phase angles can be measured in degrees or ______________.
2. Name four different types of waveforms.
3. State two advantages of AC over DC.
4. The power supplied to the end user equals ______________.

T³ RESIDENTIAL WIRING

T^3 tech tips & troubleshooting

The electric voltage supplied to your home is an AC voltage. In the United States, the most common form is nominally 120 volts RMS, 60 hertz, AC. This voltage is used to light lights and operate appliances. The power outlet or **duplex receptacle,** shown in Figure 9-18, is one of the most common means of connecting an appliance to AC voltages.

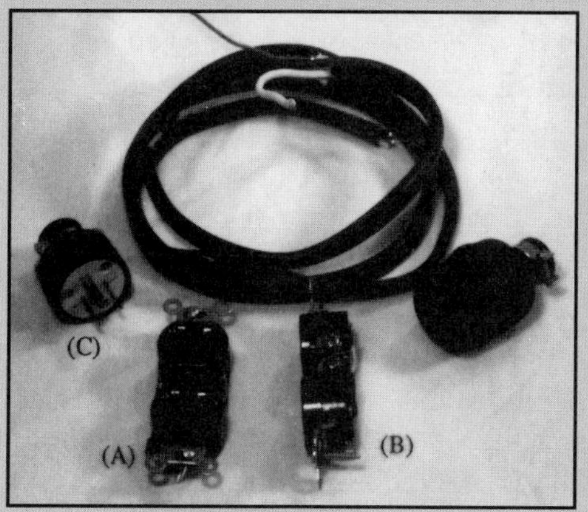

Figure 9-18 **Edison-style duplex receptacle and plug**

Note that in Figure 9-18A, the standard receptacle has two separate outlets. Hence, the name duplex. Each outlet is a three-connection device. Looking closely at the receptacle, you will notice that one opening is long, one is short, and one is round. The long, rectangular connection point or **terminal** is called the neutral or reference terminal. The short rectangular terminal is referred to as the hot terminal. This is the active terminal that supplies both the positive and negative portions of the sine wave relative to the neutral reference terminal. The hot terminal is the *dangerous* one. It can give you a harmful electric shock. The third round terminal is the ground or safety terminal. The ground terminal is always somehow physically connected to earth ground, which is truly 0 volts. Today this is accomplished by connecting a wire to a copper rod that is driven deep into the earth. Figure 9-18B shows a side view of the duplex receptacle with its connection screws. Figure 9-18C shows the mating plug of the receptacle.

! SAFETY TIP: The short rectangular terminal (hot) is the dangerous one.

Remember black to brass.

Household power cable comes in a variety of sizes and styles, as shown in Figure 9-19. The two most popular styles are known as BX and ROMEX. BX cable comes with a metal shielded jacket and ROMEX cable has a plastic outer jacket. In either style there are generally three conductors or wires. By convention, the black wire is used as the hot wire, the white wire is used as the neutral, and the bare copper wire is the ground wire.

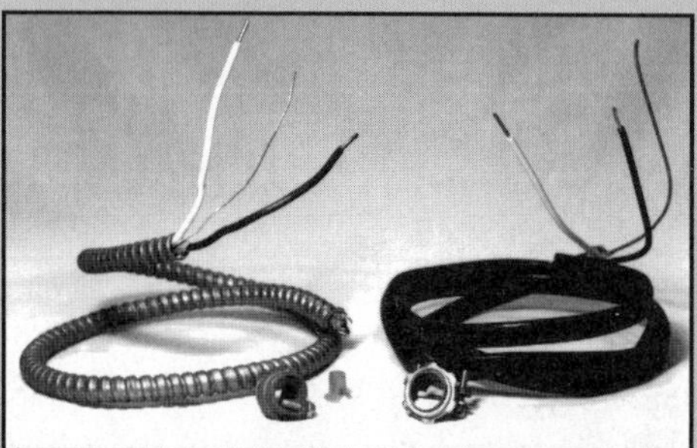

Figure 9-19 **BX and ROMEX housewire**

When connecting the wires to a duplex receptacle, always wrap the wire clockwise around the screw, as shown in Figure 9-20. In this way, when you tighten the screw you will not tend to twist the wire off. Remember that the clockwise direction generally tightens screws while the counterclockwise direction generally loosens screws.

Remember that the black wire is always connected to the brass-colored screw on the duplex receptacle. In this way, the black or hot wire will be connected to the shorter or hot rectangular terminal on the receptacle. The white wire is always connected to the nickel or silver-colored screw. Thus it will be connected to the long rectangular terminal. Finally the bare wire is always connected to the green or ground screw. It connects to the round terminal on the receptacle.

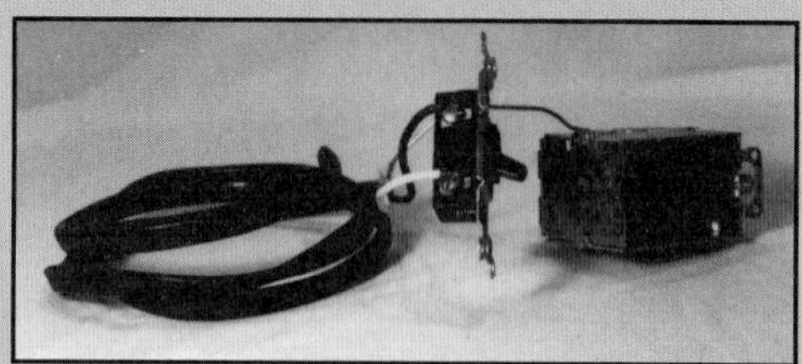

Figure 9-20 **Receptacle being wired**

SUMMARY

This chapter introduced the fundamental definitions and concepts associated with AC sinusoidal waveforms. The amplitude of a sine wave refers to its maximum or peak value. The peak-to-peak value is the peak amplitude value in the positive direction plus the absolute value of the peak amplitude value in the negative direction. The point at which a waveform repeats itself is called a cycle. The time required to complete one full cycle is called the period. The number of cycles in one second is known as the frequency.

The average value of a waveform is equal to the algebraic sum of the positive and negative areas under the waveform divided by the period. The effective or RMS value is the AC voltage or current value that will perform the same rate of work as an equivalent amount of direct current.

The phase relationship is a measurement of the difference in time between two AC waveforms. It is usually measured in degrees and referred to as the phase angle.

FORMULAS

$$f = \frac{1}{T} \qquad T = \frac{1}{f}$$

$$V_{p-p} = 2\ V_p \qquad V_p = \frac{V_{p-p}}{2}$$

$$V_{avg} = 0.637\ V_p$$ (for sine waves only)

$$I_{avg} = 0.637\ I_p$$ (for sine waves only)

$$V_{eff} = 0.707\ V_p$$ (for sine waves only)

$$I_{eff} = 0.707\ I_p$$ (for sine waves only)

$$V_P = 1.414\ V_{eff}$$ (for sine waves only)

$$I_P = 1.414\ I_{eff}$$ (for sine waves only)

$$\text{average} = \frac{\Sigma\ \text{area}}{\text{period}}$$ (for any waveform)

$$2\pi\ \text{rad} = 360°$$

$$\pi\ \text{rad} = 180°$$

$$1\ \text{rad} = 57.3°$$

$$\text{radians} = (\text{degrees})\left(\frac{\pi}{180}\right)$$

$$\text{degrees} = (\text{radians})\left(\frac{180}{\pi}\right)$$

EXERCISES

Section 9.3

1. For the three waveforms of Figure 9-21, determine the amplitude and peak-to-peak value for each waveform.
2. Determine the period of each of the three waveforms in Figure 9-21.
3. Determine the frequency of each of the three waveforms in Figure 9-21.

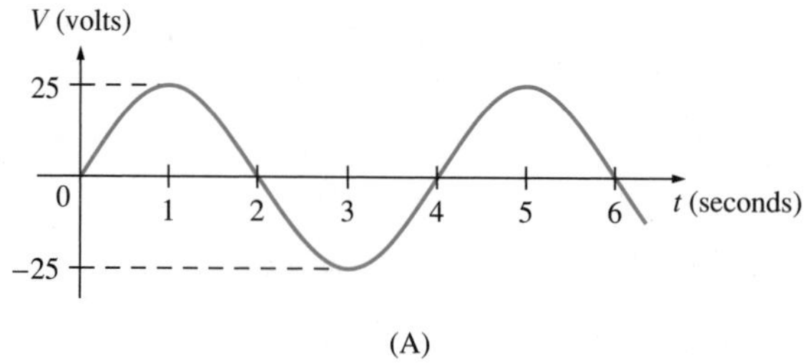

(A)

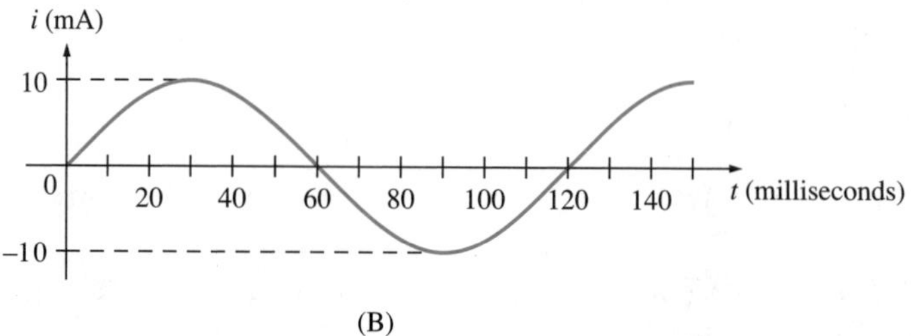

(B)

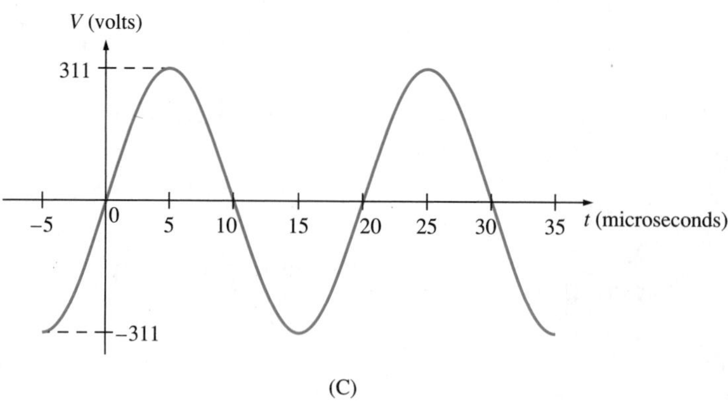

(C)

Figure 9-21

Section 9.4

4. Find the average value of the sine wave pulse in Figure 9-22A.
5. The average value of a sine wave over a half cycle is 100 volts. Find the peak value and the peak-to-peak value of the voltage waveform.
6. The average value of a sine wave over a half cycle is 250 milliamperes. Find the peak value and the peak-to-peak value of the current waveform.
7. Find the average value over a half cycle of a 100-volt peak-to-peak sine wave.
8. Find the average value over a half cycle of the waveform in Figure 9-22B.
9. Repeat problem 8 for the waveform of Figure 9-22C.
10. Repeat problem 8 for the waveform of Figure 9-22D.

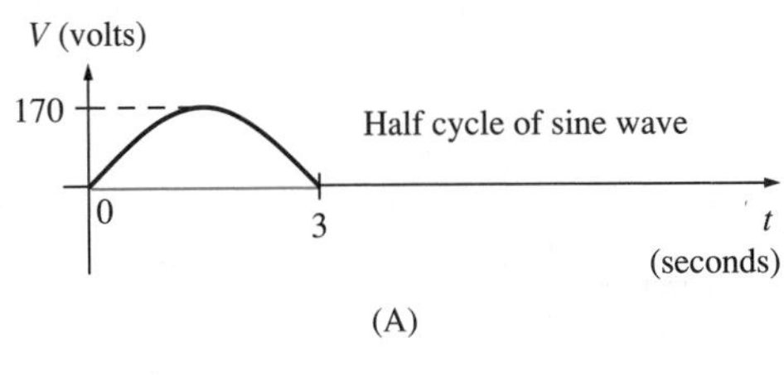

(A)

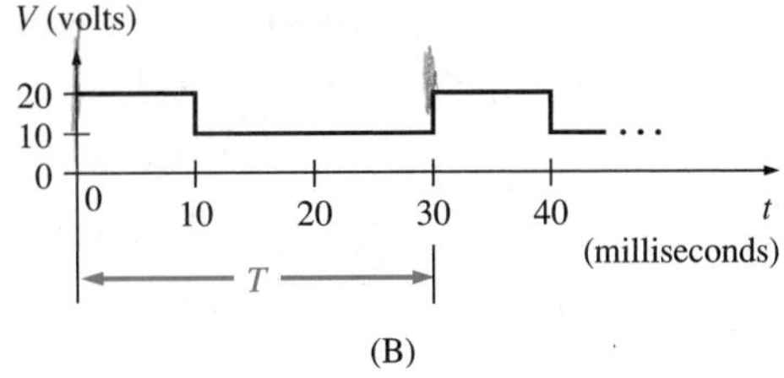

(B)

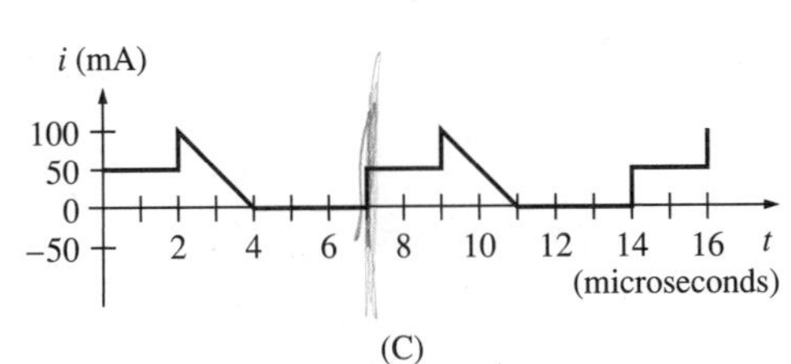

(C)

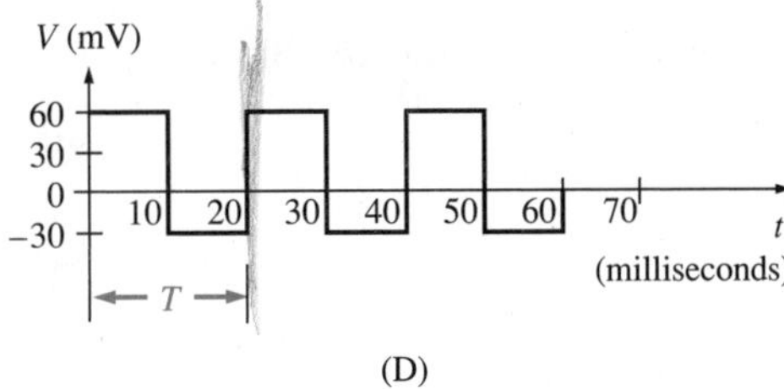

(D)

Figure 9-22

Section 9.5

11. An AC sinusoidal voltage of 208 volts RMS is applied to a circuit. Find the peak-to-peak voltage of the waveform.
12. The peak-to-peak voltage of a sinusoidal waveform is 60 volts. Determine the RMS value of the voltage.
13. The peak-to-peak current of a sinusoidal waveform is 280 milliamperes. Determine the RMS value of the current.
14. Determine the RMS value of the sinusoidal waveform of Figure 9-21A.
15. Repeat problem 14 for the waveform of Figure 9-21B.
16. Repeat problem 14 for the waveform of Figure 9-21C.

Section 9.6

17. In Figure 9-23A, what is the phase relationship between V_A and V_B? Between V_A and V_C? Between V_B and V_C?

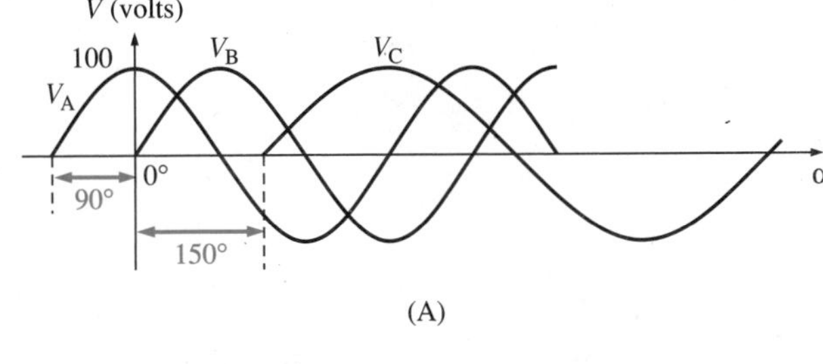

(A)

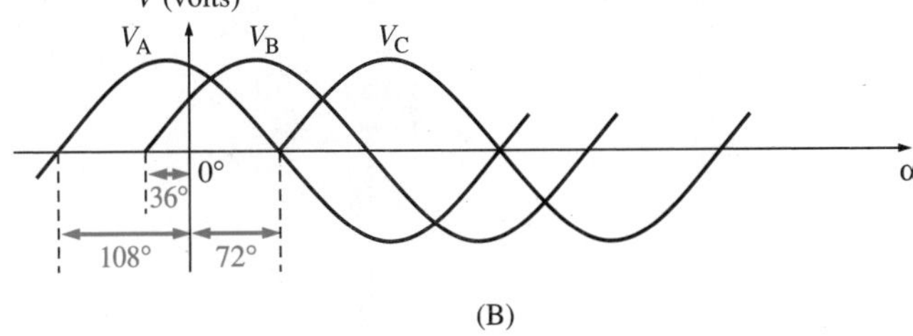

(B)

Figure 9-23

18. In Figure 9-23B, what is the phase relationship between V_A and V_B? Between V_A and V_C? Between V_B and V_C?

19. Convert the following degrees to radians:

    (A) 45° (B) 80° (C) 135°

20. Convert the following radian angles to degrees:

    (A) 1.3 rad (B) 2.1 rad (C) –0.75 rad

21. See Lab Manual for the crossword puzzle for Chapter 9.

CAREER FOCUS
On Electronics Technology

As you continue your studies of electronics technology, you will discover the applications for your knowledge are as limitless as the careers that use those applications. Certain careers might require you to mold your education in certain ways. The following pages will introduce you to some of the opportunities and educational requirements in the ever-expanding field of electronics technology.

TECHNICIANS

Electrical and electronics technicians usually hold a two-year associate's degree or a specialized technical school diploma. They can specialize their training to work in areas such as electronics technology, computer technology, power technology, and instrumentation technology. Technicians frequently work with engineers to solve technical problems.

(Photo courtesy of DeVry Institutes)

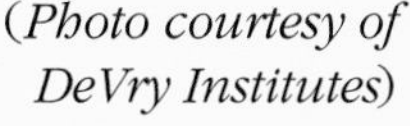

(Photo courtesy of DeVry Institutes)

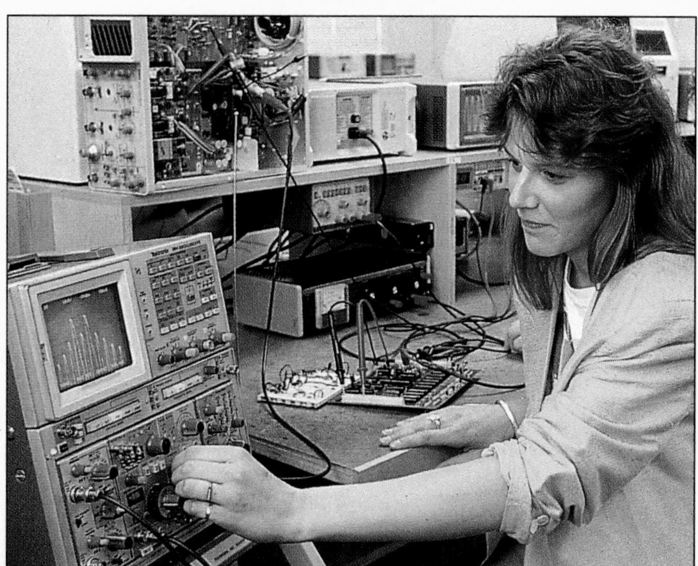

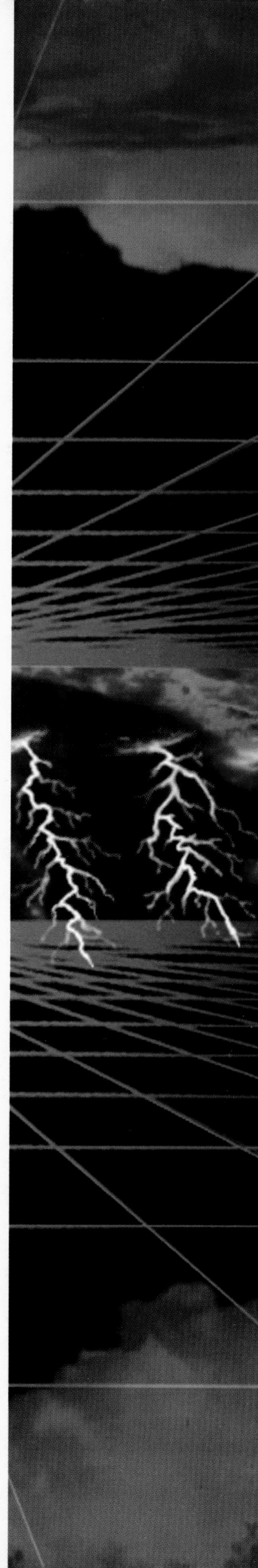

TECHNICIANS

In the electronics field, technicians are usually employed as:

- Electrical technicians
- Electronics technicians
- Electrical estimators
- Engineering technicians
- Computer maintenance technicians
- Biomedical electronics technicians
- Service representatives
- Technical representatives

Technicians working in a laser lab. (*Photo courtesy of U.S. Sprint*)

Technical estimators working at a job site. (*Photo courtesy of the Cleveland Institute of Electronics*)

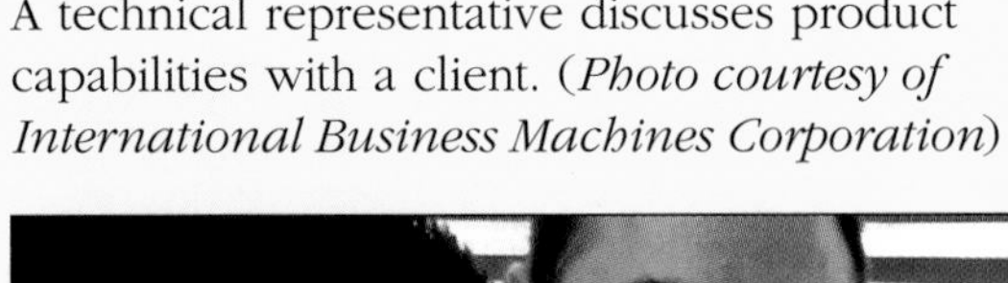

A technical representative discusses product capabilities with a client. (*Photo courtesy of International Business Machines Corporation*)

Field Service Technicians

Field service technicians install, test, adjust, service, update, and program electric and electronic equipment in commercial, industrial, and hospital facilities, as well as on ships and aircraft. Field service requires considerable practical technical experience. Two principal tasks are to solve equipment problems and to supplement customer training. Field service technicians advise customers on operating and maintenance procedures. They serve as a direct link between the company and the customer.

A field service technician locating cabling problems in a local area network (LAN). (*Photo courtesy of Hewlett-Packard Company*)

Working with a portable transceiver test system. (*Photo courtesy of Hewlett-Packard Company*)

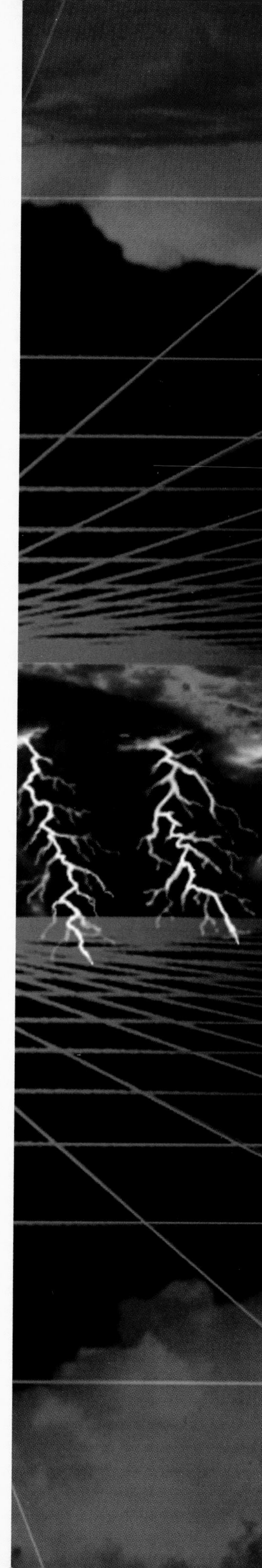

Service Representatives

Service representatives communicate between manufacturers and their customers. They may provide technical assistance for servicing equipment such as computers, X-ray equipment, and automobiles. Representatives may also schedule parts deliveries or field service visits.

Consumer products service technician. (*Photo courtesy of the Cleveland Institute of Electronics*)

TECHNICIANS

Technical Sales Representatives

Technical sales representatives assess users' needs and sell manufacturers' electric and electronic parts and equipment.

(Photo courtesy of Hewlett-Packard Company)

Technical Writers

Complicated, contemporary products need extensive and precise instructions for operation and maintenance. Technical writers develop manuals that include estimates of coverage, cost, contents, and delivery schedules. Technical illustrators prepare schematics, wiring drawings, photographs, and performance charts by working closely with engineers. This work includes preparing proposals and scientific papers.

(Photo courtesy of International Business Machines Corporation)

Production Technicians

Production technicians help construct, connect, and supervise the assembly and fabrication of electric and electronic equipment. The production technician translates product and design engineering information into production of a final product. Product technicians develop the manufacturing processes for fabricating and testing all of the detail and assembled parts.

Assembling automotive radios with electronic automation. (*Photo courtesy of Hewlett-Packard Company*)

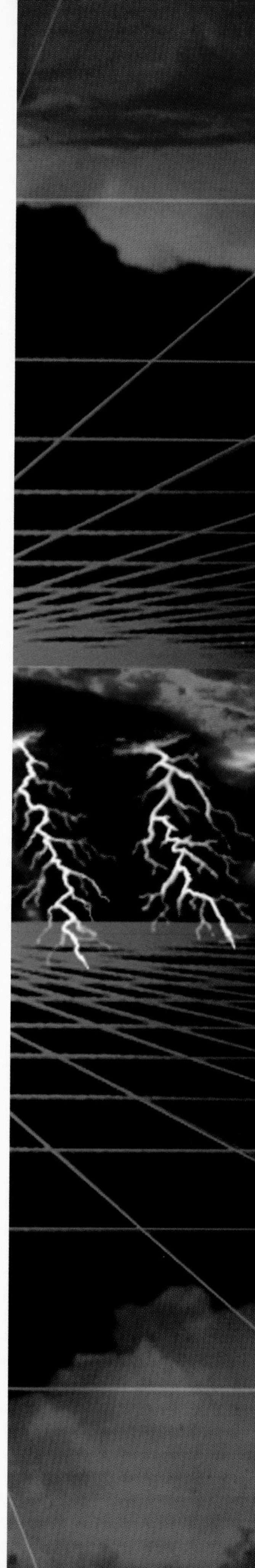

Industrial Technicians

Industrial technicians plan for effective use of people, manufacturing equipment, and floor space. These workers use analytical techniques and their knowledge of manufacturing operations to perform studies and recommend changes that will improve efficiency and reduce costs. Their work involves scheduling and creating status reports, performing work measurement studies, planning plant layout, equipment use, and materials handling, assisting in equipment and facilities acquisitions, reviewing and monitoring fixed assets and budgets, and developing manufacturing standards of operation.

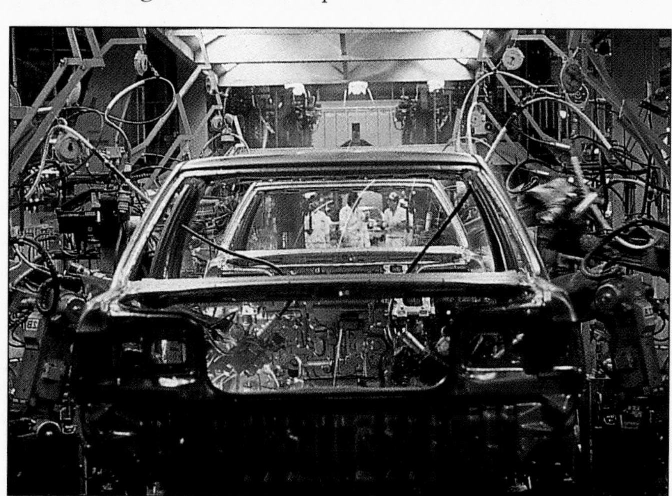

Many functions on an automotive assembly line are automated by electronics. Technicians monitor these processes. (*Photo courtesy of Honda of America Manufacturing, East Liberty, Ohio auto plant*)

TECHNICIANS

Electrical Estimators and Inspectors

Electrical estimators and inspectors work for the construction industry and for government agencies. They estimate the cost, time, and equipment needed for construction projects and for the installation of electric equipment and systems. Inspectors working for the government ensure that electrical work conforms to local codes.

An electrical estimator analyzes a construction site. (*Photo courtesy of The British Petroleum Corporation*)

Facilities Managers

Facilities management personnel evaluate, select, and design buildings, grounds, utilities, equipment, and all other installations needed to run a workplace. A close work effort between facilities engineering and facilities maintenance helps ensure that the workplace does not become obsolete. Facilities engineers help assemble construction work packages from minor rearrangements to major new construction.

(*Photo by Kim Collavo. Copyright New York State Electric and Gas*)

Biomedical Technicians

Biomedical technicians maintain, calibrate, repair, and interface electric, electronic, and computer equipment used in medical labs and hospitals. Biomedical technicians are important members of health and research teams. They apply the principles of engineering and physics to solving problems in the fields of medicine and biology.

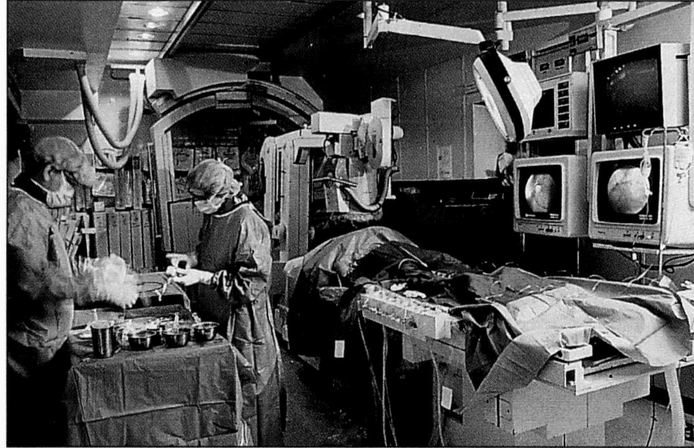

Medical diagnosis has benefited tremendously from electronics spin-off technologies. (*Photo courtesy of NASA*)

A technician calibrating a magnetic resonance imaging (MRI) system. MRI uses a superconducting magnet and radio waves to produce a computerized cross sectional (slice) of the human body. (*Photo courtesy of Siemens Medical System*)

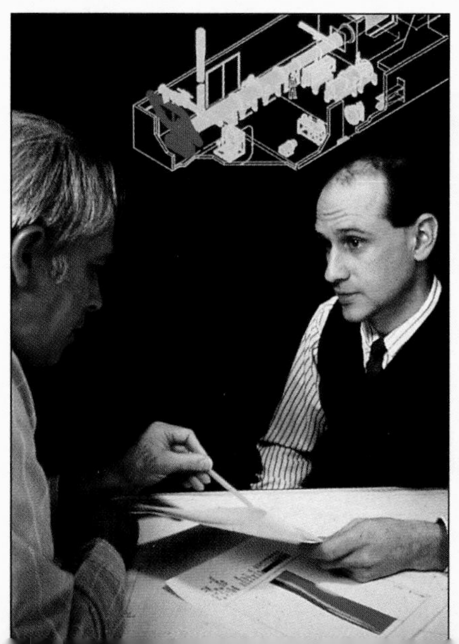

Environmental Technicians

Environmental technicians are essential to the design and testing of a new product. Equipment and people must work in a narrow band of efficiency, within the limits of the operating environment. Human factors, both physical and psychological, must be considered in controlling air conditioning, oxygen, pressurization, electronic cooling, temperature control, defogging, waste water, water purification, water management, solid waste processing, the atmosphere, the earth's ecology, and marine problems.

(*Photo courtesy of NASA*)

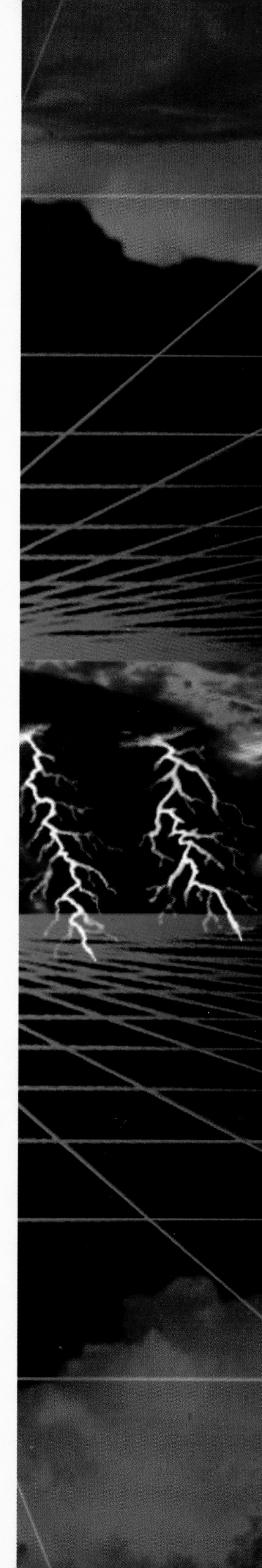

TECHNICIANS

Engineering Technicians

Engineering technicians work with engineers in research and development to prototype and test new products and systems. The scope of activity touches all aspects of the electrical and electronics disciplines throughout the industry. Engineering technicians work with design engineers and scientists to conceive electronics systems, direct the development and manufacturing of systems, and run the tests required to ensure product performance.

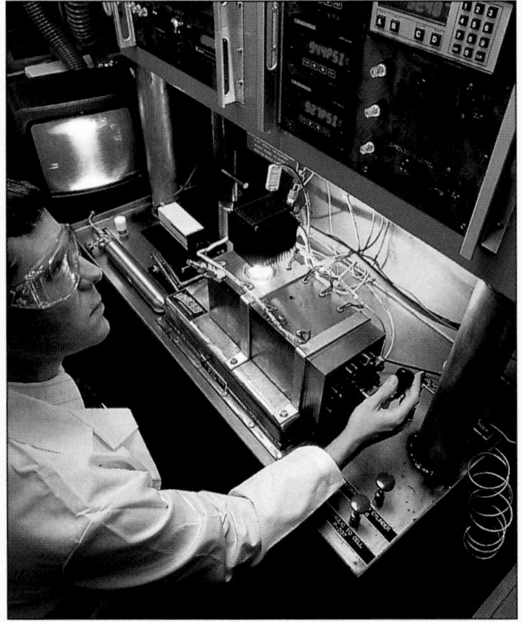

Engineering technician at work in a research lab. (*Photo copyright 1991 by Steve Kahn*)

Reliability, Maintainability, and Quality Control Technicians

Reliability goes hand in hand with maintainability; they work together in every phase of the development cycle. Technicians analyze large amounts of data gathered from operational reports and tests. Quality control applies reliability and maintainability to several distinct areas, including administration, auditing, budgets, training, inspection, procedures, and testing.

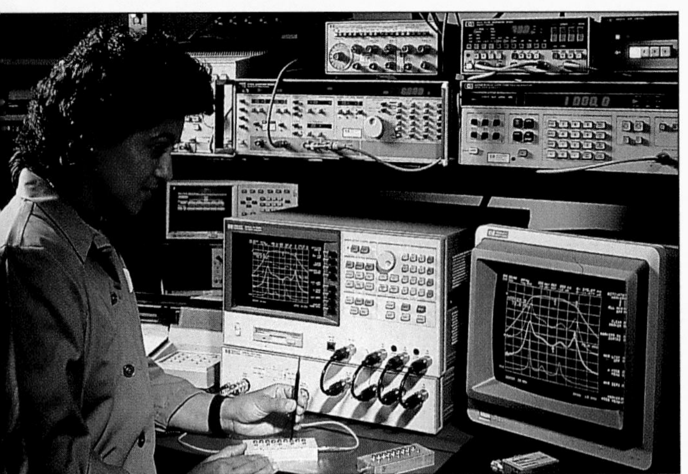

A specially designed network analyzer shortens test time and improves product accuracy. (*Photo courtesy of Hewlett-Packard Company*)

(*Photo by David Perry. Courtesy of Rockwell International*)

Instrumentation and Control Technicians

Instrumentation and control technicians work in industrial production plants and electric power company facilities. They install, maintain, upgrade, and interpret electronic instruments that monitor process controls such as aircraft engine performance, power plant output, and the production of photo film.

Power plant control room. (*Photo by Kirk Van Zandbergen. Courtesy of New York State Electric and Gas*)

Coating photographic paper with emulsion. (*Reprinted courtesy of Eastman Kodak Company. Copyright 1993 by Eastman Kodak Company*)

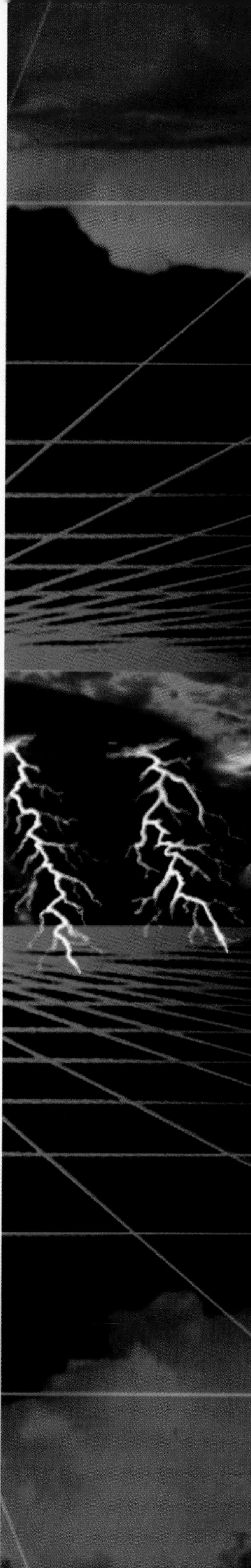

ENGINEERING TECHNOLOGISTS

Engineering technology emphasizes the understanding and application of established methods and materials of technology. An engineering technologist usually earns a four-year bachelor's degree in engineering technology. The first two years are often an associate's degree in a technical specialty. The technologist's training and job goals emphasize solving practical design problems and evaluating industrial problems.

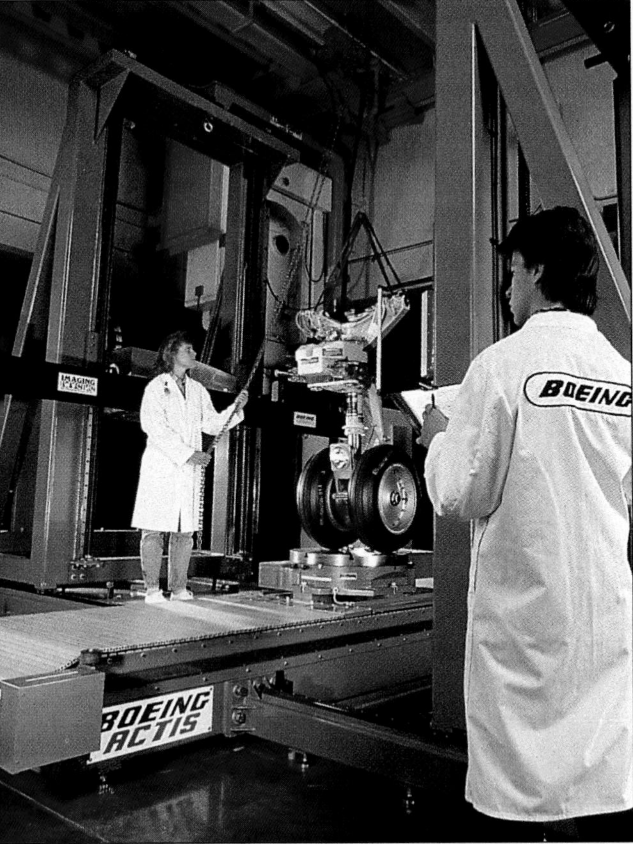

(*Photo courtesy of NASA*)

(*Photo courtesy of Texas Instruments*)

Engineering technologists often communicate between engineering and production, or sales and the customer. Typical job titles include:

- ➤ Electrical technologist
- ➤ Electronics technologist
- ➤ Sales engineer
- ➤ Service engineer
- ➤ Test engineer
- ➤ Field engineer
- ➤ Electrical buyer
- ➤ Customer engineer
- ➤ Research aide

Test engineers examine performance standards. (*Photo courtesy of Hewlett-Packard Company*)

A technician examines a "phantom" used to test an MRI system. (*Photo courtesy of Siemens Medical Systems, Inc.*)

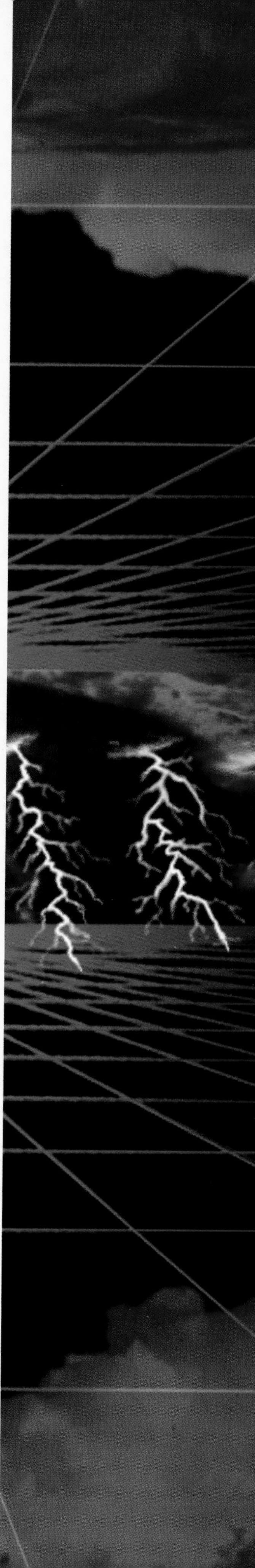

ENGINEERS

Electrical and electronics engineers perform a broad spectrum of activities in many fields and industries. An engineer's training emphasizes analytical skills and an understanding of physics. They utilize math and science skills to develop design principles that can be applied to a wide variety of problems. Typical engineering training begins with a four-year bachelor's degree in engineering and continues throughout the engineer's work experience. Many engineers complete some graduate study in engineering, computer science, economics, management, business administration, law, or medicine.

(*Photo courtesy of Rensselaer Polytechnic Institute*)

While engineering jobs stress analysis and design, many engineers work as managers overseeing other engineers and technical personnel. Engineering is a practical activity, and engineers may be found in any organization that must solve practical problems. Typical job titles include:

- Electrical engineer
- Electronics engineer
- Computer engineer
- Design engineer
- Project engineer
- Quality control engineer
- Software engineer
- Development engineer
- Research engineer
- Systems engineer
- Sales engineer
- Field engineer
- Test engineer
- Engineering manager

Inspecting electrical power systems. (*Photo by Kirk Van Zandbergen. Courtesy of New York State Electric and Gas*)

Working with a portable signal analyzer. (*Photo courtesy of Hewlett-Packard Company*)

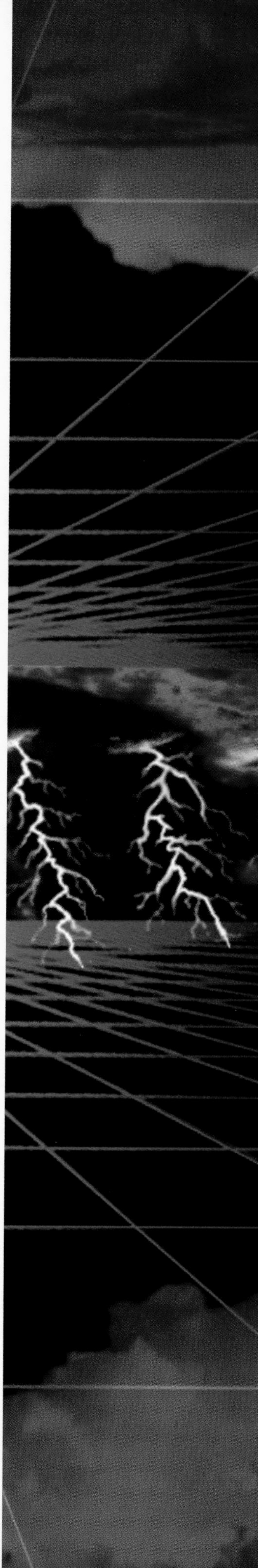

ENGINEERS

Some areas of employment in engineering include:

- *Engineering design and development.* Designs the form, composition, and operating characteristics of electronic products.
- *Production and Manufacturing.* Produces safe, economical, and high-performance products and services.
- *Field Service and Training.* Directs the set-up, operation, training, and maintenance of advanced technical products and services at the user's location.
- *Systems Analysis and Design.* Analyzes and designs interconnections of components or processes to meet required performance criteria.

Computer-aided design. (*Photo courtesy of Motorola Government Electronics Group*)

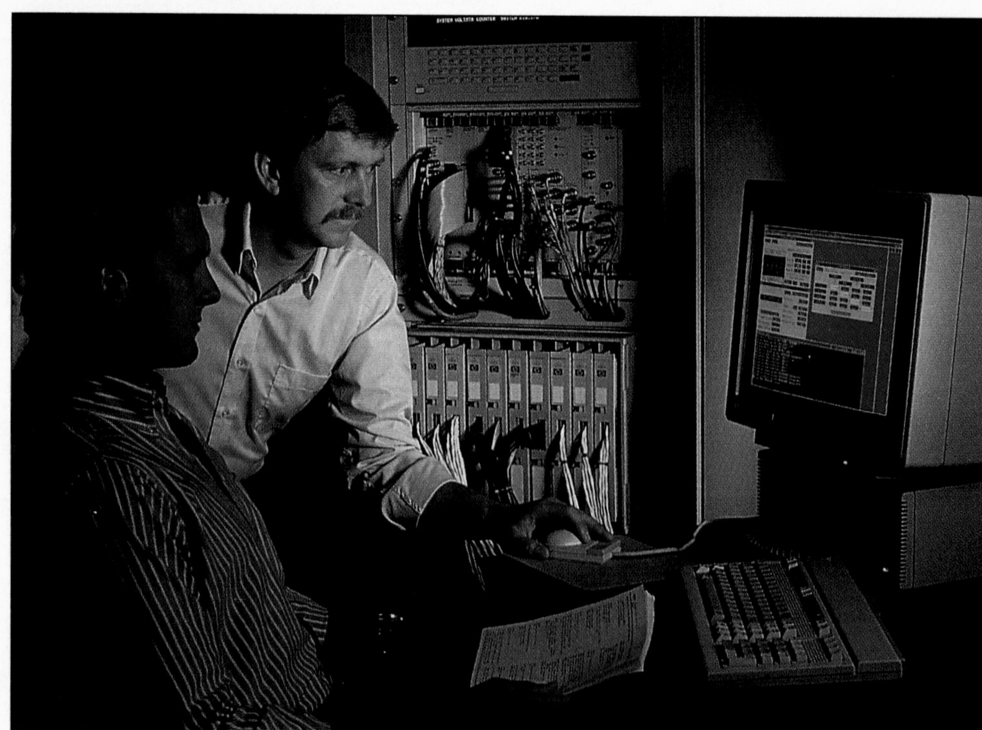

(*Photo courtesy of Hewlett-Packard Company*)

CHAPTER 10

CAPACITORS

Courtesy of Hewlett Packard

KEY TERMS

capacitance
capacitive reactance
capacitor
charged
dielectric
diodes
dynamic test
electrostatic field
farad
filter capacitor
plates
polarized
rectifier
static test
steady state
tau (τ)
time constant
working voltage

OBJECTIVES

After completing this chapter, the student should be able to:

- define capacitance.
- define dielectric.
- explain how charge is stored in a capacitor.
- calculate capacitance.
- calculate the charge of a capacitor.
- determine the total capacitance for series and parallel capacitors.
- determine the time constant for a RC circuit.
- calculate the steady state voltage across a capacitor.
- define and calculate capacitive reactance.
- describe the effects of a filter capacitor on pulsating DC.

10.1 Introduction

Capacitors store energy in an electrostatic field.

Capacitance can be defined as the ability or capacity of a device to store energy. The energy is stored in an electrostatic field in a device called a **capacitor.** The energy stored in the electrostatic field can be thought of as similar to the energy stored in a rubber band. When the rubber band is stretched, energy is stored in the elastic properties of the rubber band. When one end of the rubber band is released, the energy is returned to the source.

Capacitors are made of two plates separated by a dielectric (insulator).

A capacitor is made of two conductors separated by an insulator, as illustrated in Figure 10-1A. The conductors are called the **plates** and the insulator is called the **dielectric.** Figure 10-1B shows the schematic symbol for a capacitor.

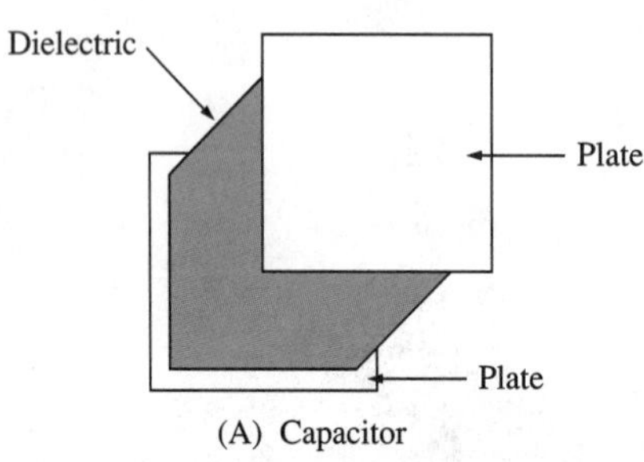

(A) Capacitor

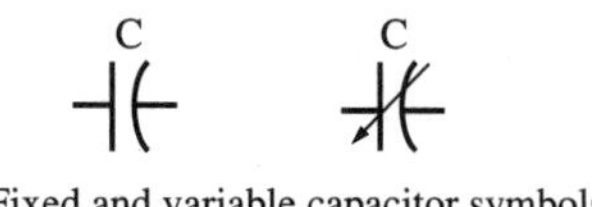

(B) Fixed and variable capacitor symbols

Figure 10-1 ***Capacitor***

The amount of energy stored in a capacitor is directly related to the size of the capacitor. The larger the capacitance (C) of the capacitor, the greater the amount of energy stored. Capacitors store energy, so even when they are removed from a circuit, care must be exercised when handling or removing capacitors from a circuit.

! SAFETY TIP: Exercise care when handling capacitors.

10.2 Capacitance

Charged capacitors store energy.

The circuit of Figure 10-2A is essentially an open circuit because the plates of the capacitor are separated by an insulator. However, when the switch is closed (Figure 10-2B), electrons from the plate connected to the positive terminal of the battery are attracted toward the positive terminal and begin to flow into the battery. Similarly, electrons from the negative terminal of the battery begin to flow into the plate connected to the negative terminal of the battery. Eventually the plates can no longer accept or give up electrons and all current flow stops. At this point we say the capacitor is fully **charged.** When the switch is opened (Figure 10-2C), the electrons in the negative plate are attracted by the positive charge (deficiency of electrons) of the positive plate but no current exists because of the insulator separating the plates. Even though the electrons cannot flow to the positive plate, a force of attraction still exists. This force is called an electrostatic force or an **electrostatic field.** The electrostatic force causes the negatively charged electrons to be attracted toward the positive plate and the more positively charged nucleus to be attracted to the negative plate. This tends to stress or distort the insulator and stores the energy similar to the

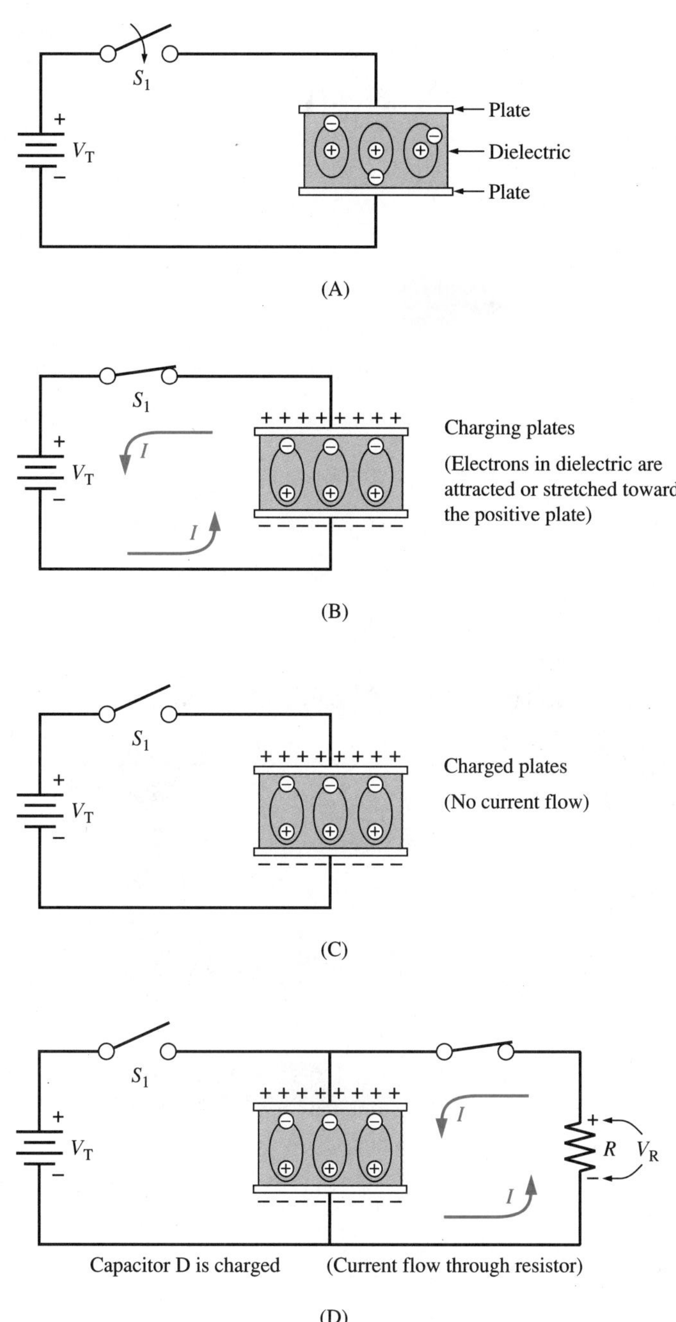

Figure 10-2 ***Capacitor charging/discharging***

way a rubber band stores energy when it is stretched. The capacitor can return the energy (Figure 10-2D) in the same way that the rubber band returns the energy when one end of the rubber band is released.

One plate is negative due to the excess of electrons and the other plate is positive due to the deficiency of electrons, thus a potential difference or voltage exists between the plates. If the capacitor is charged to a voltage that is so great that the attraction between the positive plate and the negative plate pulls the electrons through the insulator, *dielectric breakdown* occurs. This is where the resistance of the insulator is not great enough to stop charge flow (current) be-

tech tidbit

Capacitors have a maximum working voltage (VDC).

tween the plates. Thus all capacitors will have a maximum **working voltage,** which is usually stated in volts DC (VDC).

The unit of capacitance is the **farad,** however, most capacitors are rated in microfarads or picofarads. Capacitance is a measure of the capacitor's ability to store charge on its plates. The larger the area of the plates, the larger the capacitance. The greater the distance between the plates, the smaller the capacitance. The stronger the dielectric material, the greater the capacitance. Mathematically,

Equation 10-1 ▶

$$C = \frac{kA}{d}$$

where k = the dielectric constant in farads/meter

A = the surface area of one plate in square meters

d = the distance between the plates in meters

C = the capacitance in farads (F)

tech tidbit

The unit of capacitance is the farad (F).

Table 10-1 shows the dielectric constants for a number of different types of insulating materials used in capacitors.

Table 10-1 Dielectric Constants

Material	Dielectric constant (F/m)
air	8.855×10^{-12}
teflon	17.700×10^{-12}
mica	44.250×10^{-12}
glass	66.375×10^{-12}
ceramic	66375×10^{-12}

EXAMPLE 10-1

Find the capacitance of a capacitor whose plate area, A, is 1 millimeter^2 and whose plates are separated by a distance of 2 millimeters if the dielectric material is air.

Solution:

$$C = \frac{kA}{d}$$

$$= \frac{(8.855 \times 10^{-12}\ \text{F/m})(1 \times 10^{-3}\ \text{m}^2)}{}$$

$$= 4.4275 \times 10^{-12}\ \text{F}$$

$$\approx 4.4\ \text{pF}$$

EXAMPLE 10-2

If the area of a 10-microfarad capacitor is increased by four times and the distance between the plates is doubled, find the new capacitance.

Solution:

Original capacitor

$$C_0 = \frac{kA}{d}$$

$$10\ \mu F = \frac{kA}{d}$$

New capacitor

$$C_{new} = \frac{4}{2} \times \frac{kA}{d}$$

$$\text{but } \frac{kA}{d} = 10\ \mu F$$

$$\therefore C_{new} = \frac{4}{2} \times 10\ \mu F$$

$$= 20\ \mu F$$

or because the area increases four times

$$C_{new} = 4 \times 10\ \mu F = 40\ \mu F$$

but because the diameter is doubled

$$C_{new} = \frac{40\ \mu F}{2} = 20\ \mu F$$

A capacitor is said to have a capacitance of 1 farad if it holds 1 coulomb of charge when a voltage of 1 volt is across its plates. Mathematically,

$$C = \frac{Q}{V}$$

Equation 10-2

where C = the capacitance in farads

Q = the charge on the plate in coulombs

V = the voltage across the plates in volts

tech tidbit

Capacitance is directly proportional to the area of the plates.

tech tidbit

Capacitance is inversely proportional to the distance between the plates.

EXAMPLE 10-3

What is the capacitance of a capacitor with a charge of 100 microcoulombs and a voltage of 2 volts?

Solution:

$$C = \frac{Q}{V}$$

$$= \frac{100\ \mu C}{2\ V}$$

$$= 50\ \mu F$$

Capacitors like resistors are available in both fixed and variable types. The symbol for each type was given in Figure 10-1B. The curved line in the symbol indicates the plate that is usually connected to the point of the lowest voltage potential. Figure 10-3 shows a number of popular types of capacitors in use today.

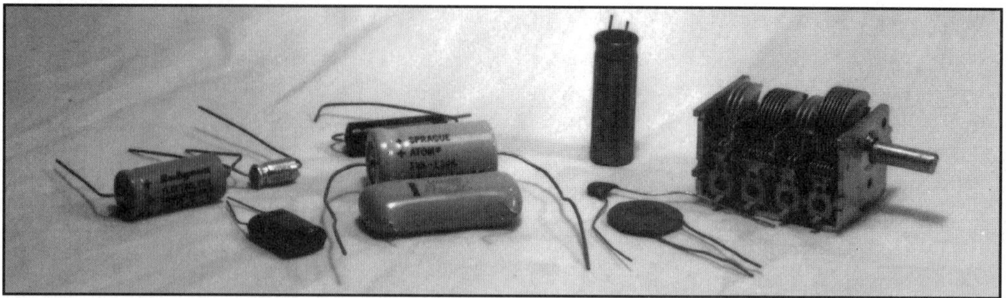

Figure 10-3 Types of capacitors

Mica capacitors have excellent temperature and high voltage characteristics with very small leakage current. Leakage current is current that leaks through the dielectric from one plate to another when the capacitor is charged. Ceramic capacitors have exceptionally high working voltages and a relatively large capacitance for their physical size. Electrolytic capacitors are used in high capacitance applications. They have lower working voltages but have the largest capacitance for their physical size. They are primarily used in DC circuits and are **polarized.** Polarized refers to the fact that the positive plate must be connected to a positive potential and the negative plate must be connected to a negative potential. Tantalum capacitors are also polarized. They are like electrolytic capacitors but are smaller and more temperature stable. Polyester-film or plastic capacitors are one of the most common capacitors used today. They are produced in molded, DIP (Duel In-line Package), or tubular styles. They are small and compact in size. Variable capacitors are often found in TV and radio tuner applications. The plate area is varied by the rotation of the shaft, which results in a change in capacitance.

> **! SAFETY TIP:** When polarized capacitors are installed backward, they can explode.

✓ Self-Test 1

1. The ability of a device to store energy is called ________________.
2. A ________________ is made of two conductors separated by an insulator.
3. The larger the capacitance of a capacitor, the ________________ the amount of energy stored.
4. The greater the distance between the plates of a capacitor, the ________________ the capacitance.

T^2 tech tidbit

When capacitors are connected in series, the total capacitance is always less than the smallest capacitor.

10.3 Capacitors in Series and Parallel

Capacitors can be connected in series and in parallel when used in electric circuits. The general rules are opposite but similar to the rules for resistors in series and in parallel. When capacitors are connected in series, the effect is similar to increasing the distance between the plates. Thus the total capacitance of two or more capacitors connected in series is reduced. Mathematically,

$$\frac{1}{C_T} = \frac{1}{C_1} + \frac{1}{C_2} + \frac{1}{C_3} + \ldots + \frac{1}{C_N}$$

◀ Equation 10-3

When capacitors are connected in parallel, effect is similar to increasing the area of the plates. Thus the total capacitance of two or more capacitors connected in parallel increases. Mathematically,

$$C_T = C_1 + C_2 + C_3 + \ldots + C_N$$

◀ Equation 10-4

Figure 10-4 illustrates the effects of capacitors in series and in parallel and shows the mathematical equations for calculating C_T.

T² tech tidbit

Capacitors in parallel: $C_T = \Sigma C$.

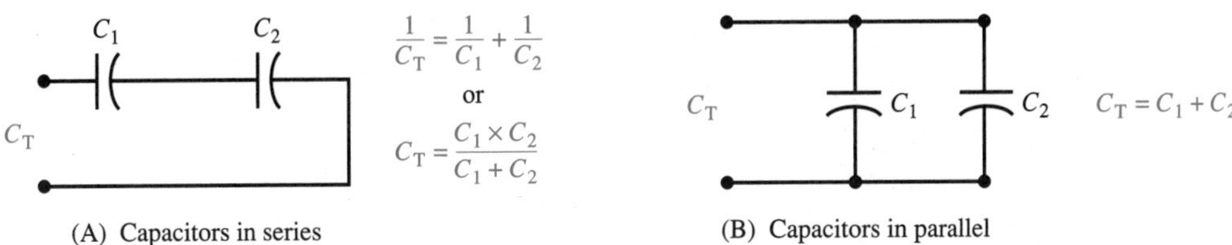

Figure 10-4 **Capacitors—series and parallel**

EXAMPLE 10-4

Find the total capacitance when a 6-farad and a 3-farad capacitor are connected in series. See Figure 10-5.

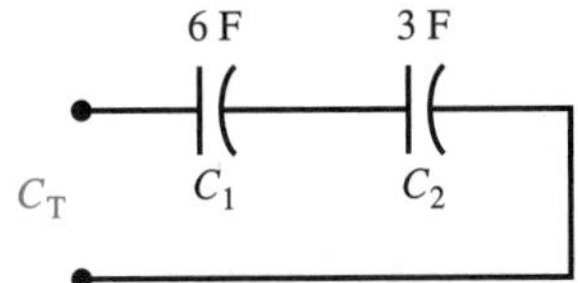

Figure 10-5

Solution:

$$C_T = \frac{C_1 \times C_2}{C_1 + C_2}$$

$$= \frac{(6\text{ F})(3\text{ F})}{(6\text{ F} + 3\text{ F})}$$

$$= \frac{18\text{ F} \cdot \text{F}}{9\text{ F}}$$

$$= 2\text{ F}$$

or

$$\frac{1}{C_T} = \frac{1}{C_1} + \frac{1}{C_2}$$

$$\frac{1}{C_T} = \frac{1}{6\text{ F}} + \frac{1}{3\text{ F}}$$

$$\frac{1}{C_T} = \frac{1}{6\text{ F}} + \frac{2}{6\text{ F}}$$

$$\frac{1}{C_T} = \frac{3}{6\ \text{F}}$$

$$\therefore \quad C_T = \frac{6\ \text{F}}{3}$$

$$= 2\ \text{F}$$

EXAMPLE 10-5

Find the total capacitance when a 6-farad and a 3-farad capacitor are connected in parallel. See Figure 10-6.

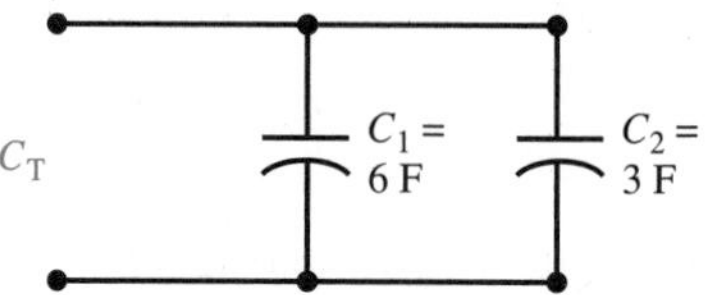

Figure 10-6

Solution:

$$C_T = C_1 + C_2$$

$$= 6\ \text{F} + 3\ \text{F}$$

$$= 9\ \text{F}$$

EXAMPLE 10-6

Calculate the total capacitance in the network of Figure 10-7A.

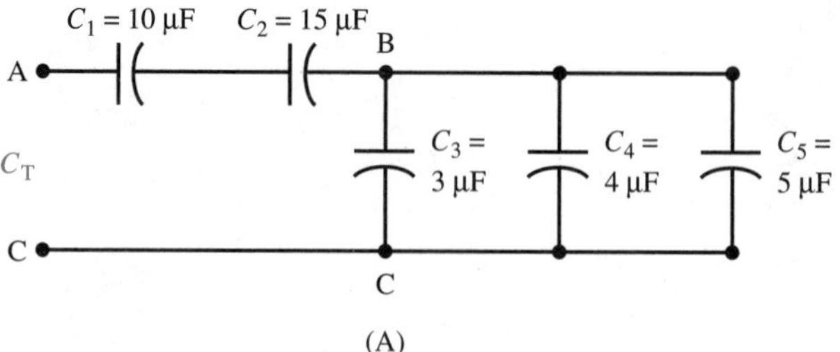

(A)

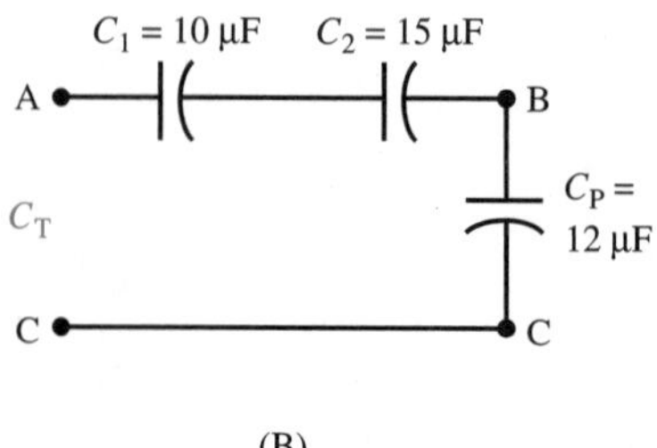

(B)

Figure 10-7

Solution:

C_3, C_4, C_5 are in parallel.

$$\therefore \quad C_P = C_3 + C_4 + C_5$$
$$= 3\ \mu F + 4\ \mu F + 5\ \mu F$$
$$= 12\ \mu F$$

Redraw circuit with C_P included. See Figure 10-7B.

$$\frac{1}{C_T} = \frac{1}{C_1} + \frac{1}{C_2} + \frac{1}{C_P}$$
$$= \frac{1}{10\ \mu F} + \frac{1}{15\ \mu F} + \frac{1}{12\ \mu F}$$
$$= \frac{0.1}{\mu F} + \frac{0.0667}{\mu F} + \frac{0.0833}{\mu F}$$
$$= \frac{0.25}{\mu F}$$
$$C_T = \frac{1\ \mu F}{0.25}$$
$$= 4\ \mu F$$

EXAMPLE 10-7

Find the total capacitance when three 15-microfarad capacitors are connected in series. See Figure 10-8.

For equal value capacitors in series: $C_T = C/N$.

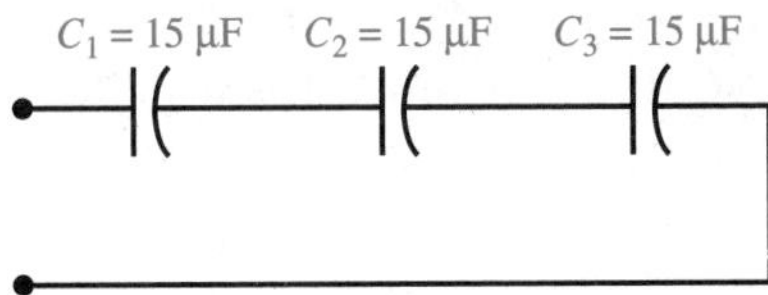

Figure 10-8

Solution:

$$C_T = \frac{C}{N}$$

where N = number of equal value capacitors

$$C_T = \frac{15\ \mu F}{3}$$
$$C_T = 5\ \mu F$$

or

$$\frac{1}{C_T} = \frac{1}{C_1} + \frac{1}{C_2} + \frac{1}{C_3}$$
$$\frac{1}{C_T} = \frac{1}{15\ \mu F} + \frac{1}{15\ \mu F} + \frac{1}{15\ \mu F}$$

$$\frac{1}{C_T} = \frac{3}{15\ \mu F}$$

$$\therefore \quad C_T = \frac{15\ \mu F}{3} = 5\ \mu F$$

10.4 RC Time Constants

When a DC source is connected to an uncharged capacitor, a quantity of charge (current) begins to flow. This surge of current will continue until the charge deposited on the plates of the capacitor produces a voltage that is equal to the DC source. Recall from Equation 10-2 ($V = Q/C$) that the voltage across the capacitor is directly related to the charge deposited on the plates.

The time required for a capacitor to fully charge is a function of the resistance of the circuit and the size of the capacitor. The larger the resistance and the larger the size of the capacitor, the longer it will take to fully charge. The smaller the resistance and the smaller the size of the capacitor, the less time it will take to fully charge. Mathematically,

Equation 10-5 ▶

$$\tau = RC,$$

where R = the resistance of the circuit in ohms

C = the size of the capacitor in farads

τ = the time constant in seconds

tech tidbit

It takes 5 τ for a capacitor to reach full charge.

The Greek letter **tau (τ)** is known as the **time constant** of the circuit. It represents the time for the capacitor to charge to 63.2% of the applied DC voltage. For all practical purposes, we can say that it takes five time constants or 5 τ for a capacitor to reach full charge or be in the steady state condition. Figure 10-9A graphically illustrates the time required for a capacitor to charge. The shape of the curve is called an *exponential curve*. For each time constant the capacitor charges 63.2% of the difference between the applied voltage and the capacitor voltage. Figure 10-9B illustrates the discharge time and Figure 10-9C is a table of τ versus the corresponding percentage of full charge.

tech tidbit

The timing properties of a capacitor are used in many timing circuits.

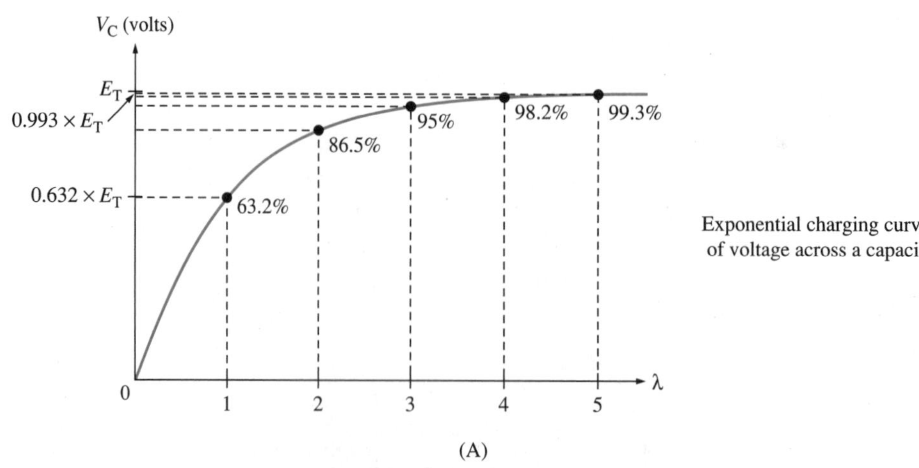

Figure 10-9 **RC time constants** (continued on page 223)

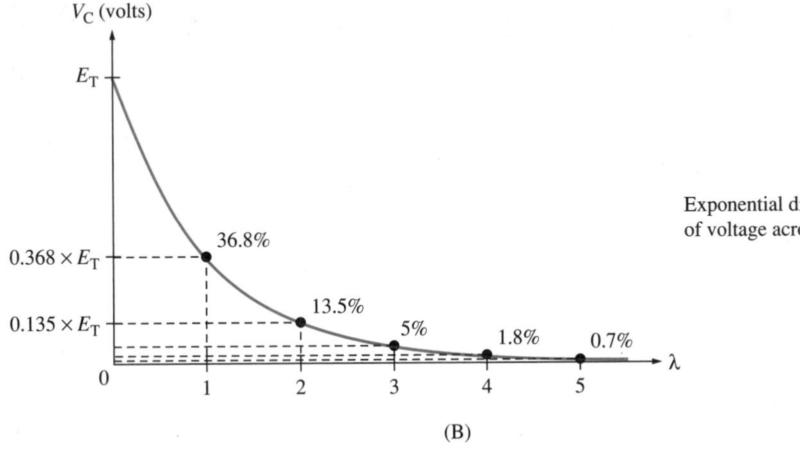

(B)

λ	V_C charging %	V_C discharging %
1	63.2	36.8
2	86.5	13.5
3	95	5
4	98.2	1.8
5	99.3	0.7

(C)

Figure 10-9 (cont.) ***RC time constants***

EXAMPLE 10-8

Calculate the time required for the capacitor to reach full charge in Figure 10-10.

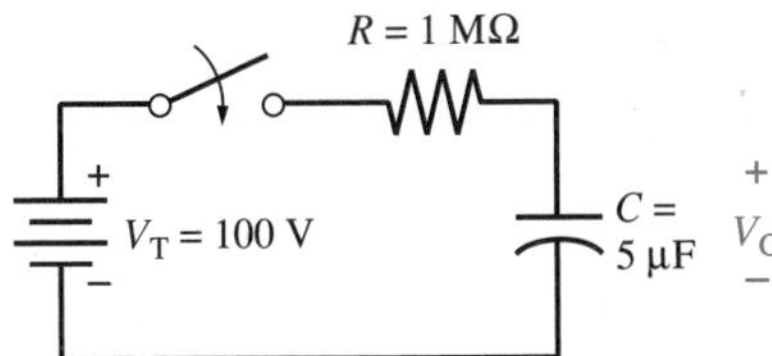

Figure 10-10

Solution:

$$\begin{aligned} \tau &= RC \\ &= (1 \times 10^6\ \Omega)(5 \times 10^{-6}\ \text{F}) \\ &= 5\ \text{s} \\ \text{full charge} &= 5\ \tau \\ &= (5)(5\ \text{s}) \\ &= 25\ \text{s} \end{aligned}$$

EXAMPLE 10-9

Find the voltage on the capacitor at 1τ, 2τ, 3τ, 4τ, and 5τ in Figure 10-10.

Solution:

for 1τ
$$V_C = 63.2\%\ V_T = (.632)(100\text{ V}) = 63.2\text{ V}$$

for 2τ
$$V_C = 86.5\%\ V_T = (.865)(100\text{ V}) = 86.5\text{ V}$$

for 3τ
$$V_C = 95\%\ V_T = (.95)(100\text{ V}) = 95\text{ V}$$

for 4τ
$$V_C = 98.2\%\ V_T = (.982)(100\text{ V}) = 98.2\text{ V}$$

for 5τ
$$V_C = 99.3\%\ V_T = (.993)(100\text{ V}) = 99.3\text{ V}$$

EXAMPLE 10-10

Find the voltage across the fully charged capacitor in Figure 10-11.

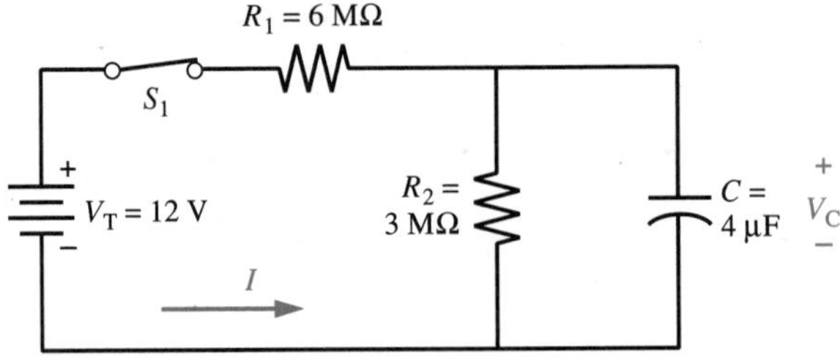

Figure 10-11

Solution:

In the *steady state,* or fully charged condition, the capacitor looks like an open circuit because no current exists in a fully charged capacitor. Therefore the voltage the capacitor will charge to is equal to the voltage across resistor R_2. Or, because the capacitor is in parallel with R_2, the voltage across the capacitor must be the same as the voltage across R_2. Recall that the voltage in a parallel circuit is the same. The voltage can be calculated by Ohm's law or by the voltage divider rule (see Figure 10-12).

At steady state capacitors look like an open circuit.

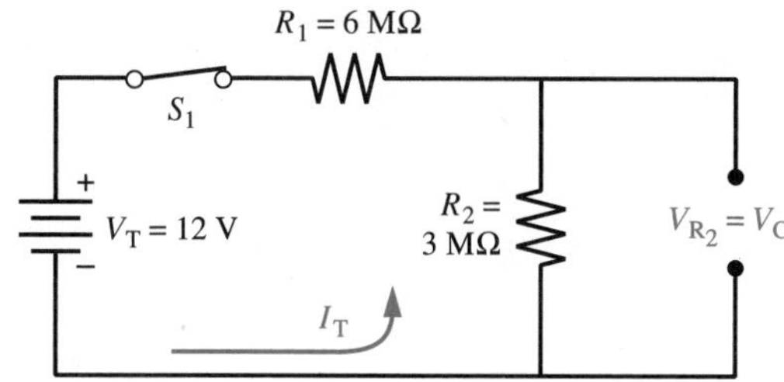

Figure 10-12

$$R_T = R_1 + R_2 = 6\ M\Omega + 3\ M\Omega = 9\ M\Omega$$

and
$$I_T = \frac{V_T}{R_T} = \frac{12\ V}{9\ M\Omega} = 1.33\ mA$$

and
$$V_{R_2} = I_T \times R_2 = (1.33\ mA)(3\ M\Omega) = 4\ V$$

$$\therefore \quad V_C = V_{R_2} = 4\ V$$

or by V.D.R.
$$V_C = V_{R_2} = \frac{R_2 \times V_T}{R_1 + R_2} = \frac{(3\ M\Omega)(12\ V)}{(3\ M\Omega + 6\ M\Omega)} = \frac{(3\ M\Omega)(12\ V)}{9\ M\Omega} = 4\ V$$

For an alternate solution, see Figure 10-13. Thevinin's theorem also has the advantage of enabling us to determine the time constant of the circuit (because the relationship $\tau = RC$ is for a series circuit and the Thevinin equivalent circuit is a series circuit).

The energy stored in the electrostatic field of a capacitor can be calculated using the following formula:

$$W_C = \frac{1}{2} CV^2$$ ◀ Equation 10-6

where
W_C = the energy stored in the capacitor in joules
C = the capacitance in farads
V = the voltage across the plates of the capacitor

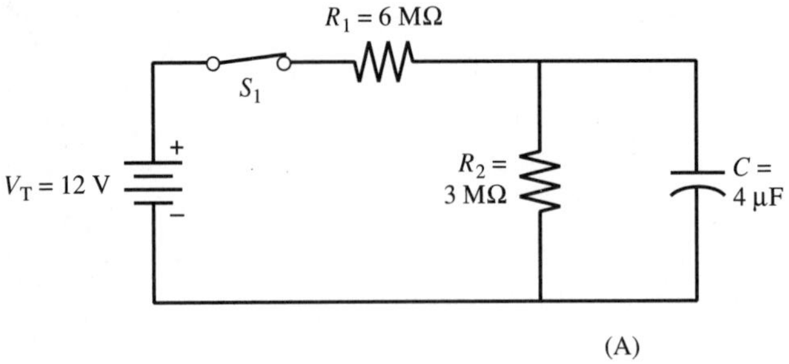

C becomes the load.
Remove the load.

(A)

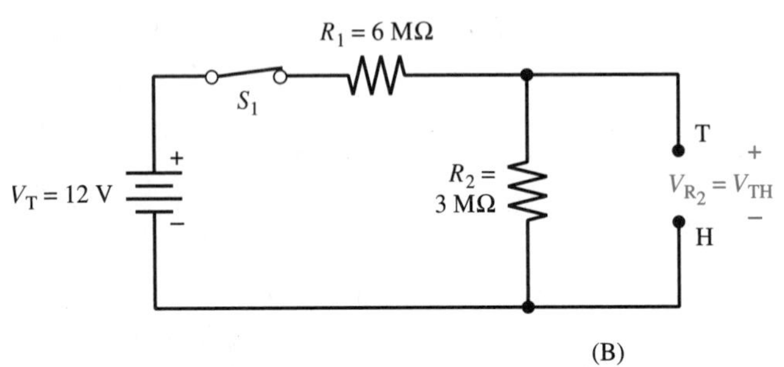

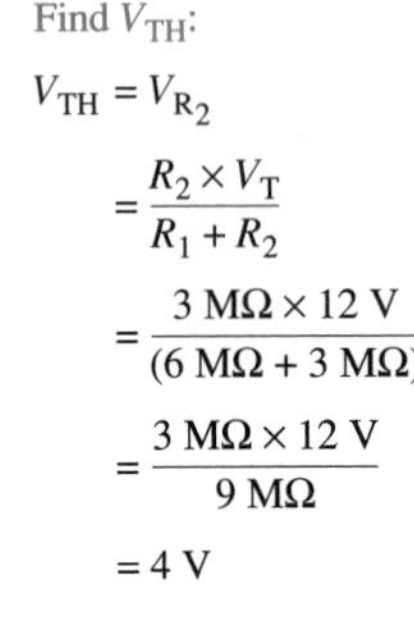

Find V_{TH}:

$$V_{TH} = V_{R_2}$$

$$= \frac{R_2 \times V_T}{R_1 + R_2}$$

$$= \frac{3\ M\Omega \times 12\ V}{(6\ M\Omega + 3\ M\Omega)}$$

$$= \frac{3\ M\Omega \times 12\ V}{9\ M\Omega}$$

$$= 4\ V$$

(B)

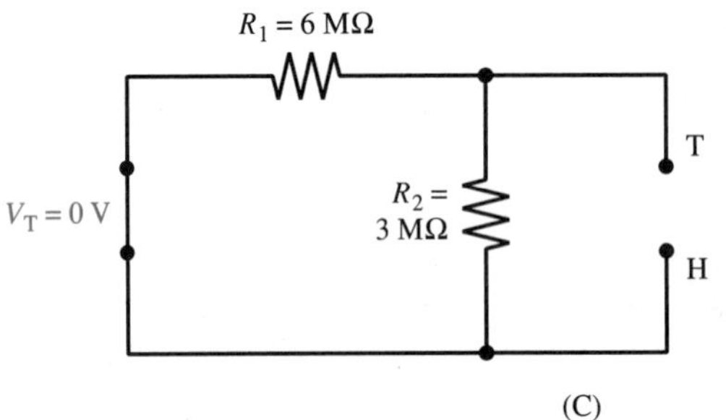

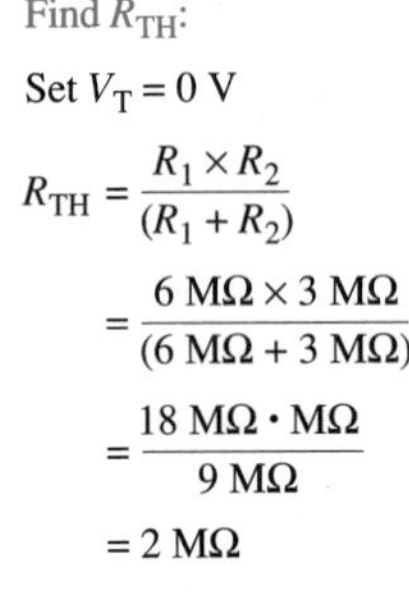

Find R_{TH}:

Set $V_T = 0\ V$

$$R_{TH} = \frac{R_1 \times R_2}{(R_1 + R_2)}$$

$$= \frac{6\ M\Omega \times 3\ M\Omega}{(6\ M\Omega + 3\ M\Omega)}$$

$$= \frac{18\ M\Omega \cdot M\Omega}{9\ M\Omega}$$

$$= 2\ M\Omega$$

(C)

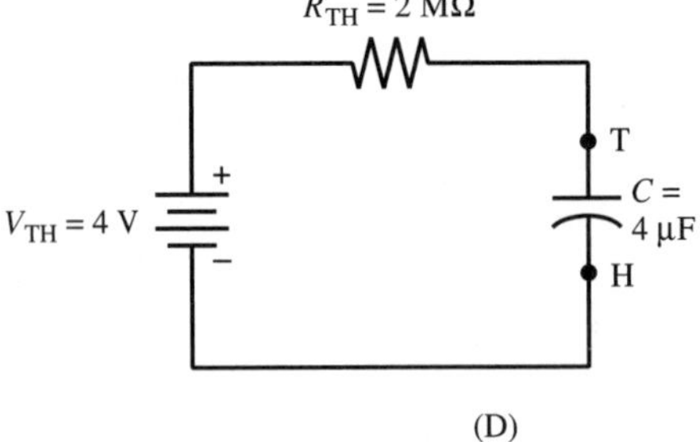

The Thevenin's equivalent circuit:

$$\tau = R_{TH}C$$

$$= 2\ M\Omega \times 4\ \mu F$$

$$= 8\ s$$

and V_C at steady state will charge to V_{TH}.

$$V_C = V_{TH} = 4\ V$$

(D)

Figure 10-13

EXAMPLE 10-11

Calculate the energy stored in a 4-microfarad capacitor that has been charged to a voltage of 10 volts.

Solution:

$$W_C = \frac{1}{2}CV^2$$
$$= (0.5)(4 \times 10^{-6}\ \text{F})(10\ \text{V})^2$$
$$= (0.5)(4 \times 10^{-6}\ \text{F})(100\ \text{V}^2)$$
$$= 200 \times 10^{-6}\ \text{J}$$
$$= 200\ \mu\text{J}$$

✓ Self-Test 2

1. When capacitors are connected in series, the effect is similar to _______________ the distance between the plates.
2. State the general equation for capacitors in series.
3. State the general equation for capacitors in parallel.
4. For each time constant, a capacitor charges _______________ of the difference between the applied voltage and the capacitor voltage.
5. The time required for a capacitor to reach its steady state voltage is said to be _______________.

10.5 Capacitors in the Steady State

Once a capacitor is fully charged (after five time constants), it is said to be in the **steady state** condition. Ideally, when a capacitor is at steady state, no current passes through it. It appears as though it were an open circuit. Furthermore, it has the effect of being a voltage source where $V = Q/C$. When capacitors are in series in a steady state condition, the charge on the plates of each capacitor is equal regardless of the size of the capacitor. This can be seen by recalling that for series circuits

Capacitors in series have equal charge on their plates.

$$I_T = I_1 = I_2 = I_3$$

but $$I = \frac{Q}{t}$$

$$\therefore \quad \frac{Q_T}{t} = \frac{Q_1}{t} = \frac{Q_2}{t} = \frac{Q_3}{t}$$

and multiplying by t $$Q_T = Q_1 = Q_2 = Q_3$$

When capacitors are in parallel in a steady state condition, the total charge will equal the sum of the charges on each individual capacitor. Recall that for parallel circuits,

Capacitors in parallel: $Q_T = \Sigma Q$.

$$I_T = I_1 + I_2 + I_3$$

but $$I = \frac{Q}{t}$$

$$\therefore \quad \frac{Q_T}{t} = \frac{Q_1}{t} + \frac{Q_2}{t} + \frac{Q_3}{t}$$

and multiplying by t $$Q_T = Q_1 + Q_2 + Q_3$$

EXAMPLE 10-12

For the circuit of Figure 10-14, calculate:
(A) the total capacitance.
(B) the charge on each plate.
(C) the voltage across each capacitor.

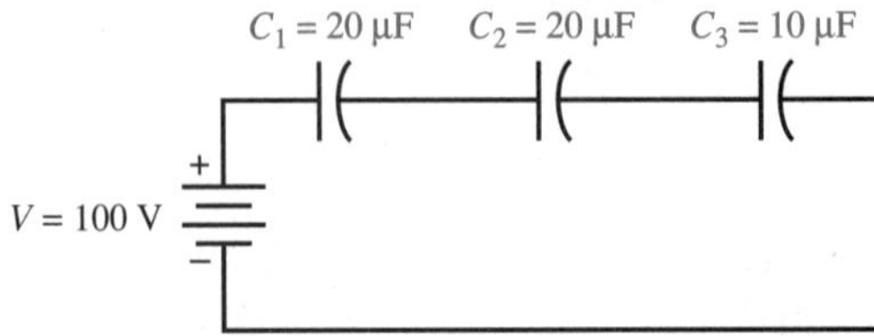

Figure 10-14

Solution:

(A)
$$\frac{1}{C_T} = \frac{1}{C_1} + \frac{1}{C_2} + \frac{1}{C_3}$$
$$\frac{1}{C_T} = \frac{1}{20\ \mu F} + \frac{1}{20\ \mu F} + \frac{1}{10\ \mu F}$$
$$\frac{1}{C_T} = \frac{1}{20\ \mu F} + \frac{1}{20\ \mu F} + \frac{2}{20\ \mu F}$$
$$\frac{1}{C_T} = \frac{4}{20\ \mu F}$$
$$C_T = \frac{20\ \mu F}{4}$$
$$C_T = 5\ \mu F$$

or
$$\frac{1}{C_T} = \frac{1}{C_1} + \frac{1}{C_2} + \frac{1}{C_3}$$
$$\frac{1}{C_T} = \frac{1}{20\ \mu F} + \frac{1}{20\ \mu F} + \frac{1}{10\ \mu F}$$
$$\frac{1}{C_T} = 0.05 \times 10^6\ F + 0.05 \times 10^6\ F + 0.1 \times 10^6\ F$$
$$\frac{1}{C_T} = 0.2 \times 10^6\ F$$
$$C_T = \frac{1}{0.2 \times 10^{-6}\ F}$$
$$C_T = 5 \times 10^{-6}\ F$$
$$C_T = 5\ \mu F$$

(B)
$$Q_T = Q_1 = Q_2 = Q_3$$
$$Q_T = C_T V_T$$
$$= (5 \times 10^{-6}\ F)(100\ V)$$
$$= 500 \times 10^{-6}\ C$$
$$= 500\ \mu C$$

(C) $V_1 = \frac{Q_1}{C_1} = \frac{Q_T}{C_1}$

$$= \frac{500 \times 10^{-6}\text{ C}}{20 \times 10^{-6}\text{ F}}$$

$$= 25\text{ V}$$

$$V_2 = \frac{Q_2}{C_2} = \frac{Q_T}{C_2}$$

$$= \frac{500 \times 10^{-6}\text{ C}}{20 \times 10^{-6}\text{ F}}$$

$$= 25\text{ V}$$

$$V_3 = \frac{Q_3}{C_3} = \frac{Q_T}{C_3}$$

$$= \frac{500 \times 10^{-6}\text{ C}}{10 \times 10^{-6}\text{ F}}$$

$$= 50\text{ V}$$

Check

$$V_T = V_1 + V_2 + V_3 \quad \text{(KVL)}$$

$$100\text{ V} = 25\text{ V} + 25\text{ V} + 50\text{ V}$$

$$100\text{ V} = 100\text{ V}$$

EXAMPLE 10-13

For the circuit of Figure 10-15, calculate:
(A) the total capacitance.
(B) the charge on each plate.
(C) the total charge.

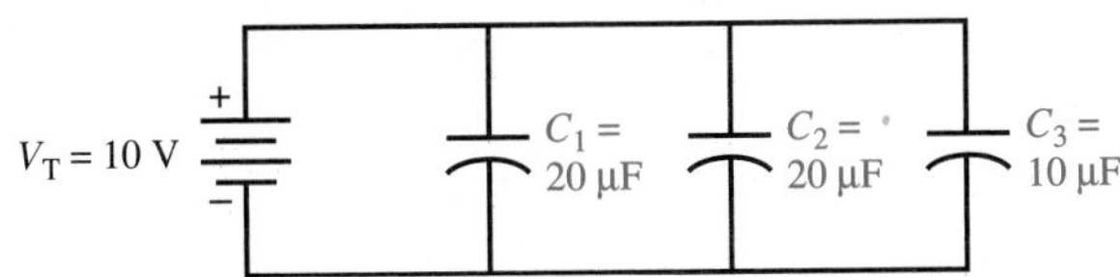

Figure 10-15

Solution:

(A) $C_T = C_1 + C_2 + C_3$

$$= 20\ \mu\text{F} + 20\ \mu\text{F} + 10\ \mu\text{F}$$

$$= 50\ \mu\text{F}$$

(B) $V_T = V_{C_1} = V_{C_2} = V_{C_3}$ (because all are in parallel)

$$Q_1 = C_1 V_T$$

$$= (20 \times 10^{-6}\text{ F})(10\text{ V})$$

$$= 200 \times 10^{-6}\text{ C}$$

$$= 200\ \mu\text{C}$$

$$Q_2 = C_2V_T$$
$$= (20 \times 10^{-6}\text{ F})(10\text{ V})$$
$$= 200\ \mu\text{C}$$

$$Q_3 = C_3V_T$$
$$= (10 \times 10^{-6}\text{ F})(10\text{ V})$$
$$= 100\ \mu\text{C}$$

(C) $$Q_T = Q_1 + Q_2 + Q_3$$
$$= 200\ \mu\text{C} + 200\ \mu\text{C} + 100\ \mu\text{C}$$
$$= 500\ \mu\text{C}$$

EXAMPLE 10-14

Find the steady state voltage across each capacitor and the total current for the circuit of Figure 10-16.

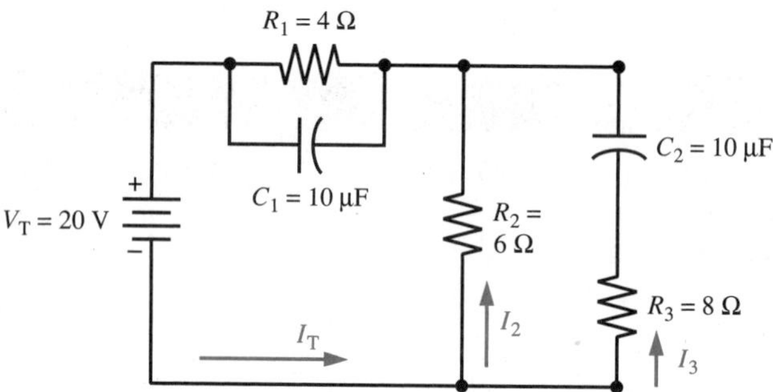

Figure 10-16

Solution:

Refer to Figure 10-17.

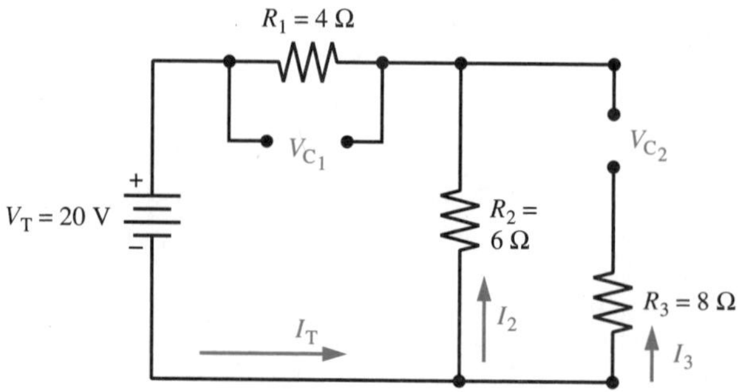

Figure 10-17

C_1 and C_2 appear as open circuits.

$$\therefore \quad I_3 = 0\ \text{A}$$

because the branch of the circuit is open, the current is zero.

$$I_2 = I_T$$

because R_1 and R_2 now form a series branch in the circuit.

$$\therefore \quad R_T = R_1 + R_2$$

$$= 4\ \Omega + 6\ \Omega$$

$$= 10\ \Omega$$

$$\text{and} \quad I_T = \frac{V_T}{R_T}$$

$$= \frac{20\ \text{V}}{10\ \Omega}$$

$$= 2\ \text{A}$$

$$\therefore \quad I_2 = I_T = 2\ \text{A}$$

V_{C_1} is equal to the voltage drop across R_1, because they are in parallel.

$$V_{C_1} = V_{R_1} = I_T R_1 = 2\ \text{A} \times 4\ \Omega = 8\ \text{V}$$

and V_{C_2} is equal to V_{R_2}, the voltage across R_2, because they are in parallel.

$$V_{C_2} = V_{R_2} = I_T R_2 = 2\ \text{A} \times 6\ \Omega = 12\ \text{V}$$

Note that the voltage divider rule could also be used directly.

$$V_{C_1} = V_{R_1} = \frac{R_1 \times V_T}{R_T}$$

$$= \frac{4\ \Omega \times 20\ \text{V}}{4\ \Omega + 6\ \Omega}$$

$$= 8\ \text{V}$$

$$\text{and} \quad V_{C_2} = V_{R_2} = \frac{R_2 \times V_T}{R_T}$$

$$= \frac{6\ \Omega \times 20\ \text{V}}{4\ \Omega + 6\ \Omega}$$

$$= 12\ \text{V}$$

Check

$$V_T = V_{R_1} + V_{R_2}$$

$$20\ \text{V} = 8\ \text{V} + 12\ \text{V}$$

$$20\ \text{V} = 20\ \text{V}$$

10.6 Capacitive Reactance

Capacitive reactance (X_C) is the AC opposition to current.

In a DC circuit the only opposition to current is the resistance of the circuit. In an AC circuit the opposition to current comes not only from the resistance but also from the capacitance. The opposition to AC current due to a capacitor is called **capacitive reactance** or X_C. Capacitive reactance is a function of frequency. As the frequency increases, the capacitive reactance will decrease. Mathematically,

$$X_C = \frac{1}{2\pi f C}$$

or

Equation 10-7 ▶

$$X_C = \frac{0.159}{fC}$$

where X_C = the capacitive reactance in ohms

f = the frequency in hertz

C = the capacitance in farads

X_C is inversely proportional to frequency.

Note that the value of X_C will decrease as the frequency or capacitance increases. X_C will increase as the frequency or capacitance of the circuit decreases.

EXAMPLE 10-15

Calculate the capacitive reactance of a 0.159-microfarad capacitor at 500 hertz, 1,000 hertz, and 2,000 hertz. Plot a graph of frequency versus X_C.

Solution:

$$X_{C_1} = \frac{1}{2\pi f_1 C}$$

$$= \frac{0.159}{f_1 C}$$

$$= \frac{0.159}{(500\ \text{Hz})(0.159 \times 10^{-6}\ \text{F})}$$

$$= 2{,}000\ \Omega$$

$$= 2\ \text{K}\Omega$$

$$X_{C_2} = \frac{0.159}{f_2 C}$$

$$= \frac{0.159}{(1{,}000\ \text{Hz})(0.159 \times 10^{-6}\ \text{F})}$$

$$= 1{,}000\ \Omega$$

$$= 1\ \text{K}\Omega$$

$$X_{C_3} = \frac{0.159}{f_3 C}$$

$$= \frac{0.159}{(2{,}000\ \text{Hz})(0.159 \times 10^{-6}\ \text{F})}$$

$$= 500\ \Omega$$

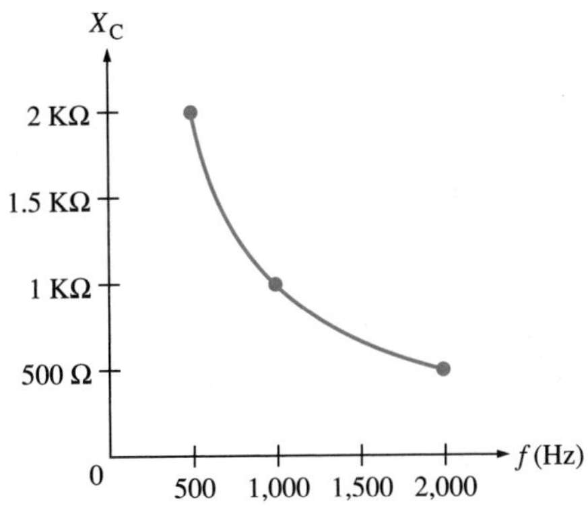

Figure 10-18

✓ Self-Test 3

1. When capacitors are in series in a steady state condition, the charge on the plates of each capacitor is ______________ regardless of the size of the capacitor.
2. When capacitors are in parallel in the steady state condition, the total charge will equal ______________ on each individual capacitor.
3. The opposition to AC current due to a capacitor is called ______________.
4. As frequency increases, X_C will ______________.

10.7 Capacitor Applications

Capacitors are used to accomplish many different types of tasks in electronics. Because of the time constant and because the reactance varies with frequency, capacitors can perform many timing and filtering applications. For

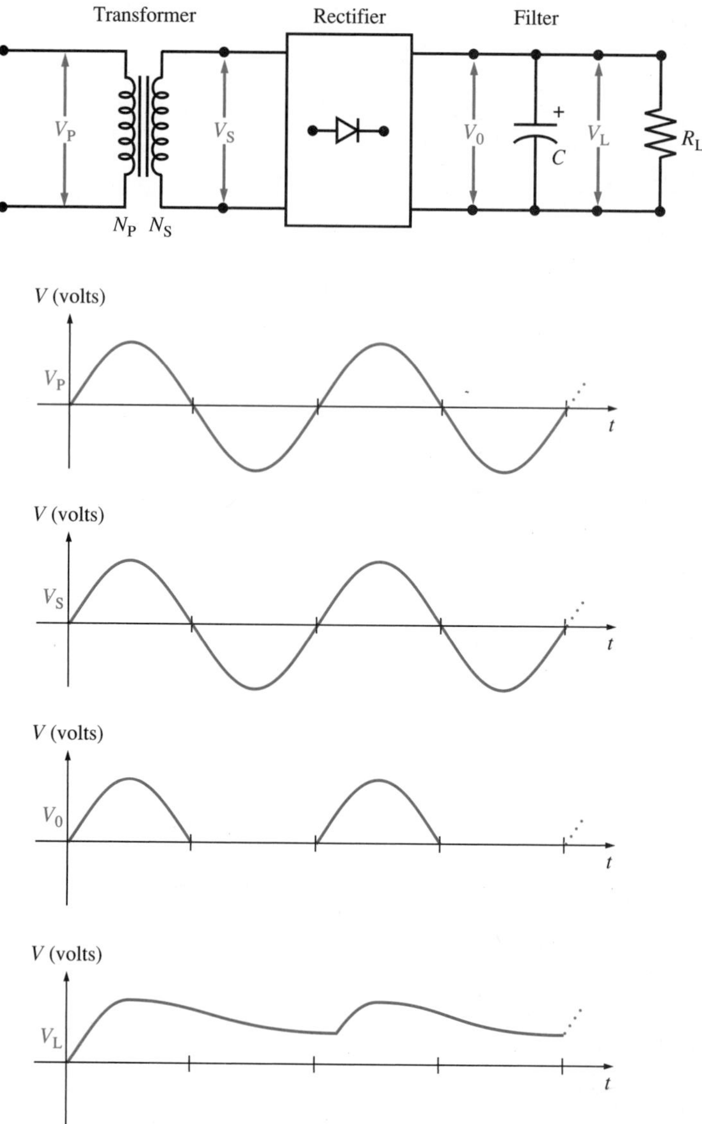

Figure 10-19 **Capacitor filter**

tech tidbit[2]

Diodes are like check valves or one way streets.

example, consider the circuit of Figure 10-19. In Figure 10-19 the transformer is used to step up or step down the voltage to a desired level. The rectifier is an electronic device that only allows current in one direction only. **Rectifiers** are made from diodes. **Diodes** are devices that have a high resistance in one direction and a low resistance in the other direction. Applying the input AC waveform of V_S to the rectifier will result in the output waveform of V_O. The negative half of the AC waveform or the reverse direction current is not allowed to flow because of the diode. This V_O waveform is called pulsating DC. Current exists for 180 degrees in a sinusoidal manner and then there is no current. The capacitor charges up during the time that current exists. During the time that there is no current, the capacitor discharges into the load resistor. Waveform V_L

shows the effect of the capacitor. The charge time is very fast and the discharge time is controlled by the load resistor. Capacitors used in this fashion are usually the electrolytic type and are referred to as **filter capacitors.** Filter capacitors are used to smooth pulsating DC voltage.

Filter capacitors are used in power supplies.

Capacitors are also often used to block DC in electronic circuits. In Figure 10-20, we have a DC and an AC source connected together. Viewing this waveform in the DC mode on an oscilloscope will show the summing effect clearly. In the DC position, the input signal is fed directly into the display amplifier. In the AC mode, the oscilloscope adds a capacitor in series with the input signal to block any DC from passing. Recall that a capacitor is an open circuit to DC signals. The opposition to AC is determined by the frequency. Thus, the capacitor blocks the DC level and only allows the AC waveform to be displayed, as shown in Figure 10-20. The AC/DC mode switch is referred to as the *vertical input coupling* switch of the oscilloscope.

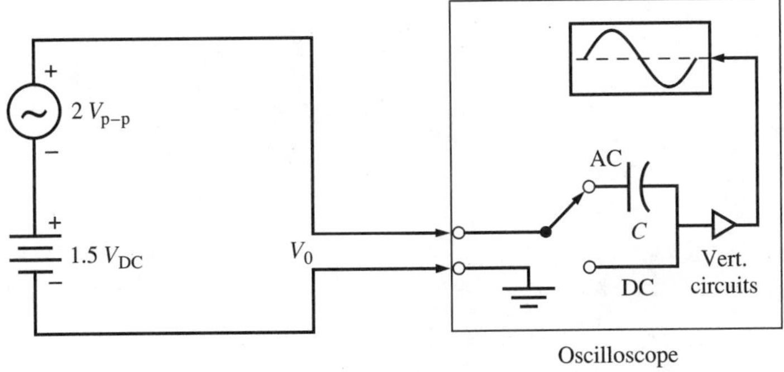

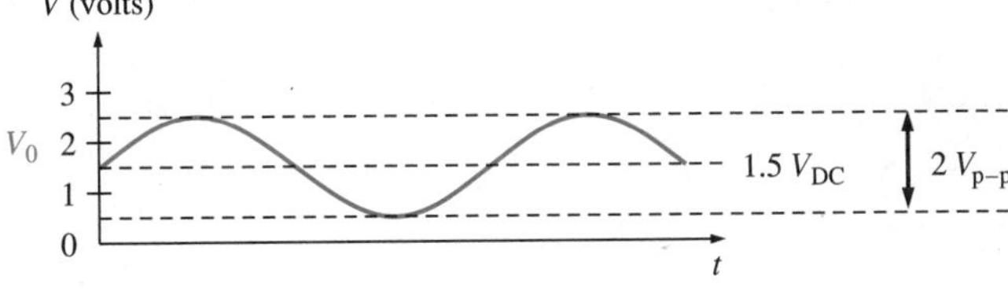

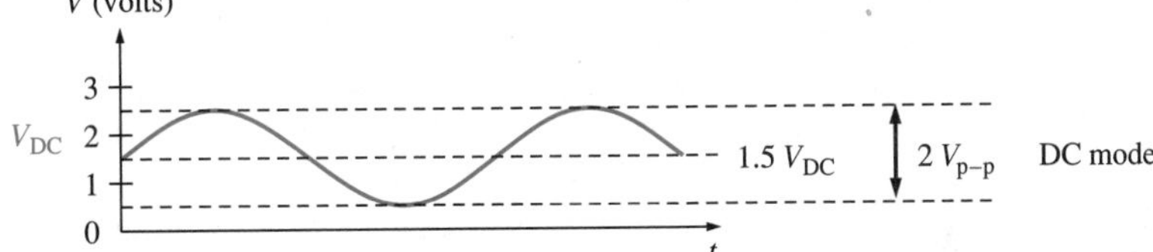

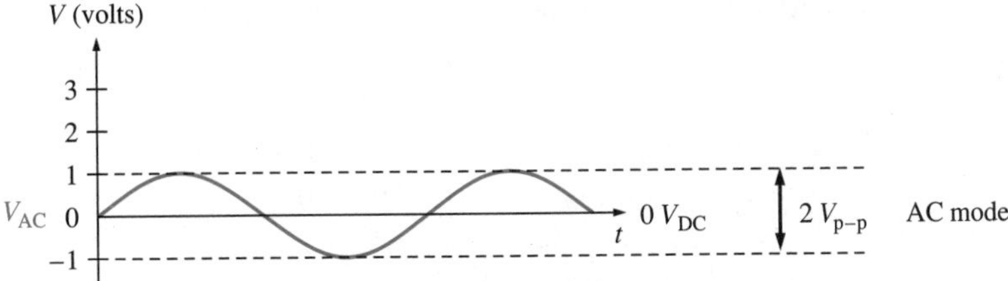

Figure 10-20 **Blocking capacitor in an oscilloscope**

T3 tech tips & trouble-shooting

T3 TESTING CAPACITORS

A capacitor is two plates separated by an insulator, so it is basically an open circuit. Therefore, we can perform some basic tests with an ohmmeter to determine whether or not a capacitor is "good." Placing the leads of an ohmmeter across a "good" capacitor should indicate an open circuit or infinity ohms (see Figure 10-21). A reading of zero or low ohms indicates that the capacitor is shorted. If the capacitor is large, the ohmmeter will give a momentary reading and then slowly increase its reading to infinity because the battery in the ohmmeter is charging the capacitor and indicating the momentary current. In either case, in the steady state the capacitor should indicate an open circuit.

T2 tech tidbit

Ohmmeters can be used to test capacitors.

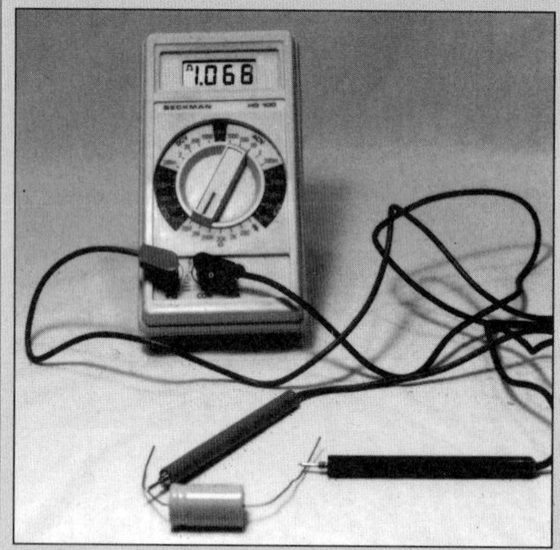

Figure 10-21 **Testing a capacitor with an ohmmeter**

Ohmmeter tests are known as **static tests.** Sometimes a capacitor will test good statically but break down and fail under circuit operation. This is known as a leaky capacitor. Leaky capacitors cannot be detected with an ohmmeter but can often be tested using a voltmeter. In this **dynamic test** a DC voltage is placed across a capacitor with a voltmeter connected, as shown in Figure 10-22. A good capacitor will show only a momentary reading on the voltmeter after which the voltmeter will read zero volts. A leaky capacitor will give a voltage reading on the voltmeter. This test should be performed at the rated voltage of the capacitor if possible.

T2 tech tidbit

A leaky capacitor is a capacitor that allows some DC current to pass. Leaky capacitors should be replaced.

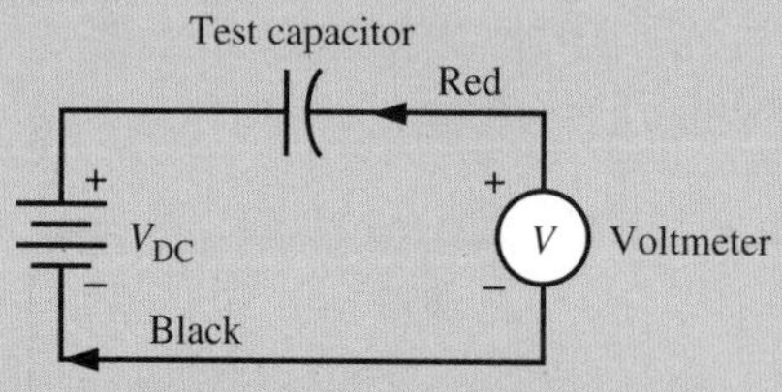

Figure 10-22 **Testing a capacitor with a voltmeter**

Recall that capacitors are used to store energy. When a capacitor is disconnected or removed from a circuit, it may still be charged. Large value capacitors when charged can store large amounts of energy. Therefore, it is important to discharge all capacitors before handling them. This can be accomplished by connecting a low value resistor to an insulated test lead and shorting the terminals of the capacitor after it has been removed from the circuit or the energy source disconnected. Be careful to hold the insulated portion of the test lead. This procedure can be very dangerous when working on high voltage. For example, when discharging the capacitance of a TV picture tube. Do not attempt to discharge a picture tube until you have been instructed on the proper safety precautions.

! **SAFETY TIP:** Always discharge capacitors before handling them.

SUMMARY

A capacitor is made of two conductors separated by an insulator called a dielectric. The ability of a capacitor to store energy is called capacitance. The unit of capacitance is the farad.

Capacitors can be connected in series and in parallel. The general rules for calculating the total capacitance are opposite but similar to the rules for resistors in series and in parallel. The time required for a capacitor to reach 63.2% of its full charge is called the time constant (τ). Five time constants are required to reach full charge.

The opposition to AC current due to a capacitor is called capacitive reactance or X_C. Capacitive reactance is a function of frequency. As the frequency increases, the capacitive reactance decreases. As the frequency decreases, the capacitive reactance increases.

FORMULAS

$$C = \frac{kA}{d}$$

$$C = \frac{Q}{V} \qquad Q = CV \qquad V = \frac{Q}{C}$$

$$\frac{1}{C_T} = \frac{1}{C_1} + \frac{1}{C_2} + \frac{1}{C_3} + \ldots + \frac{1}{C_N}$$ (capacitors in series)

$$C_T = \frac{C_1 \times C_2}{C_1 + C_2}$$ (two capacitors in series)

$$C_T = \frac{C}{N}$$ (equal capacitors in series)

$$C_T = C_1 + C_2 + C_3 + \ldots + C_N$$ (capacitors in parallel)

$$\tau = RC$$

$$\text{full charge} = 5\tau$$

$$W_C = \frac{1}{2}CV^2$$

$$Q_T = Q_1 = Q_2 = Q_3 = \ldots = Q_N$$ (capacitors in series)

$$Q_T = Q_1 + Q_2 + Q_3 + \ldots + Q_N$$ (capacitors in parallel)

$$X_C = \frac{1}{2\pi f C} \qquad f = \frac{1}{2\pi X_C C} \qquad C = \frac{1}{2\pi f X_C}$$

EXERCISES

Section 10.2

1. Find the capacitance of a capacitor whose plate area, A, is 2 millimeters^2 if the plates are separated by a distance of 4 millimeters and the dielectric material is air.

2. Find the capacitance of a capacitor whose plate area, *A*, is 3 millimeters^2 if the plates are separated by a distance of 5 millimeters and the dielectric material is mica.
3. If the plate area, *A*, of a 5-microfarad capacitor is doubled, find the new capacitance.
4. If the distance between the plates of a 20-microfarad capacitor is halved, what is the new capacitance?
5. What is the capacitance of a capacitor with a charge of 200 microcoulombs and a voltage of 25 volts?
6. What is the capacitance of a capacitor with a charge of 50 millicoulombs and a voltage of 10 volts?
7. How much charge is deposited on a 12-microfarad capacitor with a voltage of 6 volts across its plates?

Section 10.3

8. Find the total capacitance when a 10-microfarad and a 15-microfarad capacitor are connected in series.
9. Find the total capacitance when a 6-microfarad, a 12-microfarad, and a 4-microfarad capacitor are connected in series.
10. Find the total capacitance when a 5-microfarad and a 10-microfarad capacitor are connected in series.
11. Find the total capacitance when six 18-microfarad capacitors are connected in series.
12. Find the total capacitance when a 15-microfarad and a 20-microfarad capacitor are connected in parallel.
13. Find the total capacitance when a 100-microfarad, a 200-microfarad, and a 300-microfarad capacitor are connected in parallel.
14. Find the total capacitance for the network of Figure 10-23A.
15. Find the total capacitance for the network of Figure 10-23B.

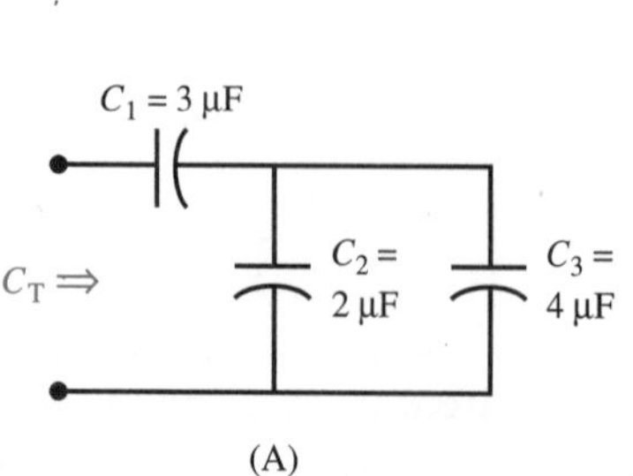

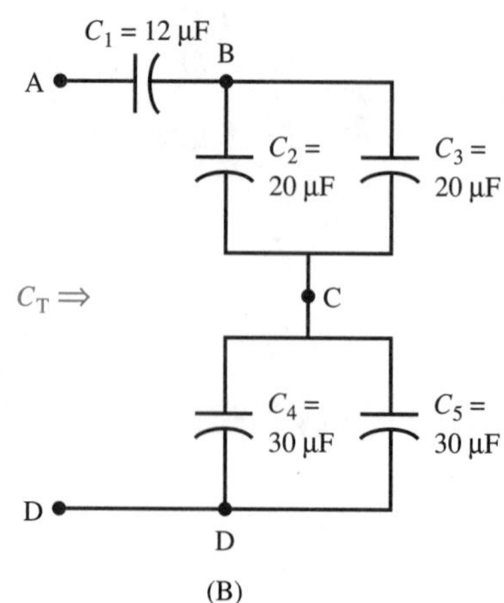

Figure 10-23

Section 10-4

16. Calculate the time required for a 20-microfarad capacitor connected in series with a 3-megohm resistor to reach full charge.
17. Calculate the time required for a 3-microfarad capacitor connected in series with a 50-kilohm resistor to reach full charge.
18. A 20-volt battery is connected in series with a 2-microfarad capacitor and a 6-megohm resistor. Calculate the voltage across the capacitor at 1τ, 2τ, 3τ, 4τ, and 5τ.
19. Find the steady state voltage across the capacitor in Figure 10-24.
20. Find the time constant for the circuit of Figure 10-24.

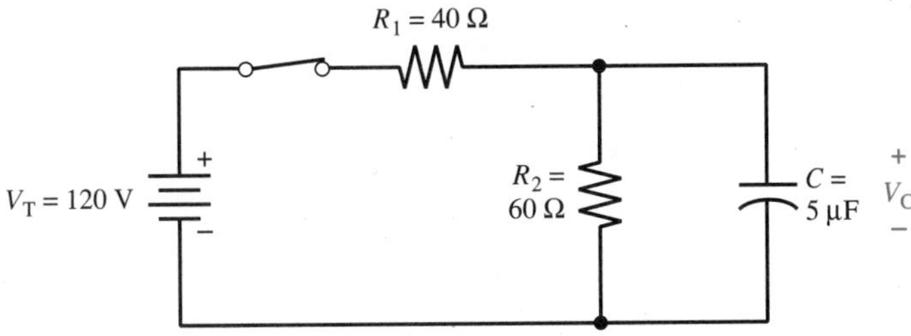

Figure 10-24

21. Calculate the energy stored in a 12-microfarad capacitor that has been charged to a voltage of 6 volts.
22. A capacitor that is charged to 10 volts stores 400 microjoules of energy. Find the value of the capacitor.

Section 10.5

23. A 60-volt battery, a 6-microfarad capacitor, a 3-microfarad capacitor, and a 2-microfarad capacitor are all connected in series. Calculate
    (A) the total capacitance.
    (B) the charge on each plate of each capacitor.
    (C) the voltage across each capacitor.
24. A 60-volt battery, a 6-microfarad capacitor, a 3-microfarad capacitor, and a 2-microfarad capacitor are all connected in parallel. Calculate
    (A) the total capacitance.
    (B) the charge on each plate.
    (C) the total charge.
25. Find the steady state voltage across the capacitor and the total current for the circuit of Figure 10-25.

Section 10.6

26. Calculate the capacitive reactance of a 5-microfarad capacitor at 100 hertz, a 10-microfarad capacitor at 1,000 hertz, and a 15-microfarad capacitor at 1,000 hertz.

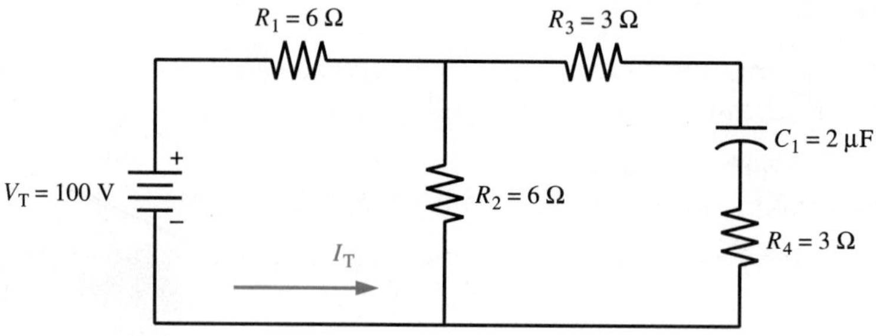

Figure 10-25

27. Calculate the capacitive reactance of a 10.6-microfarad capacitor at 1,000 hertz, 2,000 hertz, and 4,000 hertz. Plot a graph of frequency versus X_C.
28. See Lab Manual for the crossword puzzle for Chapter 10.

CHAPTER 11

INDUCTORS

Courtesy of Hewlett Packard

KEY TERMS

autotransformer
center tap transformer
choke
copper losses
counter EMF
eddy currents
efficiency
exponential
henry
hysteresis
ideal inductor
impedance
inductance
inductive reactance
inductor
iron losses
isolation
laminated
mutual inductance
primary
quality
reflected impedance
secondary
steady state
step-down transformer
step-up transformer
tapped
time constant
transformer
transients
turns ratio
variac
wiper

OBJECTIVES

After completing this chapter, the student should be able to:

- define inductance.
- identify types of inductors.
- state Lenz's law.
- calculate the total inductance of a series, parallel, and series-parallel network.
- define steady state.
- calculate the time constant for an RL circuit.
- use the universal time constant curve to solve problems.
- calculate the quality of an inductor.
- calculate inductive reactance.
- describe transformer action.
- classify types of transformers.
- calculate transformer ratios.
- compute the efficiency of transformers.
- explain the safety applications of a transformer.

11.1 Introduction

Inductance—ability to oppose the change in current.

Inductance (L) is the physical characteristic of an electrical conductor that affects current. It is a measure of the ability of a material to produce a magnetic field. Inductance is defined as the property or ability of a conductor to oppose any change in current through the conductor. A straight piece of wire or conductor will have very little inductance and will have little affect on changes in current magnitude. However, a coil of wire will have more inductance and will oppose changes in current magnitude because a changing magnetic field will induce a voltage or current in the opposite direction from the applied voltage or current. It is important to note that the effect of inductance is only due to a change in current. Therefore the value of the effects of inductance will be zero for direct current (DC), because the magnitude of the current is constant and does not change. In alternating current (AC), the magnitude of the current is always changing due to the nature of the sine wave. Therefore, the inductance of the circuit must be considered when analyzing alternating current circuits. This inductance will oppose the magnitude of current in a manner similar to the way that a resistor opposes current in a DC circuit. Furthermore, the faster the current is changing (frequency), the greater the opposition to the current.

11.2 Inductors

The iron core increases the strength of the magnetic field.

A length of wire wound into a coil creates an **inductor.** The inductance is due to the magnetic effects that are increased by the coiling of the wire. The greater the number of turns, the greater the inductance. If the coil of wire is wound around an iron core, the inductance will also be increased. This is because the iron core increases the strength of the magnetic field around the coil. Sometimes the coil is wound around a nonmagnetic core. This is done for support and will have no additional effect on the inductance. If the coil has no core, it is sometimes referred to as an air-core inductor. Figure 11-1 illustrates the schematic symbols for an air-core or a nonmagnetic core (Figure 11-1A) inductor and an iron core inductor (Figure 11-1B). The unit of inductance is the **henry** (H), named after the United States scientist Joseph Henry (1797–1878).

The unit of inductance is henries (H).

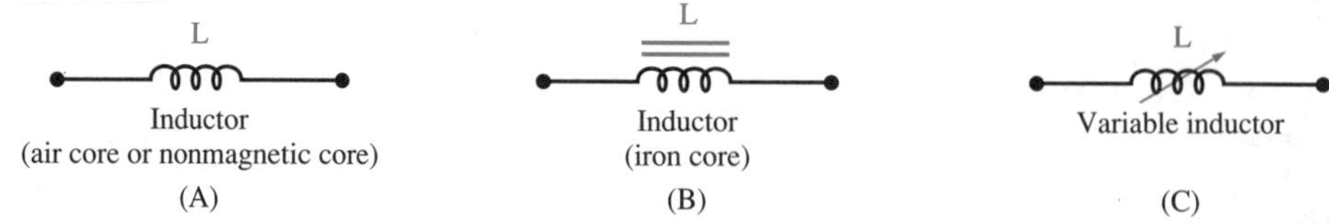

Figure 11-1 ***Inductor symbols***

Inductance is determined by the type, cross-sectional area, and length of the core and the number of turns of wire.

Figure 11-1C shows the schematic symbol for a variable inductor. Variable inductors are made by moving a magnetic core in and out of the coil of wire. The position of the core determines the strength of the field and the value of the inductance.

The amount of inductance that an inductor has is determined by the type and cross-sectional area of the core, the length of the core, and the number of turns of wire. Mathematically,

$$L = \frac{KN^2}{l}$$

Equation 11-1

where K = a constant determined by the type of material and cross-sectional area of the core

N = the number of turns of wire of the coil

l = the length of the core

L = the inductance in henries (H)

Note that the inductance, L, varies directly as the square of the number of turns and inversely as the length of the core. Therefore, the number of turns will have the greatest affect on the inductance of an inductor. Thus, if we double the number of turns, we will increase the inductance four times. If we halve the number of turns, we decrease the inductance by one-fourth.

Inductors are also called chokes or coils.

Figure 11-2 shows a number of different types of inductors used in electric circuits today. Inductors are also referred to as **chokes** or coils.

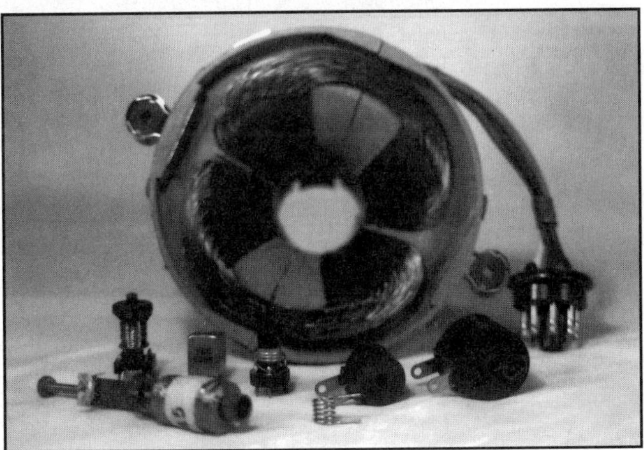

Figure 11-2 Types of inductors

11.3 Lenz's Law

Whenever a magnetic field expands or collapses across a coil of wire, a voltage will be induced in the coil of wire. The induced voltage will always be in the opposite direction to the current that created the magnetic field. The induced voltage is referred to as a counter electromotive force or **counter EMF** because it is in the opposite direction. This phenomenon is known as Lenz's law and is illustrated in Figure 11-3. It was named after its discoverer, Emil Khristianovich Lenz (1804–1865).

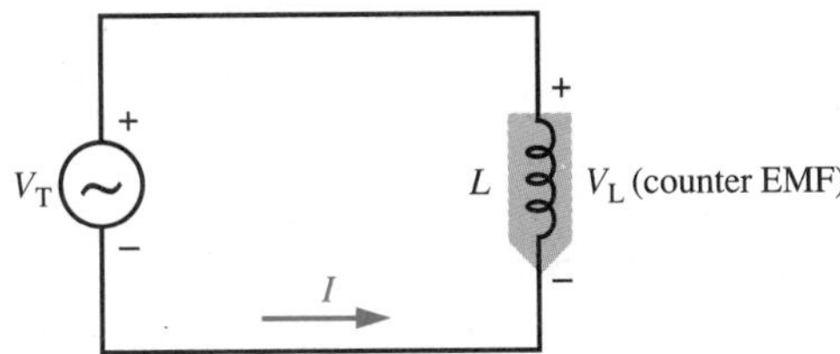

Figure 11-3 Lenz's Law

The induced voltage or counter EMF can be calculated mathematically by use of the following equation:

Equation 11-2

$$V_L = L \times \frac{\Delta i}{\Delta t}$$

where V_L = the induced voltage in volts

L = the inductance in henries

Δi = the change in current in amperes

Δt = the time it took the current to change in seconds

Note that the magnitude of the induced voltage or counter EMF is very dependent on the rate of change of the current.

T² tech tidbit

The magnitude of the induced voltage is mostly dependent on the rate of change of the current.

EXAMPLE 11-1

The ignition system in Mr. Dunnigan's car increases from 0 to 2 amperes in 1 millisecond. Find the voltage delivered to the spark plugs if the inductance of the ignition coil is 10 henries.

Solution:

$$V_L = L\frac{\Delta i}{\Delta t}$$

$$\Delta i = 2\text{ A} - 0\text{ A} = 2\text{ A}$$

$$\Delta t = 1 \times 10^{-3}\text{ s} - 0\text{ s} = 1 \times 10^{-3}\text{ s}$$

$$V_L = L\frac{\Delta i}{\Delta t}$$

$$= 10\text{ H}\left(\frac{2\text{ A}}{\quad}\right)$$

$$= \frac{20\text{ H} \cdot \text{A}}{\quad}$$

$$= 20 \times 10^{3}\text{ V}$$

$$= 20\text{ KV}$$

✓ Self-Test 1

1. The ______________ the number of turns, the greater the inductance.
2. If the number of turns in an inductor is doubled, its inductance will increase ______________ times.
3. The induced voltage in a coil is referred to as ______________.
4. ______________ states that whenever a magnetic field expands or collapses across a coil of wire, a voltage will be induced in the coil of wire.

11.4 Inductors in Series and Parallel

For inductors in series, $L_T = \sum L$.

Inductors can be connected in series and in parallel when used in electric circuits. The total inductance is calculated in the same way that the total resistance is calculated in series and in parallel. Figure 11-4A illustrates that the total inductance of three inductors connected in series is equal to the sum of the individual inductors. In general, for inductors connected in series,

$$L_T = L_1 + L_2 + L_3 + \ldots + L_N$$

Equation 11-3

Figure 11-4B illustrates how to calculate the total inductance of three inductors connected in parallel. In general, for inductors connected in parallel,

$$\frac{1}{L_T} = \frac{1}{L_1} + \frac{1}{L_2} + \frac{1}{L_3} + \ldots + \frac{1}{L_N}$$

Equation 11-4

For inductors in parallel, L_T will always be less than the smallest inductor.

Note that just as with parallel resistors, the special cases of the following formulas:

For N equal inductors in parallel

$$L_T = \frac{L}{N}$$

and

For 2 inductors in parallel

$$L_T = \frac{L_1 L_2}{L_1 + L_2}$$

will still be valid.

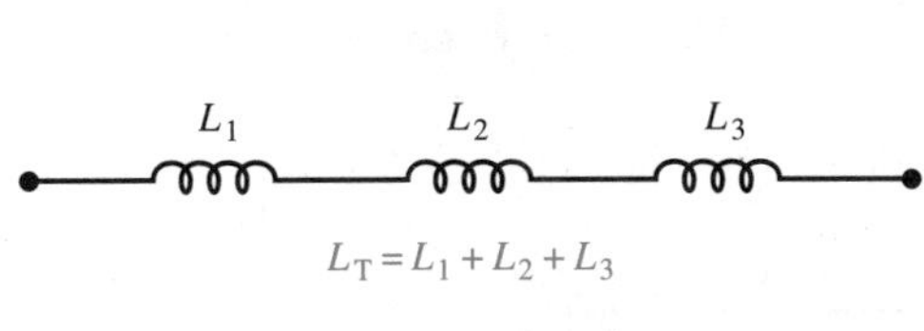

(A) Inductors in series

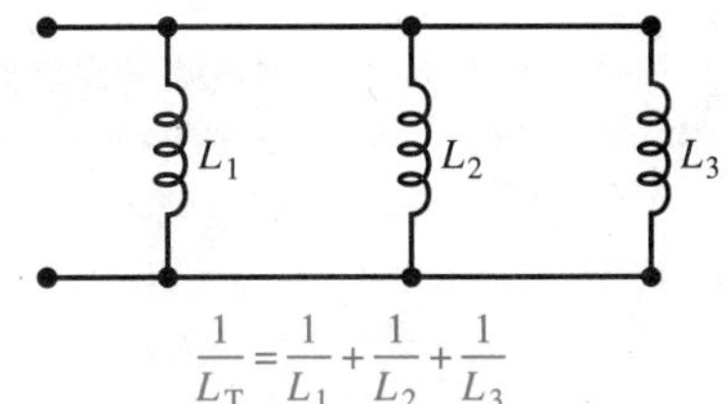

(B) Inductors in parallel

Figure 11-4 ***Inductors in series and in parallel***

EXAMPLE 11-2

Find the total inductance of two 10-henry coils connected in series. See Figure 11-5.

Figure 11-5

Solution:

$$\begin{aligned} L_{AB} &= L_1 + L_2 \\ &= 10\text{ H} + 10\text{ H} \\ &= 20\text{ H} \end{aligned}$$

EXAMPLE 11-3

A 3-millihenry coil, a 6-millihenry coil, and a 2-millihenry coil are connected in series. Find the total inductance. See Figure 11-6.

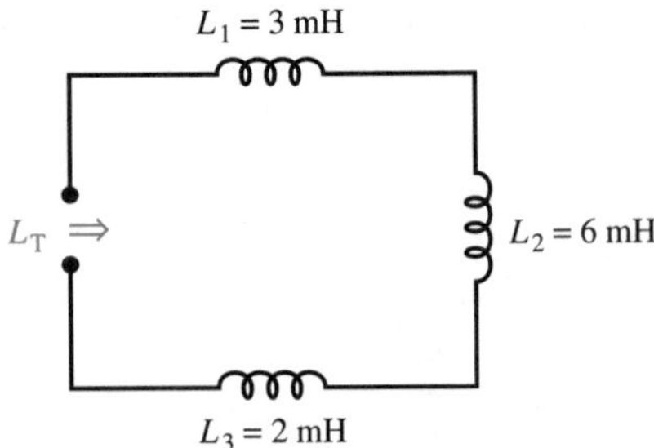

Figure 11-6

Solution:

$$L_T = L_1 + L_2 + L_3$$
$$= 3\text{ mH} + 6\text{ mH} + 2\text{ mH}$$
$$= 11\text{ mH}$$

EXAMPLE 11-4

A 12-henry coil, a 4-henry coil, and a 3-henry coil are connected in parallel. Find the total inductance. See Figure 11-7.

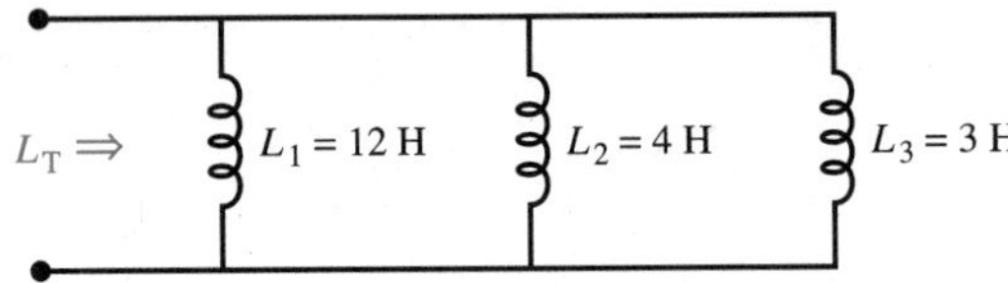

Figure 11-7

Solution:

$$\frac{1}{L_T} = \frac{1}{L_1} + \frac{1}{L_2} + \frac{1}{L_3}$$
$$\frac{1}{L_T} = \frac{1}{12\text{ H}} + \frac{1}{4\text{ H}} + \frac{1}{3\text{ H}}$$
$$\frac{1}{L_T} = \frac{1}{12\text{ H}} + \frac{3}{12\text{ H}} + \frac{4}{12\text{ H}}$$
$$\frac{1}{L_T} = \frac{8}{12\text{ H}}$$
$$\therefore \quad L_T = \frac{12\text{ H}}{8}$$
$$= 1.5\text{ H}$$

or $$\frac{1}{L_T} = \frac{1}{L_1} + \frac{1}{L_2} + \frac{1}{L_3}$$

$$\frac{1}{L_T} = \frac{1}{12\text{ H}} + \frac{1}{4\text{ H}} + \frac{1}{3\text{ H}}$$

$$\frac{1}{L_T} = \frac{0.083}{\text{H}} + \frac{0.25}{\text{H}} + \frac{0.33}{\text{H}} = \frac{0.67}{\text{H}}$$

$$\therefore \quad L_T = \frac{1}{0.67}\text{ H}$$

$$= 1.5\text{ H}$$

EXAMPLE 11-5

A 6-millihenry coil and a 3-millihenry coil are connected in parallel. Find the total inductance. See Figure 11-8.

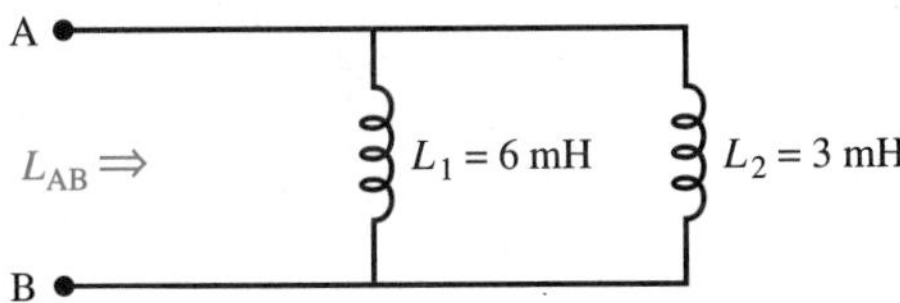

Figure 11-8

Solution:

$$L_{AB} = \frac{L_1 \times L_2}{L_1 + L_2}$$

$$= \frac{(6\text{ mH})(3\text{ mH})}{6\text{ mH} + 3\text{ mH}}$$

$$= \frac{18\text{ m}\cancel{\text{H}} \cdot \text{mH}}{9\text{ m}\cancel{\text{H}}}$$

$$= 2\text{ mH}$$

EXAMPLE 11-6

Five 100-microhenry coils are connected in parallel. Find the total inductance. See Figure 11-9.

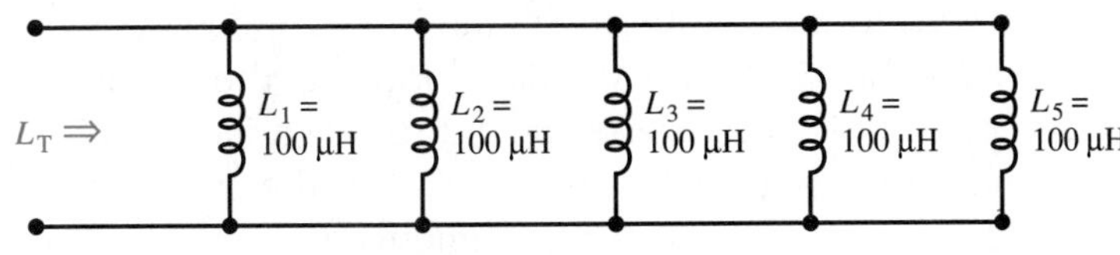

Figure 11-9

Solution:

$$N = 5 \text{ (equal inductors of 100 mH)}$$

$$L_T = \frac{L}{N} = \frac{100\ \mu\text{H}}{5} = 20\ \mu\text{H}$$

EXAMPLE 11-7

Find the total inductance of the circuit of Figure 11-10.

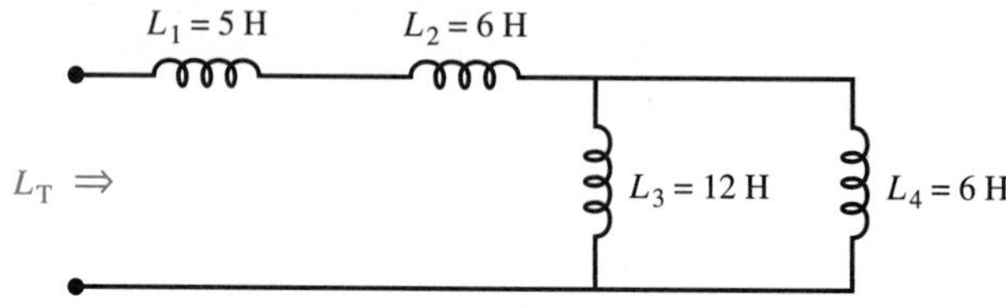

Figure 11-10

Solution:

$$L_{34} = \frac{L_3 L_4}{L_3 + L_4} = \frac{(12\ \text{H})(6\ \text{H})}{12\ \text{H} + 6\ \text{H}} = \frac{72\ \text{H} \bullet \text{H}}{18\ \text{H}} = 4\ \text{H}$$

$$L_T = L_1 + L_2 + L_{34} = 5\ \text{H} + 6\ \text{H} + 4\ \text{H} = 15\ \text{H}$$

11.5 RL Time Constants

A steady state condition exists after all transients have settled down.

Recall that in DC circuits the current is constant. That is, it does not change in value. This is true under steady state conditions. **Steady state** is defined as the time or condition after all changes (**transients**) have settled down. If a circuit contains only resistance, the steady state condition occurs instantaneously. However, if the circuit contains inductance, the current cannot change instantaneously. This is due to Lenz's law. When the switch is closed, the current tries to increase from zero to its steady state value instantaneously. However, the current is opposed by the counter EMF developed in the inductor. Therefore, it takes a definite period of time for the current to reach its steady state or its maximum value. The more inductance in the circuit, the greater the counter EMF that will be produced, and the longer the time for the current to reach its steady state value.

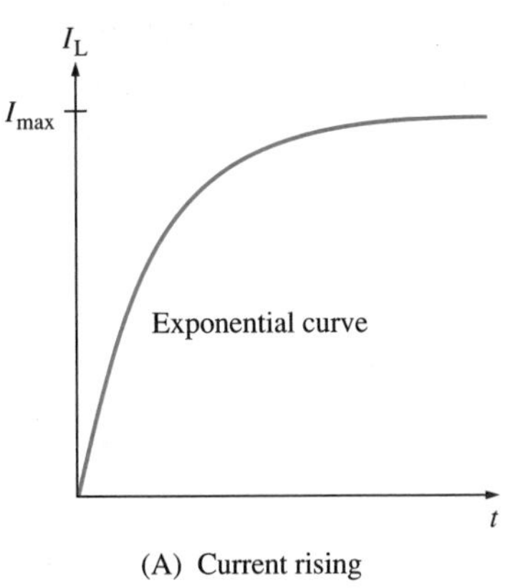

(A) Current rising

I_L

I_{max}

Exponential curve

t

(B) Current dropping

Figure 11-11 ***Inductor current***

When the switch is opened, a similar situation exists. This time the counter EMF opposes the change so that the current decreases to zero gradually. Figure 11-11 illustrates the shape of the current waveform for a current rising from zero to a maximum value and for a current dropping from a maximum value to zero. This waveform is called an **exponential** waveform.

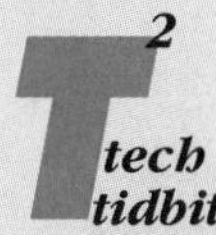

The exponential curve is broken up into five equal periods of time called time constants.

Regardless of the amount of resistance and inductance in the circuit, the shape of the current curve will always follow the exponential form. The exponential curve is broken up into five equal periods of time called **time constants.** The actual time in seconds is solely related to the amount of inductance and resistance in the circuit. Mathematically, the time constant (τ) for an inductive circuit can be calculated by the following equation:

$$\tau = \frac{L}{R}$$

Equation 11-5

where τ = the time constant in seconds

L = the inductance in henries

R = the resistance in ohms

Steady state = 5τ

The total practical time to reach steady state is considered to be five time constants. This is because the current will rise 63.2% of the change in every time constant. Table 11-1 illustrates the percentage of the steady state values for five time constants.

Table 11-1 TIME CONSTANTS

Time constant	% of steady state for current rising	% of steady state for current dropping
1	63.2	36.8
2	86.5	13.5
3	95.1	4.9
4	98.1	1.9
5	99.3	0.7

Figure 11-12 illustrates the graphing of Table 11-1 and is sometimes referred to as the universal time constant curve. It is based on the evaluation of the mathematical function ε^{-x}. The symbol ε, the Greek letter Epsilon, equals 2.718.

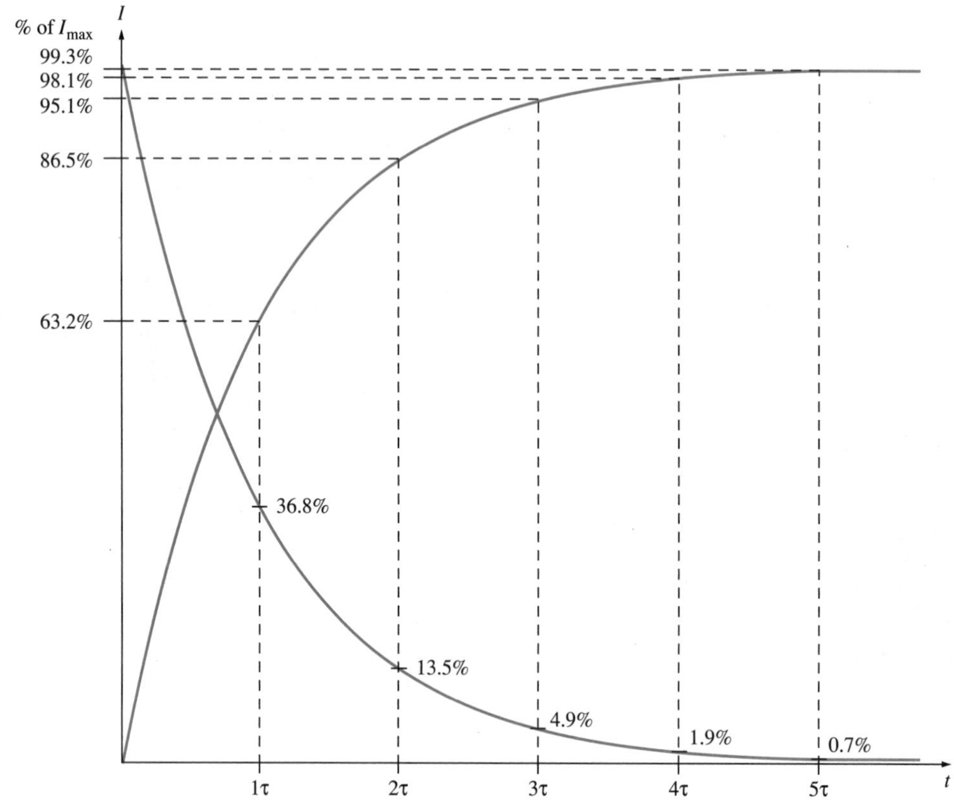

Figure 11-12 **Universal time constant exponential curves**

EXAMPLE 11-8

Find the time constant for a 0.2-henry inductor in series with a 100-ohm resistor.

Solution:

$$\tau = \frac{L}{R}$$

$$= \frac{0.2\ \text{H}}{100\ \Omega}$$

$$= 0.002\ \text{s}$$

$$= 2 \times 10^{-3}\ \text{s}$$

$$= 2\ \text{ms}$$

EXAMPLE 11-9

For the circuit of Figure 11-13, find the time required to reach steady state and the value of the current at steady state.

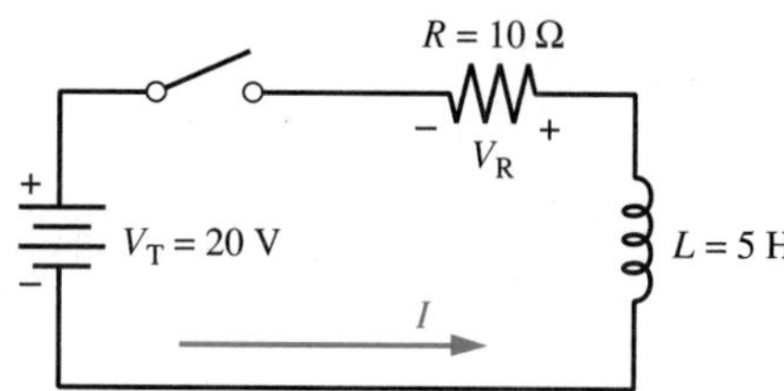

Figure 11-13

Solution:

$$\begin{aligned}
\tau &= \frac{L}{R} \\
&= \frac{5\text{ H}}{10\ \Omega} \\
&= 0.5\text{ s} \\
\text{Steady state time} &= 5\,\tau \\
&= (5)(0.5\text{ s}) \\
&= 2.5\text{ s}
\end{aligned}$$

At steady state the inductor can be considered a short circuit because it is a coil of wire with negligible resistance.

$$\begin{aligned}
I_{SS} &= \frac{V_T}{R} \\
&= \frac{20\text{ V}}{10\ \Omega} \\
&= 2\text{ A}
\end{aligned}$$

EXAMPLE 11-10

Find the value of the current three time constants after the switch in Figure 11-14 is closed.

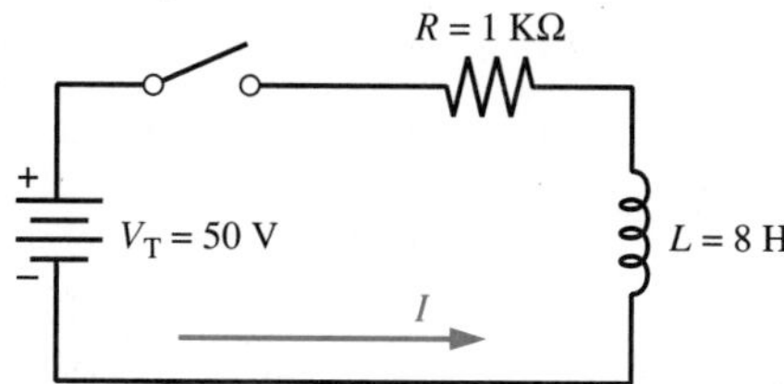

Figure 11-14

Solution:

The steady state current is

$$I_{SS} = \frac{V_T}{R} = \frac{50\text{ V}}{1{,}000\ \Omega} = \frac{50\text{ V}}{1\text{ K}\Omega} = 50\text{ mA}$$

From the universal time constant curve of Figure 11-12 or Table 11-1:

$$3\tau = 95.1\%$$

$$I_{3\tau} = (95.1\%)\ I_{SS} = (0.951)(50\text{ mA}) = 47.55\text{ mA}$$

tech tidbit

Energy is measured in joules.

The energy stored in the magnetic field of an inductor can be calculated using the following formula:

Equation 11-6

$$W_L = \frac{1}{2}LI^2$$

where W_L = the energy in joules
L = the inductance in henries
I = the current in amperes

EXAMPLE 11-11

Find the energy stored in a 5-henry coil with a steady state current of 2 amperes flowing through it.

Solution:

$$W_L = \frac{1}{2}LI^2 = (0.5)(5\text{ H})(2\text{ A})^2 = 10\text{ J}$$

✓ Self-Test 2

1. What is the total inductance of five 10-henry inductors in series?
2. What is the total inductance of five 10-henry inductors in parallel?
3. If a circuit contains inductance, the _______________ cannot change instantaneously.
4. The time constant of an RL circuit is directly proportional to its _______________ and inversely proportional to its _______________.

11.6 Inductive Reactance

In a DC circuit the only opposition to the flow of current is the resistance of the circuit. In an AC circuit, the opposition to current comes not only from the resistance but also from the inductance. For an inductor, the amount of current that flows is determined by the amount of counter EMF (Lenz's law). The counter EMF opposes current in the same way that a resistor opposes current. Although counter EMF is rated in volts, the effect or opposition of counter EMF can be equated to ohms. This effect is called **inductive reactance,** or X_L. The inductive reactance is dependent on the amount of inductance and the frequency of the circuit. Mathematically,

$$X_L = 2\pi f L$$

Equation 11-7

where X_L = the inductive reactance in ohms

f = the frequency in hertz

L = the inductance in henries

X_L increases when f increases.

Note that the value of X_L will increase as the frequency or the inductance of the circuit increases. Note also that if the frequency is 0 hertz, as in DC, X_L will equal 0 ohms. Thus there is no X_L in DC circuits.

EXAMPLE 11-12

Calculate the inductive reactance of a 3.2-henry coil at 10 hertz, 100 hertz, and 1,000 hertz. Plot a graph of frequency versus X_L.

Solution:

See Figure 11-15.

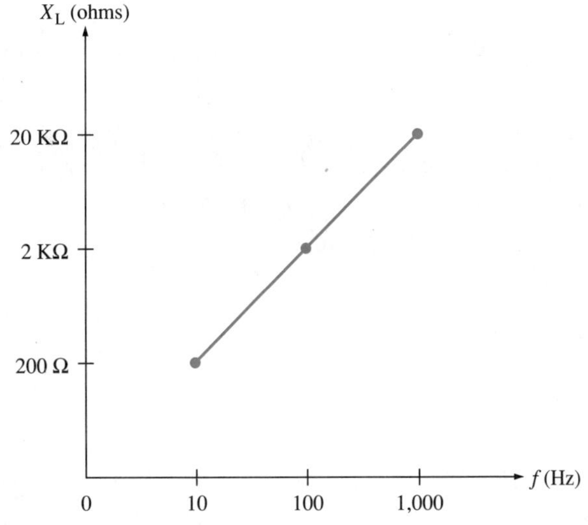

Figure 11-15

$$X_{L_1} = 2\pi f_1 L \qquad \text{at 10 Hz}$$
$$= 6.28 \times 10 \text{ Hz} \times 3.2 \text{ H}$$
$$= 201\ \Omega$$

$$X_{L_2} = 2\pi f_2 L \qquad \text{at 100 Hz}$$
$$= 6.28 \times 100 \text{ Hz} \times 3.2 \text{ H}$$
$$= 2{,}010\ \Omega$$
$$= 2\ \text{K}\Omega$$

$$X_{L_3} = 2\pi f_3 L \qquad \text{at 1,000 Hz}$$
$$= 6.28 \times 1{,}000 \text{ Hz} \times 3.2 \text{ H}$$
$$= 20{,}100\ \Omega$$
$$= 20\ \text{K}\Omega$$

11.7 Quality of Inductors

tech tidbit

An ideal inductor has a Q of infinity.

Up to this point, we have assumed that inductors have no resistance. This is referred to as an **ideal inductor.** However, real inductors are made of many turns of relatively fine wire. Therefore, real inductors have resistance. The resistance of an inductor affects how the coil will work in an AC circuit. The less the resistance compared to the inductive reactance, the more the coil will act like an ideal inductor. The quantity that measures the ability of a coil to respond like an ideal inductor is known as the **quality** (Q) of the coil. Quality is the ratio of the inductive reactance to the resistance of the coil. Mathematically,

Equation 11-8

$$Q = \frac{X_L}{R_L}$$

where Q = the quality of the coil (unitless)

X_L = the inductive reactance in ohms

R_L = the resistance of the coil in ohms

tech tidbit

Q is unitless.

EXAMPLE 11-13

An 8-henry inductor has a resistance of 10 ohms. Calculate the Q of the coil (A) at 60 hertz and (B) at 1 kilohertz.

Solution:

(A) at 60 Hz

$$X_L = 2\pi f L$$
$$= (2)(3.14)(60 \text{ Hz})(8 \text{ H})$$
$$= 3{,}014\ \Omega$$

(B) at 1 KHz

$$X_L = 2\pi f L$$
$$= (2)(3.14)(1{,}000 \text{ Hz})(8 \text{ H})$$
$$= 50{,}240\ \Omega$$

$$Q = \frac{X_L}{R}$$
$$= \frac{3{,}014\ \Omega}{10\ \Omega}$$
$$= 301.4$$

$$Q = \frac{X_L}{R}$$
$$= \frac{50{,}240\ \Omega}{10\ \Omega}$$
$$= 5{,}024$$

✓ Self-Test 3

1. The unit of inductive reactance is the ______________.
2. Calculate the inductive reactance of a 5-henry coil operating at 400 hertz.
3. An ideal inductor has no ______________.
4. The ______________ of a coil measures its ability to respond like an ideal inductor.

11.8 Transformers

When two coils of wire are positioned close to each other, the flux lines from the expanding and contracting magnetic field of one coil cut across the windings of the second coil and induce a voltage in the second coil. The effective inductance of the two coils is known as the **mutual inductance** (M). Mutual inductance makes it possible to transfer energy magnetically. When mutual inductance exists between two coils, a changing current in one coil induces a voltage into the second coil. This is also referred to as transformer action. Devices that operate on this principle are called **transformers.** Figure 11-16 illustrates the effects of mutual inductance.

Mutual inductance makes it possible to transfer energy magnetically.

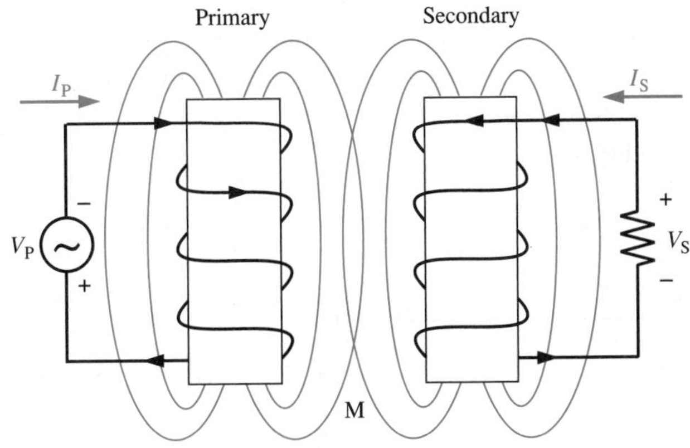

Figure 11-16 **Mutual inductance (M)**

The schematic symbols used for transformers are shown in Figure 11-17. The symbol consists basically of two coils of wire. A **center tap transformer** is a transformer that has a coil split or tapped into two or more pieces to produce different voltage levels, as shown in Figure 11-17.

Transformer design is dependent on three main factors: the operating voltage, the operating frequency, and the power it must handle. Air core transformers (Figure 11-17A) are used in high-frequency applications. Iron core transformers (Figure 11-17B) are used in low frequency and high power applications.

The input coil of a transformer is called the primary coil.

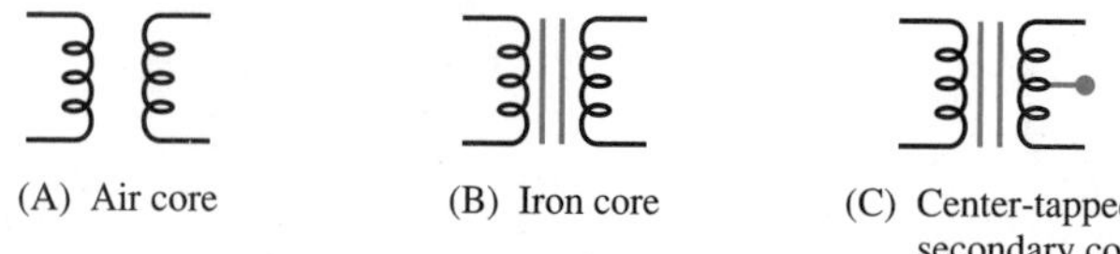

Figure 11-17 Transformer symbols

Referring to Figure 11-18, the input coil of a transformer is called the **primary** coil. The output coil of a transformer is called the **secondary** coil. The phase relationship between the primary and secondary can be either in phase or 180 degrees out of phase, as shown in Figure 11-18. A dot is used to indicate the phase relationship (polarity) between the primary and secondary coils. When the dots are on the same side, the output voltage is in phase with the input voltage. When the dots are on diagonally opposite sides, the output voltage will be 180 degrees out of phase with the input voltage.

The output coil of a transformer is called the secondary coil.

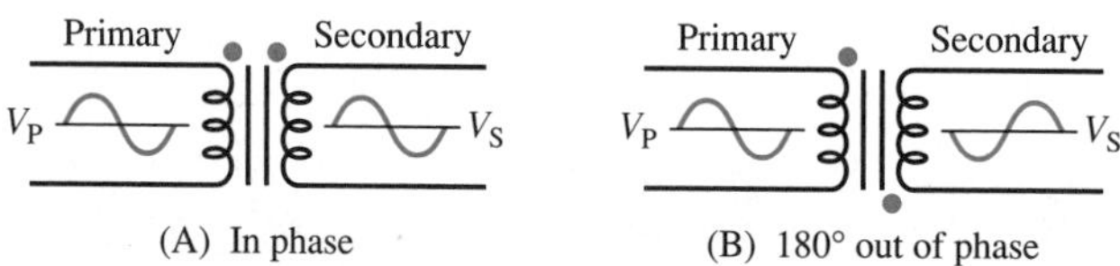

Figure 11-18 Phase relationships in transformers

11.9 Transformer Ratios

Step-down transformers lower voltages.

Transformers are often used to change voltage levels in a circuit. When the primary coil has more turns than the secondary coil, the voltage induced in the secondary will be less than the primary voltage. This is referred to as a **step-down transformer** and is illustrated in Figure 11-19A. When the primary coil has less turns than the secondary coil, the voltage induced in the secondary will be more than the primary voltage. This is referred to as a **step-up transformer** and is illustrated in Figure 11-19B.

Step-up transformers raise voltages.

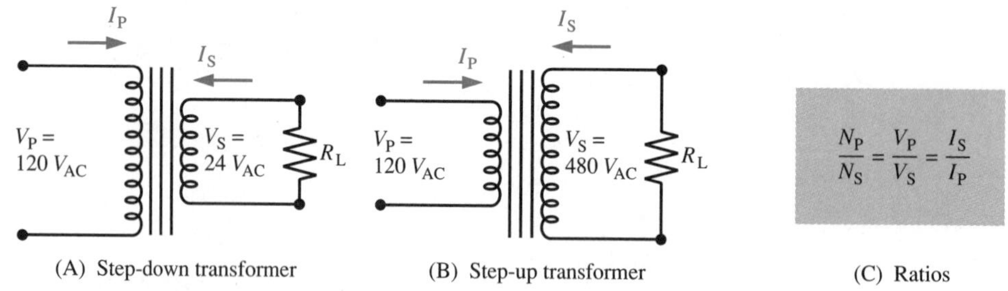

Figure 11-19 Transformer ratios

Changing voltage levels does not mean we are creating energy. Transformers actually transfer power from the primary to the secondary through the magnetic field. The voltage levels may change, but the power in the primary will always equal the power in the secondary (neglecting any losses). Therefore, in an ideal transformer, the power in the primary equals the power in the secondary. Recall that power equals voltage times current. Mathematically,

$$P_{pri} = P_{sec}$$

$$V_P I_P = V_S I_S$$

or

$$\frac{V_P}{V_S} = \frac{I_S}{I_P}$$

Equation 11-9

Furthermore, the voltage ratio of the primary coil voltage (V_P) to the secondary coil voltage (V_S) varies directly with the turns ratio of the number of primary turns (N_P) and the number of secondary turns (N_S) because the flux linking both coils is essentially the same. Mathematically,

$$\frac{N_P}{N_S} = \frac{V_P}{V_S}$$

Equation 11-10

From Equations 11-9 and 11-10, it can be seen that

$$\frac{N_P}{N_S} = \frac{I_S}{I_P}$$

Equation 11-11

The **turns ratio** is sometimes referred to as the transformation ratio and uses the symbol a. Mathematically,

$$a = \frac{N_P}{N_S}$$

Equation 11-12

Note that the side with the high voltage has the low current and vice versa. Thus, if a transformer steps up the voltage, it will step down the current for the same power.

$$V_P = \frac{V_S I_S}{I_P}, \; V_S = \frac{V_P I_P}{I_S}$$

EXAMPLE 11-14

A step-down transformer has 500 turns in the primary coil and 100 turns in the secondary coil. If 120 volts are applied to the primary, calculate the voltage induced into the secondary. See Figure 11-20.

$$V_P = \frac{V_S N_P}{N_S}, \; V_S = \frac{V_P N_S}{N_P}$$

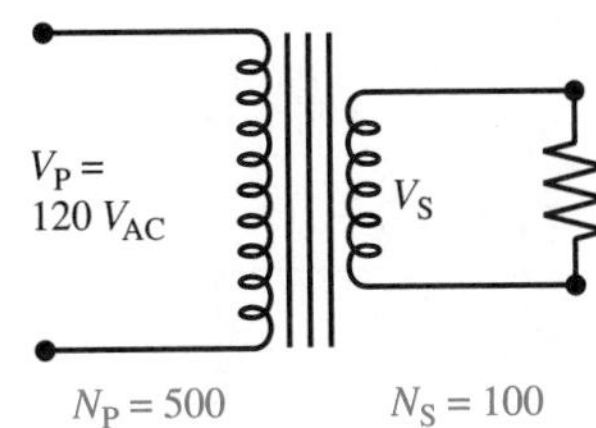

Figure 11-20

Solution:

$$\frac{N_P}{N_S} = \frac{V_P}{V_S}$$

$$\frac{500}{100} = \frac{120\text{ V}}{V_S}$$

$$5 = \frac{120\text{ V}}{V_S}$$

$$V_S = \frac{120\text{ V}}{5}$$

$$V_S = 24\text{ V}$$

$$I_P = \frac{I_S N_S}{N_P}, \; I_S = \frac{I_P N_P}{N_S}$$

$$\frac{V_P}{V_S} = \frac{I_S}{I_P} = \frac{N_P}{N_S} = a$$

EXAMPLE 11-15

A 120 V/480 V step-up transformer is used to supply power to a load. Find the turns ratio of the transformer. See Figure 11-21.

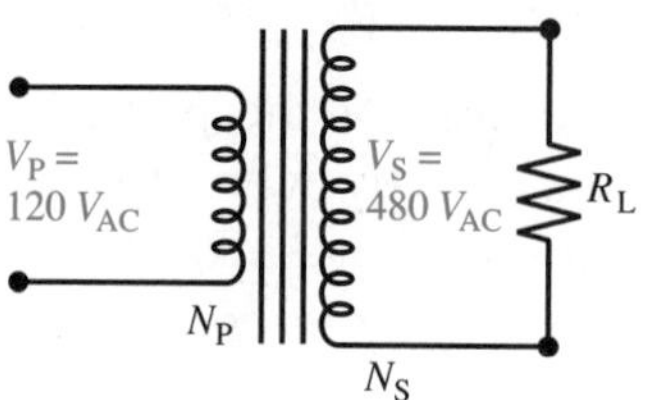

Figure 11-21

Solution:

$$\frac{N_P}{N_S} = \frac{V_P}{V_S}$$

$$\frac{N_P}{N_S} = \frac{120\ V_{AC}}{480\ V_{AC}}$$

$$\frac{N_P}{N_S} = \frac{1}{4}$$

$$a = \frac{N_P}{N_S} = \frac{1}{4}$$

$$a = 1{:}4$$

EXAMPLE 11-16

A 120 V/12 V step-down transformer is connected to a 3-ohm resistive load. Calculate the current in the primary coil. See Figure 11-22.

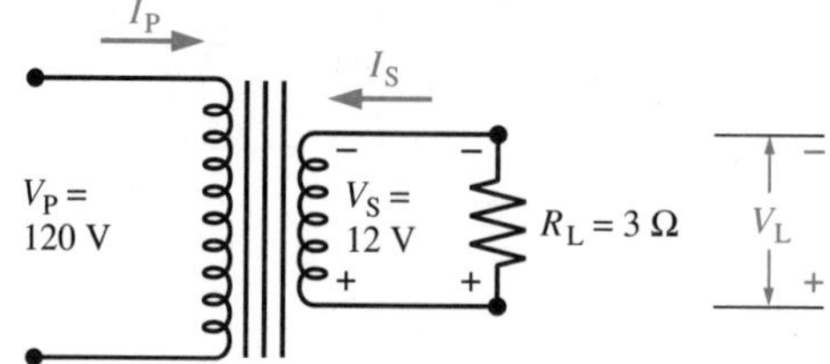

Figure 11-22

Solution:

$$V_S = V_L = 12\ V$$

$$I_S = \frac{V_L}{R_L} = \frac{12\ V}{3\ \Omega} = 4\ A$$

$$\frac{V_P}{V_S} = \frac{I_S}{I_P}$$

$$\frac{120\text{ V}}{12\text{ V}} = \frac{4\text{ A}}{I_P}$$

$$10 = \frac{4\text{ A}}{I_P}$$

$$I_P = \frac{4\text{ A}}{10}$$

$$I_P = 0.4\text{ A}$$

11.10 Impedance Matching

The term **impedance** in AC circuits refers to the total opposition to the current. Impedance includes both resistance and reactance. The units of impedance are ohms and the symbol is "Z." For right now, we can consider impedance to be the AC resistance of a circuit. Transformers can change impedance in much the same way as they change voltage or current. Recall from Thevenin's theorem that maximum power is delivered to a load when the Thevenin resistance or impedance of the circuit equals the resistance or impedance of the load. A transformer can be a useful tool for accomplishing impedance matching for maximum power transformation. Thus, a transformer can make a load appear to be either larger or smaller than it really is. A step-up transformer will make the load (ohms) appear *smaller.* A step-down transformer will make the load (ohms) appear *larger.* The load is connected to the secondary and the apparent load seen in the primary circuit is called the **reflected impedance** of the load, as shown in Figure 11-23.

Impedance is the total opposition to AC current.

Reflected impedance is the apparent load seen at the primary.

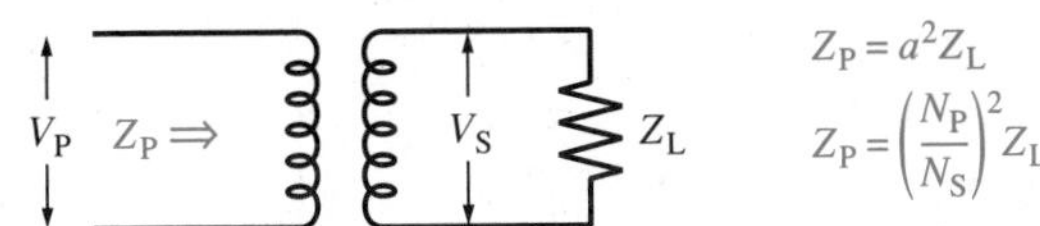

Figure 11-23 **Reflected impedance**

The reflected impedance is a function of the square of the turns ratio. Mathematically,

$$Z_P = a^2 Z_L \qquad \text{Equation 11-13}$$

where Z_P = the reflected primary impedance in ohms

Z_L = the secondary load impedance in ohms

a = the turns ratio

$$Z_P = \left(\frac{N_P}{N_S}\right)^2 Z_L$$

EXAMPLE 11-17

An 8-ohm speaker is connected to an amplifier using a step-down impedance matching transformer. If the primary coil has 1,000 turns and the secondary coil has 10 turns, calculate the reflected impedance to the amplifier. See Figure 11-24.

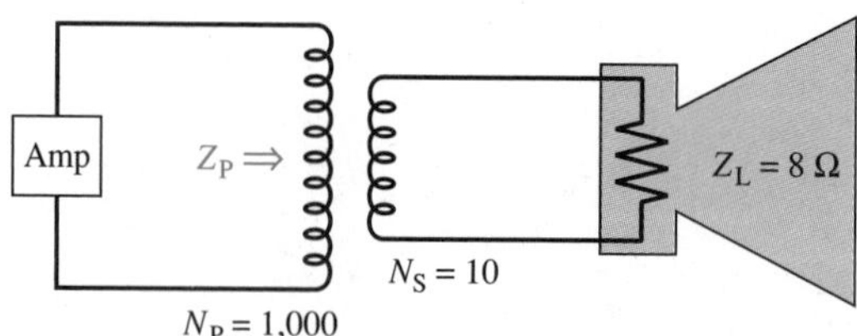

Figure 11-24

Solution:

$$Z_P = a^2 Z_L$$

$$Z_P = \left(\frac{N_P}{N_S}\right)^2 \times Z_L$$

$$Z_P = \left(\frac{1,000}{10}\right)^2 (8\ \Omega)$$

$$Z_P = 80,000\ \Omega$$

$$Z_P = 80\ \text{K}\Omega$$

EXAMPLE 11-18

Mr. Dunnigan wishes to connect an 8-ohm speaker to his amplifier. Determine the turns ratio required to achieve maximum power transfer if the amplifier has an output impedance of 5,000 ohms. See Figure 11-25.

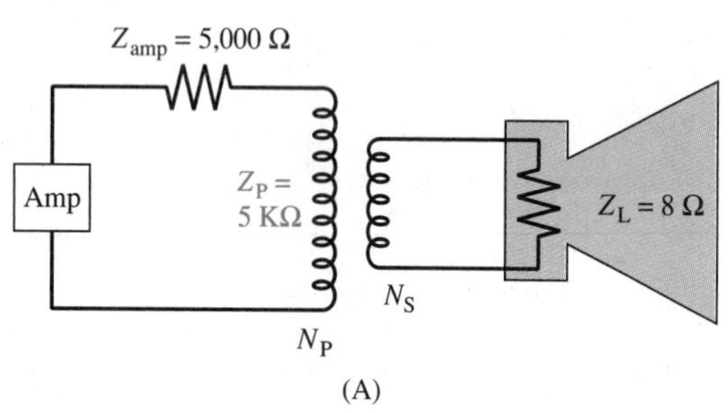

Figure 11-25

Solution:

$$Z_P = a^2 Z_L$$

$$5,000\ \Omega = a^2 (8\ \Omega)$$

$$a^2 = \frac{5,000\ \Omega}{8\ \Omega}$$

$$a^2 = 625$$

$$\sqrt{a^2} = \sqrt{625}$$

$$a = 25 = \left(\frac{N_P}{N_S}\right)$$

turns ratio of 25 to 1

✓ Self-Test 4

1. When mutual inductance exists a change in current in one coil induces a ______________ in the other coil.
2. Transformer design is dependent on ______________, ______________, and ______________.
3. Transformers transfer power from the ______________ to the ______________.
4. The unit of impedance is the ______________.
5. A step-up transformer will make the reflected impedance appear ______________.

11.11 Transformer Applications

tech tidbit

Transformers are used to step up and step down voltage.

Transformers are used in power transmission because they can easily and efficiently raise and lower voltage and current. Recall that the energy into a transformer primary is approximately equal to the energy out of the transformer secondary. $V_P I_P = V_S I_S$, so if the voltage from the primary to the secondary is stepped up, the current must be stepped down. This is very important in power transmission. The power lost as heat due to the resistance of the power line is equal to I^2R. By stepping up the voltage for long-distance power transmission, the current is reduced. This causes a tremendous reduction in the amount of power lost due to heat, because the power lost is a function of the current squared. Thus $P_{loss} = I^2R$, where I is the current in the transmission line and R is the resistance of the line.

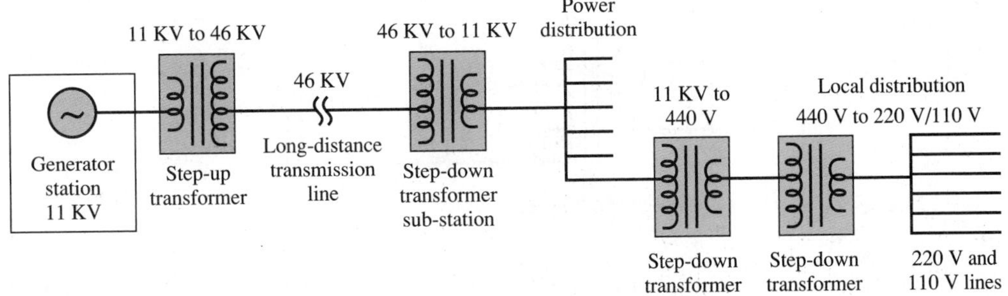

Figure 11-26 **Power distribution using transformers**

tech tidbit

The voltage is stepped up for long-distance power transmission in order to lower the current and the power loss.

In Figure 11-26 a generating station produces 11,000 volts for transmission to a city many miles away. The voltage is first stepped up to 46,000 volts, which has the effect of reducing the current in the long-distance power line. As we approach the city, the voltage is stepped down to a lower voltage (11,000 volts) for distribution at a nearby substation. The stepped-down voltage is then sent to a local distribution station and further stepped down to 440 volts. Finally, it is sent on local utility poles and stepped down once more to the 220-volt and 110-volt service we use in our homes.

tech tidbit

High voltage produces electromagnetic radiation.

The high voltage (as high as 1,000,000 volts) used in long-distance power transmission reduces the power loss due to the resistance of the wire because of the low current. However, very high voltages produce electromagnetic radiation, which poses health and environmental concerns. For this and other safety reasons, as we approach a city or populated area, voltage is stepped down to a lower level.

Transformers sometimes have multiple secondaries to provide different voltage outputs simultaneously. Sometimes a single secondary winding is **tapped** or divided into two or more parts to provide different voltage outputs. Figure 11-27 illustrates both a multiple secondary and a tapped secondary transformer.

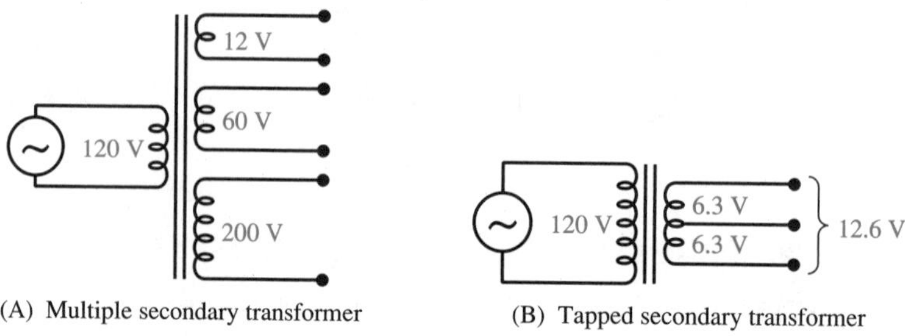

Figure 11-27 **Multioutput transformer configurations**

tech tidbit

Secondaries in parallel and in phase will provide more current.

Multiple secondary transformers can be used to raise and lower voltages and currents by summation. Connecting two secondaries (with the same voltage) in parallel and in phase has the affect of providing more current to the load. Figure 11-28 illustrates how the connections are made. Note that the dots are connected together to provide the in-phase connection. Note also that the secondary voltage does not change because both secondaries are in parallel.

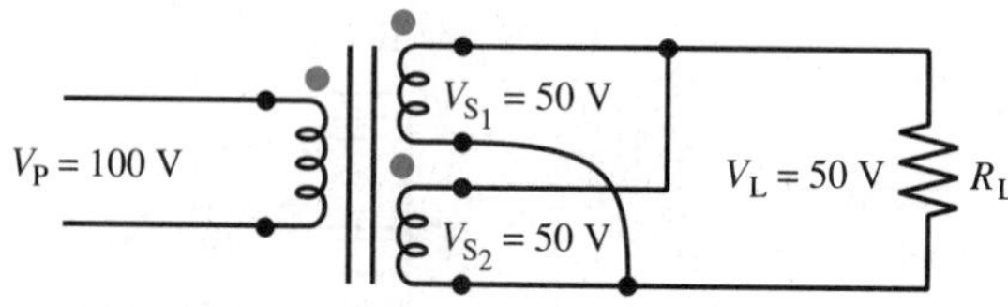

Figure 11-28 **Parallel secondaries**

tech tidbit

When secondaries are in series, the total voltage will add or subtract.

When secondaries are connected in series, the total voltage can be the sum of the two secondary voltages or the difference between the two secondary voltages. The result depends on the phase connections. In Figure 11-29A, the secondaries are connected in phase (series aiding), which results in the addition of the secondary voltages. In Figure 11-29B, the secondaries are connected out of phase (series opposing), which results in the subtraction of the secondary voltages.

Note that the dot determines whether the polarity is in phase or out of phase. If we consider the half cycle where the dot is positive, then all dots will be positive at the same time. Thus, by Kirchhoff's voltage law, we can see the addition or subtraction of the voltage.

Transformers can be made variable just as resistors can be made variable. Figure 11-30 shows the symbol for a variable transformer. As the **wiper** or variable arm is moved up and down, it contacts the turns on the secondary and results in a different output voltage. Variable transformers can also be made by controlling the position of the iron core. Moving the core in and out of the magnetic field results in a variable output voltage. Another way of making a variable

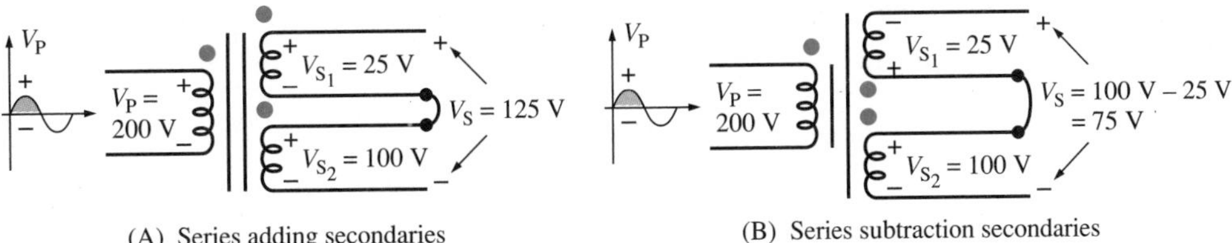

(A) Series adding secondaries

(B) Series subtraction secondaries

Figure 11-29 **Transformer secondaries in series**

transformer is to change the position or angle of the secondary winding with respect to the primary winding. Some variable transformers are also known as **variacs.**

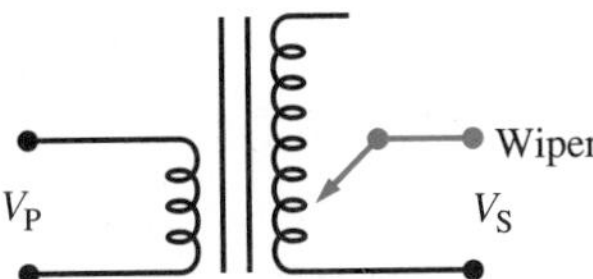

Figure 11-30 **Variable transformer**

A transformer with a common primary and secondary winding is called an **autotransformer.** Autotransformers have the primary and secondary windings connected both magnetically and electrically. These transformers function like normal transformers but do not provide secondary isolation. This is because the secondary is electrically connected to the primary. Autotransformers are generally cheaper to manufacture. The high voltage coil (induction coil) or transformer in an automobile is an example of an autotransformer. Figure 11-31 shows the schematic symbol for an autotransformer.

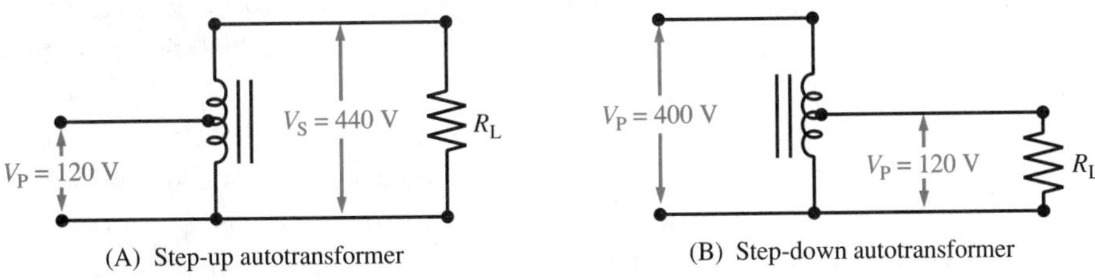

(A) Step-up autotransformer

(B) Step-down autotransformer

Figure 11-31 **Autotransformers**

11.12 Transformer Efficiency

Efficiency— ratio of power out to power in.

Efficiency is defined as the ratio of the power out to the power in. Up to this point, we have assumed that transformers are 100% efficient. However, real transformers have losses between the primary or input power and the secondary or output power. Mathematically,

$$\eta_{(\%)} = \frac{P_{out}}{P_{in}} \times 100\%$$

Equation 11-14

where $\eta_{(\%)}$ = the percent efficiency

P_{out} = the secondary output power in watts

P_{in} = the primary input power in watts

It should also be noted that the power out is equal to the power in minus the power lost. Mathematically,

$$P_{out} = P_{in} - P_{lost}$$

EXAMPLE 11-19

The input power to the primary of a transformer is 1,000 watts. Calculate the efficiency if the output power is 950 watts.

Solution:

$$\eta_{(\%)} = \frac{P_{out}}{P_{in}} \times 100\%$$

$$= \frac{950\ W}{1{,}000\ W} \times 100\%$$

$$= 0.95 \times 100\%$$

$$= 95\%$$

tech tidbit

Transformer heat losses are known as copper losses.

tech tidbit

Eddy currents can be reduced by laminating the iron core.

Transformers get warm to the touch, so it follows that some of the input power is being converted to heat. The heat energy is power lost by the transformer. It is due to the resistance of the windings themselves and can be calculated from the power formula $P = I^2R$. Heat energy losses are also known as **copper losses** because they are a result of the resistance of the copper wire used to make the windings. In addition to copper losses, there are two types of **iron losses** that result from the iron core of the transformer. These two types of iron losses are called eddy currents and hysteresis.

Eddy currents are small circulating currents induced by the AC voltage in the core. In a transformer, voltage is induced into the secondary winding and also into the iron core itself. This is because the iron core is a conductor. The induced voltage in the iron core results in small currents called eddy currents that heat up the core. This heat is another form of lost power. To reduce the effects of eddy currents, transformer cores are **laminated.** A laminated core is a core made up of many thin layers (insulated from each other) instead of one solid thick piece of iron. Laminating a core reduces the power lost due to eddy currents. Often the laminations vibrate and can be heard when a transformer operates. Figure 11-32 illustrates the effects of eddy currents and core lamination of a transformer.

Hysteresis is the power lost as a result of the need to reverse the magnetic field in an iron core of the transformer. The faster the AC current reverses, the greater the hysteresis losses. Figure 11-33 shows a graph of flux density versus the magnetizing force in a transformer. As the molecules set up the magnetic field in one direction, they lag behind the magnetizing force. The curve in Figure 11-33 is sometimes referred to as a hysteresis loop. The greater the area of the loop, the greater the power lost.

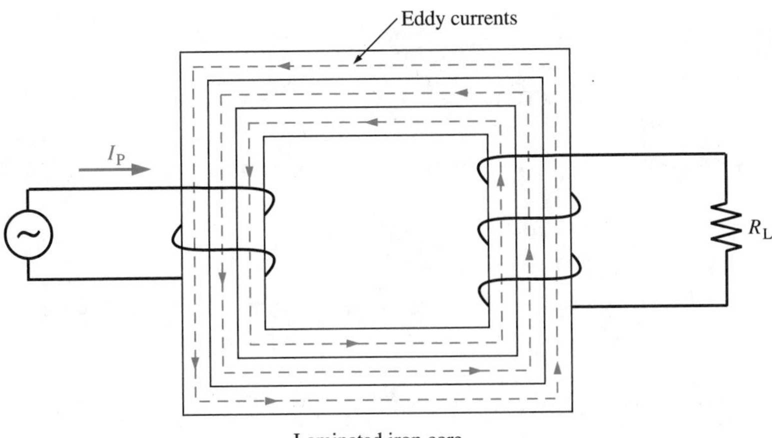

Figure 11-32 **Eddy currents**

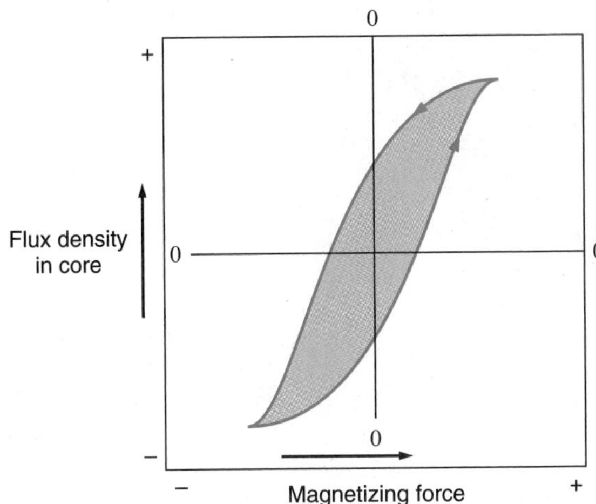

Figure 11-33 **Hysteresis curve**

✓ Self-Test 5

1. The power lost as heat due to the resistance of a wire is equal to _______________.
2. When the secondary of a transformer is divided into two or more parts, it is referred to as a _______________ winding.
3. The movable arm in a variable transformer is called the _______________.
4. _______________ transformers have the primary and secondary windings connected magnetically and electrically.
5. Two types of iron core losses in a transformer are called _______________ and _______________.

3 tech tips & trouble-shooting

T3 ISOLATION TRANSFORMERS

Transformers are often used to isolate a device from ground. **Isolation** means that the device has no physical electrical connection to ground. Isolation provides safety from electrical shock. Figure 11-34 shows a technician working on a TV set. One side of the power receptacle is connected to earth ground. If the technician is standing on a grounded surface like a cement floor, and touches a component connected to the hot side of the receptacle, the technician will get an electric shock. Current can flow through the technician's body, through the floor to ground. An isolation transformer removes the direct connection to ground and helps to protect against electric shock.

Isolation transformers are also useful when making electrical measurements. In Figure 11-35, the ground side of the oscilloscope shorts out resistors R_2 and R_3. Using an isolation transformer removes the direct connection to ground and allows correct measurements to be made.

2 tech tidbit

Transformer isolation provides safety from electric shock.

2 tech tidbit

Damp or moist cement is a good conductor.

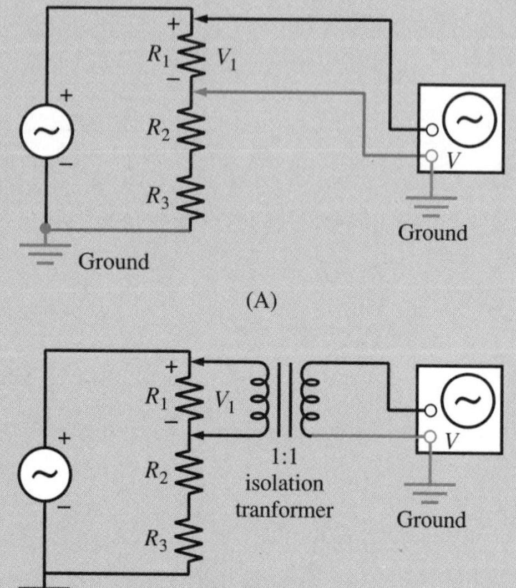

Figure 11-35 Isolation measurements

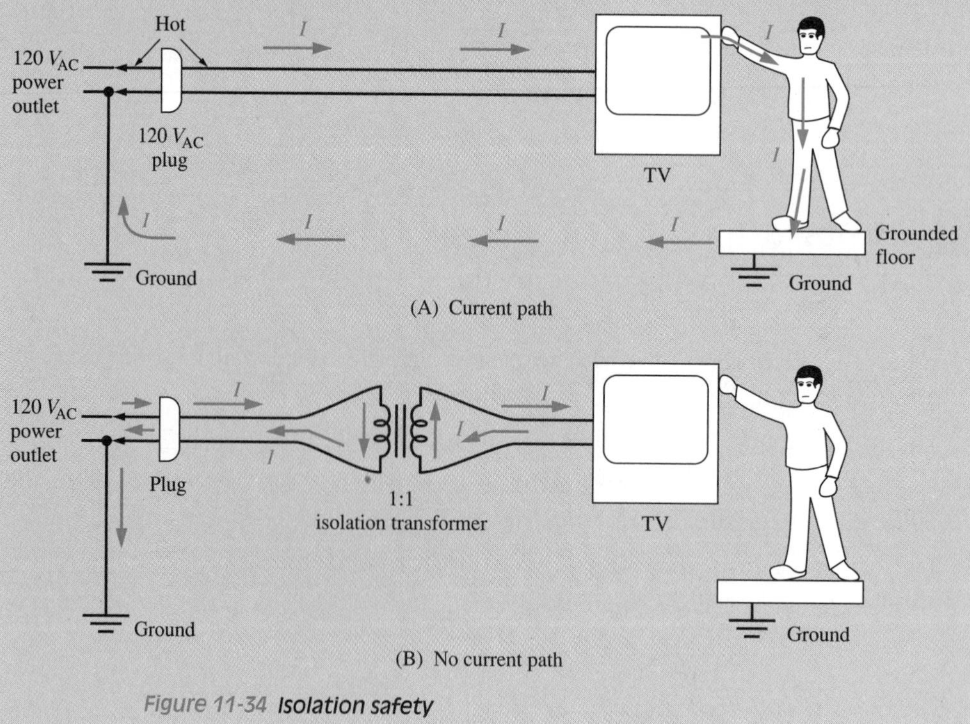

Figure 11-34 Isolation safety

SUMMARY

Inductance is defined as the property or ability of a conductor to oppose a change in current through the conductor. The amount of inductance that an inductor has is determined by the type and cross-sectional area of the core, the length of the core, and the number of turns of wire. The unit of inductance is the henry.

Lenz's law states that whenever a magnetic field expands or collapses across a coil of wire, a voltage will be induced in the coil of wire. The induced voltage will always be in the opposite direction to the current that created the magnetic field.

When inductors are in series or in parallel, the total inductance is calculated in the same way that the total resistance is calculated when resistors are in series or parallel. The greater the amount of inductance in a circuit, the greater the time required for the current or voltage to reach its steady-state value. One time constant, τ, is the time required for the current or voltage to reach 63.2% of its steady state value. Five time constants are required to reach steady state.

The effect or opposition of the induced voltage in an inductor can be equated to ohms and is called inductive reactance, X_L. The ratio of the inductive reactance of a coil to the resistance of the coil is referred to as the quality, Q, of the coil.

Transformers are often used to change voltage, current, or impedance levels in a circuit. In a step-up transformer, the secondary winding has more turns than the primary winding. In a step-down transformer, the secondary winding has less turns than the primary winding. Transformer efficiency is the ratio of the output power to the input power of the transformer. Power losses result from copper losses and from iron losses.

FORMULAS

$$L = \frac{KN^2}{l}$$

$$V_L = L \times \frac{\Delta i}{\Delta t}$$

$$L_T = L_1 + L_2 + L_3 + \ldots + L_N$$ (inductors in series)

$$\frac{1}{L_T} = \frac{1}{L_1} + \frac{1}{L_2} + \frac{1}{L_3} + \ldots + \frac{1}{L_N}$$ (inductors in parallel)

$$L_T = \frac{L_1 L_2}{L_1 + L_2}$$ (two inductors in parallel)

$$L_T = \frac{L}{N}$$ (equal inductors in parallel)

$$\tau = \frac{L}{R}$$ Steady state = 5τ

$$W_L = \frac{1}{2}LI^2$$

$$X_L = 2\pi fL$$

$$Q = \frac{X_L}{R}$$

$$\frac{V_P}{V_S} = \frac{I_S}{I_P}$$

$$\frac{N_P}{N_S} = \frac{V_P}{V_S}$$

$$\frac{N_P}{N_S} = \frac{I_S}{I_P}$$

$$a = \frac{N_P}{N_S}$$

$$Z_P = a^2 Z_L \qquad Z_P = \left(\frac{N_P}{N_S}\right)^2 Z_L$$

$$\frac{V_P}{V_S} = \frac{I_S}{I_P} = \frac{N_P}{N_S} = a$$

$$\eta_{(\%)} = \frac{P_{out}}{P_{in}} \times 100\%$$

EXERCISES

Section 11.3

1. The current through a 5-henry inductor increases from 0 to 3 amperes in 10 milliseconds. Find the voltage induced across the coil.
2. Find the induced voltage across a 10-henry coil if the rate of change of current through the coil is 0.1 ampere/second.
3. Find the inductance of a coil if the induced voltage across the coil is 10 volts when the current change is 0.5 ampere/second.

Section 11.4

4. Find the total inductance when a 2-henry, 8-henry, 5-henry, and a 10-henry coil are all connected in series.
5. Find the total inductance when a 5-millihenry, 12-millihenry, and an 8-millihenry coil are all connected in series.
6. Find the total inductance when a 1-millihenry, a 500-microhenry, and a 1,500-microhenry coil are all connected in series.
7. Four 10-henry coils are connected in parallel. Find the total inductance.
8. A 3-henry coil and a 6-henry coil are connected in parallel. Find the total inductance.
9. A 12-henry coil, a 6-henry coil, and a 4-henry coil are connected in parallel. Find the total inductance.

10. Find the total inductance for the network of Figure 11-36.

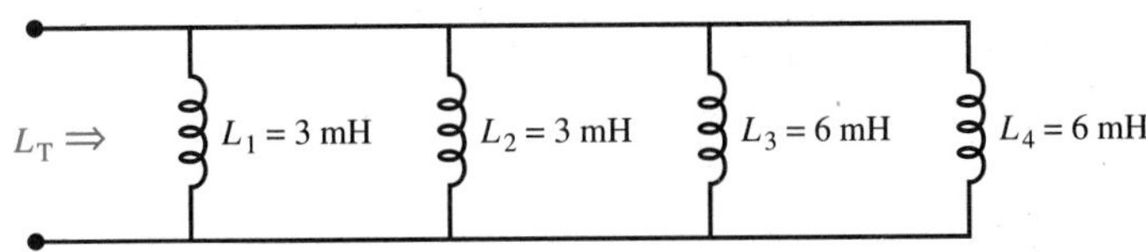

Figure 11-36

11. Find the total inductance of the network of Figure 11-37.

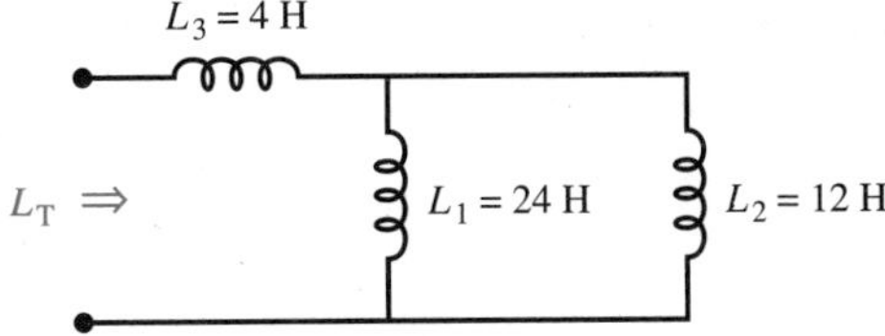

Figure 11-37

12. Find the total inductance of the network of Figure 11-38.

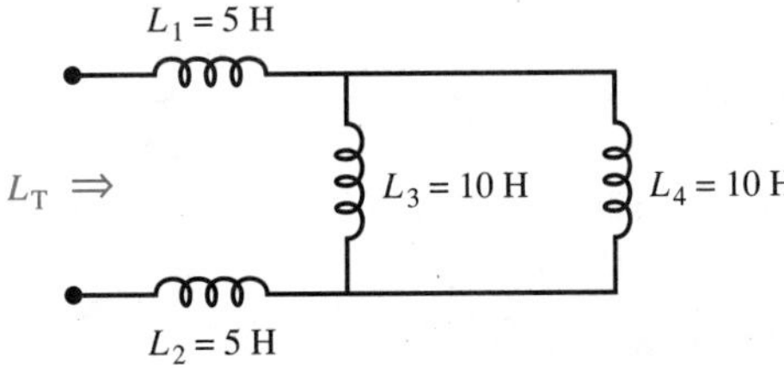

Figure 11-38

Section 11.5

13. Find the time constant for an 8-henry inductor in series with a 1-kilohm resistor.

14. Find the time constant for a 10-millihenry inductor in series with a 50-ohm resistor.

15. For the circuit of Figure 11-39, $L = 10$ H, $R = 100\ \Omega$, and $V = 150$ V. Find the time required to reach steady state and the value of the current at steady state. Calculate the energy stored in the coil at steady state.

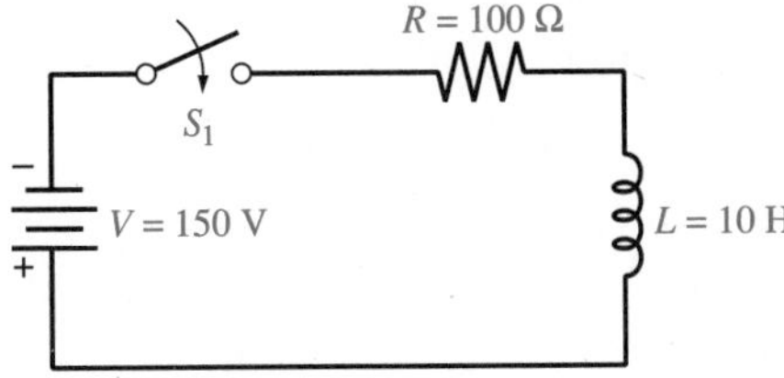

Figure 11-39

16. For the circuit of problem 15, find the value of the current two time constants after the switch is closed.
17. For the circuit of Figure 11-39, L = 50 mH, R = 25 Ω, and V = 100 V. Find the time constant of the circuit and the current at one time constant after the switch is closed.

Section 11.6

18. Find the inductive reactance of a 10-henry coil at 60 hertz.
19. Find the inductive reactance of a 10-millihenry coil at 60 hertz and at 60 kilohertz.
20. Calculate the inductive reactance of a 5-henry coil at 10 hertz, 100 hertz, and 1,000 hertz. Plot a graph of frequency versus X_L.

Section 11.7

21. Find the quality (Q) of a 10-ohm, 4-henry coil at 100 hertz.
22. Find the quality (Q) of a 100-ohm, 7-millihenry coil at 60 KHz.
23. A highly selective radio tuner at 100 megahertz requires a coil with a Q of 5,000. If the X_L required for maximum power transfer is 15 kilohms, find L and R_L of the coil.

Section 11.9

24. A step-up transformer has 50 turns in the primary coil and 500 turns in the secondary coil. If 120 volts are applied to the primary, calculate the voltage induced into the secondary.
25. Find the applied voltage to a transformer with a secondary voltage of 440 volts when N_P = 120 turns and N_S = 480 turns.
26. Find the turns ratio of a 120 V/600 V step-up transformer.
27. A 240 V/24 V step-down transformer is connected to a 6-ohm load. Calculate the primary current.
28. If V_L = 120 V, R_L = 20 Ω, I_P = 0.5 A, and N_S = 100 turns, find the number of turns in the primary of the transformer.

Section 11.10

29. A 4-ohm speaker is connected to an amplifier using a step-down transformer. If the primary coil has 600 turns and the secondary coil has 30 turns, calculate the reflected impedance to the amplifier.
30. A 4-ohm speaker is to be connected to an amplifier whose output impedance is 64 ohms. Find the turns ratio of a transformer required to deliver maximum power to the speaker.

Section 11.11

31. Find the output voltage V_S in Figure 11-40A.
32. Find the output voltage V_S in Figure 11-40B.
33. Find the output voltage V_S in Figure 11-40C.

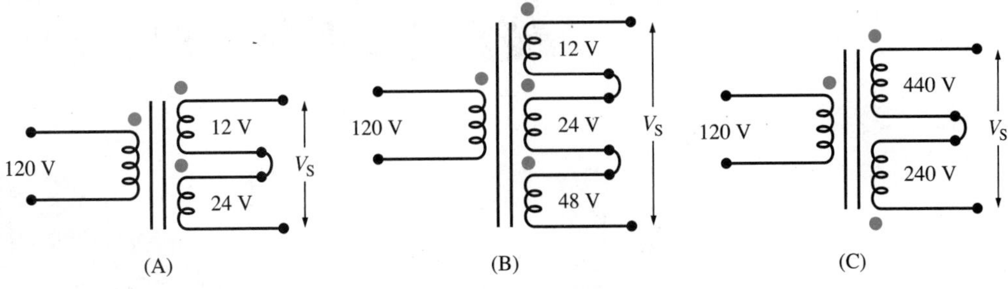

Figure 11-40

Section 11.12

34. The input power to the primary of a transformer is 1,500 watts. Calculate the efficiency if the output power is 1,425 watts.
35. A transformer has a primary voltage of 120 volts and a primary current of 15 amperes. The transformer loses 100 watts in copper losses and 80 watts in core losses. Find the efficiency of the transformer.

Section T^3

36. An isolation transformer is being used to measure a 40,000-volt power line. Find the voltage measured on the secondary if the transformer has a turns ratio of 1,000 to 1.
37. A voltmeter can measure up to 100 volts. Find the turns ratio of a transformer required to measure an 11,000-volt power line for full-scale deflection of the meter.
38. See Lab Manual for the crossword puzzle for Chapter 11.

CHAPTER 12

RLC CIRCUITS

Courtesy of Motorola, Inc.

KEY TERMS

attenuate
band pass filter
band stop filter
bandwidth
bass
cut-off frequency
direction
ELI the ICE man
filter
half-power points
high pass filter
impedance
low pass filter
magnitude
notch filter
phasor
quality (*Q*)
resonance
resonant frequency
selectivity
tank circuit
treble

OBJECTIVES

After completing this chapter, the student should be able to:

- analyze RLC circuits in the DC steady state condition.
- express AC quantities in phasor notation.
- calculate the total impedance in an AC circuit.
- analyze RL circuits.
- analyze RC circuits.
- analyze RLC circuits.
- calculate the resonant frequency of an RLC circuit.
- determine the properties of a series resonant circuit.
- determine the properties of a parallel resonant circuit.
- identify and analyze filter circuits.

12.1 Introduction

Most electric circuits include combinations of resistors, inductors, and capacitors. Recall that a resistor is something that offers opposition to the current and results in the transformation of electric energy into heat. For a pure resistance this opposition is the same in DC and in AC. An inductor's opposition to AC current is known as inductive reactance, or X_L. In DC, because an inductor is a coil of wire, it normally appears as a low resistance. For this reason, an inductor can be considered a short circuit (because its resistance is practically zero ohms) in the steady state. A capacitor's opposition to AC current is known as capacitive reactance, or X_C. In DC, because a capacitor is two plates (conductors) separated by an insulator, it normally appears as a high resistance. For this reason, a capacitor can be considered as an open circuit (infinite ohms) in the steady state. Thus, when analyzing RLC circuits in the DC steady state condition, we can replace all inductors with short circuits and all capacitors with open circuits, as shown in Figure 12-1.

tech tidbit

Inductive reactance (X_L)— An inductor's opposition to AC current.

tech tidbit

Capacitive reactance (X_C)— A capacitor's opposition to AC current.

tech tidbit

In the steady state, replace inductors with shorts; replace capacitors with opens.

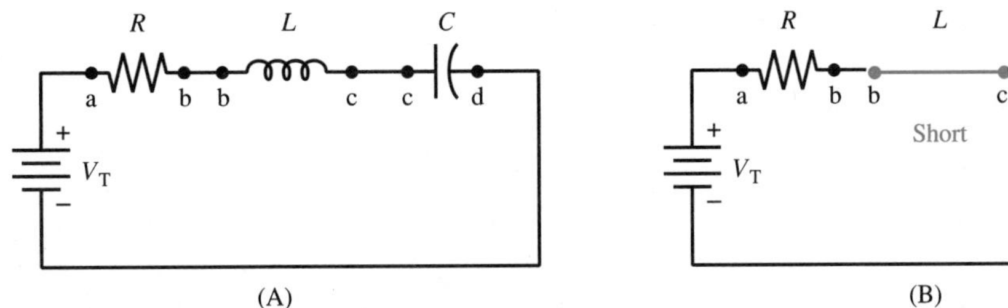

Figure 12-1 RLC steady state DC circuit

EXAMPLE 12-1

Find (A) the steady state current I_T and (B) the voltage drop across each resistor in Figure 12-2.

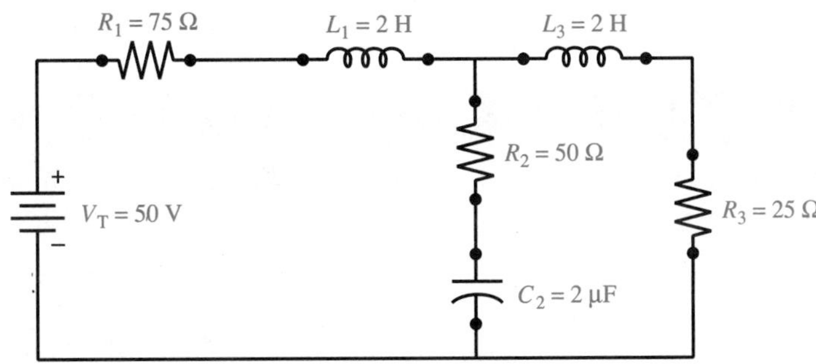

Figure 12-2

Solution:

See Figure 12-3.

(A) $R_T = R_1 + R_3$

$= 75\ \Omega + 25\ \Omega$

$= 100\ \Omega$

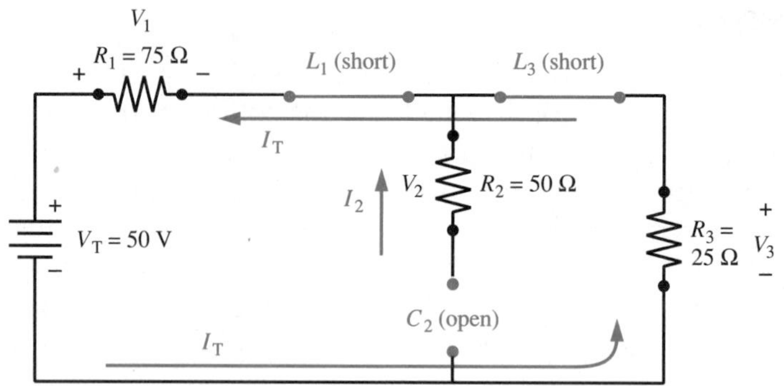

Figure 12-3

$$I_T = \frac{V_T}{R_T}$$

$$= \frac{50\ \text{V}}{100\ \Omega}$$

$$= 0.5\ \text{A}$$

(B) $V_1 = I_T R_1$

$$= (0.5\ \text{A})(75\ \Omega)$$

$$= 37.5\ \text{V}$$

$$V_3 = I_T R_3$$

$$= (0.5\ \text{A})(25\ \Omega)$$

$$= 12.5\ \text{V}$$

because the middle branch is an open circuit, $I_2 = 0$

$$V_2 = I_2 R_2$$

$$= (0\ \text{A})(50\ \Omega)$$

$$= 0\ \text{V}$$

EXAMPLE 12-2

Find (A) the total current in Figure 12-4 and (B) the steady state branch currents.

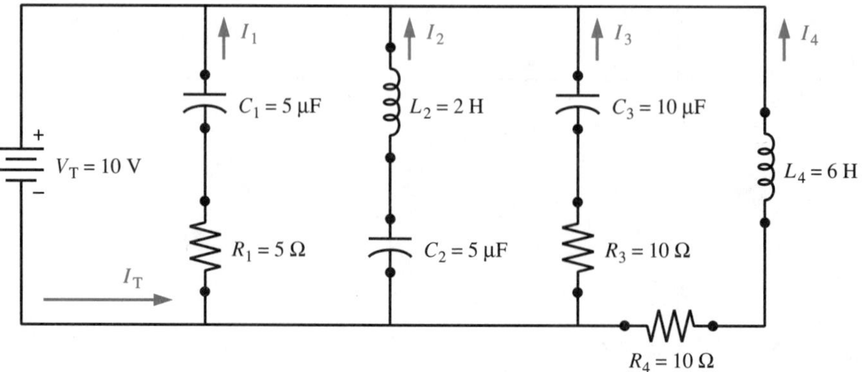

Figure 12-4

Solution:

See Figure 12-5.

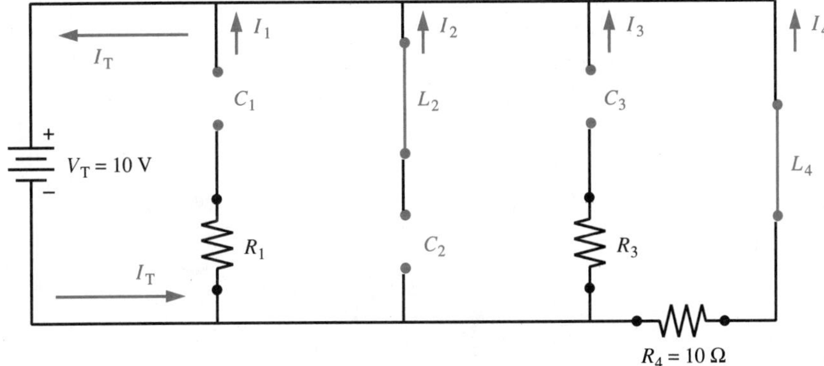

Figure 12-5

(A) $R_T = R_4$

$= 10\ \Omega$

Note: R_1 and R_3 are in "open" branches.

$$\therefore\ I_T = \frac{V_T}{R_T} = \frac{10\text{ V}}{10\ \Omega} = 1\text{ A}$$

(B) $I_1 = 0$ A (C₁ opens branch)

$I_2 = 0$ A (L_2 = short, C_2 = open)

$I_3 = 0$ A (C_3 opens branch)

$I_4 = I_T$ (total current goes to R_4)

12.2 Phasors

Phasors, like vectors in mathematics, are quantities that have magnitude and direction. A phasor is a type of graph used to illustrate the action of opposing forces. The **magnitude** of the phasor is represented by the length of the phasor line. The **direction** is represented by the angle at which the phasor line is drawn. Figure 12-6 illustrates how sinusoidal waveforms can be represented by a phasor diagram. The RMS value of the voltage is used to represent the magnitude of the phasor. The phase angle is used to represent the direction or its relationship to time. Phase angles always increase in value as the phasor is rotated counterclockwise on the phasor diagram. The reference (0 degrees) is always taken as the positive horizontal axis.

Phasors have magnitude and direction.

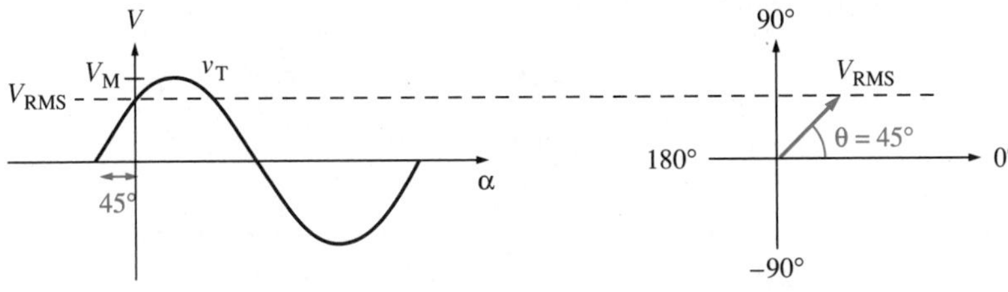

Figure 12-6 Phasors

Figure 12-7 shows how phasors can be used to describe the time relationship between two sinusoidal waveforms. In Figure 12-7, V_1 leads V_2 by 60 degrees. Notice that on the phasor diagram of Figure 12-7, V_1 is rotated 60 degrees counterclockwise from V_2. This is because V_1 leads V_2 on the sine wave diagram. V_2 is drawn on the reference line because it starts at 0 degrees. Hence when referring to either diagram, we can say that V_1 leads V_2 by 60 degrees or V_2 lags V_1 by 60 degrees.

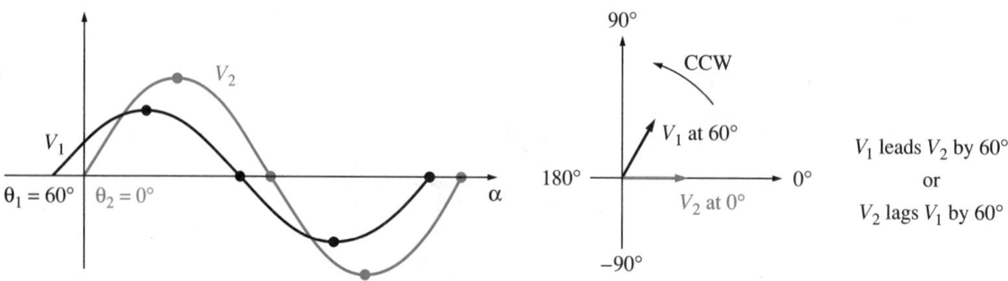

Figure 12-7 **Phase relationship**

In a purely resistive AC circuit, the voltage and current are in phase.

When an AC voltage is applied to a purely resistive circuit, the voltage and current will remain in phase. That is, as the voltage increases, the current will increase accordingly at the exact same instant in time (or the maximums occur at the same time or angle). Figure 12-8A illustrates the sinusoidal and phasor diagrams for a purely resistive circuit. Note that because the voltage and current are in phase they lie along the same line on the phasor diagram. In this case, the line is the positive horizontal axis or 0 degrees. If the voltage had started at 60 degrees, the current would also have been plotted at 60 degrees.

In a purely inductive AC circuit, *V* leads *I* by 90°.

In a purely inductive AC circuit, the voltage will lead the current by 90 degrees. Recall that the current through an inductor cannot change instantaneously because of the counter EMF induced. However the applied voltage does appear instantaneously across the inductor. It can be shown experimentally that the voltage applied to a pure inductor will always lead the current by 90 degrees or the current will lag the voltage by 90 degrees. Figure 12-8B illustrates the sinusoidal and phasor diagrams for a purely inductive circuit.

In a purely capacitive AC circuit, *I* leads *V* by 90°.

In a purely capacitive AC circuit, the current will lead the voltage by 90 degrees or the voltage will lag the current by 90 degrees. Recall that the voltage cannot change instantaneously across a capacitor due to the electrostatic field. Thus, it takes time for the capacitor to charge. The current, however, immediately begins to flow and deposit charge on the plates of the capacitor. It can be shown experimentally that the current will always lead the voltage in a purely capacitive circuit by 90 degrees. Figure 12-8C illustrates the sinusoidal and phasor diagrams for a purely capacitive circuit.

ELI—Inductor—V_L leads I_L.

One way of remembering the phase relationship for inductive and capacitive circuits is with the familiar saying

ICE—Capacitor—I_C leads V_C.

ELI the ICE man

where *ELI* stands for the voltage, *E*, across an inductor, *L*, that leads the current, *I*, by 90 degrees. And, *ICE* stands for the current, *I*, through a capacitor, *C*, that leads the voltage, *E*, by 90 degrees. Note that the symbol for voltage, *E*, is sometimes used interchangeably with *V* and is used as the symbol for electromotive force (source voltage).

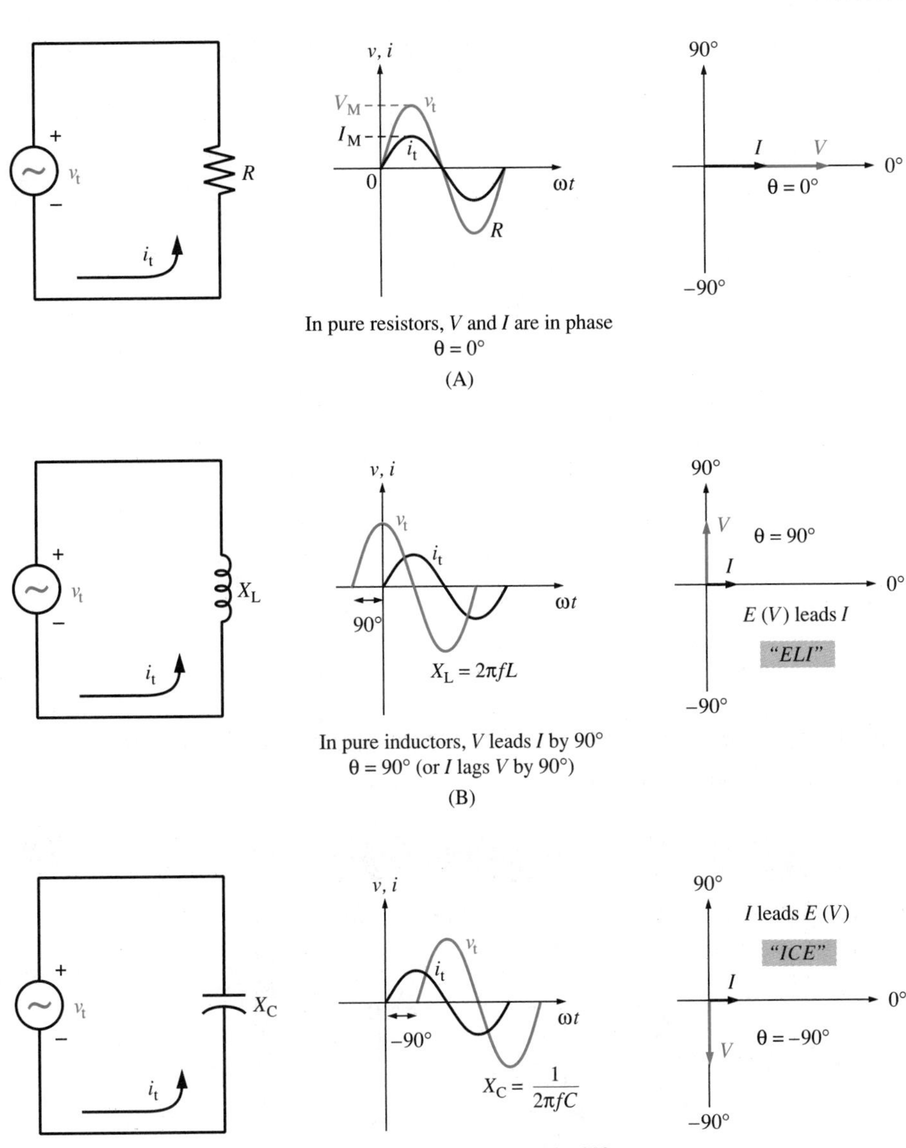

Figure 12-8 **Phasor representation**

EXAMPLE 12-3

Find the current in the circuit of Figure 12-9. Sketch the waveform and phasor diagram of the voltage and current.

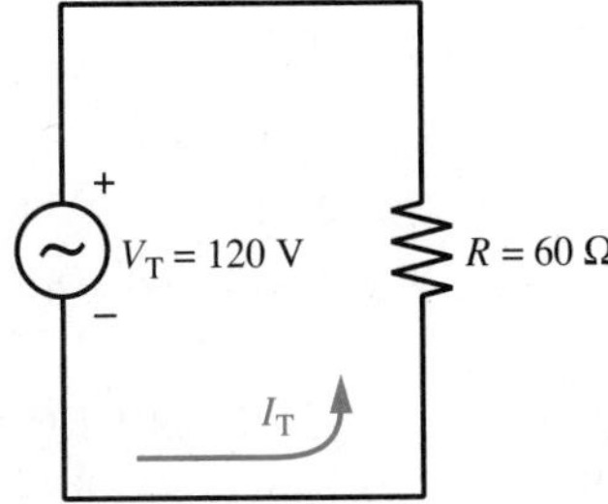

Figure 12-9

Solution:

See Figure 12-10.

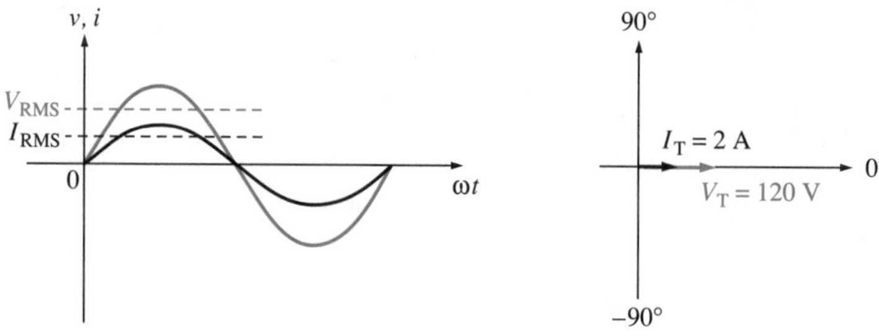

Figure 12-10

$$I_T = \frac{V_T}{R} = \frac{120 \text{ V}}{60 \ \Omega} = 2 \text{ A}$$

Note: V_T and I_T are in phase. This can be seen from the waveform diagram and phasor diagram.

EXAMPLE 12-4

Find the current in the circuit of Figure 12-11. Sketch the waveform and phasor diagram of the voltage and current.

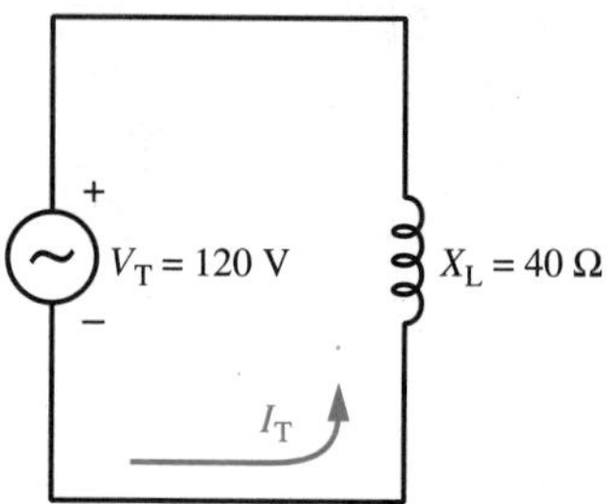

Figure 12-11

Solution:

See Figure 12-12.

$$I_T = \frac{V_T}{X_L} = \frac{120 \text{ V}}{40 \ \Omega} = 3 \text{ A}$$

Note: V_T (*E*) leads I_T by 90° (*ELI*). This can be seen from the waveform diagram and phasor diagram. (Also, *I* lags V_T by 90°.)

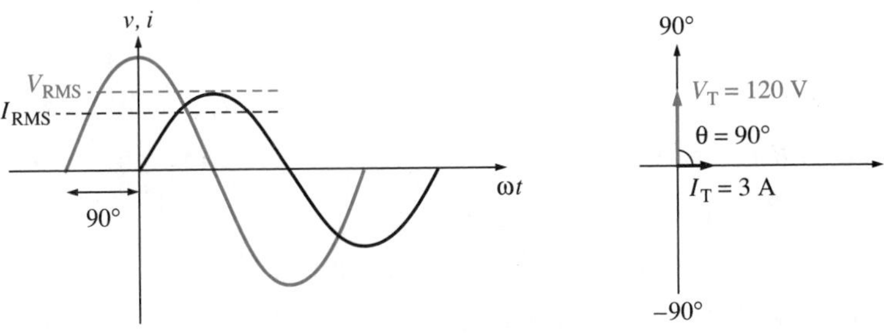

Figure 12-12

EXAMPLE 12-5

Find the current in the circuit of Figure 12-13. Sketch the waveform and phasor diagram of the voltage and current.

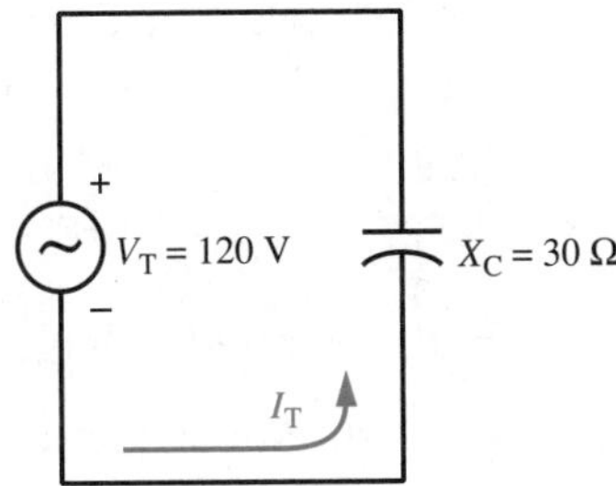

Figure 12-13

Solution:

See Figure 12-14.

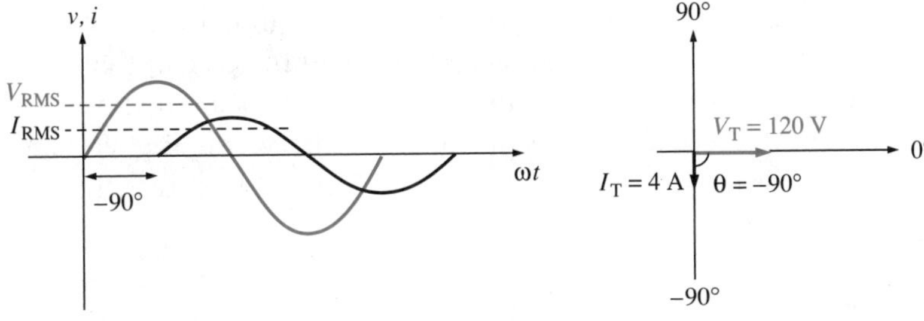

Figure 12-14

$$I_T = \frac{V_T}{X_C}$$

$$= \frac{120\ \text{V}}{30\ \Omega}$$

$$= 4\ \text{A}$$

Note: I_T leads V_T (*E*) by 90° (*ICE*). This can be seen from the waveform diagram and phasor diagram.

✓ Self-Test 1

1. An inductor can be considered a ______________ in the steady state.
2. A capacitor can be considered an ______________ in the steady state.
3. Phasors are quantities that have ______________ and ______________.
4. In purely resistive AC circuits, the voltage and current are ______________.
5. In purely inductive AC circuits, the voltage will ______________ the current by 90 degrees.
6. In purely capacitive AC circuits, the current will ______________ the voltage by 90 degrees.

Impedance, Z_T, is the total opposition to AC current.

12.3 Impedance

Impedance is the total opposition to current in an AC circuit. The impedance of an AC circuit is expressed in ohms and uses the letter Z as its symbol. Impedance can be defined using Ohm's law as

Equation 12-1 ▶

$$Z_T = \frac{V_T}{I_T}$$

where V_T = the applied voltage in volts

I_T = the total current in amperes

Z_T = the total impedance in ohms

Using algebra we can rewrite the formulas for impedance as follows:

Equation 12-2 ▶

$$V_T = I_T Z_T$$

Equation 12-3 ▶

$$I_T = \frac{V_T}{Z_T}$$

$Z_R = R < 0°$

The impedance of a resistor is equal to the value of the resistor (R) in ohms. It can be expressed as Z_R on the phasor diagram. Recall that a phasor has magnitude (size) and direction. The magnitude is equal to the resistance (R) in ohms. The direction is equal to the phase angle between the voltage and the current (θ). For a resistor, the phase angle is 0 degrees because the voltage and current are in phase, as shown in Figure 12-15.

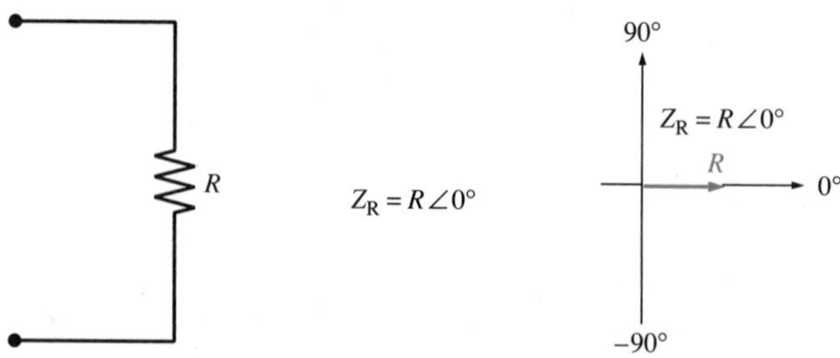

$Z_R = R\angle 0°$ is read as the impedance Z_R is equal to the resistance R at a phase angle of zero degrees.

Figure 12-15 **Impedance of a resistor**

The impedance of an inductor is equal to the value of the inductive reactance (X_L) in ohms. It is expressed as Z_L in our phasor diagram. The magnitude is equal to the inductive reactance (X_L) in ohms. The direction is equal to the phase angle between the voltage and the current (θ). Recall that for a pure inductor, the voltage leads the current by 90 degrees (ELI). Therefore, Z_L is equal to X_L at a phase angle of +90 degrees, as shown in Figure 12-16.

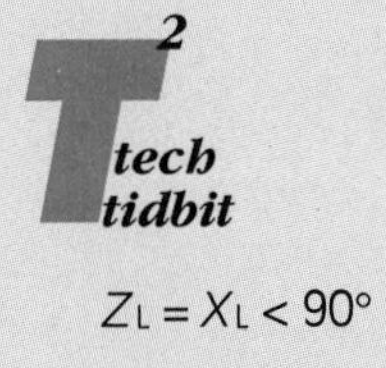

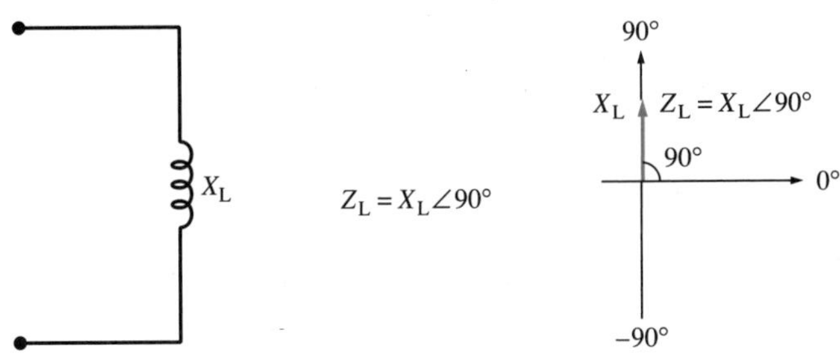

$Z_L = X_L \angle 90°$ is read as the impedance Z_L is equal to the reactance X_L at a phase angle of +90 degrees.

Figure 12-16 **Impedance of an inductor**

The impedance of a capacitor is equal to the value of the capacitive reactance (X_C) in ohms. It is expressed as Z_C on the phasor diagram. The magnitude is equal to the capacitive reactance (X_C) in ohms. The direction is equal to the phase angle between the voltage and the current (θ). Recall that for a pure capacitor, the current leads the voltage by 90 degrees (*ICE*). Therefore, Z_C is equal to X_C at a phase angle of −90 degrees, as shown in Figure 12-17.

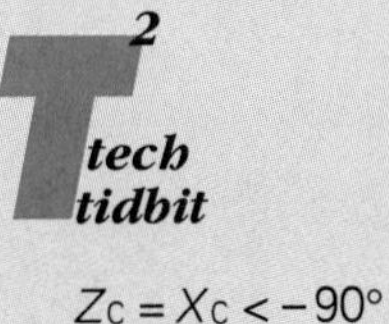

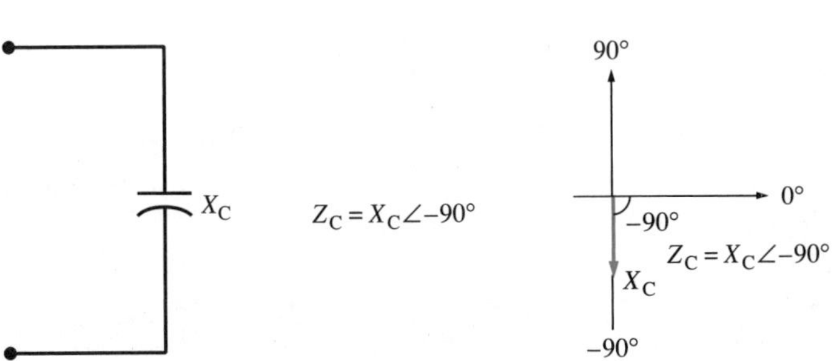

$Z_C = X_C \angle -90°$ is read as the impedance Z_C is equal to the reactance X_C at a phase angle of −90 degrees.

Figure 12-17 **Impedance of a capacitor**

EXAMPLE 12-6

For the circuit of Figure 12-18, find (A) Z_T, (B) I_T, (C) V_1, and (D) V_2.

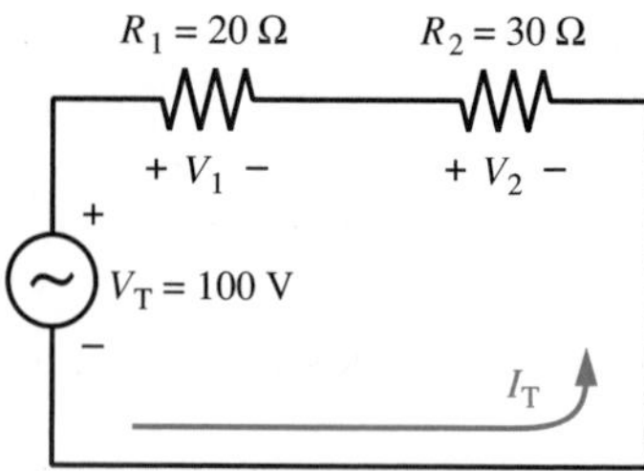

Figure 12-18

Solution:

(A) $Z_T = Z_{R_1} + Z_{R_2}$

$= 20\ \Omega + 30\ \Omega$

$= 50\ \Omega$

(B) $I_T = I_1 = I_2$ (series circuit)

$$I_T = \frac{V_T}{Z_T}$$

$$= \frac{100\text{ V}}{50\ \Omega}$$

$= 2$ A

(C) $V_1 = I_1 Z_{R_1}$

$= (2\text{ A})(20\ \Omega)$

$= 40$ V

(D) $V_2 = I_2 Z_{R_2}$

$= (2\text{ A})(30\ \Omega)$

$= 60$ V

Note that in Example 12-6, the magnitude of the impedance $Z_{R_1} = R_1$. The magnitude of the impedance $Z_{R_2} = R_2$. The circuit is purely resistive, so the total impedance Z_T is the arithmetic sum of $Z_{R_1} + Z_{R_2}$.

EXAMPLE 12-7

For the circuit of Figure 12-19, find (A) Z_T, (B) I_T, (C) V_1, and (D) V_2. (E) Sketch the waveform of V_T versus I_T and show the equivalent phasor diagram.

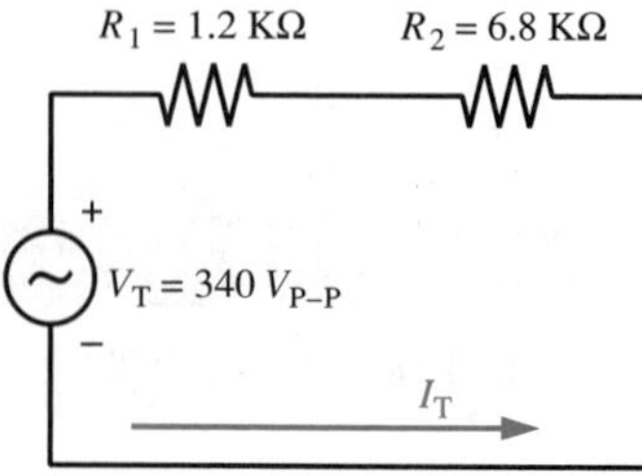

Figure 12-19

Solution:

(A) $Z_T = Z_{R_1} + Z_{R_2}$

$= 1.2\ K\Omega + 6.8\ K\Omega$

$= 8\ K\Omega$

(B) $V_P = \dfrac{V_{P-P}}{2} = \dfrac{340\ V_{P-P}}{2} = 170\ V_P$

$V_{T_{RMS}} = (0.707)(V_P)$

$= (0.707)(170\ V_P)$

$= 120\ V$

$I_T = \dfrac{V_T}{Z_T}$

$= \dfrac{120\ V}{8\ K\Omega}$

$= 15\ mA$

(C) $V_1 = I_T Z_{R_1}$

$= (15\ mA)(1.2\ K\Omega)$

$= 18\ V$

(D) $V_2 = I_T Z_{R_2}$

$= (15\ mA)(6.8\ K\Omega)$

$= 102\ V$

(E) See Figure 12-20.

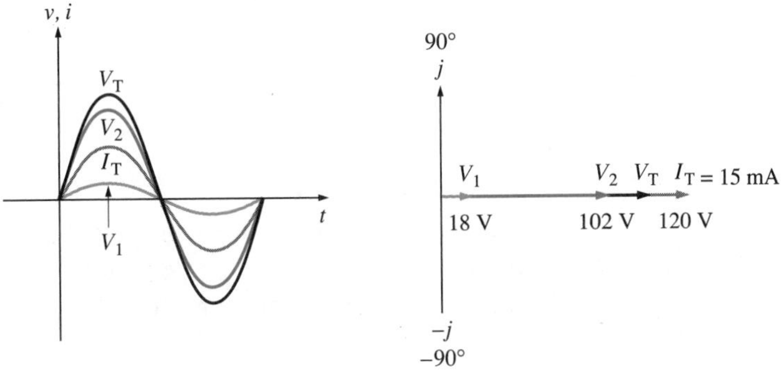

Figure 12-20

AC voltages and currents are assumed to be RMS unless otherwise specified.

The letter *j* can be used to represent 90°.

Note that in Example 12-7 the value of I_T was computed as the RMS or effective value. As indicated previously, by industry convention, whenever an AC current or voltage is specified, it is assumed to be the RMS (effective) value unless otherwise stated and is usually written as just V or I without stating RMS or effective or AC.

EXAMPLE 12-8

For the circuit of Figure 12-21, find (A) I_1, (B) I_2, (C) I_T, and (D) Z_T.

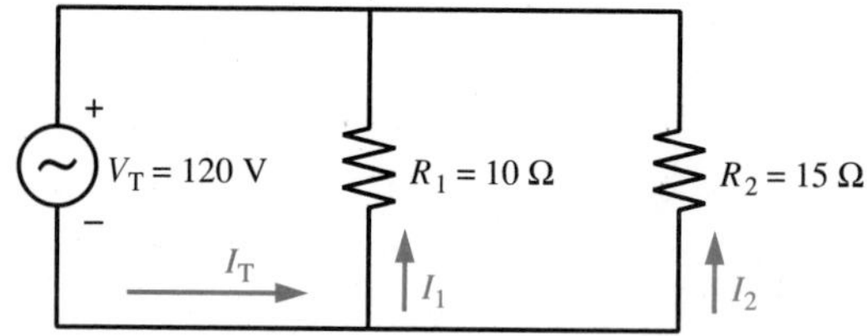

Figure 12-21

Solution:

Method I

(A) $I_1 = \frac{V_T}{Z_{R_1}} = \frac{120\ \text{V}}{10\ \Omega} = 12\ \text{A}$

(B) $I_2 = \frac{V_T}{Z_{R_2}} = \frac{120\ \text{V}}{15\ \Omega} = 8\ \text{A}$

(C) $I_T = I_1 + I_2$

$= 12\ \text{A} + 8\ \text{A}$

$= 20\ \text{A}$

(D) $Z_T = \frac{V_T}{I_T} = \frac{120\ \text{V}}{20\ \text{A}} = 6\ \Omega$

Method 2

(D) $Z_T = \frac{R_1 R_2}{R_1 + R_2} = \frac{(10\ \Omega)(15\ \Omega)}{10\ \Omega + 15\ \Omega} = 6\ \Omega$

(C) $I_T = \frac{V_T}{Z_T} = \frac{120\ \text{V}}{6\ \Omega} = 20\ \text{A}$

(A) $I_1 = \frac{V_T}{Z_{R_1}} = \frac{120\ \text{V}}{10\ \Omega} = 12\ \text{A}$

(B) $I_2 = \frac{V_T}{Z_{R_2}} = \frac{120\ \text{V}}{15\ \Omega} = 8\ \text{A}$

It is important to note that in each of the three preceding examples, the voltage and current were always in phase because the circuits are purely resistive. Thus we are able to add voltages and currents directly.

12.4 RL Circuits

The combination of a resistor and an inductor in a circuit is called an RL circuit. Figure 12-22A illustrates an RL circuit. Kirchhoff's voltage law tells us that $V_T = V_R + V_L$. However, because V_R and V_L are 90 degrees out of phase (Figure 12-22B), we cannot simply add them together. V_R and V_L can be added by the use of phasors. Referring to Figure 12-22C, we scale the magnitude of V_R (3 volts) in the X direction. V_L (4 volts) is scaled in the Y direction because it leads by 90 degrees. Completing the rectangle, the diagonal becomes V_T. V_T can be measured with a ruler to be 5 units or in this case 5 volts. The angle between V_T and V_R is called the phase angle (θ_T) of the entire circuit. It could be measured graphically with a protractor as 53 degrees.

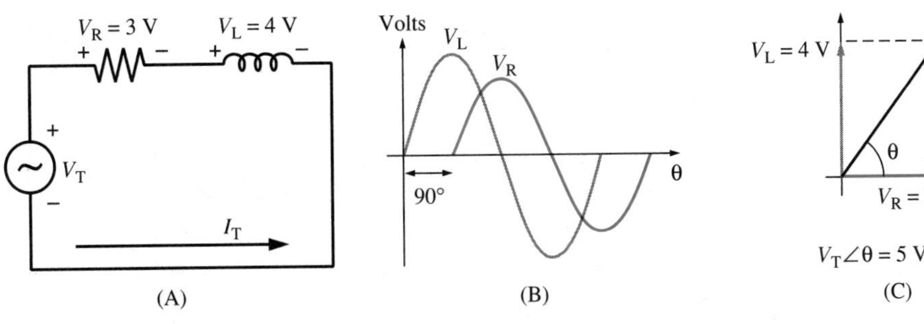

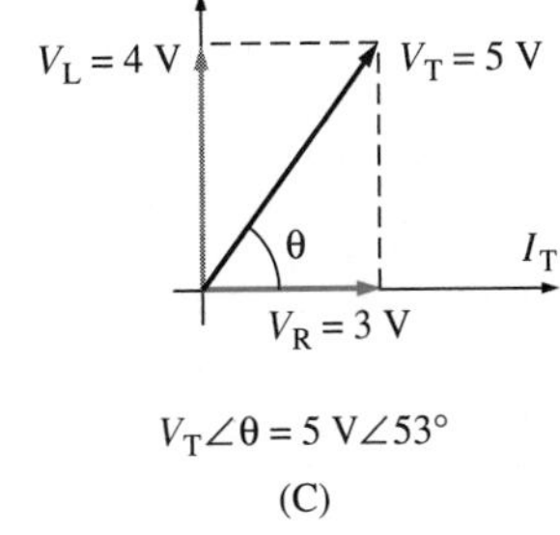

Figure 12-22 **Voltage in an RL series circuit**

$V_T^2 = V_R^2 - V_L^2$
$V_R^2 = V_T^2 - V_L^2$
$V_L^2 = V_T^2 - V_R^2$

Mathematically the solution can be calculated using the Pythagorean theorem where

$$V_T^2 = V_R^2 + V_L^2$$

or

$$V_T = \sqrt{V_R^2 + V_L^2}$$

Equation 12-4

In this case,

$$\begin{aligned} V_T &= \sqrt{(3\text{ V})^2 + (4\text{ V})^2} \\ &= \sqrt{9\text{ V}^2 + 16\text{ V}^2} \\ &= \sqrt{25\text{ V}^2} \\ &= 5\text{ V} \end{aligned}$$

$$\begin{aligned} \cos\theta_T &= \frac{\text{adjacent side}}{\text{hypotenuse}} \\ &= \frac{V_R}{V_T} \\ &= \frac{3\text{ V}}{5\text{ V}} \\ &= 0.6 \\ \therefore \quad \theta_T &= \cos^{-1} 0.6 \\ &= 53° \end{aligned}$$

The phase angle, θ_T, can be obtained on a calculator by entering 0.6 and pressing the $\cos^{-1}$ key. The phase angle could also be calculated using the sin or tan trigonometric functions. The impedance, Z_T, of an RL circuit, is the total opposition to current offered by both the resistor and the inductor. Recall that according to Ohm's law

$$Z_T = \frac{V_T}{I_T}$$

Referring to the phasor diagram of Figure 12-23, the impedance is obtained in the same manner as V_T in the previous diagram. The resistance (R) is scaled in the X (horizontal) direction (10 units). The inductive reactance X_L is scaled in the Y (vertical) direction (15 units). Completing the rectangle, the impedance,

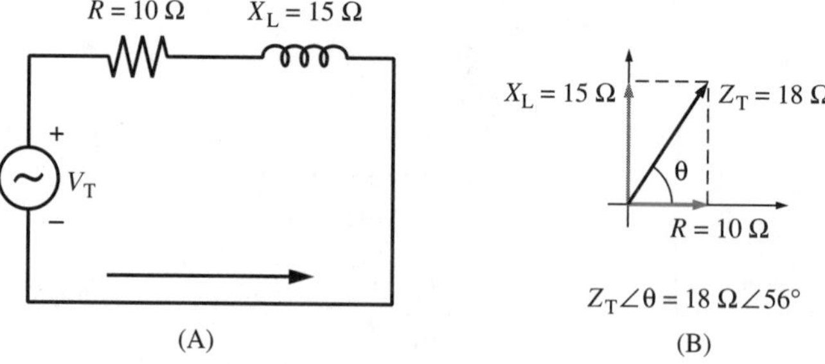

Figure 12-23 **Impedance of an RL series circuit**

Z_T, becomes the diagonal of the rectangle. It can be measured with a ruler as 18 units or in this case 18 ohms. The phase angle (θ_T) for the entire circuit is the angle measured between the horizontal and the diagonal. This can be measured using a protractor as 56 degrees. When Z_T is in the counterclockwise direction from the horizontal reference, it is said to be positive. Thus, θ_T = +56 degrees. Mathematically the impedance can be calculated from the Pythagorean theorem using the following equations:

$$Z_T^2 = R_2 + X_L^2$$

or

Equation 12-5 ▶

$$Z_T = \sqrt{R^2 + X_L^2}$$

In this case,

$$Z_T = \sqrt{(10\ \Omega)^2 + (15\ \Omega)^2}$$

$$= \sqrt{100\ \Omega^2 + 225\ \Omega^2}$$

$$= \sqrt{325\ \Omega^2}$$

$$= 18\ \Omega$$

The phase angle can be calculated using trigonometry as follows:

$$\cos\theta_T = \frac{R}{Z_T}$$

$$= \frac{10\ \Omega}{18\ \Omega}$$

$$= 0.555$$

$$\therefore \quad \theta_T = \cos^{-1} 0.555$$

$$= 56°$$

EXAMPLE 12-9

For the circuit of Figure 12-24, find (A) Z_T, (B) θ_T, (C) I_T, (D) V_R, and (E) V_L. Verify your answer using Kirchhoff's voltage law.

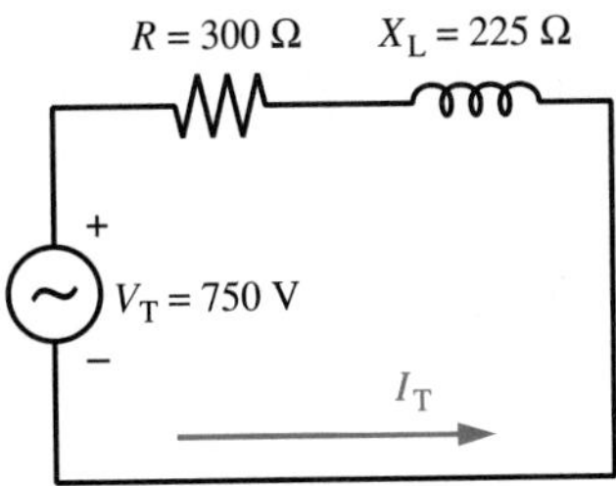

Figure 12-24

Solution:

(A)
$$Z_T = \sqrt{R^2 + X_L{}^2}$$
$$= \sqrt{(300\ \Omega)^2 + (225\ \Omega)^2}$$
$$= \sqrt{90{,}000\ \Omega^2 + 50{,}625\ \Omega^2}$$
$$= \sqrt{140{,}625\ \Omega^2}$$
$$= 375\ \Omega$$

(B)
$$\cos\theta_T = \frac{R}{Z_T}$$
$$= \frac{300\ \Omega}{375\ \Omega}$$
$$= 0.8$$
$$\therefore\ \theta_T = \cos^{-1} 0.8$$
$$= 37°$$

(C)
$$I_T = \frac{V_T}{Z_T}$$
$$= \frac{750\ \text{V}}{375\ \Omega}$$
$$= 2\ \text{A}$$

(D)
$$V_R = I_T R$$
$$= (2\ \text{A})(300\ \Omega)$$
$$= 600\ \text{V}$$

(E)
$$V_L = I_T X_L$$
$$= (2\ \text{A})(225\ \Omega)$$
$$= 450\ \text{V}$$

See Figure 12-25.

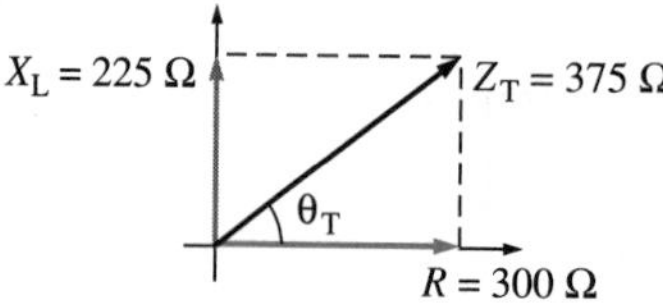

$Z_T\angle\theta = 375\ \Omega\angle 37°$

(A)

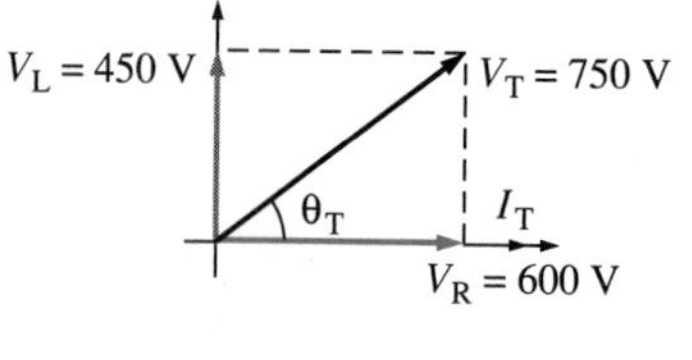

$V_T\angle\theta = 750\ \text{V}\angle 37°$

(B)

Figure 12-25

Check

$$V_T = \sqrt{V_R^2 + V_L^2}$$

$$= \sqrt{(600\ \text{V})^2 + (450\ \text{V})^2}$$

$$= 750\ \text{V}$$

Note that if the horizontal component ($R = 300\ \Omega$) is greater than the vertical component ($X_L = 225\ \Omega$) then θ_T will always be less than 45 degrees. When the horizontal component is less than the vertical component, θ_T will always be greater than 45 degrees. If the horizontal component equals the vertical component, then θ_T will equal 45 degrees.

EXAMPLE 12-10

A 120-volt, 60-hertz voltage source is connected in series with a 2.5-kilohm resistor, a 2-henry choke, and a 3-henry choke. Find (A) Z_T, (B) θ_T, (C) V_R, (D) V_{L_1}, and (E) V_{L_2}.

Solution:

See Figure 12-26.

(A)

$$X_{L_1} = 2\pi f L_1$$

$$= (2)(3.14)(60\ \text{Hz})(2\ \text{H})$$

$$= 753.6\ \Omega$$

$$X_{L_2} = 2\pi f L_2$$

$$= (2)(3.14)(60\ \text{Hz})(3\ \text{H})$$

$$= 1{,}130.4\ \Omega$$

$$X_{L_T} = X_{L_1} + X_{L_2}$$

$$= 753.6\ \Omega + 1{,}130.4\ \Omega$$

$$= 1{,}884\ \Omega$$

$$Z_T = \sqrt{R^2 + X_{L_T}^2}$$

$$= \sqrt{(2{,}500\ \Omega)^2 + (1{,}884\ \Omega)^2}$$

$$= 3{,}130\ \Omega$$

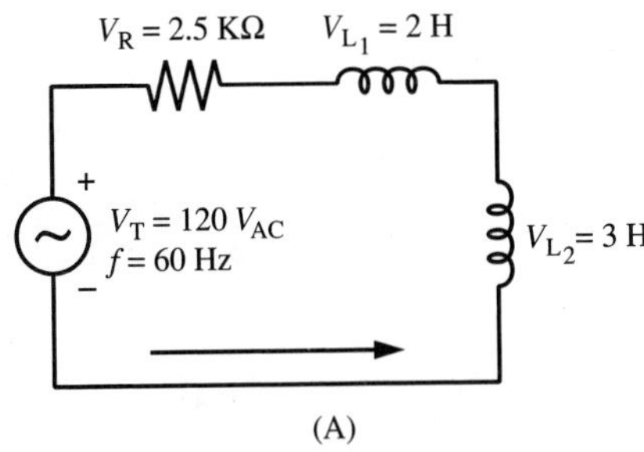

(A)

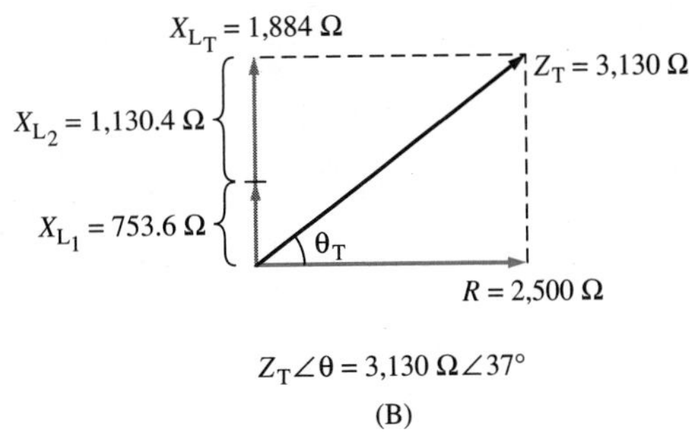

(B)

Figure 12-26

(B) $\cos\theta_T = \dfrac{R}{Z_T}$

$= \dfrac{2,500\ \Omega}{3,130\ \Omega}$

$= 0.798$

$\therefore\quad \theta_T = \cos^{-1} 0.798$

$= 37°$

(C) $I_T = \dfrac{V_T}{Z_T}$

$= \dfrac{120\ V}{3,130\ \Omega}$

$= 0.038\ A$

$= 38.3\ mA$

$V_R = I_T R$

$= (38.3\ mA)(2,500\ \Omega)$

$= (38.3\ mA)(2.5\ K\Omega)$

$= 95.8\ V$

(D) $V_{L_1} = I_T X_{L_1}$

$= (38.3\ mA)(753.6\ \Omega)$

$= (38.3\ mA)(0.7536\ K\Omega)$

$= 28.9\ V$

(E) $V_{L_2} = I_T X_{L_2}$

$= (38.3\ mA)(1,130.4\ \Omega)$

$= (38.3\ mA)(1.1304\ K\Omega)$

$= 43.3\ V$

Check

$$V_T = \sqrt{V_R^2 + (V_{L1} + V_{L2})^2}$$
$$= \sqrt{(95.8\text{ V})^2 + (28.9\text{ V} + 43.3\text{ V})^2}$$
$$= \sqrt{(95.8\text{ V})^2 + (72.2\text{ V})^2}$$
$$= 120\text{ V}$$

T^2 tech tidbit

For parallel circuits, use *V* as the reference. For series circuits, use *I* as the reference.

When an inductor and a resistor are in parallel, all of the previous rules for parallel circuits apply. That is, the source voltage and the voltage across each parallel branch are the same, as shown in Figure 12-27. The total current at any instant in time will equal the phasor sum of the currents through each branch. Note that in Figure 12-27 the current I_L and the current I_T are lagging I_R. This is consistent with ELI. In a parallel circuit, the voltage is the same and is therefore used as the reference. Recall that in a series circuit, the current is the same and was always used as the reference.

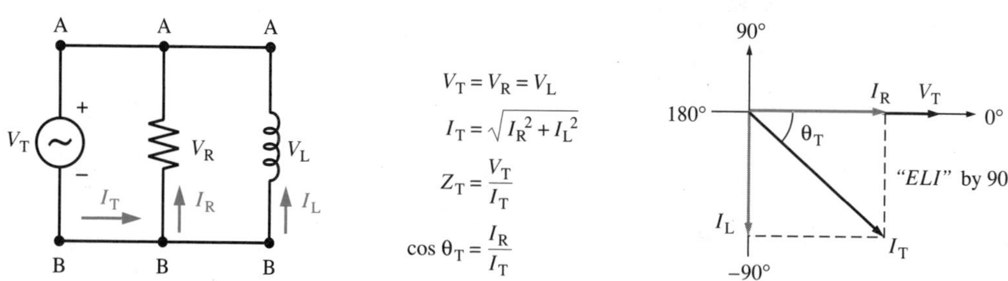

Figure 12-27 **Parallel RL circuits**

EXAMPLE 12-11

For the circuit of Figure 12-28, find (A) I_R, (B) I_L, (C) I_T, (D) Z_T, and (E) the circuit phase angle θ_T.

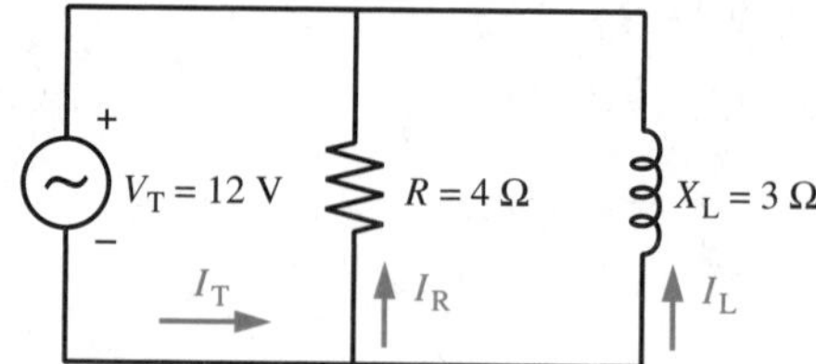

Figure 12-28

Solution:

(A) $$I_R = \frac{V_T}{Z_R} = \frac{12\text{ V}}{4\ \Omega} = 3\text{ A}$$

(B) $I_L = \frac{V_T}{Z_L}$

$= \frac{12\text{ V}}{3\ \Omega}$

$= 4\text{ A}$

(C) $I_T = \sqrt{I_R^2 + I_L^2}$

$= \sqrt{(3\text{ A})^2 + (4\text{ A})^2}$

$= 5\text{ A}$

(D) $Z_T = \frac{V_T}{I_T}$

$= \frac{12\text{ V}}{5\text{ A}}$

$= 2.4\ \Omega$

(E) $\cos\theta_T = \frac{I_R}{I_T}$

$= \frac{3\text{ A}}{5\text{ A}}$

$= 0.6$

$\therefore\ \theta_T = \cos^{-1} 0.6$

$= 53°$ lagging (I_T lags V_T)

EXAMPLE 12-12

For the circuit of Figure 12-29, find (A) I_R, (B) I_L, (C) I_T, (D) Z_T, and (E) the circuit phase angle θ_T.

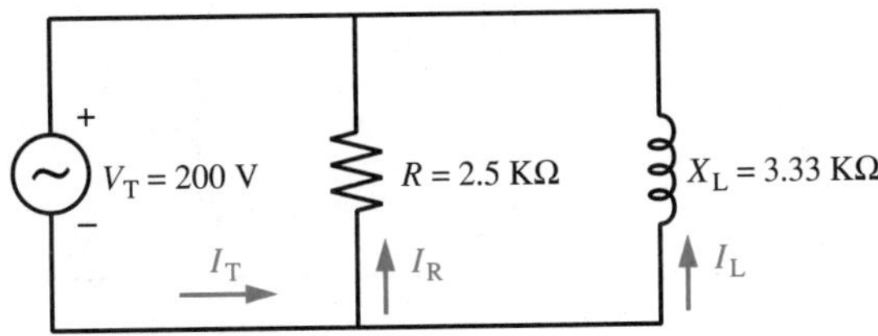

Figure 12-29

Solution:

(A) $I_R = \frac{V_T}{Z_R}$

$= \frac{200\text{ V}}{2.5\text{ K}\Omega}$

$= 80\text{ mA}$

(B) $$I_L = \frac{V_T}{Z_L}$$

$$= \frac{200 \text{ V}}{3.33 \text{ K}\Omega}$$

$$= 60 \text{ mA}$$

(C) $$I_T = \sqrt{I_R^2 + I_L^2}$$

$$= \sqrt{(80 \text{ mA})^2 + (60 \text{ mA})^2}$$

$$= 100 \text{ mA}$$

(D) $$Z_T = \frac{V_T}{I_T}$$

$$= \frac{200 \text{ V}}{100 \text{ mA}}$$

$$= 2 \text{ K}\Omega$$

(E) $$\cos \theta_T = \frac{I_R}{I_T}$$

$$= \frac{80 \text{ mA}}{100 \text{ mA}}$$

$$= 0.8$$

$$\therefore \quad \theta_T = \cos^{-1} 0.8$$

$$= 37° \text{ lagging } (I_T \text{ lags } V_T)$$

EXAMPLE 12-13

For the circuit of Figure 12-30, find (A) I_R, (B) I_{L_1}, (C) I_{L_2}, (D) I_T, (E) Z_T, and (F) the circuit phase angle θ_T.

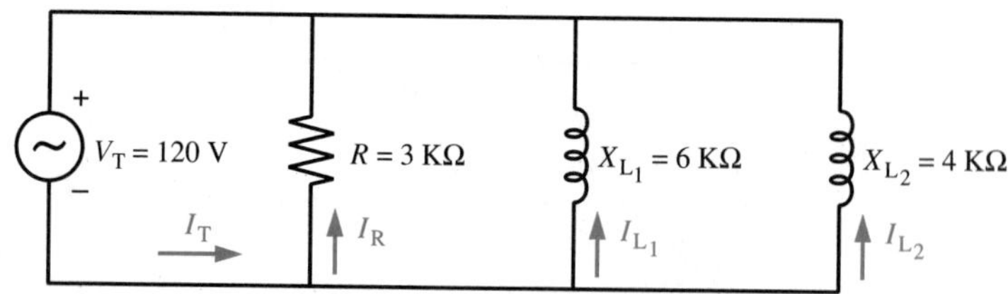

Figure 12-30

Solution:

See Figure 12-31.

(A) $$I_R = \frac{V_T}{Z_R}$$

$$= \frac{120 \text{ V}}{3 \text{ K}\Omega}$$

$$= 40 \text{ mA}$$

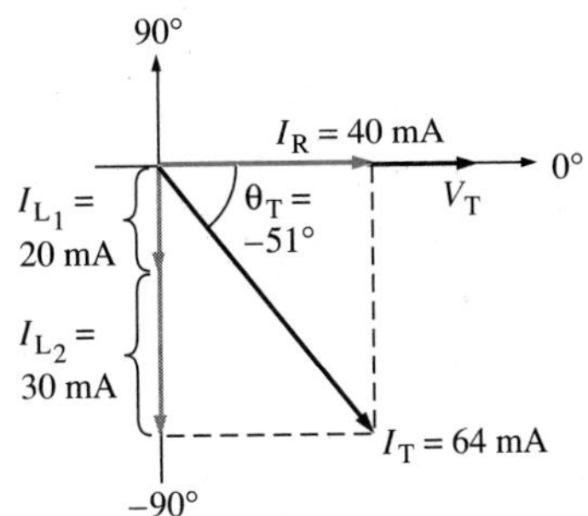

Figure 12-31

(B)
$$I_{L_1} = \frac{V_T}{Z_{L_1}} = \frac{120 \text{ V}}{6 \text{ K}\Omega} = 20 \text{ mA}$$

(C)
$$I_{L_2} = \frac{V_T}{Z_{L_2}} = \frac{120 \text{ V}}{4 \text{ K}\Omega} = 30 \text{ mA}$$

(D)
$$I_T = \sqrt{I_R^2 + (I_{L1} + I_{L2})^2}$$
$$= \sqrt{(40 \text{ mA})^2 + (20 \text{ mA} + 30 \text{ mA})^2}$$
$$= \sqrt{(40 \text{ mA})^2 + (50 \text{ mA})^2}$$
$$= 64 \text{ mA}$$

(E)
$$Z_T = \frac{V_T}{I_T} = \frac{120 \text{ V}}{64 \text{ mA}} = 1.875 \text{ K}\Omega$$

(F)
$$\cos \theta_T = \frac{I_R}{I_T} = \frac{40 \text{ mA}}{64 \text{ mA}} = 0.625$$

$$\therefore \quad \theta_T = \cos^{-1} 0.625$$
$$= 51° \text{ lagging } (I_T \text{ lags } V_T)$$

Note that in the previous three examples, the method of solution does not require us to calculate Z_T directly from Z_R and Z_L. The total impedance, Z_T, can be calculated using the product over sum rule for parallel circuits. Mathematically, this results in the following formulas:

Equation 12-6 ▶

$$Z_T = \frac{(R)(X_L)}{\sqrt{R^2 + X_L^2}}$$

or

$$Z_T = \frac{(R)(Z_L)}{\sqrt{R^2 + Z_L^2}}$$

✓ Self-Test 2

1. The total opposition to current in an AC circuit is called ______________.
2. Find the total impedance when a 3-ohm resistor and a 4-ohm inductive reactance are connected in series.
3. Referring to question 2, find the total current if the applied voltage is 100 volts.
4. Referring to question 2, find the phase angle, θ_T, for the circuit.

12.5 RC Circuits

The combination of a resistor and a capacitor in a circuit is called a RC circuit. Figure 12-32A illustrates an RC series circuit. Kirchhoff's voltage law tells us that $V_T = V_R + V_C$. However, because V_R and V_C are 90 degrees out of phase (Figure 12-32B), we cannot simply add them together. V_R and V_C are added using phasors in a similar manner to that used for V_R and V_L in the case of the inductor. Referring to Figure 12-32C, we scaled V_R (3 volts) in the X (horizontal) direction. V_C (4 volts) is scaled in the $-Y$ (vertical) direction, because it lags by 90 degrees. Completing the rectangle, the diagonal becomes V_T. V_T can be measured using a ruler to be 5 units or 5 volts. The angle between V_T and V_R is called the phase angle of the entire circuit. It can be measured graphically with a protractor as 53 degrees lagging (V_T lags I_T, $\theta_T = -53°$).

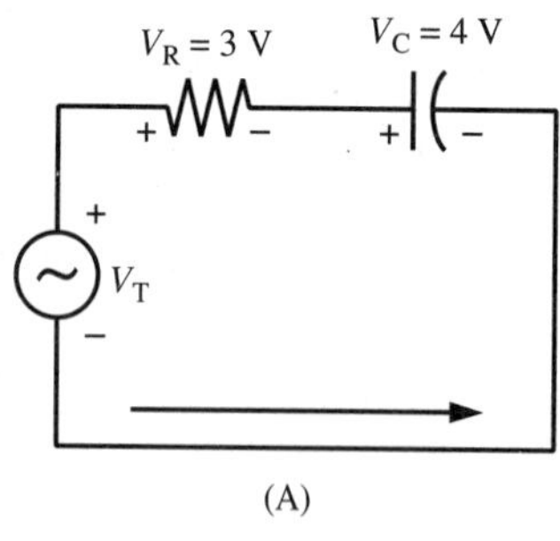

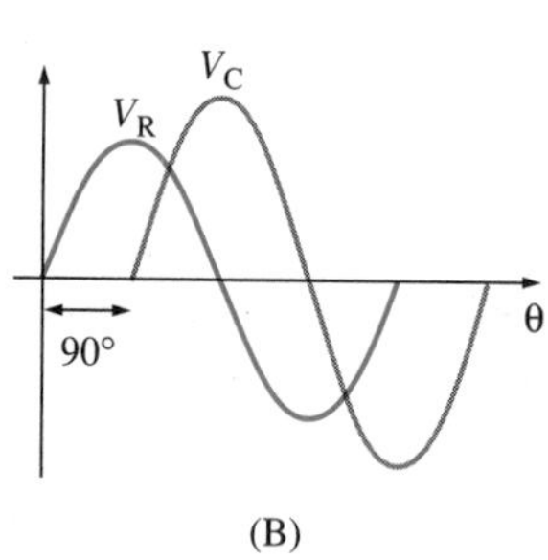

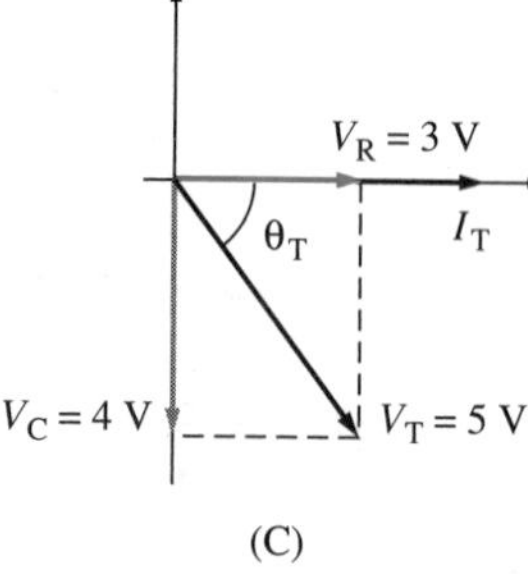

Figure 12-32 **RC series circuit**

T² tech tidbit

$V_T^2 = V_R^2 + V_C^2$
$V_R^2 = V_T^2 - V_L^2$
$V_L^2 = V_T^2 - V_R^2$

Mathematically, the solution can be calculated using the Pythagorean theorem:

$$V_T^2 = V_R^2 + V_C^2$$

or

Equation 12-7 ▶

$$V_T = \sqrt{V_R^2 + V_C^2}$$

In this case,

$$V_T = \sqrt{(3\text{ V})^2 + (4\text{ V})^2}$$
$$= 5\text{ V}$$
$$\cos\theta_T = \frac{\text{adjacent side}}{\text{hypotenuse}}$$
$$= \frac{V_R}{V_T}$$
$$= \frac{3\text{ V}}{5\text{ V}}$$
$$= 0.6$$
$$\therefore \quad \theta_T = \cos^{-1} 0.6$$
$$= 53° \text{ lagging } (V_T \text{ lags } I_T)$$

Note that lagging is also indicated by writing $\theta_T = -53$ degrees.

The total impedance (Z_T) of an RC circuit is the total opposition to current offered by the resistor and the capacitor. Recall that according to Ohm's law,

$$Z_T = \frac{V_T}{I_T}$$

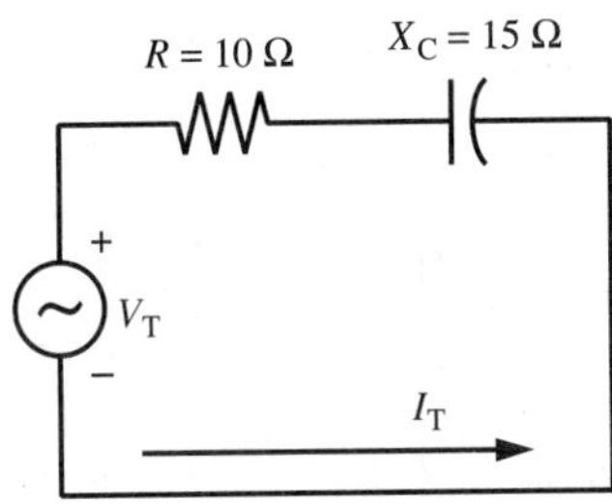

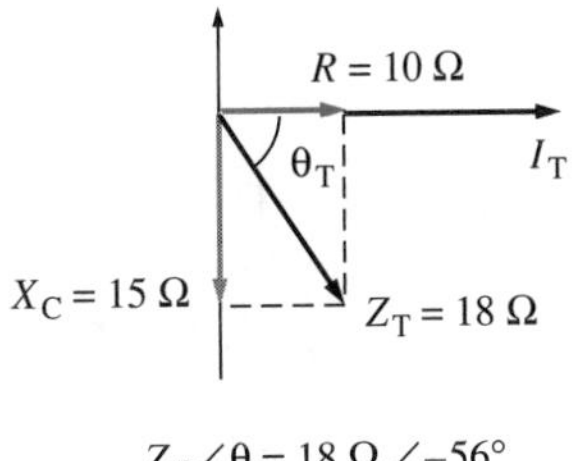

Figure 12-33 Impedance in an RC series circuit

Referring to the phasor diagram of Figure 12-33, the impedance can be obtained by the use of phasors. The resistance (R) is scaled in the X direction (10 units). The capacitive reactance (X_C) is scaled in the $-Y$ direction (15 units). Completing the rectangle, the impedance (Z_T) becomes the diagonal of the rectangle. It can be measured with a ruler as 18 units or in this case 18 ohms. The phase angle (θ_T) for the entire circuit is the angle measured between the horizontal and the diagonal. This can be measured using a protractor as -56 degrees. Mathematically, the impedance can be calculated from the Pythagorean theorem using the following equation:

$$Z_T^2 = R^2 + X_C^2$$

or

$$Z_T = \sqrt{R^2 + X_C^2}$$

Equation 12-8

In this case,

$$Z_T = \sqrt{(10\ \Omega)^2 + (15\ \Omega)^2}$$
$$= 18\ \Omega$$

The phase angle can be calculated using trigonometry as follows:

$$\cos \theta_T = \frac{R}{Z_T}$$
$$= \frac{10\ \Omega}{18\ \Omega}$$
$$= 0.555$$
$$\therefore \quad \theta_T = \cos^{-1} 0.555$$
$$= 56° \text{ lagging}$$
$$= -56° \ (V_T \text{ lags } I_T)$$

EXAMPLE 12-14

For the circuit of Figure 12-34, find (A) Z_T, (B) I_T, (C) V_R, (D) V_C, and (E) the circuit phase angle θ_T.

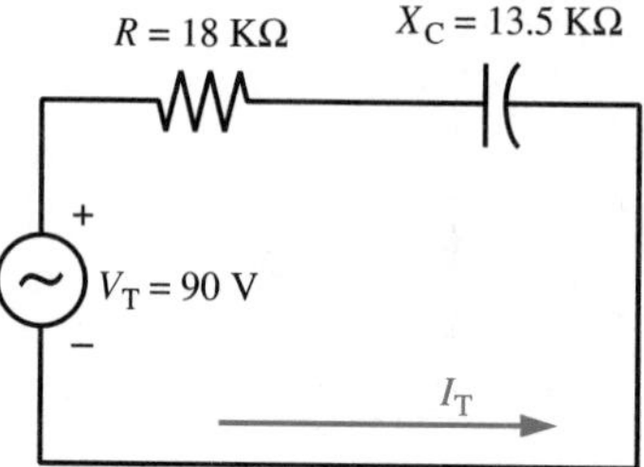

Figure 12-34

Solution:

See Figure 12-35.

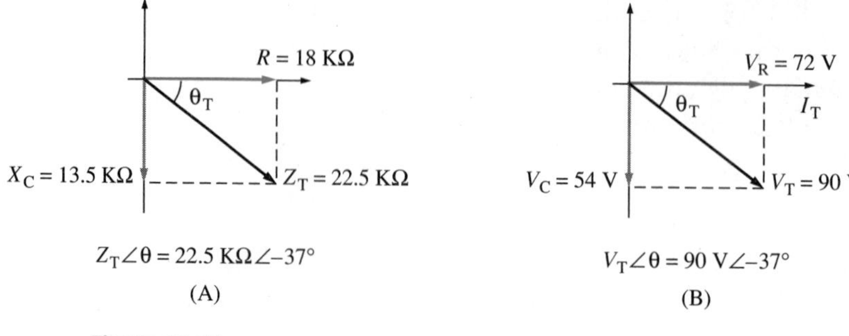

Figure 12-35

(A)
$$Z_T = \sqrt{R^2 + X_C^2}$$
$$= \sqrt{(18\ \text{K}\Omega)^2 + (13.5\ \text{K}\Omega)^2}$$
$$= 22.5\ \text{K}\Omega$$

(B)
$$I_T = \frac{V_T}{Z_T}$$
$$= \frac{90\ \text{V}}{22.5\ \text{K}\Omega}$$
$$= 4\ \text{mA}$$

(C)
$$V_R = I_T R$$
$$= (4\ \text{mA})(18\ \text{K}\Omega)$$
$$= 72\ \text{V}$$

(D)
$$V_C = I_T X_C$$
$$= (4\ \text{mA})(13.5\ \text{K}\Omega)$$
$$= 54\ \text{V}$$

(E)
$$\cos\theta_T = \frac{R}{Z_T}$$
$$= \frac{18\ \text{K}\Omega}{22.5\ \text{K}\Omega}$$
$$= 0.8$$
$$\therefore\ \theta_T = \cos^{-1} 0.8$$
$$= 37°\ \text{lagging}$$
$$= -37°\ (V_T\ \text{lags}\ I_T)$$

Check

$$V_T = \sqrt{V_R^2 + V_C^2}$$
$$= \sqrt{(72\ \text{V})^2 + (54\ \text{V})^2}$$
$$= 90\ \text{V}$$

When a capacitor and a resistor are in parallel, all of the previous rules for parallel circuits apply. That is, the source voltage and the voltage across each parallel branch are the same, as shown in Figure 12-36. The total current at any instant in time will equal the phasor sum of the currents through each branch. Note that in Figure 12-36, the current I_C and the current I_T are leading I_R. This is consistent with *ICE.* In a parallel circuit, the voltage is the same and is therefore used as the reference.

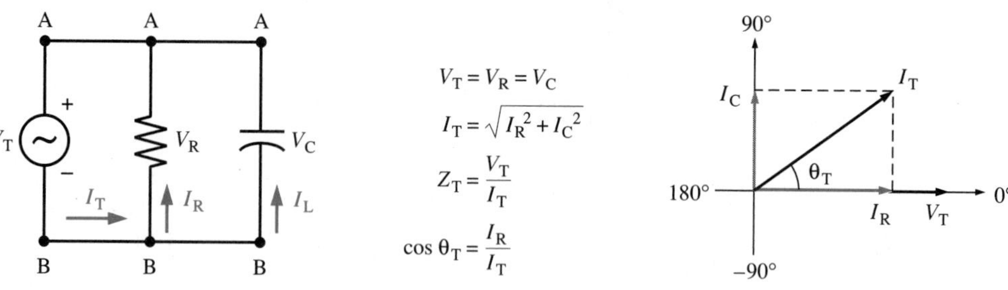

Figure 12-36 **Parallel RC circuits**

EXAMPLE 12-15

For the circuit of Figure 12-37, find (A) I_R, (B) I_C, (C) I_T, (D) Z_T, and (E) the circuit phase angle θ_T.

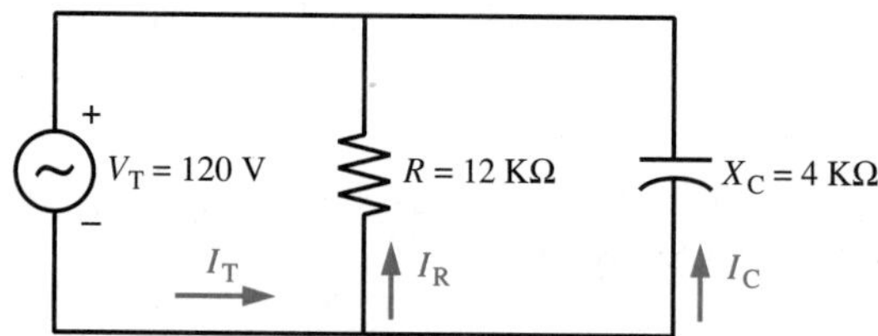

Figure 12-37

Solution:

(A)
$$I_R = \frac{V_T}{Z_R} = \frac{120 \text{ V}}{12 \text{ K}\Omega} = 10 \text{ mA}$$

(B)
$$I_C = \frac{V_T}{Z_C} = \frac{120 \text{ V}}{4 \text{ K}\Omega} = 30 \text{ mA}$$

(C)
$$I_T = \sqrt{I_R^2 + I_C^2} = \sqrt{(10 \text{ mA})^2 + (30 \text{ mA})^2} = 31.6 \text{ mA}$$

(D)
$$Z_T = \frac{V_T}{I_T} = \frac{120 \text{ V}}{31.6 \text{ mA}} = 3.8 \text{ K}\Omega$$

(E) $\cos\theta_T = \dfrac{I_R}{I_T}$

$= \dfrac{10 \text{ mA}}{31.6 \text{ mA}}$

$= 0.316$

$\therefore \quad \theta_T = \cos^{-1} 0.316$

$= 71.6°$ leading (I_T leads V_T)

EXAMPLE 12-16

For the circuit of Figure 12-38, find (A) I_R, (B) I_{C_1}, (C) I_{C_2}, (D) I_T, (E) Z_T, and (F) the circuit phase angle θ_T.

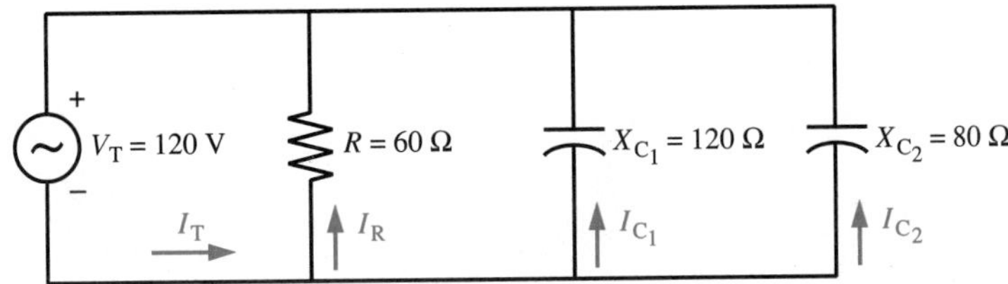

Figure 12-38

Solution:

See Figure 12-39.

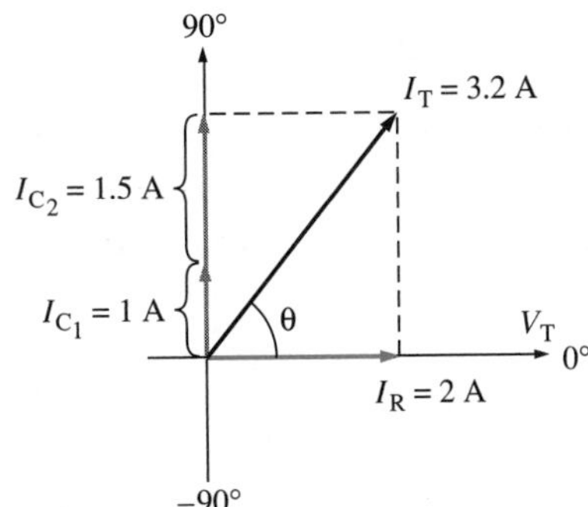

Figure 12-39

(A) $I_R = \dfrac{V_T}{Z_R}$

$= \dfrac{120 \text{ V}}{60\ \Omega}$

$= 2 \text{ A}$

(B) $I_{C_1} = \dfrac{V_T}{Z_{C_1}}$

$= \dfrac{120 \text{ V}}{120\ \Omega}$

$= 1 \text{ A}$

(C) $I_{C_2} = \frac{V_T}{Z_{C_2}}$

$= \frac{120\text{ V}}{80\ \Omega}$

$= 1.5\text{ A}$

(D) $I_T = \sqrt{I_R^2 + (I_{C1} + I_{C2})^2}$

$= \sqrt{(2\text{ A})^2 + (1\text{ A} + 1.5\text{ A})^2}$

$= \sqrt{(2\text{ A})^2 + (2.5\text{ A})^2}$

$= 3.2\text{ A}$

(E) $Z_T = \frac{V_T}{I_T}$

$= \frac{120\text{ V}}{3.2\text{ A}}$

$= 37.5\ \Omega$

(F) $\cos\theta_T = \frac{I_R}{I_T}$

$= \frac{2\text{ A}}{3.2\text{ A}}$

$= 0.625$

$\therefore \theta_T = \cos^{-1} 0.625$

$= 51.3°$ leading (I_T leads V_T)

Note that in the previous examples, the method of solution does not require us to calculate Z_T directly from Z_R and Z_C. The total impedance, Z_T, can be calculated using the product over sum rule for parallel circuits. Mathematically, this results in the following formulas:

Equation 12-9 ▶

$$Z_T = \frac{(R)(X_C)}{\sqrt{R^2 + X_C^2}}$$

or

$$Z_T = \frac{(R)(Z_C)}{\sqrt{R^2 + Z_C^2}}$$

✓ Self-Test 3

1. Find the total impedance when a 5-ohm resistor and a 12-ohm capacitive reactance are connected in series.
2. Referring to question 1, find the total current if the applied voltage is 52 volts.
3. Referring to question 1, find the phase angle, θ_T, for the circuit.
4. Referring to question 2, find the magnitude of the voltage across the capacitor.

12.6 RLC Circuits

The combination of a resistor, an inductor, and a capacitor in a circuit is called an RLC circuit. Figure 12-40 illustrates an RLC series circuit. Recall that the general formula for impedance is

$$Z_T = \sqrt{R^2 + X^2}$$

When the reactance is caused by an inductor, the formula becomes

$$Z_T = \sqrt{R^2 + X_L^2}$$

When the reactance is caused by a capacitor, the formula becomes

$$Z_T = \sqrt{R^2 + X_C^2}$$

Referring to the phasor diagram of Figure 12-40, we see that X_L is plotted at +90 degrees. This is due to the fact that for an inductor, the voltage leads the current by 90 degrees (ELI). We also note that X_C is plotted at −90 degrees. This is because for a capacitor the current leads the voltage by 90 degrees (ICE). R is plotted at 0 degrees because for a resistor the voltage and current are in phase.

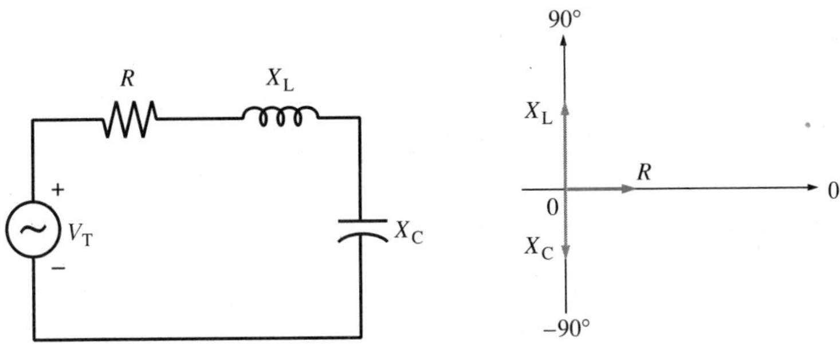

Figure 12-40 **Impedance in an RLC series circuit**

It can be seen through vector addition that because X_L and X_C are 180 degrees out of phase and are therefore opposed to each other, the general formula for impedance becomes

$$Z_T = \sqrt{R^2 + (X_L - X_C)^2}$$

◄ Equation 12-10

EXAMPLE 12-17

Calculate the impedance of the RLC series circuit of Figure 12-41.

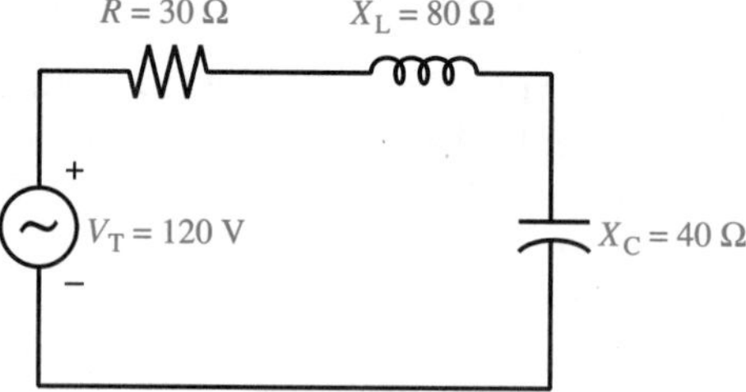

Figure 12-41

Solution:

See Figure 12-42.

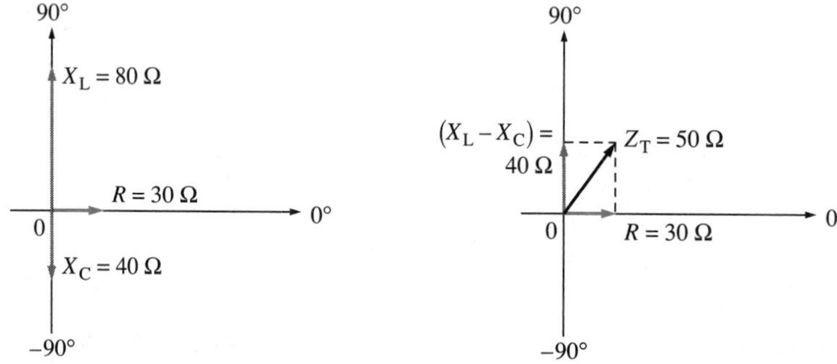

Figure 12-42

$$
\begin{aligned}
Z_T &= \sqrt{R^2 + (X_L - X_C)^2} \\
&= \sqrt{(30\ \Omega)^2 + (80\ \Omega - 40\ \Omega)^2} \\
&= \sqrt{(30\ \Omega)^2 + (40\ \Omega)^2} \\
&= \sqrt{900\ \Omega^2 + 1{,}600\ \Omega^2} \\
&= \sqrt{2{,}500\ \Omega^2} \\
&= 50\ \Omega
\end{aligned}
$$

EXAMPLE 12-18

Calculate (A) the total impedance, (B) total current, (C) voltage across each element, and (D) the circuit phase angle in the series circuit of Figure 12-43.

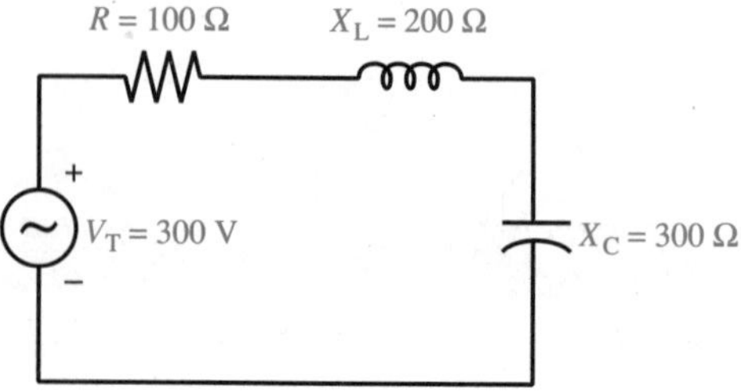

Figure 12-43

Solution:

See Figure 12-44.

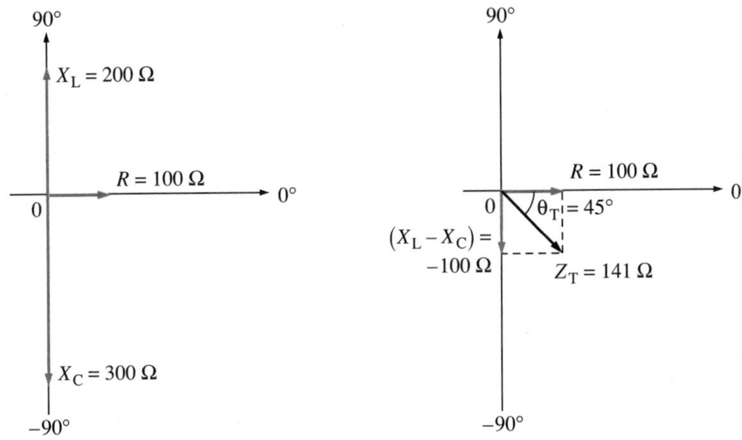

Figure 12-44

(A)
$$Z_T = \sqrt{R^2 + (X_L - X_C)^2}$$
$$= \sqrt{(100\ \Omega)^2 + (200\ \Omega - 300\ \Omega)^2}$$
$$= \sqrt{(100\ \Omega)^2 + (-100\ \Omega)^2}$$
$$= \sqrt{10{,}000\ \Omega^2 + 10{,}000\ \Omega^2}$$
$$= \sqrt{20{,}000\ \Omega^2}$$
$$= 141\ \Omega$$

(B)
$$I_T = \frac{V_T}{Z_T}$$
$$= \frac{300\ \text{V}}{141\ \Omega}$$
$$= 2.13\ \text{A}$$

(C)
$$V_R = I_T R = (2.13\ \text{A})(100\ \Omega) = 213\ \text{V}$$
$$V_L = I_T X_L = (2.13\ \text{A})(200\ \Omega) = 426\ \text{V}$$
$$V_C = I_T X_C = (2.13\ \text{A})(300\ \Omega) = 639\ \text{V}$$

(D)
$$\cos\theta_T = \frac{R}{Z_T} = \frac{100\ \Omega}{141\ \Omega} = 0.707$$
$$\therefore\ \theta_T = \cos^{-1} 0.707$$
$$= 45°\ \text{lagging}\ (I_T \text{ lags } V_T)$$
$$\text{or}\ \theta_T = -45°$$

Note that the voltage across the inductor and the capacitor can be greater than the applied voltage in a series circuit. This is due to the phase relationship.

Check

See Figure 12-45.

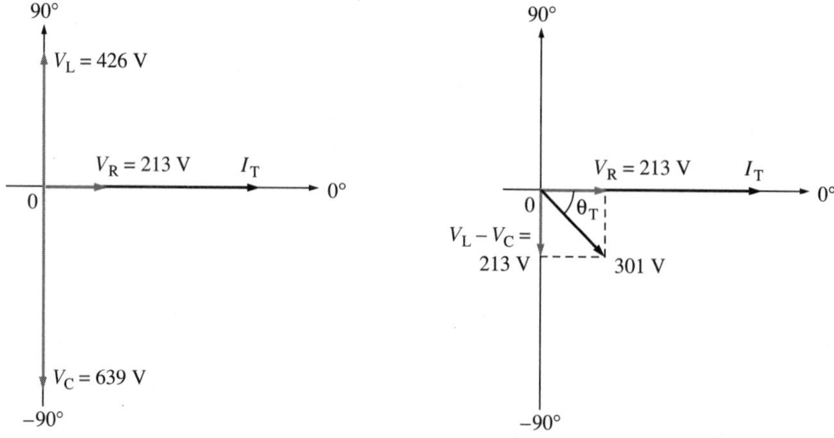

Figure 12-45

$$V_T = \sqrt{V_R^2 + (V_L - V_C)^2}$$

$$= \sqrt{(213\text{ V})^2 + (426\text{ V} - 639\text{ V})^2}$$

$$= \sqrt{(213\text{ V})^2 + (-213\text{ V})^2}$$

$$= \sqrt{45{,}369\text{ V}^2 + 45{,}369\text{ V}^2}$$

$$= \sqrt{90{,}738\text{ V}^2}$$

$$= 301\text{ V}$$

$$\approx 300\text{ V}$$

Note that V_T is actually 300 volts. The 1-volt difference is due to a round-off error in calculating V_R, V_L, and V_C.

Figure 12-46 shows a RLC parallel circuit. In a parallel circuit, the voltage across each component is the same. However, the currents are out of phase with each other. The general formula that describes this relationship is

$$I_T = \sqrt{I_R^2 + I_X^2}$$

For an RLC parallel circuit the relationship becomes

$$I_T = \sqrt{I_R^2 + (I_L - I_C)^2}$$

EXAMPLE 12-19

For the circuit of Figure 12-46, calculate the (A) current through each component, (B) the total current, (C) the total impedance, and (D) the circuit phase angle.

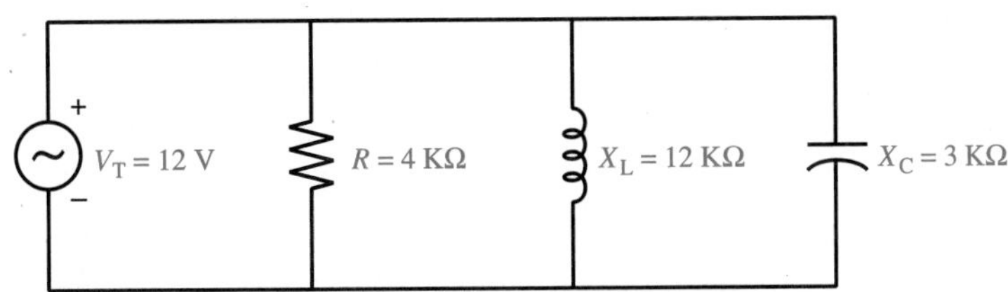

Figure 12-46

Solution:

(A)

$$I_R = \frac{V_T}{R} = \frac{12\text{ V}}{4\text{ K}\Omega} = 3\text{ mA}$$

$$I_L = \frac{V_T}{X_L} = \frac{12\text{ V}}{12\text{ K}\Omega} = 1\text{ mA}$$

$$I_C = \frac{V_T}{X_C} = \frac{12\text{ V}}{3\text{ K}\Omega} = 4\text{ mA}$$

(B)

$$I_T = \sqrt{I_R^{\,2} + (I_L - I_C)^2}$$

$$= \sqrt{(3\text{ mA})^2 + (1\text{ mA} - 4\text{ mA})^2}$$

$$= \sqrt{(3\text{ mA})^2 + (-3\text{ mA})^2}$$

$$= \sqrt{9\text{ mA}^2 + 9\text{ mA}^2}$$

$$= \sqrt{18\text{ mA}^2}$$

$$= 4.24\text{ mA}$$

Refer to Figure 12-47.

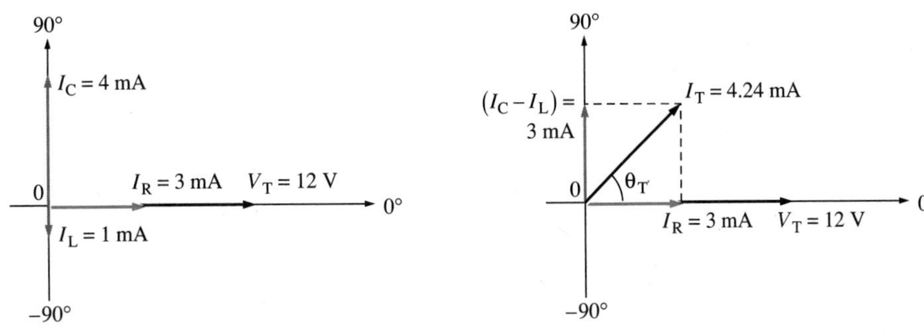

Figure 12-47

(C) $Z_T = \frac{V_T}{I_T}$

$= \frac{12\text{ V}}{4.24\text{ mA}}$

$= 2.83\text{ K}\Omega$

$= 2{,}830\ \Omega$

(D) $\cos\theta_T = \frac{I_R}{I_T}$

$= \frac{3\text{ mA}}{4.24\text{ mA}}$

$= 0.707$

$\therefore\ \theta_T = \cos^{-1} 0.707$

$= 45°$ leading (I_T leads V_T)

Note that in a parallel circuit the voltage is used as the reference, therefore, I_C is plotted at +90 degrees (ICE) and I_L is plotted at −90 degrees (ELI).

✓ Self-Test 4

1. A 1.5-kilohm resistor, an inductor with a reactance of 4 kilohms, and a capacitor with a reactance of 2 kilohms are all connected in series. Find Z_T.
2. Referring to question 1, find the total current if the applied voltage is 100 volts.
3. For a series AC circuit, the ______________ is used as the reference on the phasor diagram.
4. For a parallel AC circuit, the ______________ is used as the reference on the phasor diagram.

12.7 Resonance

T2 tech tidbit

X_L and X_C are always 180° out of phase.

The phasor diagrams for RLC series circuits show that X_L and X_C are always 180 degrees out of phase. Therefore, we can subtract X_L and X_C algebraically. However, when the value of X_L equals the value of X_C, the resultant, X_T, is zero. The point at which X_L equals X_C is called **resonance.** As a result, the only opposition to current is caused by the resistance. Thus, at resonance, the current in the circuit is at a maximum value. Figure 12-48 shows the phasor relationship at resonance.

The formula for X_L is

$$X_L = 2\pi f L$$

The formula for X_C is

$$X_C = \frac{1}{2\pi f L}$$

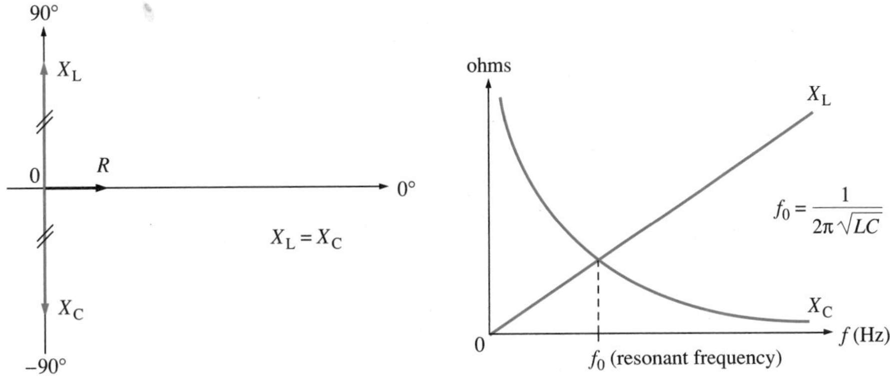

Figure 12-48 **Phasor relationship at resonance**

X_L equals X_C at resonance, so it follows that

$$X_L = X_C$$

$$2\pi f_0 L = \frac{1}{2\pi f_0 C}$$

where f_0 equals the frequency at which X_L will equal X_C. This frequency is called the **resonant frequency.** Solving for f_0 we obtain

tech tidbit

At resonance $X_L = X_C$.

$$f_0 = \frac{1}{2\pi\sqrt{LC}}$$

◀ *Equation 12-11*

where f_0 = the resonant frequency in hertz

L = the inductance in henries

C = the capacitance in farads

tech tidbit

For any given L and C, resonance occurs at only one frequency.

EXAMPLE 12-20

(A) Find the resonant frequency for a 5-microfarad capacitor and a 50-microhenry coil. Calculate (B) X_L and (C) X_C at this frequency.

Solution:

tech tidbit

f_0 is sometimes written as f_r.

(A) $$f_0 = \frac{1}{2\pi\sqrt{LC}}$$

$$= \frac{1}{(2)(3.14)\sqrt{(50 \times 10^{-6}\text{ H})(5 \times 10^{-6}\text{ F})}}$$

$$= 10{,}071 \text{ Hz}$$

(B) $$X_L = 2\pi f L$$

$$= (2)(3.14)(10{,}071 \text{ Hz})(50 \times 10^{-6}\text{ H})$$

$$= 3.16\ \Omega$$

$$\text{(C)}\quad X_C = \frac{1}{2\pi f L}$$

$$= \frac{1}{(2)(3.14)(10{,}071\text{ Hz})(5 \times 10^{-6}\text{ F})}$$

$$= 3.16\ \Omega$$

Checks because at resonance $X_L = X_C$.

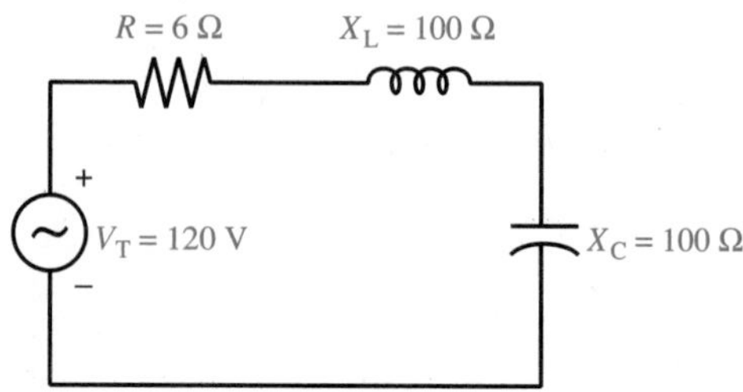

Figure 12-49

At series resonance, $Z_T = R$.

Figure 12-49 shows a series circuit consisting of a resistor, an inductor, and a capacitor. X_L equals X_C, so the series circuit is in resonance. Therefore,

$$Z_{T_0} = \sqrt{R^2 + (X_L - X_C)^2}$$

$$= \sqrt{R^2 + (0)^2}$$

$$= R$$

The current is then

$$I = \frac{V_T}{Z_{T_0}}$$

$$= \frac{120\text{ V}}{6\ \Omega}$$

$$= 20\text{ A}$$

and $V_R = I_T R = (20\text{ A})(6\ \Omega) = 120\text{ V}$ (Note: $V_R = V_T$ at resonance)

$V_L = I_T X_L = (20\text{ A})(100\ \Omega) = 2{,}000\text{ V}$ (Note: $V_L = V_C$ at resonance)

$V_C = I_T X_C = (20\text{ A})(100\ \Omega) = 2{,}000\text{ V}$

At resonance, V_L and V_C can be greater than V_T.

Notice that the voltage across *L* and *C* is much higher than the applied voltage to the circuit. The ability to produce a voltage higher than the applied voltage is one of the most remarkable characteristics of the series resonant circuit. It is possible because of the ability of the inductor and the capacitor to store energy. This is illustrated in the phasor diagram of Figure 12-50.

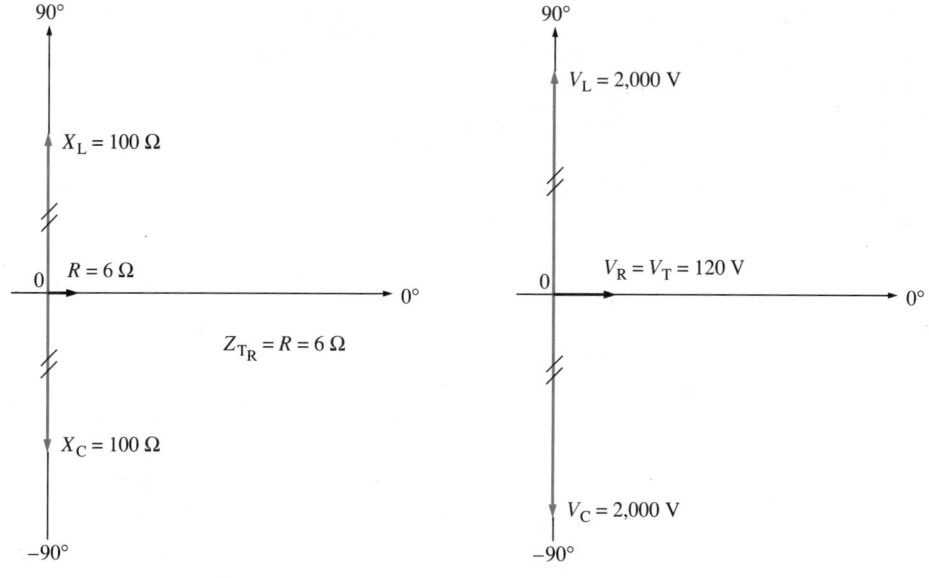

Figure 12-50

tech tidbit

Quality (Q) is the ratio of reactance to resistance.

The magnification of the voltage across L and C is determined by the **quality (Q)** of the circuit. Quality is defined as the ratio of the reactance of the coil to the resistance. Mathematically,

$$Q = \frac{X}{R}$$

Equation 12-12

Or because at resonance, $X_L = X_C$

$$Q = \frac{X_L}{R}$$

or

$$Q = \frac{X_C}{R}$$

tech tidbit

The greater the value of R in a circuit, the lower the Q.

Thus, if 1 volt is applied to a circuit with a Q of 10, the voltage across L and C at resonance will equal

$$\begin{aligned} V_L &= QV_{in} \\ &= (10)(1\text{ V}) \\ &= 10\text{ V} \end{aligned}$$

or

$$\begin{aligned} V_C &= QV_{in} \\ &= (10)(1\text{ V}) \\ &= 10\text{ V} \end{aligned}$$

It should also be noted that the current at resonance is the highest value (maximum) because the impedance is at its lowest value and equal to the resistance, R. Therefore the power (watts) at resonance is also at a maximum value (I^2R).

Bandwidth is determined by Q.

Resonant circuits are selective. That is, they respond more readily to their resonant frequency (f_0) than to other frequencies. Although the resonant frequency will have the greatest affect on the circuit, the frequencies close to the resonant frequency will also have a large effect on the circuit. The frequencies with the greatest effect on a circuit are called a band of frequencies, or the **bandwidth,** of the circuit. Figure 12-51 illustrates the effect of Q on the bandwidth of a circuit.

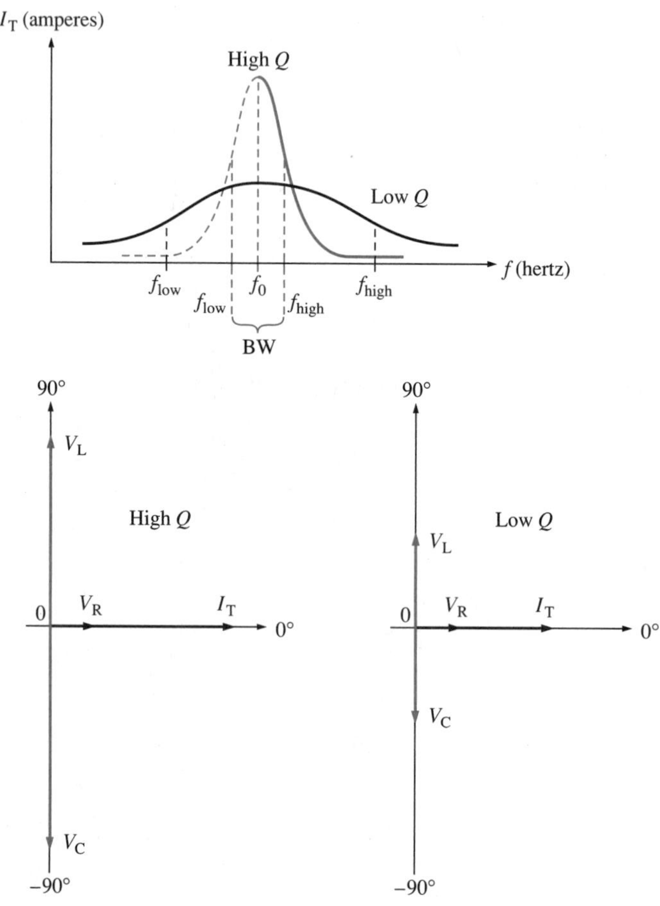

Figure 12-51 ***Effect of Q on the bandwidth of a circuit***

Mathematically,

Equation 12-13 ▶

$$\text{bandwidth} = BW = \frac{f_0}{Q}$$

Referring to Figure 12-51, we note that the bandwidth is centered about the resonant frequency, f_0. Furthermore, the bandwidth can be defined as the range of frequencies over which at least 50% of the maximum power is delivered to the circuit. These points are also the points at which at least 70.7% of the maximum current or voltage will be delivered to the circuit. These points are called the **half-power points.** Mathematically,

$$
\begin{aligned}
P_{max} &= I^2_{max}R \\
&= (0.707\ I_{max})^2R \\
&= 0.5\ I_{max}{}^2R \\
&= \frac{1}{2}P_{max}
\end{aligned}
$$

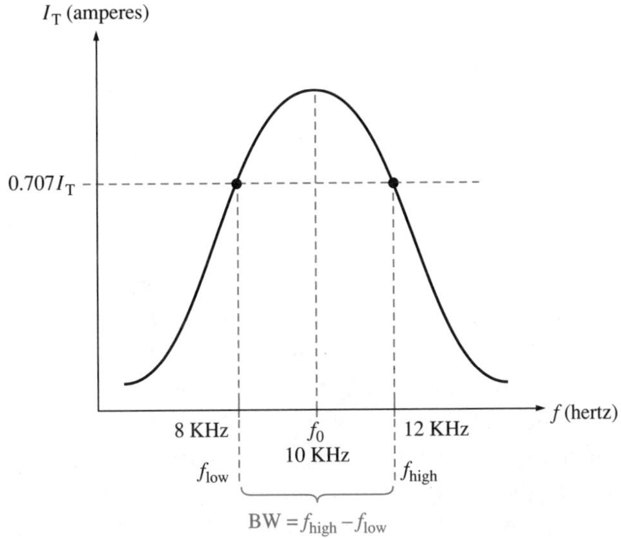

Figure 12-52 ***Half-power points***

The bandwidth can also be calculated from the half-power points as follows:

$$BW = f_H - f_L$$ ◀ *Equation 12-14*

Referring to Figure 12-52,

$$
\begin{aligned}
BW &= f_H - f_L \\
&= 12\ \text{KHz} - 8\ \text{KHz} \\
&= 4\ \text{KHz}
\end{aligned}
$$

EXAMPLE 12-21

The circuit of Figure 12-53 is at resonance. Calculate the Q and the BW of the circuit when (A) R = 20 ohms and (B) R = 2 ohms.

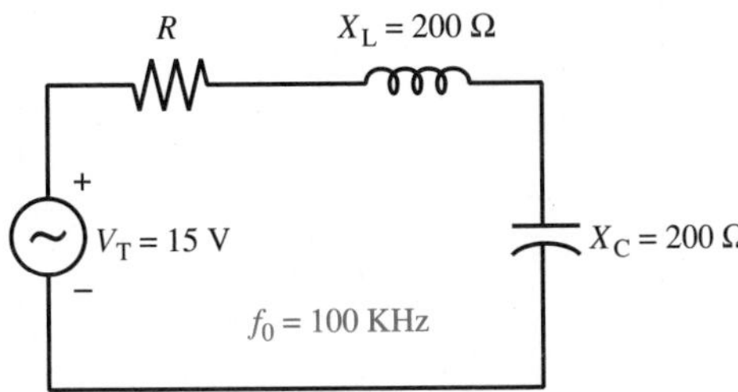

Figure 12-53

Solution:

(A)

$R = 20\ \Omega$

$$Q = \frac{X_L}{R} = \frac{200\ \Omega}{20\ \Omega} = 10$$

$$BW = \frac{f_0}{Q} = \frac{100\ \text{KHz}}{10} = 10\ \text{KHz}$$

(B)

$R = 2\ \Omega$

$$Q = \frac{X_L}{R} = \frac{200\ \Omega}{2\ \Omega} = 100$$

$$BW = \frac{f_0}{Q} = \frac{100\ \text{KHz}}{100} = 1\ \text{kHz}$$

tech tidbit

The higher the Q, the narrower the bandwidth. The lower the Q, the wider the bandwidth.

Notice that in Example 12-21, the amount of resistance affects both the Q and the bandwidth of the circuit and not the resonant frequency. The higher the resistance, the smaller the Q and the greater the bandwidth.

tech tidbit

The narrower the bandwidth, the more selective the circuit.

The bandwidth of the circuit determines the **selectivity** of the circuit. The selectivity of the circuit is the ability of a circuit to select one desired frequency out of a group of available frequencies. For example, a radio receives all the signals broadcast in a given area. Each station's signal is at a different frequency. The resonant circuit in the radio is set to select the frequency of the one station we wish to hear and reject all other frequencies. Therefore, the circuit requires a small or narrow bandwidth, hence, a high Q circuit. If the bandwidth is too high (wide), the radio would receive more than one station. In the case of a circuit like a radar detector, we would want a high bandwidth (wide) or low Q circuit. This would enable us to receive all radar signals in the area.

tech tidbit

Radios and TV sets use high Q very selective circuits.

Figure 12-54 summarizes the affects of resonance on a series circuit. At series resonance, $Z = R$ and is at a minimum value. The current is therefore at a maximum value. The bandwidth is the range of frequencies over which at least half of the maximum power (at resonance) is available.

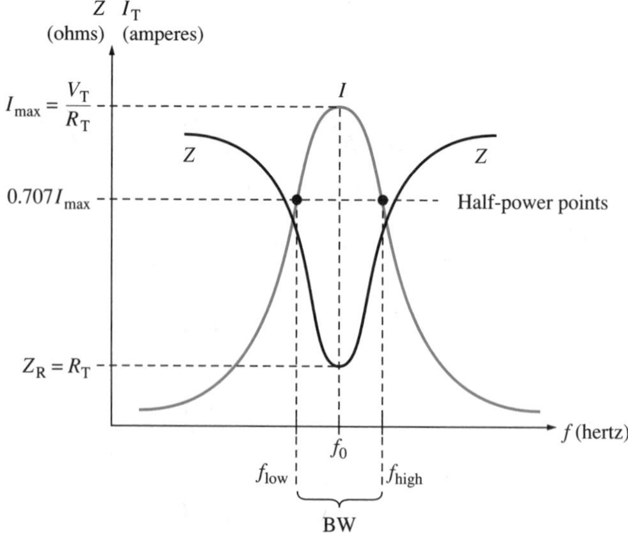

Figure 12-54 Series resonant effects

Parallel circuits can also be at resonance. Figure 12-55 illustrates a parallel resonant circuit. This circuit is sometimes referred to as a **tank circuit.** In an ideal sense, the circuit operates like a seesaw. During one half cycle, the capacitor is charged by the collapsing magnetic field in the inductor. During the next cycle, the capacitor discharges into the inductor. This sets up the magnetic field again and the process repeats. If the inductor and capacitor were both ideal (no resistance), the process would continue indefinitely because no losses would occur. However, everything has some resistance. Therefore the process will eventually end if the energy lost due to resistance in the circuit was not replaced.

tech tidbit

Tank circuits are parallel resonant circuits.

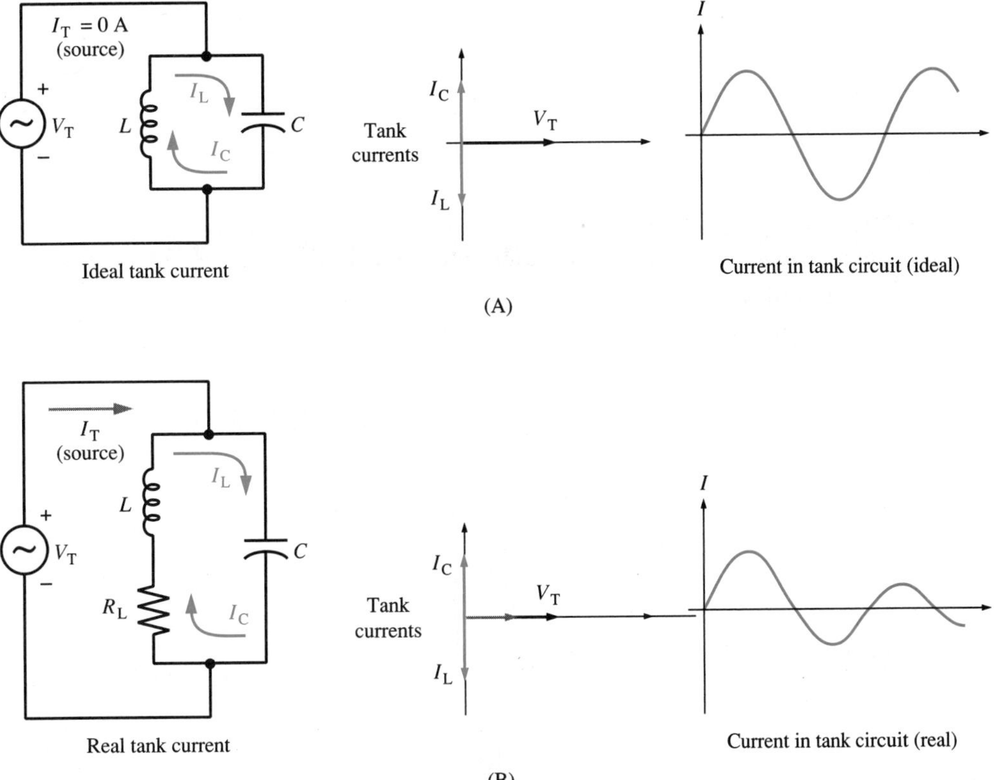

Figure 12-55 ***Parallel resonant (tank) circuit***

As with series resonant circuits, Q can be expressed as

$$Q = \frac{X}{R}$$

This equation is valid as long as the source resistance R_S is a lot greater than the resistance of the tank R_P. In actuality,

$$Q = \frac{I_{tank}}{I_{source}}$$

Equation 12-15

In a tank circuit, the source current is low while the circulating current (I_{tank}) is high. This is because the source current is used to replace the energy lost due to resistance in the circuit. Recall that Q can be thought of as a magnification factor. However, in this case, it is not the voltage that is being magnified,

At parallel resonance, Z_T approaches infinity.

it is the impedance that is magnified. The source current is low, so it follows that the impedance is high. In fact, at parallel resonance, the impedance of the tanks approaches infinity. For a parallel resonant circuit with a high Q, the impedance can be shown to be,

$$Z = QX$$

because $X_L = X_C$

$$Z = QX_L$$

or $Z = QX_C$

As with series resonant circuits, the bandwidth of a parallel resonant circuit is

Equation 12-16

$$BW = \frac{f_0}{Q}$$

EXAMPLE 12-22

For the circuit of Figure 12-56, find (A) the resonant frequency (f_0), (B) Q, (C) I_{tank}, (D) Z, and (E) BW.

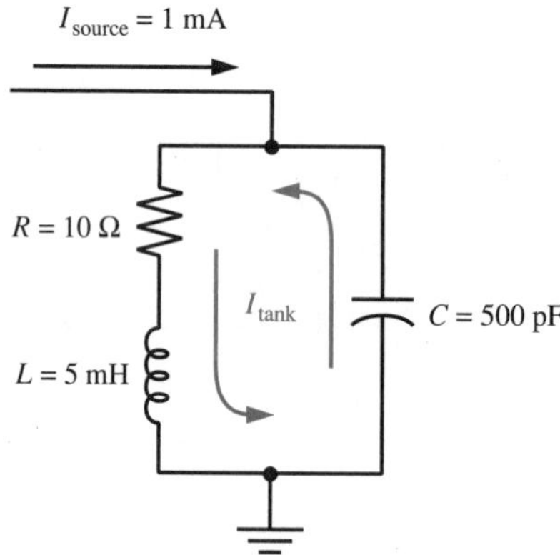

Figure 12-56

Solution:

(A)
$$f_0 = \frac{1}{2\pi\sqrt{LC}}$$

$$= \frac{1}{(2)(3.14)\sqrt{(5\text{ mH})(500\text{ pF})}}$$

$$= \frac{1}{(6.28)\sqrt{(5 \times 10^{-3}\text{ H})(500 \times 10^{-12}\text{ F})}}$$

$$= 100{,}700\text{ Hz}$$

$$= 100.7\text{ KHz}$$

(B) $X_L = 2\pi f L$

$= (2)(3.14)(100.7\text{ KHz})(5\text{ mH})$

$= 3{,}160\ \Omega$

$= 3.16\text{ K}\Omega$

$Q = \dfrac{X_L}{R}$

$= \dfrac{3.16\text{ K}\Omega}{10\ \Omega} = \dfrac{3{,}160\ \Omega}{10\ \Omega}$

$= 316$

(C) at resonance

$X_C = X_L = 3.16\text{ K}\Omega$

$Q = \dfrac{I_{\text{tank}}}{I_{\text{source}}}$

$I_{\text{tank}} = QI_{\text{source}}$

$= (316)(1\text{ mA})$

$= 316\text{ mA}$

$I_{\text{tank}} = I_L = I_C = 316\text{ mA}$

(D) $Z = QX_C$

$= (316)(3.16\text{ K}\Omega)$

$= 998.6\text{ K}\Omega$

$= 0.9986\text{ M}\Omega$

$\approx 1\text{ M}\Omega$

(E) $BW = \dfrac{f_0}{Q}$

$= \dfrac{100.7\text{ KHz}}{316}$

$= 318.7\text{ Hz}$

12.8 Filters

A **filter** is a RL or RC circuit especially designed to be frequency selective. RLC circuits are also frequency selective and, therefore, can be used as filters. There are several basic types of filter circuits. A **low pass filter** is used to **attenuate** or drop the output voltage of all frequencies above a certain **cut-off frequency.** Thus, it passes low frequencies and blocks high frequencies. A **high pass filter** passes signals above the cut-off frequency but blocks those below the cut-off frequency. Thus, it attenuates signals below the cut-off frequency. A **band pass filter** is designed to pass a small band of frequencies while rejecting both higher and lower frequencies. A **band stop** (notch) **filter** is designed to stop a small band of frequencies while passing higher and lower frequencies. It is the opposite of a band pass filter and is used to block out any annoying or interfering frequencies.

A low pass filter passes low frequencies and blocks high frequencies.

The high pass filter passes high frequencies and blocks low frequencies.

tech tidbit

A band pass filter passes a range of frequencies and blocks lower and higher frequencies.

A band stop filter blocks a range of frequencies and passes all others.

Figure 12-57 shows the characteristics of a low pass filter and Figure 12-58 shows the characteristics of a high pass filter.

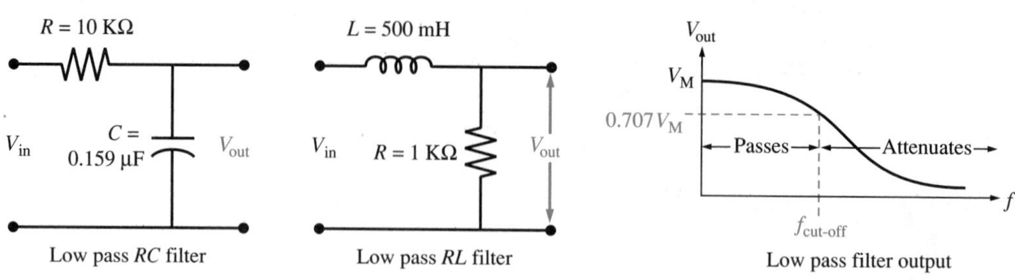

Figure 12-57 **Low pass filters**

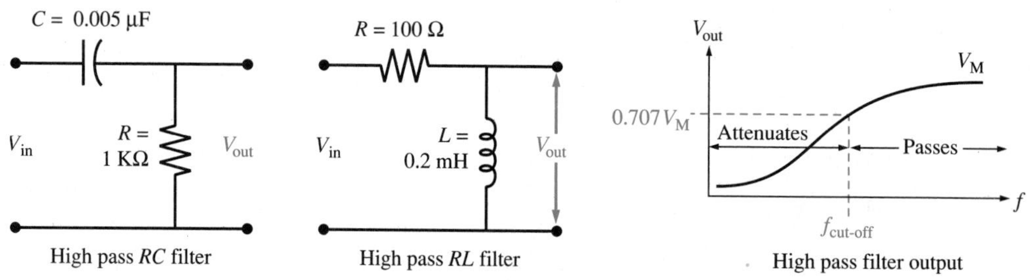

Figure 12-58 **High pass filters**

In either case, the cut-off frequency can be obtained using the following equations. For an RC filter,

Equation 12-17 ▶

$$f_{cut\text{-}off} = \frac{1}{2\pi RC}$$

where $f_{cut\text{-}off}$ = the cut-off frequency in hertz

R = the resistance in ohms

C = the capacitance in farads

For an RL filter,

Equation 12-18 ▶

$$f_{cut\text{-}off} = \frac{1}{2\pi\left(\frac{L}{R}\right)}$$

where $f_{cut\text{-}off}$ = the cut-off frequency in hertz

R = the resistance in ohms

L = the inductance in henries

T^2 tech tidbit

$\tau = RC$, $\tau = L/R$

Note that the only difference between a low pass filter and a high pass filter is the placement of the components. The cut-off frequency is defined as the frequency at which the output voltage is 70.7% of the input voltage.

f_{cutoff} occurs when $V_{out} = 0.707\ V_{in}$.

EXAMPLE 12-23

Calculate the low pass cut-off frequency for the (A) RC and (B) RL circuit of Figure 12-57.

Solution:

(A) Low pass RC filter

$$f_{\text{cut-off}} = \frac{1}{2\pi RC}$$

$$= \frac{1}{(2)(3.14)(10{,}000\ \Omega)(0.159 \times 10^{-6}\ \text{F})}$$

$$= 100\ \text{Hz}$$

(B) Low pass RL filter

$$f_{\text{cut-off}} = \frac{1}{2\pi\left(\frac{L}{R}\right)}$$

$$= \frac{1}{(2)(3.14)\left(\frac{500 \times 10^{-3}\ \text{H}}{}\right)}$$

$$= 318.5\ \text{Hz}$$

EXAMPLE 12-24

Calculate the high pass cut-off frequency for the (A) RC and (B) RL circuits of Figure 12-58.

Solution:

(A) High pass RC filter

$$f_{\text{cut-off}} = \frac{1}{2\pi RC}$$

$$= \frac{1}{(2)(3.14)(1 \times 10^{3})(0.005 \times 10^{-6}\ \text{F})}$$

$$= 31{,}800\ \text{Hz}$$

$$= 31.8\ \text{KHz}$$

(B) High pass RL filter

$$f_{\text{cut-off}} = \frac{1}{2\pi\left(\frac{L}{R}\right)}$$

$$= \frac{1}{(2)(3.14)\left(\frac{0.2 \times 10^{-3}\ \text{H}}{100\ \Omega}\right)}$$

$$= 79{,}600\ \text{Hz}$$

$$= 79.6\ \text{KHz}$$

Figure 12-59 shows the characteristics of a band pass filter. Note that resonant frequency circuit analysis is used to design band pass filter circuits. These circuits can be either series resonant or parallel resonant. The bandwidth determines the band pass capabilities.

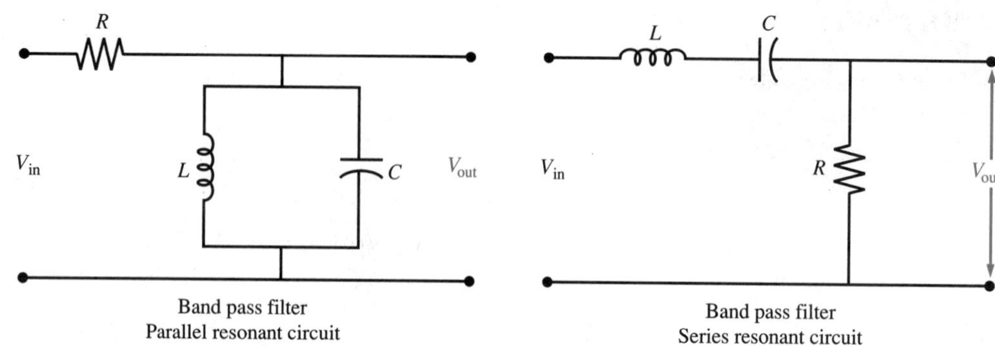

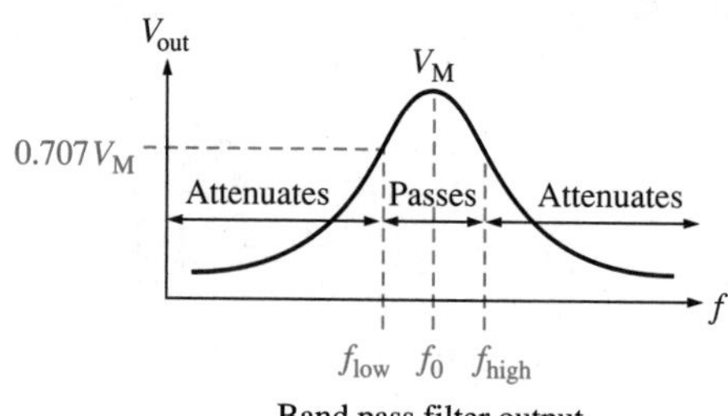

Figure 12-59 **Band pass circuits**

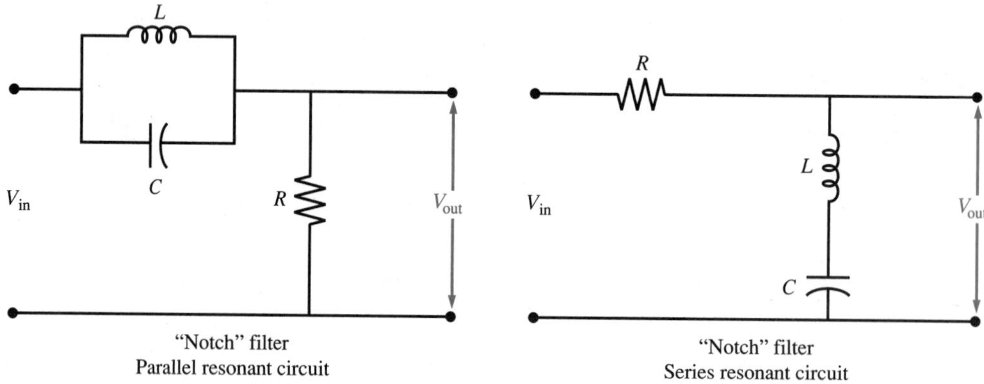

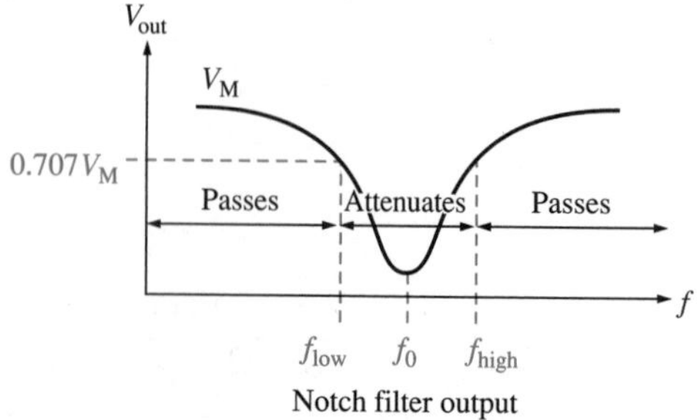

Figure 12-60 **Band stop "notch" filter circuit**

Figure 12-60 illustrates the characteristics of a band stop filter. It is also known as a **notch filter**. Again, resonant circuits are used. However, the placement of the components are opposite to that of the band pass filter. The band stop filter is the opposite of that of the band pass filter. The bandwidth is used to determine the notch or stop frequencies.

Notch filters are band stop filters.

✓ Self-Test 5

1. The point at which $X_L = X_C$ is called ____________.
2. Quality is defined as the ratio of ____________ to ____________.
3. The higher the Q of a circuit, the ____________ the bandwidth.
4. Name four types of filter circuits.

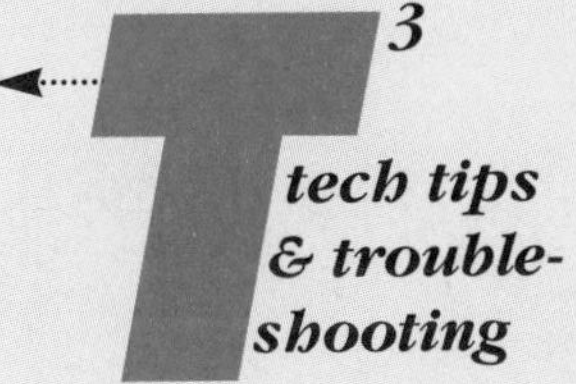

T³ TONE CONTROL

A tone control provides us with a method of allowing high or low frequencies to be amplified or attenuated. The human ear is not equally responsive to all frequencies. It does not receive high and low frequencies as well as midrange frequencies. Thus, high and low frequencies do not appear as loud as midrange frequencies. The tone control gives us a method of amplifying high and low frequencies without increasing the overall volume of the amplifier. Figure 12-61 illustrates the operation of a simple tone control.

Capacitor C_1 controls the **treble** frequencies. The treble frequencies are the high frequencies. Capacitor C_3 controls the base frequencies. The **bass** frequencies are the low frequencies. Capacitor C_2 controls the midrange frequencies. The higher the reactance of the capacitor, the greater the signal that will be allowed to pass to the amplifier. The lower the reactance, the smaller the output signal. For example, at 20,000 hertz, the high end of the audio band, C_1 has a reactance of approximately 4 kilohms, C_2 is about 0.8 kilohms, and C_3 will be about 150 ohms. Thus, most frequencies are allowed to pass when the switch is in position C_1. In position C_2, frequencies above 5,000 hertz are

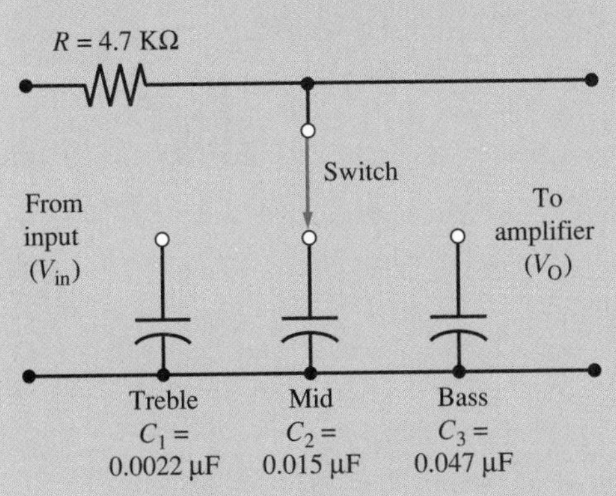

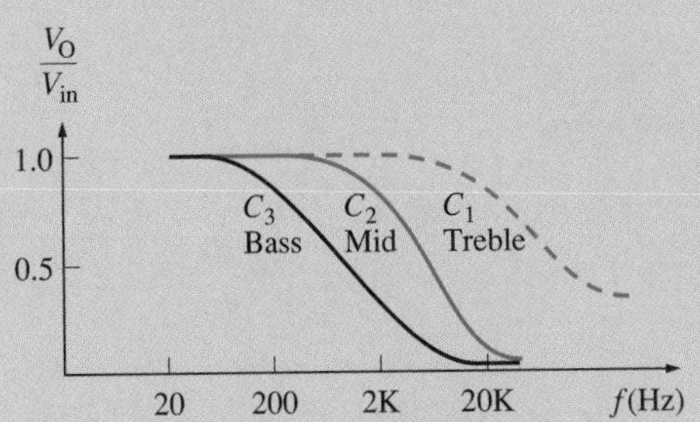

Figure 12-61 Simple tone control

attenuated. In position C_3, frequencies above 1,000 hertz are attenuated.

Figure 12-62A illustrates the circuit of an actual tone control. The circuit utilizes separate controls for bass and treble. For low frequencies, the reactance of the capacitors will be very high and the capacitors will look like open circuits. Figure 12-62B shows the equivalent circuit at low frequencies. Note that the treble control has no affect because capacitor C_3 reacts like an open circuit to low frequencies.

At high frequencies, all of the capacitors look like short circuits. Capacitors C_1 and C_2 effectively short out the bass control potentiometer. Figure 12-62C shows the equivalent circuit at high frequencies. Note that resistors R_1 and R_6 are effectively in parallel with the treble control R_5.

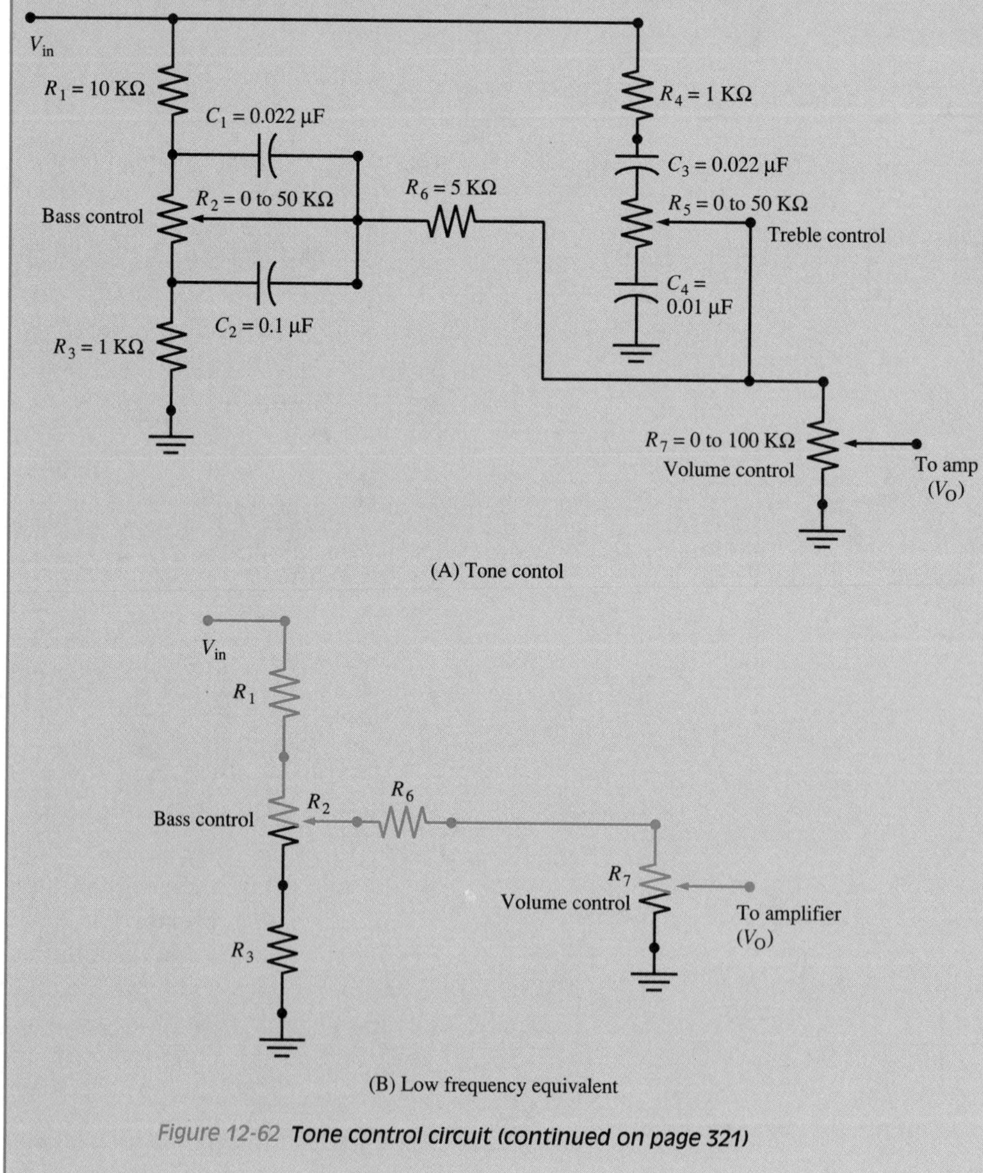

Figure 12-62 **Tone control circuit** **(continued on page 321)**

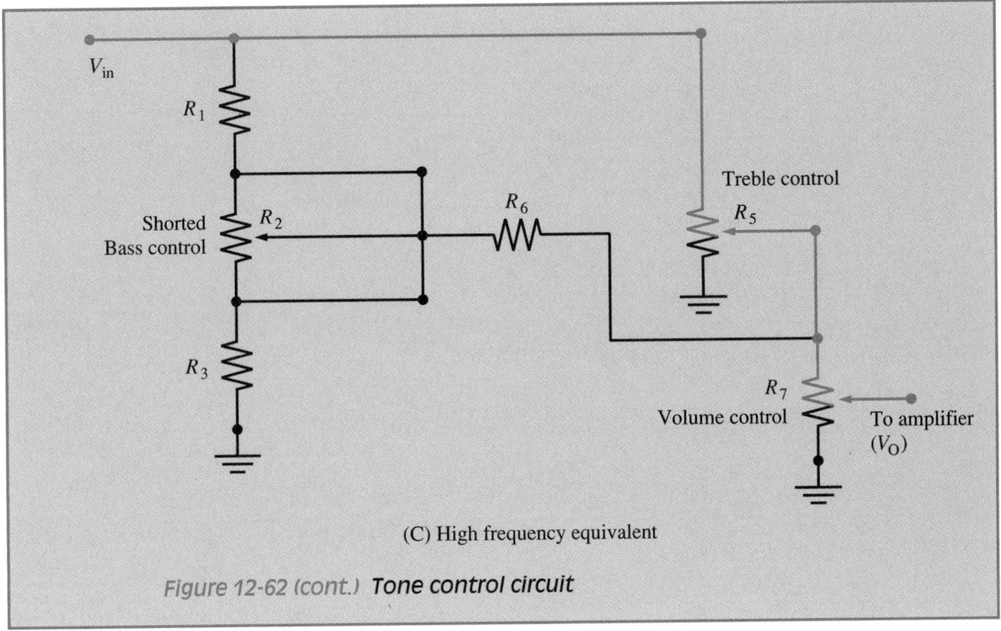

(C) High frequency equivalent

Figure 12-62 (cont.) **Tone control circuit**

SUMMARY

In this chapter we analyzed RLC circuits with DC and with AC energy sources. You should note the similarities in the analysis of DC and AC circuits. Ohm's law uses its same basic format. Basically, the current is equal to the voltage divided by the opposition.

Phasors are a type of graph that displays magnitude and direction. In a purely resistive circuit, the voltage and current are said to be in phase. In a purely inductive circuit, the voltage will lead the current by 90 degrees. In a purely capacitive circuit, the current will lead the voltage by 90 degrees.

Impedance is the total opposition to current in an AC circuit. It is the combined effect of R, X_L, and X_C. Impedance uses the letter Z as its symbol and is expressed in ohms. The point at which X_L equals X_C is called resonance. At series resonance, $Z_T = R$.

Resonant circuits are selective. That is, they respond more readily to their resonant frequency and to frequencies close to their resonant frequency. The frequencies with the greatest effect on a circuit are called a band of frequencies or the bandwidth of a circuit. Bandwidth is affected by the quality, Q, of the circuit. The bandwidth can also be calculated from the half-power points. Filters are RL or RC circuits specially designed to be frequency selective.

FORMULAS

$$Z_T = \frac{V_T}{I_T} \qquad V_T = I_T Z_T \qquad I_T = \frac{V_T}{Z_T}$$

$$X_L = \frac{V_L}{I_L} \qquad V_L = I_L X_L \qquad I_L = \frac{V_L}{X_L}$$

$$X_C = \frac{V_C}{I_C} \qquad V_C = I_C X_C \qquad I_C = \frac{V_C}{X_C}$$

$$V_T = \sqrt{V_R^2 + (V_L - V_C)^2} \qquad \text{(series RLC circuits)}$$

$$I_T = \sqrt{I_R^2 + (I_L - I_C)^2} \qquad \text{(parallel RLC circuits)}$$

$$\cos\theta_T = \frac{\text{adjacent side}}{\text{hypotenuse}} = \frac{V_R}{V_T} = \frac{R}{Z_T} = \frac{I_R}{I_T}$$

$$Z_T = \sqrt{R^2 + (X_L - X_C)^2} \qquad \text{(series RLC circuits)}$$

$$Z_T = \frac{(R)(X_L)}{\sqrt{R^2 + X_L^2}} \qquad \text{(parallel RL circuits)}$$

$$Z_T = \frac{(R)(X_C)}{\sqrt{R^2 + X_C^2}} \qquad \text{(parallel RC circuits)}$$

$$Z_T = \frac{V_T}{I_T} \qquad \text{(parallel RLC circuits)}$$

$$f_0 = \frac{1}{2\pi\sqrt{LC}} \qquad \text{(resonance)}$$

$$Z_{T_0} = R \qquad \text{(series resonance)}$$

$$Q = \frac{X_L}{R} \qquad Q = \frac{X_C}{R} \qquad Q = \frac{I_{tank}}{I_{source}}$$

$$V_L = QV_{in} \qquad V_C = QV_{in}$$

$$BW = \frac{f_0}{Q} \qquad BW = f_H - f_L$$

$$Z = QX_L \qquad Z = QX_C \qquad \text{(parallel resonance)}$$

$$f_{cut\text{-}off} = \frac{1}{2\pi RC} \qquad \text{(RC filter)}$$

$$f_{cut\text{-}off} = \frac{1}{2\pi\left(\frac{L}{R}\right)} \qquad \text{(RL filter)}$$

EXERCISES

Section 12.1

1. Find the steady state current, I_T, and the voltage drop across resistors R_1, R_2, and R_3 in Figure 12-63A.
2. In Figure 12-63B, find I_T, V_1, and V_2 in the steady state.
3. For the steady state current condition, find, I_T, V_1, V_2, V_3, V_4, and V_5 in the circuit of Figure 12-63C.

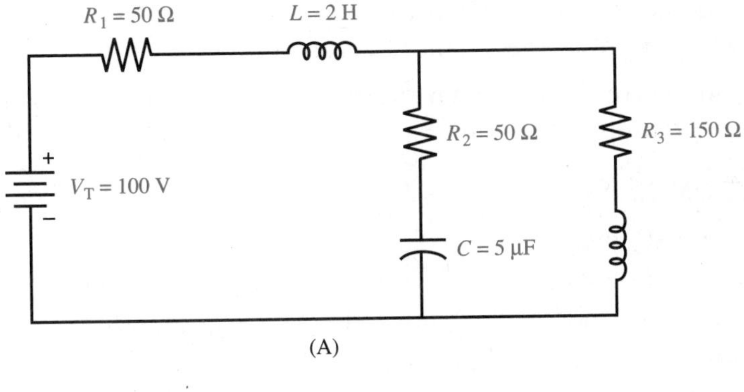

(A)

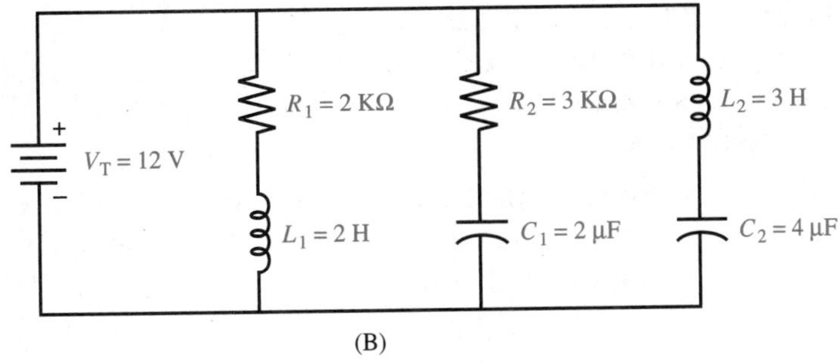

(B)

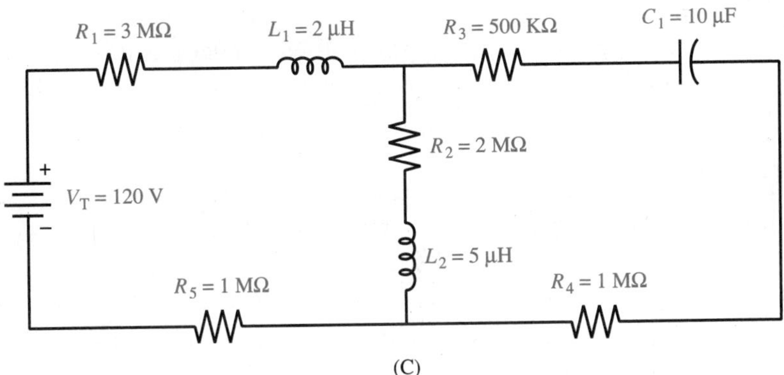

(C)

Figure 12-63

Section 12.2

4. For the circuit of Figure 12-64A, find the current and sketch the waveform and phasor diagram of the voltage and current.

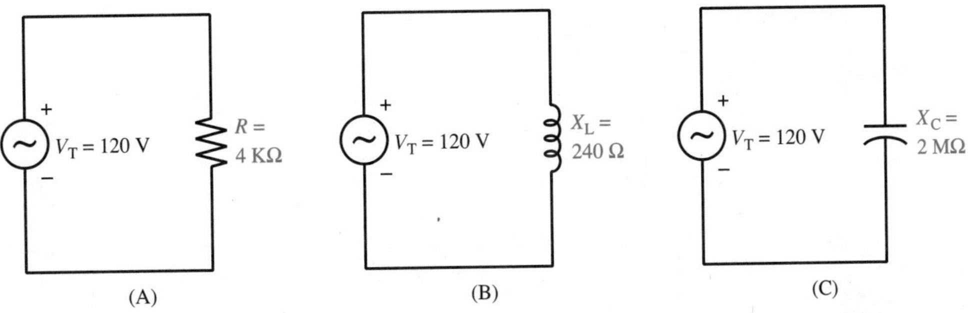

Figure 12-64

5. Find the current and sketch the waveform and phasor diagram of the voltage and current in the circuit of Figure 12-64B.

6. Find the current and sketch the waveform and phasor diagram of the voltage and current in the circuit of Figure 12-64C.

Section 12.3

7. Three resistors, 2 kilohms, 3 kilohms, and 5 kilohms, are connected in series with a 120-volt AC source. Find Z_T, I_T, V_{2K}, V_{3K}, and V_{5K}.

8. Two resistors, 300 ohms and 600 ohms, are connected in parallel with a 100-volt AC source. Find $I_{300\ \Omega}$, $I_{600\ \Omega}$, I_T, and Z_T.

Section 12.4

9. A 400-ohm resistor and an inductor with a reactance of 1,200 ohms are connected in series with a 120-volt AC source. Find Z_T, θ_T, I_T, V_R, and V_L. Verify your answers using Kirchhoff's voltage law.

10. A 4-kilohm resistor and an inductor with a reactance of 3 kilohms are connected in series with a 120-volt AC source. Find Z_T, θ_T, I_T, V_R, and V_L. Verify your answers using Kirchhoff's voltage law.

11. A 25-ohm resistor and an inductor with a reactance of 50 ohms are connected in parallel with a 100-volt AC source. Find I_R, I_L, I_T, Z_T, and the circuit phase angle θ_T.

12. A 10-kilohm resistor and an inductor with a reactance of 15 kilohms are connected in parallel with a 120-volt AC source. Find I_R, I_L, I_T, Z_T, and the circuit phase angle θ_T.

Section 12.5

13. A 12-kilohm resistor and a capacitor with a reactance of 4 kilohms are connected in series with a 48-volt AC source. Find Z_T, θ_T, I_T, V_R, and V_C. Verify your answers using Kirchhoff's voltage law.

14. A 20-ohm resistor and a capacitor with a reactance of 30 ohms are connected in series with a 100-volt AC source. Find Z_T, θ_T, I_T, V_R, and V_C. Verify your answers using Kirchhoff's voltage law.

15. A 3-kilohm resistor and a capacitor with a reactance of 4 kilohms are connected in parallel with a 12-volt AC source. Find I_R, I_C, I_T, Z_T, and the circuit phase angle θ_T.

16. A 330-ohm resistor and a capacitor with a reactance of 200 ohms are connected in parallel with a 10-volt AC source. Find I_R, I_C, I_T, Z_T, and the circuit phase angle θ_T.

Section 12.6

17. A 50-ohm resistor, a capacitor with a reactance of 40 ohms, and an inductor with a reactance of 160 ohms are connected in series. Find the total impedance.

18. A 64-ohm resistor, a capacitor with a reactance of 60 ohms, and an inductor with a reactance of 12 ohms are connected in series. Find the total impedance.

19. A 66-ohm resistor, a capacitor with a reactance of 32 ohms, and an inductor with a reactance of 120 ohms are connected in series with a 120-volt AC source. Find Z_T, I_T, the voltage across each element, and the phase angle of the circuit.

20. A 85-ohm resistor, a capacitor with a reactance of 68 ohms, and an inductor with a reactance of 272 ohms are connected in series with a 440-volt AC source. Find Z_T, I_T, the voltage across each element, and the phase angle of the circuit.

21. A 50-ohm resistor, a capacitor with a reactance of 150 ohms, and an inductor with a reactance of 100 ohms are connected in parallel with a 300-volt AC source. Find I_R, I_L, I_C, I_T, Z_T, and the circuit phase angle θ_T.

22. A 60-ohm resistor, a capacitor with a reactance of 25 ohms, and an inductor with a reactance of 50 ohms are connected in parallel with a 300-volt AC source. Find I_R, I_L, I_C, I_T, Z_T, and the circuit phase angle θ_T.

Section 12.7

23. Find the resonant frequency for an 8-microfarad capacitor and a 30-microhenry coil.

24. Find the resonant frequency for a 20-microfarad capacitor and a 20-millihenry coil. Find X_L and X_C at this frequency.

25. A 50-ohm resistor is connected in series with an inductor whose reactance is 1,000 ohms and a capacitor whose reactance is 1,000 ohms. If the source voltage is 100 VAC, find Z_T, I_T, V_R, V_L, and V_C.

26. A 25-ohm resistor and an inductor whose reactance is 500 ohms are connected in series with a 100 VAC 10-kilohertz source. Find Q and the bandwidth of the circuit.

27. Repeat problem 26 when $R = 5$ ohms.

28. For the circuit of Figure 12-65, find the resonant frequency, f_0, Q, I_{tank}, Z, and BW.

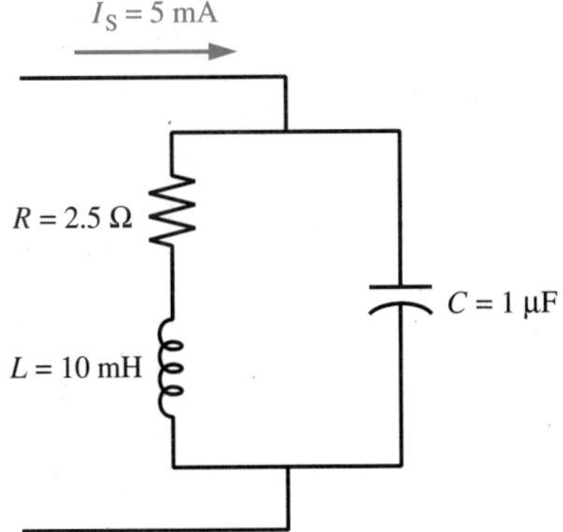

Figure 12-65

Section 12.8

29. Calculate the low pass cut-off frequency for the RC and RL circuit of Figure 12-66.

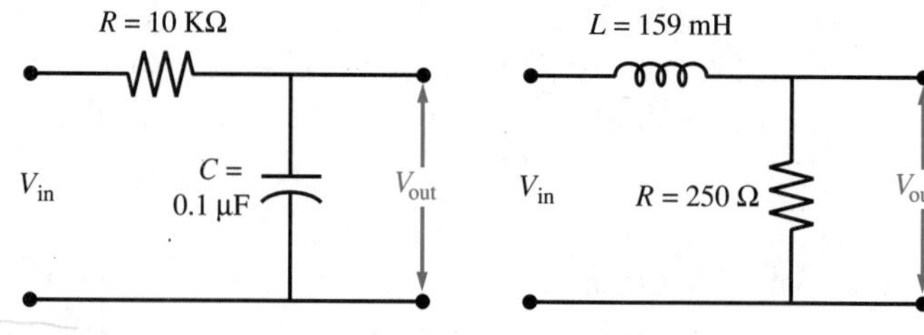

Figure 12-66

30. Calculate the high pass cut-off frequency for the RC and RL circuit of Figure 12-67.

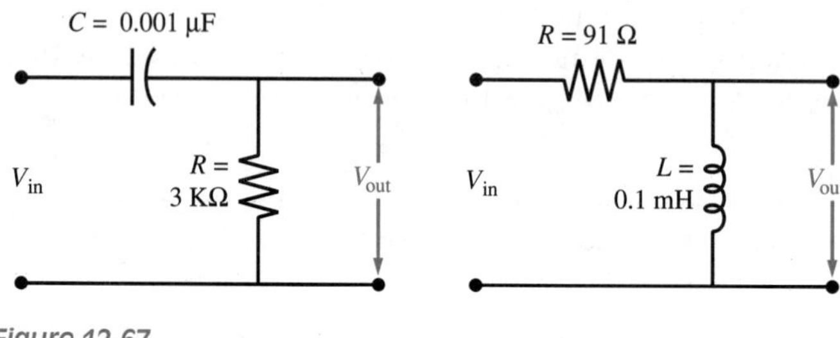

Figure 12-67

31. See Lab Manual for the crossword puzzle for Chapter 12.

CHAPTER 13

POWER, MOTORS, AND GENERATORS

Courtesy of Hewlett Packard

KEY TERMS

apparent power
armature
asynchronous motor
balanced system
capacitor start motor
carbon brushes
commutator
compound motor
copper losses
core losses
delta
efficiency
electric generator
horsepower
induction motor
line voltage
mechanical losses
motor
nameplate
phase voltage
power factor
power triangle
reactive power
real power
rotor
series motor
shaded pole motor
shunt motor
slip
split phase motor
stator
synchronous motor
synchronous speed
three-phase
torque
universal motor
VARS
volt-amperes
watts
wye

OBJECTIVES

After completing this chapter, the student should be able to:

- define power in AC circuits.
- calculate reactive, apparent, and real power.
- use the power triangle.
- perform power factor corrections.
- analyze power in three-phase circuits.
- identify the parts of a motor.
- define motor characteristics.
- calculate horsepower, efficiency, and torque.
- calculate the properties of DC machines.
- calculate the properties of AC machines.
- interpret nameplate data.

13.1 Introduction

Power is the rate of performing work.

Recall from our study of DC electric circuits that power is the rate of performing work. Mathematically, it is equal to the voltage times the current. However, in AC electric circuits, the voltage and current are always changing. Therefore, the instantaneous power must equal the product of the instantaneous value of the voltage times the instantaneous value of the current at any point in time. For resistive circuits, the voltage and current are in phase. The power is simply the product of the voltage times the current as it was in DC. All three forms of the power formulas still apply.

$$P = VI \quad \text{(watts)}$$

$$P = I^2R \quad \text{(watts)}$$

$$P = \frac{V^2}{R} \quad \text{(watts)}$$

Referring to Figure 13-1, we can see that the peak power will equal the peak voltage times the peak current. The power at 45 degrees will equal the voltage at 45 degrees times the current at 45 degrees. Note that at 270 degrees the power is still positive because a negative voltage times a negative current equals a positive power.

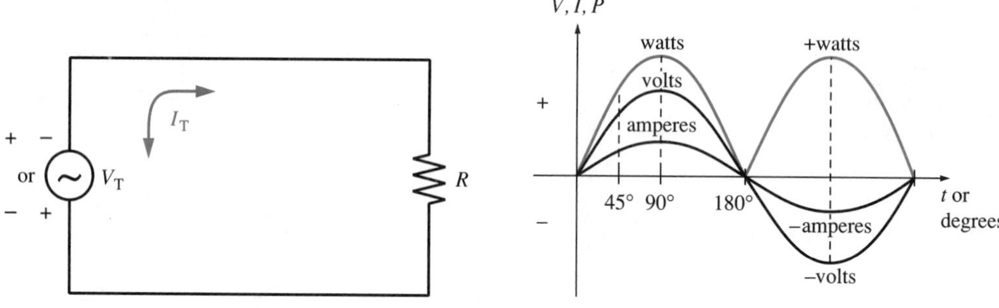

Figure 13-1 **AC Power**

EXAMPLE 13-1

For the circuit of Figure 13-2, find (A) the power dissipated by each resistor and (B) the total power used by the circuit.

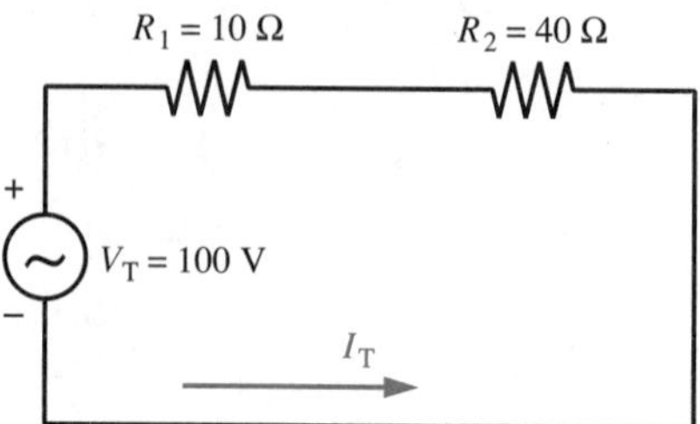

Figure 13-2

Solution:

(A) $I_T = \dfrac{V_T}{R_T}$

$= \dfrac{100\text{ V}}{10\ \Omega + 40\ \Omega}$

$= 2\text{ A}$

$P_1 = I_T^2 R_1$

$= (2\text{ A})^2(10\ \Omega)$

$= 40\text{ W}$

$P_2 = I_T^2 R_2$

$= (2\text{ A})^2(40\ \Omega)$

$= 160\text{ W}$

(B) $P_T = V_T I_T$

$= (100\text{ V})(2\text{ A})$

$= 200\text{ W}$

Check

$P_T = P_1 + P_2$

$= 40\text{ W} + 160\text{ W}$

$= 200\text{ W}$

Note that the total **real power** (power transformed by a resistor into heat) in **watts** will always equal the sum of the power dissipated by each resistor. This will be true regardless of how the circuit is constructed. Therefore, three 100-watt light bulbs will consume 300 watts of power regardless of how the bulbs are connected (series, parallel, or series-parallel) provided that the rated voltage is across each bulb.

Real power is measured in watts.

When reactance is added to a circuit, the total voltage and total current are out of phase, as shown in Figure 13-3. To compute the power, we cannot simply multiply the voltage times the current because they are no longer in phase. For example, in Figure 13-3, we note that the peak voltage and the peak current occur at different times. The power is computed by considering the phase angle using the following equation:

$$P = VI \cos \theta$$

Equation 13-1

where P = the power in watts

V = the RMS voltage in volts

I = the RMS current in amperes

θ = the phase angle between the voltage and the current

$\cos \theta = \dfrac{P}{VI}$

Note that Equation 13-1 is only valid for sine waves.

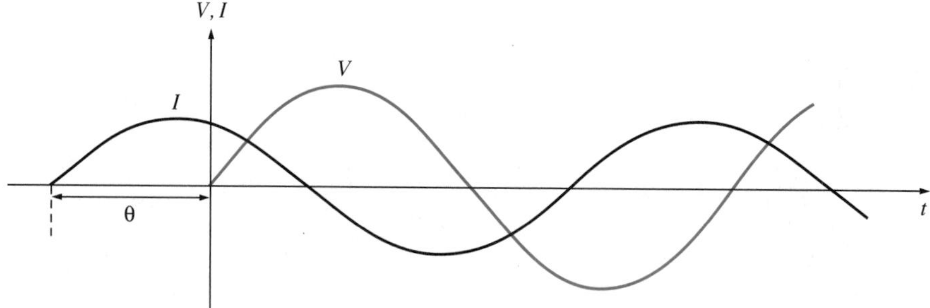

Figure 13-3 **Out-of-phase AC circuit waveform**

EXAMPLE 13-2

Calculate (A) the power consumed by the circuit of Figure 13-4 and (B) the power delivered to the circuit.

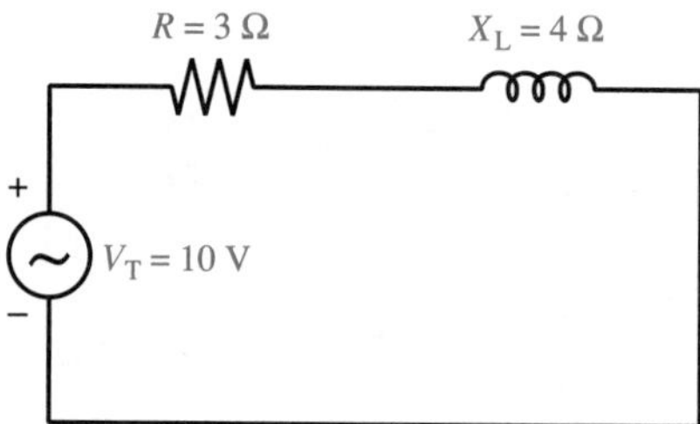

Figure 13-4

Solution:

(A)
$$Z_T = \sqrt{R^2 + X_L^2} = \sqrt{(3\ \Omega)^2 + (4\ \Omega)^2} = 5\ \Omega$$

$$I_T = \frac{V_T}{Z_T} = \frac{10\ \text{V}}{5\ \Omega} = 2\ \text{A}$$

$$P_R = I_T^2 R = (2\ \text{A})^2(3\ \Omega) = 12\ \text{W (power consumed)}$$

(B)
$$\cos \theta_T = \frac{R}{Z_T} = \frac{3\ \Omega}{5\ \Omega} = 0.6$$

$$\therefore \quad \theta_T = \cos^{-1} 0.6$$
$$= 53.13°$$
$$P_T = V_T I_T \cos \theta_T$$
$$= (10 \text{ V})(2 \text{ A})(\cos 53.13°)$$
$$= (10 \text{ V})(2 \text{ A})(0.6)$$
$$= 12 \text{ W (power delivered)}$$

Check

$$P_{consumed} = P_{delivered}$$
$$12 \text{ W} = 12 \text{ W}$$

✓ Self-Test 1

1. Power is defined as ________________.
2. The power consumed by a resistor in watts is called ________________.
3. To calculate the power in watts, the voltage and current must be multiplied by the ________________.
4. Calculate the power when $V = 100$ V, $I = 2$ A, and $\theta = 60°$.

13.2 Reactive Power

The rate of change of energy in a reactive element like an inductor or a capacitor is called **reactive power.** Reactive power differs from real power in that the energy is not dissipated in heat. Instead, the energy is stored during one-half of the cycle and returned during the next half cycle. Capacitors and inductors are 180 degrees out of phase, so when one is storing the energy, the other will be returning the energy and vice versa. This is similar to the action of the parallel resonant tank circuit. The symbol used for reactive power is P_Q or simply Q. The units of reactive power are **VARS** (volt-amperes reactive). Reactive power can be computed using the following equations:

tech tidbit

Reactive power is measured in VARS.

$$P_Q = I^2 X_L \text{ (inductive VARS)}$$ ◀ *Equation 13-2*

or

$$P_Q = I^2 X_C \text{ (capacitive VARS)}$$ ◀ *Equation 13-3*

where P_Q = the reactive power in VARS
I = the RMS current in amperes
X_L = the inductive reactance in ohms
X_C = the capacitive reactance in ohms

In circuits with both inductance and capacitance, the total reactive power will subtract (because they are 180 degrees out of phase), as shown in Figure 13-5.

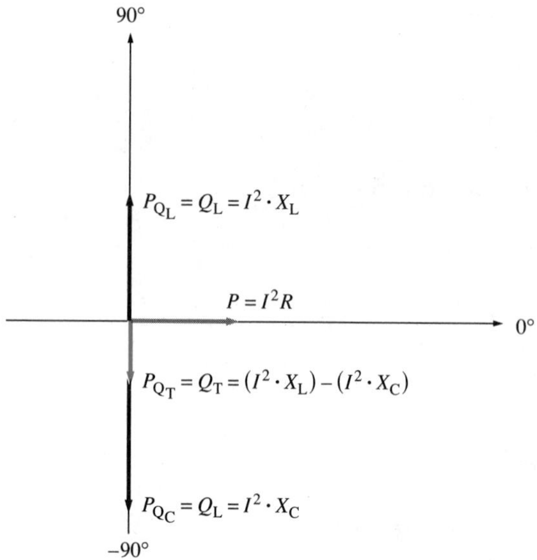

Figure 13-5 **Reactive power (P_Q or Q)**

EXAMPLE 13-3

Calculate (A) the real power and (B) the reactive power in the circuit of Figure 13-6.

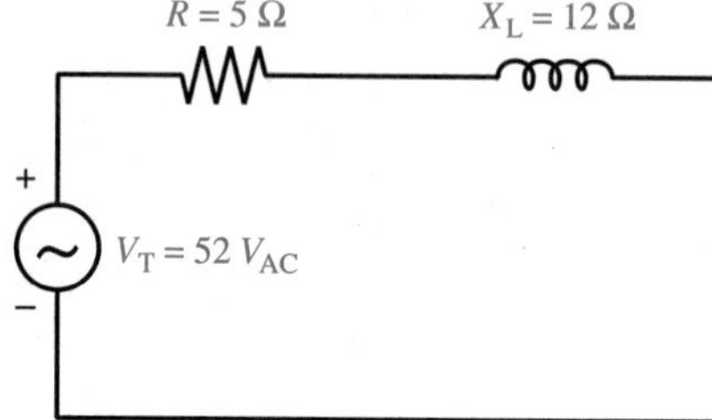

Figure 13-6

Solution:

(A)
$$Z_T = \sqrt{R^2 + X_L{}^2}$$
$$= \sqrt{(5\ \Omega)^2 + (12\ \Omega)^2}$$
$$= 13\ \Omega$$

$$I_T = \frac{V_T}{Z_T}$$
$$= \frac{52\ \text{V}}{13\ \Omega}$$
$$= 4\ \text{A}$$

$$\cos \theta_T = \frac{R}{Z_T}$$

$$= \frac{5\ \Omega}{13\ \Omega}$$

$$= 0.385$$

$$\therefore \quad \theta_T = \cos^{-1} 0.385$$

$$= 67.38°$$

$$P_T = V_T I_T \cos \theta_T$$

$$= (52\ V)(4\ A)(\cos 67.38°)$$

$$= (52\ V)(4\ A)(0.385)$$

$$= 80\ W$$

or

$$P_T = I_T{}^2 R$$

$$= (4\ A)^2(5\ \Omega)$$

$$= 80\ W$$

(B) $P_Q = I_T{}^2 X_L$

$$= (4\ A)^2(12\ \Omega)$$

$$= 192\ \text{VARS (inductive type)}$$

EXAMPLE 13-4

Calculate (A) the real power and (B) the reactive power in the circuit of Figure 13-7.

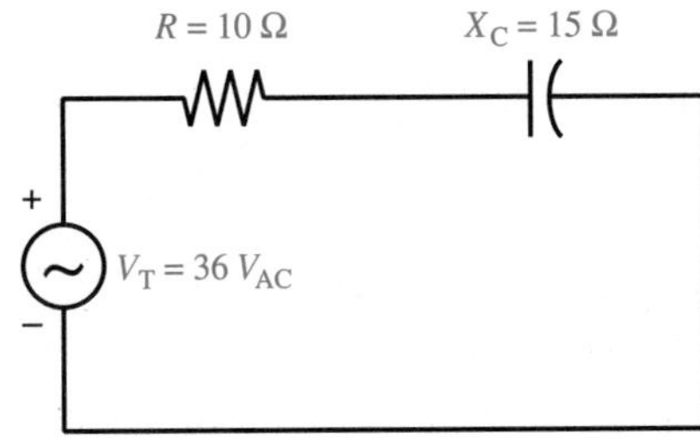

Figure 13-7

Solution:

(A) $Z_T = \sqrt{R^2 + X_C{}^2}$

$$= \sqrt{(10\ \Omega)^2 + (15\ \Omega)^2}$$

$$= 18\ \Omega$$

$$I_T = \frac{V_T}{Z_T}$$

$$= \frac{36 \text{ V}}{18 \ \Omega}$$

$$= 2 \text{ A}$$

$$\cos \theta_T = \frac{R}{Z_T}$$

$$= \frac{10 \ \Omega}{18 \ \Omega}$$

$$= 0.555$$

$$\therefore \quad \theta_T = \cos^{-1} 0.555$$

$$= 56.25°$$

$$P_T = V_T I_T \cos \theta_T$$

$$= (36 \text{ V})(2 \text{ A})(\cos 56.25°)$$

$$= (36 \text{ V})(2 \text{ A})(0.555)$$

$$= 40 \text{ W}$$

or

$$P_T = I_T{}^2 R$$

$$= (2 \text{ A})^2(10 \ \Omega)$$

$$= 40 \text{ W}$$

(B) $$P_Q = I_T{}^2 X_C$$

$$= (2 \text{ A})^2(15 \ \Omega)$$

$$= 60 \text{ VARS (capacitive type)}$$

EXAMPLE 13-5

Calculate (A) the total real power and (B) the reactive power in the circuit of Figure 13-8.

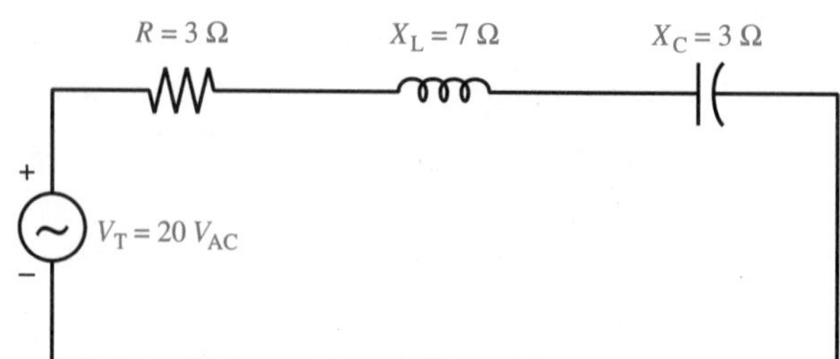

Figure 13-8

Solution:

(A) $Z_T = \sqrt{R^2 + (X_L - X_C)^2}$

$= \sqrt{(3\ \Omega)^2 + (7\ \Omega - 3\ \Omega)^2}$

$= \sqrt{(3\ \Omega)^2 + (4\ \Omega)^2}$

$= 5\ \Omega$

$I_T = \dfrac{V_T}{Z_T}$

$= \dfrac{20\ \text{V}}{5\ \Omega}$

$= 4\ \text{A}$

$\cos \theta_T = \dfrac{R}{Z_T}$

$= \dfrac{3\ \Omega}{5\ \Omega}$

$= 0.6$

$\therefore\ \theta_T = \cos^{-1} 0.6$

$= 53.13°$

Note: That in a RLC circuit, Z_T, the total impedance, may result in a value less than that of a reactive element, X ($Z_T = 5\ \Omega$, $X_L = 7\ \Omega$).

$P_T = V_T I_T \cos \theta_T$

$= (20\ \text{V})(4\ \text{A})(\cos 53.13°)$

$= (20\ \text{V})(4\ \text{A})(0.6)$

$= 48\ \text{W}$

or

$P_T = I_T{}^2 R$

$= (4\ \text{A})^2(3\ \Omega)$

$= 48\ \text{W}$

(B) $P_{Q_L} = I_T{}^2 X_L$

$= (4\ \text{A})^2(7\ \Omega)$

$= 112$ VARS (lagging inductive type)

$P_{Q_C} = I_T{}^2 X_C$

$= (4\ \text{A})^2(3\ \Omega)$

$= 48$ VARS (leading capacitive type)

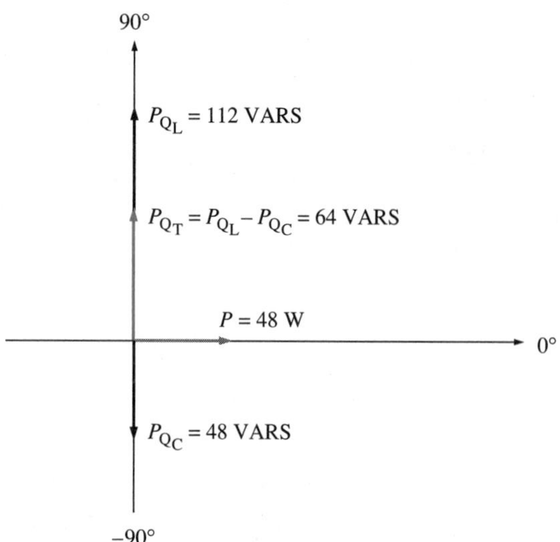

Figure 13-9

$$P_{Q_T} = P_{Q_L} - P_{Q_C}$$
$$= 112 \text{ VARS} - 48 \text{ VARS}$$
$$= 64 \text{ VARS (lagging inductive type)}$$
$$X_T = X_L - X_C$$
$$= 7\ \Omega - 3\ \Omega$$
$$= 4\ \Omega \text{ (inductive)}$$
$$P_{Q_T} = I_T{}^2 X_T$$
$$= (4\text{ A})^2(4\ \Omega)$$
$$= 64 \text{ VARS (lagging inductive type)}$$

See Figure 13-9.

EXAMPLE 13-6

Calculate (A) the total real power and (B) the reactive power in the circuit of Figure 13-10.

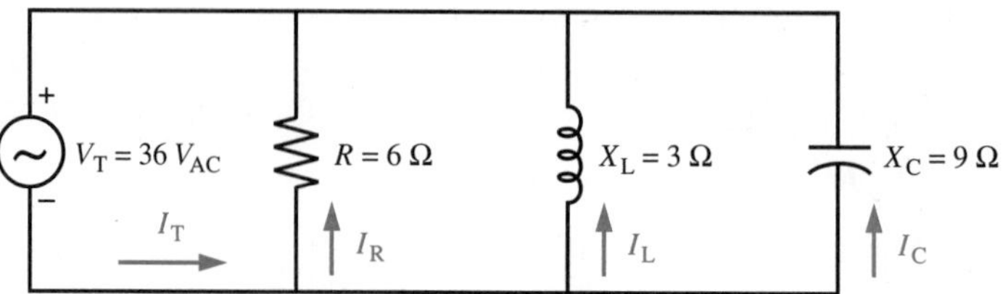

Figure 13-10

Solution:

(A)
$$I_R = \frac{V_T}{R} = \frac{36\text{ V}}{6\ \Omega} = 6\text{ A}$$

$$I_L = \frac{V_T}{X_L} = \frac{36\text{ V}}{3\ \Omega} = 12\text{ A}$$

$$I_C = \frac{V_T}{X_C} = \frac{36\text{ V}}{9\ \Omega} = 4\text{ A}$$

$$I_T = \sqrt{I_R^2 + (I_L - I_C)^2} = \sqrt{(6\text{ A})^2 + (12\text{ A} - 4\text{ A})^2} = \sqrt{(6\text{ A})^2 + (8\text{ A})^2} = 10\text{ A}$$

$$Z_T = \frac{V_T}{I_T} = \frac{36\text{ V}}{10\text{ A}} = 3.6\ \Omega$$

$$\cos\theta_T = \frac{I_R}{I_T} = \frac{6\text{ A}}{10\text{ A}} = 0.6$$

$$\therefore\ \theta_T = \cos^{-1} 0.6 = 53.13^\circ$$

$$P_T = V_T I_T \cos\theta_T = (36\text{ V})(10\text{ A})(\cos 53.13^\circ) = (36\text{ V})(10\text{ A})(0.6) = 216\text{ W}$$

or
$$P_T = I_T^2 R = (6\text{ A})^2(6\ \Omega) = 216\text{ W}$$

Checks because power (watts) can only be dissipated by a resistor.

(B)
$$P_{Q_L} = I_L^2 X_L = (12\text{ A})^2(3\ \Omega) = 432\text{ VARS (inductive type)}$$

$$P_{Q_C} = I_C^2 X_C = (4\text{ A})^2(9\ \Omega) = 144\text{ VARS (capacitive type)}$$

$$P_{Q_T} = P_{Q_L} - P_{Q_C} = 432\text{ VARS} - 144\text{ VARS} = 288\text{ VARS (inductive type)}$$

13.3 Power Triangle

Reactive power and real power can be related using a technique known as the **power triangle.** Figure 13-11 illustrates the Pythagorean relationship of the power triangle. The *real power, P,* equals $VI\cos\theta$ and the reactive power, P_Q, is equal to $VI\sin\theta$. The hypotenuse of the triangle is known as the **apparent power** and is equal to V times I. Apparent power is measured in the units of **volt-amperes** and uses the symbol P_a or S. Mathematically the equations for power can be summarized as

Equation 13-4 ▶

$$P = VI\cos\theta \quad \text{(watts)}$$
$$= P_a\cos\theta \qquad \text{real power}$$

Equation 13-5 ▶

$$P_Q = VI\sin\theta \quad \text{(VARS)}$$
$$= P_a\sin\theta \qquad \text{reactive power}$$

Equation 13-6 ▶

$$P_a = S = VI \quad \text{(VA)} \qquad \text{apparent power}$$

Equation 13-7 ▶

$$P_a = S = \sqrt{P^2 + P_Q^2} \quad \text{(VA)}$$

It should be noted that the previous equations are only valid for sine waves.

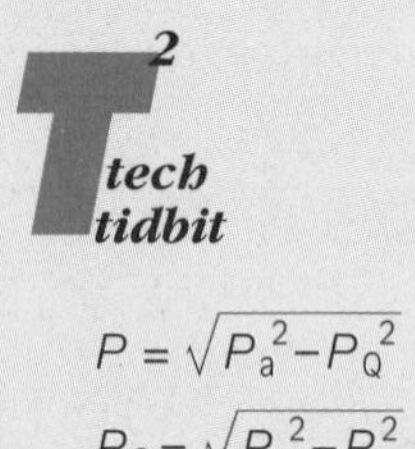

$$P = \sqrt{P_a^2 - P_Q^2}$$
$$P_Q = \sqrt{P_a^2 - P^2}$$

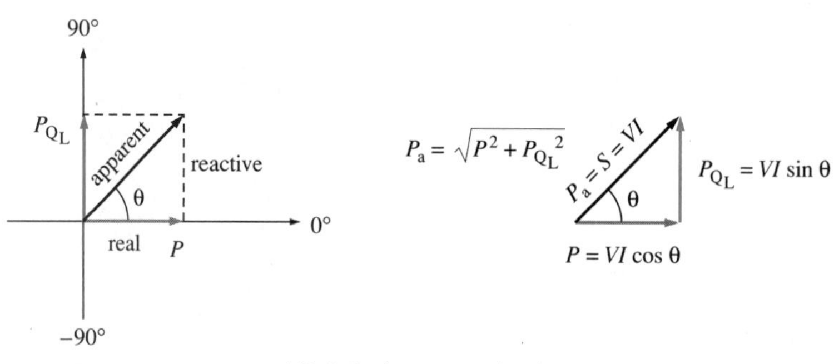

(A) Inductive power triangle

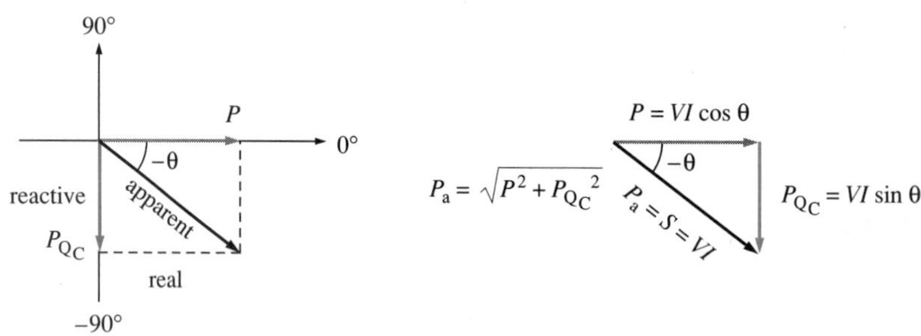

(B) Capacitive power triangle

Figure 13-11 **Power triangle**

EXAMPLE 13-7

Find (A) the real power, (B) the apparent power, and (C) the reactive power for the circuit of Figure 13-12.

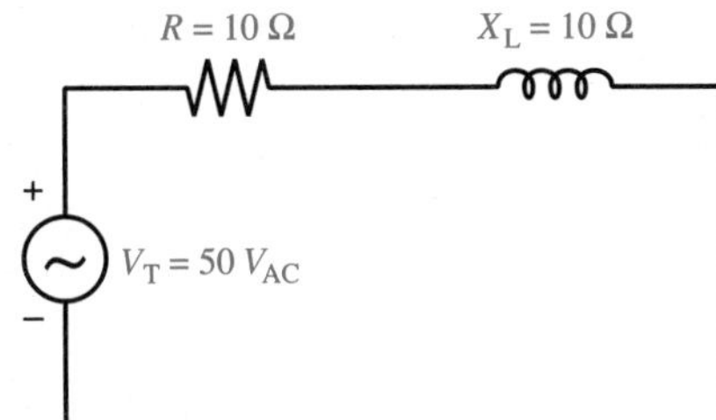

Figure 13-12

Solution:

(A) $Z_T = \sqrt{R^2 + X_L^2}$

$= \sqrt{(10\ \Omega)^2 + (10\ \Omega)^2}$

$= 14.14\ \Omega$

$I_T = \dfrac{V_T}{Z_T}$

$= \dfrac{50\ \text{V}}{14.14\ \Omega}$

$= 3.54\ \text{A}$

$\cos\theta_T = \dfrac{R}{Z_T}$

$= \dfrac{10\ \Omega}{14.14\ \Omega}$

$= 0.707$

$\therefore\ \theta_T = \cos^{-1} 0.707$

$= 45°$

$P_T = V_T I_T \cos\theta_T$

$= (50\ \text{V})(3.54\ \text{A})(\cos 45°)$

$= (50\ \text{V})(3.54\ \text{A})(0.707)$

$= 125.16\ \text{W}$

or

$P_T = I_T^2 R$

$= (3.54\ \text{A})^2(10\ \Omega)$

$= 125.16\ \text{W}$

(B) $P_a = S = V_T I_T$

$= (50\text{ V})(3.54\text{ A})$

$= 177\text{ VA}$

(C) $P_{Q_L} = V_T I_T \sin\theta_T$

$= (50\text{ V})(3.54\text{ A})(\sin 45°)$

$= (50\text{ V})(3.54\text{ A})(0.707)$

$= 125.16\text{ VARS}$

or

$P_{Q_L} = I_T^2 X_L$

$= (3.54\text{ A})^2(10\ \Omega)$

$= 125.16\text{ VARS}$

Check

$P_a = S = \sqrt{P_T^2 + P_Q^2}$

$= \sqrt{(125.16\text{ W})^2 + (125.16\text{ VARS})^2}$

$= 177\text{ VA}$

Power factor is the ratio of real power to apparent power.

13.4 Power Factor Correction

The ratio of real power to apparent power is known as the **power factor.** The power factor is also equal to the cosine of θ, where θ is the angle between the total voltage and the total current. Mathematically,

Equation 13-8 ▶

$$PF = \cos\theta = \frac{P}{P_a} = \frac{P}{VI}$$

In a purely resistive circuit: $PF = 1$.

In a purely reactive circuit: $PF = 0$.

The power factor is always a number between 0 and 1. In a purely resistive circuit, the phase angle, θ, is 0 degrees. Thus, the $PF = \cos\theta = \cos 0° = 1$. In a purely reactive circuit, the voltage and current are 90 degrees out of phase. Thus, the $PF = \cos\theta = \cos 90° = 0$. The power factor is sometimes expressed as a percentage. It can range from 0% to 100%.

It is interesting to note that when we use electric energy, we pay for the real power used ($P = VI\cos\theta$). However, the power company must produce the apparent power ($P_a = VI$). Therefore, particularly in industrial applications, the power company will try to make the power factor as close to 1 as possible. The power factor of New York City (and most cities), as a load on the power system, is typically about 0.8 lagging.

EXAMPLE 13-8

A 440-volt AC electric motor draws 20 amperes. A wattmeter connected to the circuit indicates 1,160 watts. Calculate the power factor of the circuit.

Solution:

$$PF = \frac{P}{P_a} = \frac{P}{VI}$$

$$= \frac{1{,}160 \text{ W}}{(440 \text{ V})(20 \text{ A})}$$

$$= 0.7$$

or

$$= (0.7)(100\%)$$

$$= 70\%$$

Note that an AC electric motor can normally be represented by a resistor and an inductor. Therefore, the power factor is lagging (I_T lags V_T).

EXAMPLE 13-9

(A) Find the power factor in the circuit of Figure 13-13A. (B) What effect does the addition of capacitor C have on the power factor (Figure 13-13B)?

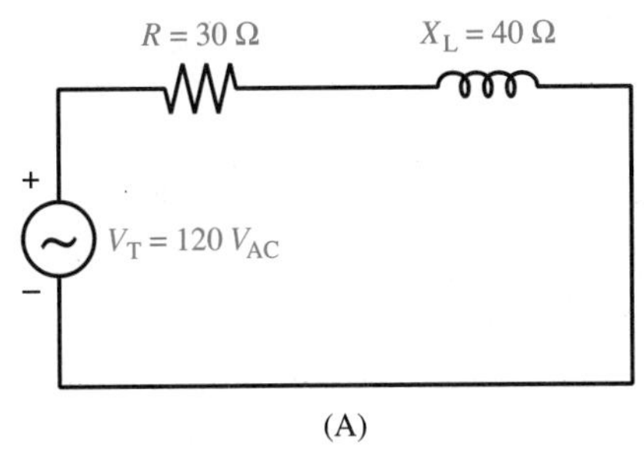

(A)

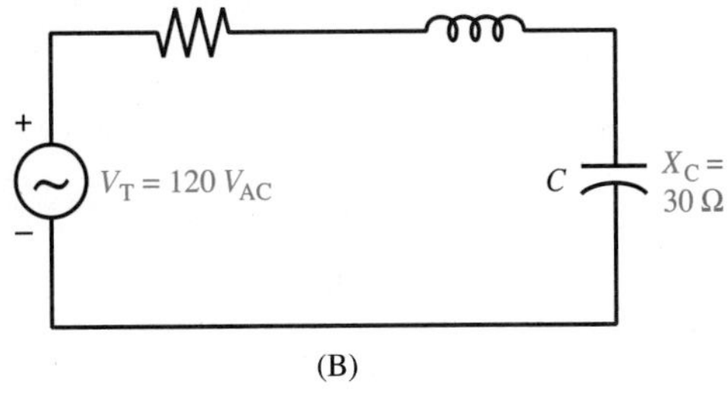

(B)

Figure 13-13

Solution:

$$\text{(A)} \quad Z_T = \sqrt{R^2 + X_L^2}$$

$$= \sqrt{(30\ \Omega)^2 + (40\ \Omega)^2}$$

$$= 50\ \Omega$$

$$PF = \cos \theta_T = \frac{R}{Z_T}$$

$$= \frac{30\ \Omega}{50\ \Omega}$$

$$= 0.6$$

(B) *with capacitor*

$$Z_T = \sqrt{R^2 + (X_L - X_C)^2}$$

$$= \sqrt{(30\ \Omega)^2 + (40\ \Omega - 30\ \Omega)^2}$$

$$= \sqrt{(30\ \Omega)^2 + (10\ \Omega)^2}$$

$$= 31.62\ \Omega$$

$$PF = \cos\theta_T = \frac{R}{Z_T}$$

$$= \frac{30\ \Omega}{31.62\ \Omega}$$

$$= 0.95$$

Capacitor C raises the power factor to 0.95.

Example 13-10

A 6-horsepower motor with a 0.7 lagging power factor and an efficiency of 80% is connected to a 220-volt, 60-hertz supply. How much capacitance must be added in a parallel with the motor to raise the power factor of the combined system to unity?

Solution:

$$P_O = 6\ \text{HP}$$

$$= (6\ \text{HP})(746\ \text{W/HP})$$

$$= 4{,}476\ \text{W}$$

$$P_{in} = \frac{P_O}{\eta}$$

$$= \frac{4{,}476\ \text{W}}{0.8}$$

$$= 5{,}595\ \text{W}$$

$$\theta = \cos^{-1}(0.7)$$

$$= 45.57°$$

$$P = VI\cos\theta$$

$$5{,}595\ \text{W} = (220\ \text{V})(I)(0.7)$$

$$I = \frac{5{,}595\ \text{W}}{(220\ \text{V})(0.7)}$$

$$= 36.33\ \text{A}$$

$$P_{QM} = VI\sin 45.57$$

$$= (220\ \text{V})(36.33\ \text{A})(\sin 45.57°)$$

$$= 5{,}708\ \text{VARS}$$

$$P_a = \sqrt{P^2 + P_{QM}^2}$$
$$= \sqrt{(5{,}595 \text{ W})^2 + (5{,}708 \text{ VARS})^2}$$
$$= 7{,}993 \text{ VA}$$

For unity power factor $P_{QC} = P_{QM} = 5{,}708$ VARS

$$P_{QC} = \frac{V^2}{X_C}$$
$$X_C = \frac{V^2}{P_{QC}}$$
$$= \frac{(220 \text{ V})^2}{5{,}708}$$
$$= 8.48\ \Omega$$
$$X_C = \frac{1}{2\pi f C}$$
$$C = \frac{1}{2\pi f X_C}$$
$$= \frac{1}{(6.28)(60 \text{ Hz})(8.48\ \Omega)}$$
$$= 313\ \mu\text{F}$$

✓ Self-Test 2

1. The energy in a reactive element is called ______________.
2. The three sides of the power triangle represent ______________, ______________, and ______________.
3. The ratio of real power to apparent power is known as ______________.
4. The power company wants the power factor to be as close to ______________ as possible.

13.5 Three-Phase Power

An electric power plant produces electricity in the form of **three-phase** AC voltages. In three-phase systems, three sine waves are produced. Each sine wave is displaced by 120 degrees in time. Each sine wave is called a phase and is referred to using the symbol ϕ. Three-phase is written as 3ϕ. Figure 13-14 illustrates a typical 3ϕ waveform. Phase 1 is shown as the reference phase because it starts at 0 degrees. Phase 2 starts 120 degrees after phase 1 but is 120 degrees before phase 3. Phase 3 is 240 degrees after phase 1 and is also 120 degrees after phase 2. Mathematically, the three-phase voltages can be expressed as

Three-phase is written as 3ϕ.

$$V_{\phi_1} = V_M \sin \alpha$$
$$V_{\phi_2} = V_M \sin (\alpha - 120°)$$

$$V_{\phi_3} = V_M \sin(\alpha - 240°) = V_{\phi_3} = V_M \sin(\alpha + 120°)$$

It is interesting to note that the algebraic sum of the voltages of each phase

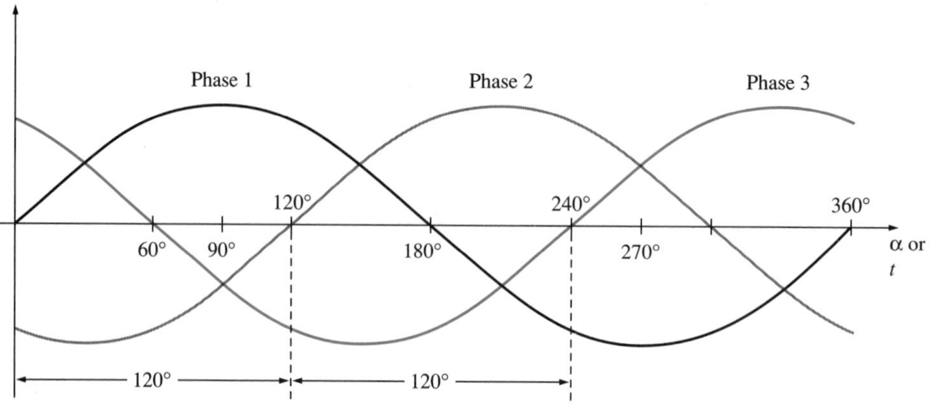

Figure 13-14 **Three-phase AC waveform**

at any instant in time will always equal zero.

Figure 13-15 shows the phasor representation of a three-phase waveform. It illustrates the phase sequence, rotation, and magnitude of the voltage. For example in Figure 13-15, V_{ϕ_1} is the reference phase and is plotted at 0 degrees. Continuing *clockwise,* V_{ϕ_2} is phase 2, which lags phase 1 by 120 degrees. V_{ϕ_3} is phase 3, which lags phase 1 by 240 degrees. Note that phase 3 can also be con-

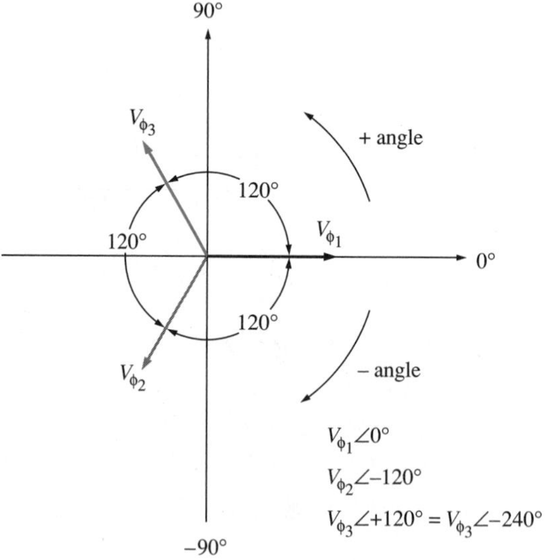

Figure 13-15 **Phasor diagram of a three-phase waveform**

sidered as leading phase 1 by 120 degrees because the rotation is continuous.

Three-phase AC generators (alternators) are constructed in either **wye** or **delta** configurations (although wye is more common). In three-phase alternators, the stator houses the conductors that produce the three-phase voltages and the rotor has the field windings that are supplied with DC from an external source through brushes on slip rings. Figure 13-16 illustrates a wye generator. Wye connected systems are normally four-wire systems. The fourth wire is the neutral and is connected to the common point of the wye. The **phase voltage** (V_ϕ) is measured from the neutral to each phase. The **line voltage** (V_L) is the voltage between any two line leads. The line voltage will be the phasor sum of two phase voltages and is related by the square root of 3. Mathematically,

tech tidbit

WYE connected systems are four wire systems.

Equation 13-9

$$V_L = \sqrt{3}\, V_\phi$$

Note that Equation 13-9 is only valid for a balanced 3ϕ system. A balanced 3ϕ system is one in which the three loads are identical. Equation 13-9 calculates only the magnitude of the line voltage.

tech tidbit

$$V_\phi = \frac{V_L}{\sqrt{3}}$$

The line voltage in a wye connection is always greater (in magnitude only) than the phase voltage by the $\sqrt{3}$ because it is the sum of two phases that are 120° apart. The $\sqrt{3}$ is equal to 1.732, which can be remembered as the year of George Washington's birthday (1732). The phase current in a wye connection is the same as the line current because the current in any line comes directly from a phase. Mathematically,

Equation 13-10

$$I_L = I_\phi$$

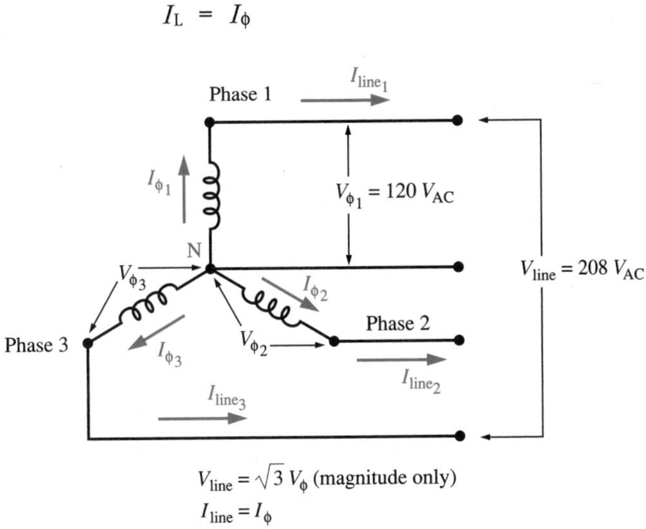

Figure 13-16 **Three-phase wye generator**

One of the world's largest 3ϕ wye connected AC generators, called "Big Allis," is part of Consolidated Edison's power system in New York. It is rated at 1,000,000 kilowatts (1,000 MW) and supplies approximately one-seventh of the load requirements of New York City. It has two field poles on the rotor and rotates at a speed of 3,600 revolutions per minute to produce a frequency of 60 hertz.

Figure 13-17 illustrates the delta configuration. The delta system is a three-wire system. Referring to Figure 13-17, we note that the line voltage and the phase voltage are equal. However, the line current (I_L) comes from two phases (I_ϕ) and is related by the $\sqrt{3}$ (for a balanced system). Mathematically, for a delta configuration,

tech tidbit

Delta connected systems are three wire systems.

Equation 13-11 ▶

$$V_L = V_\phi$$

Equation 13-12 ▶

$$I_L = \sqrt{3}\, I_\phi$$

$$I_\phi = \frac{I_L}{\sqrt{3}}$$

Note that Equation 13-12 is only valid for a balanced 3ϕ system. A balanced 3ϕ system is one in which the three loads are identical. Equation 13-12 calculates only the magnitude of the line current.

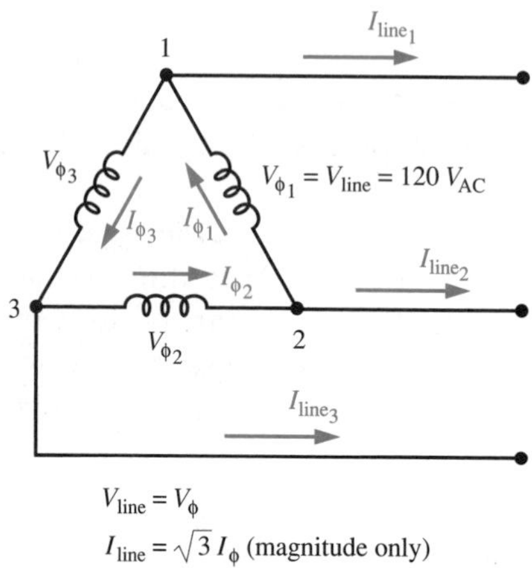

Figure 13-17 **Three-phase delta generator**

EXAMPLE 13-11

Find (A) the voltage across R_1 and (B) the current I_1 in the 3ϕ load of Figure 13-18.

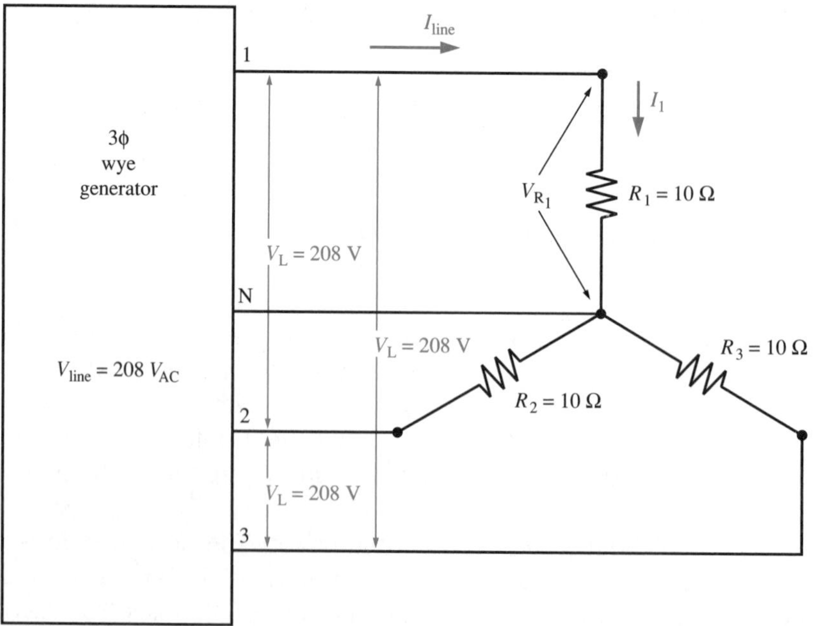

Figure 13-18

Solution:

(A) $V_L = \sqrt{3}\,V_\phi$

$$V_\phi = \frac{V_L}{\sqrt{3}} = \frac{208\text{ V}}{1.732} = 120\text{ V}$$

$$V_\phi = V_{R_1} = 120\text{ V}$$

(B) $$I_1 = \frac{V_{R_1}}{R_1} = \frac{120\text{ V}}{10\ \Omega} = 12\text{ A}$$

$$I_L = I_1 = 12\text{ A}$$

Note that in Example 13-11, all of the load resistors are the same value. This is referred to as a **balanced system.** In a balanced system, all of the load voltages and currents will have the same magnitude but will be 120 degrees out of phase with each other.

In a balanced three-phase system, all load voltages and currents are the same magnitude and 120° out of phase.

EXAMPLE 13-12

Find (A) the voltage across R_1, (B) the current through R_1, and (C) the line current for the system of Figure 13-19.

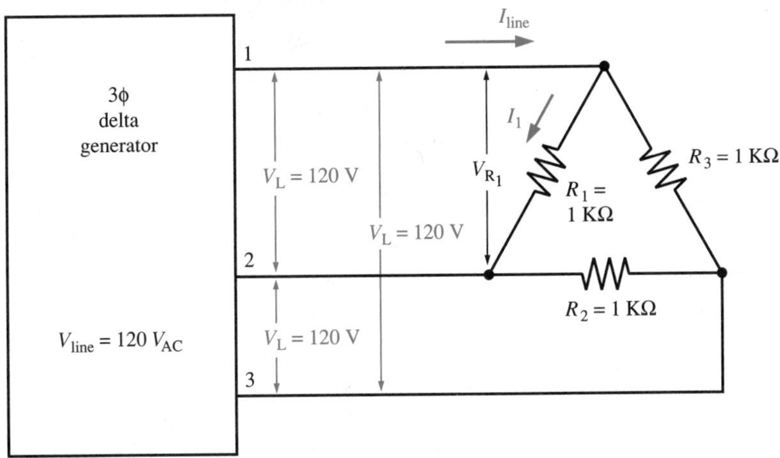

Figure 13-19

Solution:

(A) $V_L = V_\phi = V_{R_1} = 120\text{ V}$

(B) $$I_{\phi_1} = \frac{V_{R_1}}{R_1} = \frac{120\text{ V}}{1\text{ K}\Omega} = 120\text{ mA}$$

(C) $$I_L = \sqrt{3}\,I_\phi$$

$$= (1.732)(120\text{ mA})$$

$$= 207.84\text{ mA}$$

The total power in a balanced 3ϕ wye or delta system is equal to three times the power delivered or dissipated by each individual phase. Mathematically,

Equation 13-13 ▶

$$P_T = 3P_\phi$$

Asynchronous motors—speed can vary with load.

✓ Self-Test 3

1. An electric power plant produces electricity in the form of a _______________ AC voltage.
2. Three-phase generators can be constructed in either _______________ or _______________ configurations.
3. The line voltage in a wye connection is always _______________ than the phase voltage by $\sqrt{3}$ for a balanced system.
4. In a delta system, the phase voltage _______________ the line voltage.

13.6 Motors

Synchronous motors—constant speed.

The electric **motor** is a transducer because it changes electric energy (input) to mechanical energy of rotation (output). Two-thirds of the total electric power supplied by power companies is used to run electric motors used in industry, commercial buildings, and residential homes. All electric motors are classified as asynchronous or synchronous. An **asynchronous motor** is one in which the speed can vary depending on the load. A **synchronous motor** is one in which the speed is constant and does not vary with the load. All DC motors are asynchronous but AC motors may be either asynchronous or synchronous.

Universal motors—operate on DC or AC.

The input to a motor may be DC or AC. The speed of DC motors can be varied more easily than AC motors. However, the maintenance for DC motors may be more expensive than for AC motors. A **universal motor** is a motor that operates on either DC or AC.

tech tidbit

Motors are rated in units of horsepower.

All electric motors consist of a stator (fixed part) and a rotor (the rotating member). The output of a motor is rated in units of **horsepower.** The term *horsepower* comes from the average amount of work a draft horse can perform in a 12-hour day. Mathematically, one horsepower equals 550 foot-pounds per second or 33,000 foot-pounds per minute. Electric power is usually measured in the unit of watts. Watts can be related to horsepower using the following equation:

Equation 13-14 ▶

$$1 \text{ HP} = 746 \text{ watts}$$

The efficiency of a motor is the ratio of the power output (mechanical) to the power input (electrical). It is often expressed as a percentage. Mathematically,

Equation 13-15 ▶

$$\% \text{ efficiency} = \frac{P_{out}}{P_{in}} \times 100\%$$

or

Equation 13-16 ▶

$$\% \text{ efficiency} = \frac{P_{in} - P_{loss}}{P_{in}} \times 100\%$$

Motor losses can be grouped into three general categories:

1. mechanical losses
2. core losses
3. copper losses.

P_{out} = (% eff) (P_{in})
$P_{in} = P_{out} + P_{loss}$

Mechanical losses are power losses due to windage and friction. As the shaft of the motor turns, a fan that is often attached for cooling uses energy to move the air. The energy lost in this process is called windage and includes the losses due to the rotation of the rotor in air. The parts of the motor must rotate in order to convert electrical energy to mechanical energy. The turning of the rotating parts causes friction to occur. Friction dissipates power in the form of heat and is known as friction losses.

Core losses include eddy currents and hysteresis. Eddy currents and hysteresis are the same losses discussed in transformers. These losses are due to the physical and magnetic makeup of the motor.

Copper losses or I^2R losses are losses due to the resistance of the windings of the motor. These include the field windings and the armature windings.

EXAMPLE 13-13

A one-quarter horsepower (power output) DC motor operates on 12 volts and draws 20 amperes. Calculate the efficiency of the motor.

Solution:

$$\begin{aligned}\text{\% efficiency} &= \frac{P_{out}}{P_{in}} \times 100\% \\ &= \frac{(0.25\ \text{HP})(746\ \text{W/HP})}{(12\ \text{V})(20\ \text{A})} \times 100\% \\ &= \frac{186.5\ \text{W}}{240\ \text{W}} \times 100\% \\ &= 71.3\%\end{aligned}$$

The relationship between speed, horsepower, and torque can be expressed by the following formula:

$$HP = \frac{TS}{5{,}252}$$ Equation 13-17

where T = the torque in pound-feet

S = the speed in rpm

HP = horsepower

EXAMPLE 13-14

Calculate the output torque of a 3-horsepower 1,800-revolutions-per-minute motor.

Solution:

$$HP = \frac{TS}{5{,}252}$$

$$T = \frac{(5{,}252)(HP)}{S}$$

$$= \frac{(5{,}252)(3\ HP)}{1{,}800\ \text{rpm}}$$

$$= 8.75\ \text{lb-ft}$$

13.7 DC Machines

The DC motor consists of a stator section and a rotor section. The **stator** section contains the magnetic poles with windings wrapped around them (field windings) and the frame. The **rotor** section (also called the **armature**) contains armature coils in which voltage is induced, a **commutator,** which is a mechanical rectifier, and **carbon brushes,** which pass electricity to the commutator and rotor. The armature coils are embedded into an iron core on the rotating shaft. Figure 13-20 illustrates the basic parts of a DC motor.

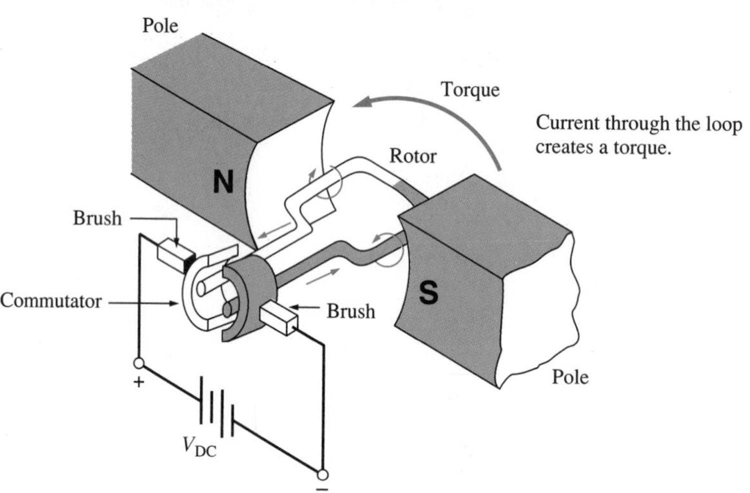

Figure 13-20 **Basic DC motor parts**

tech tidbit

The rotor's magnetic field interacts with the magnetic field of the stationary poles to cause the armature to rotate.

DC voltage is fed into the DC motor through the brushes and commutator. The brushes act as an interface to the rotating commutator shaft. This allows current into the coils of the rotating armature and creates a magnetic field. The magnetic field interacts with the magnetic field of the stationary field poles to cause the armature to rotate. The direction of current determines the direction of the magnetic field in the rotor. The magnetic field around the rotor interacts with the main magnetic field produced by the field windings around the poles in the stator. The interaction of the two magnetic fields causes rotary motion, as shown in Figure 13-21.

To change the direction of rotation, the direction of current can be reversed through the rotor or the direction of current through the windings of the stationary field poles can be reversed. This can be observed using the righthand

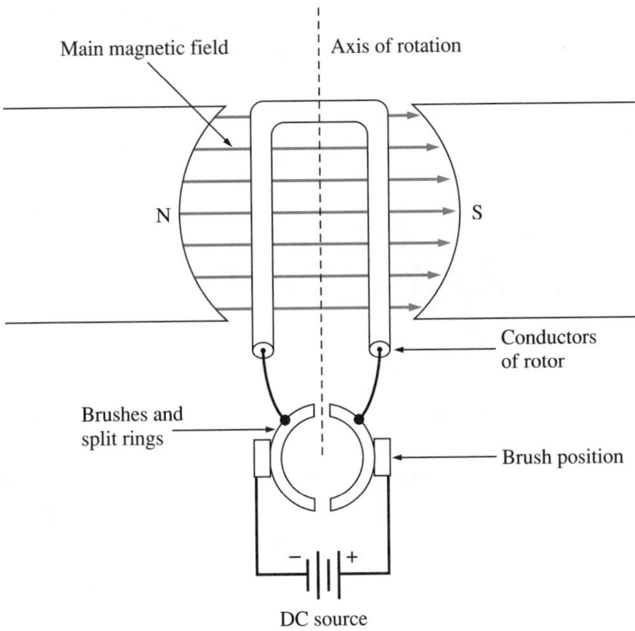

Figure 13-21 **Rotation in a DC motor**

rule, as shown in Figure 13-22. The righthand rule will determine the direction of motion from the thumb and the first two fingers of the right hand, as shown in Figure 13-22. The rotary motion produced in this manner is called **torque** or motor action. The amount of torque produced by a motor depends on the amount of current through the rotor and the strength of the main magnetic field. If either is increased, the amount of torque will be increased.

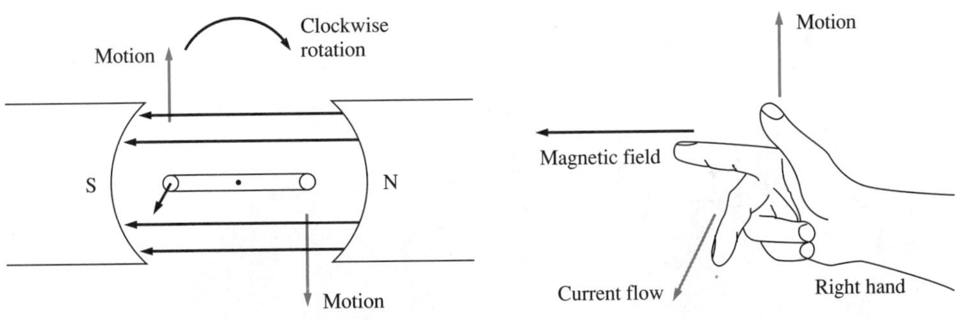

Figure 13-22 **Righthand rule**

The essential parts of a real motor vary from type to type. However, all motors have a stator, a rotor, and some means of supporting the rotor to keep it from touching the stator poles as it rotates. In Figure 13-23, the armature or the rotor section is supported so that it will not contact the stator or housing by the two bearing assemblies on the front and back ends of the armature. The stator shown in Figure 13-23 houses the stationary field pole windings.

Figure 13-24 is a photograph showing the various parts of a DC motor.

DC motors can be classified by the electrical connection of the field windings to the armature windings. The general configurations are the shunt motor, the series motor, and the compound motor.

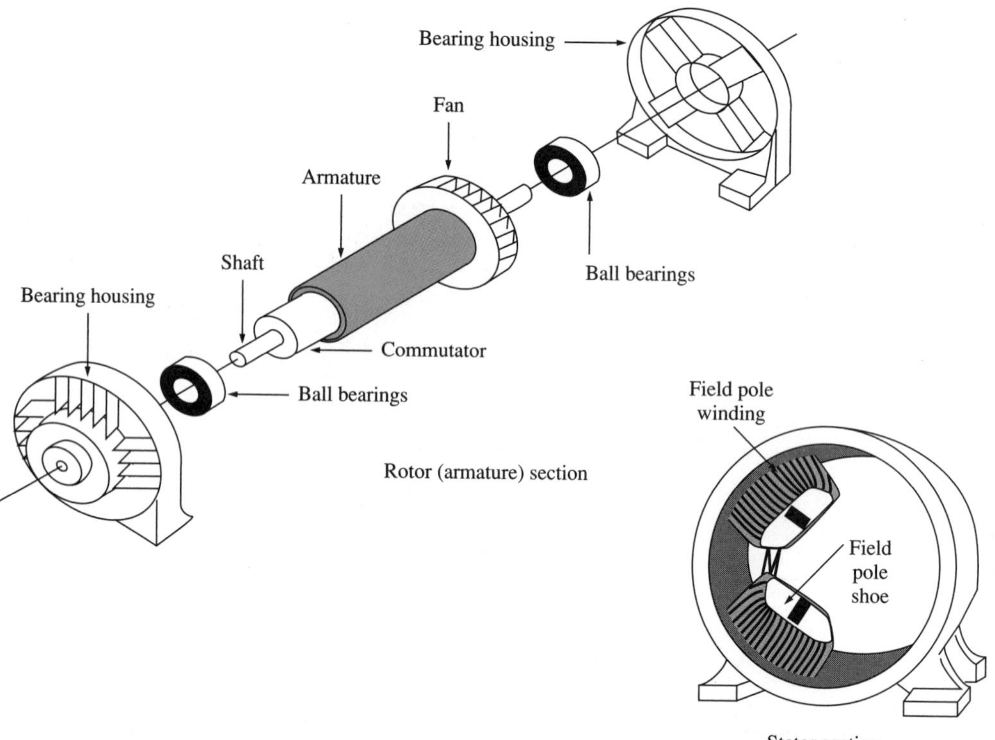

Figure 13-23 **DC motor parts**

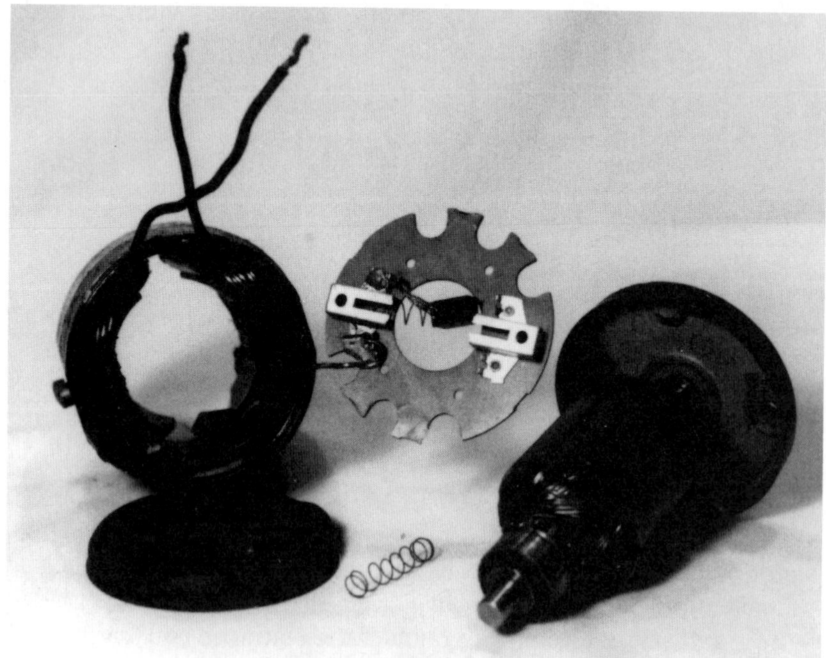

Figure 13-24 **Parts of a DC motor**

In the **shunt motor,** the field winding is in parallel with the armature winding. In the **series motor,** the field winding is in series with the armature winding. In the compound motor, there are two field windings, one is in series with the armature winding and one is in parallel (shunt) with the armature winding. There are two types of compound motors. The *cumulative compound motor,* in which the shunt field aids the series field winding in producing flux, is used more commonly in industrial applications. The *differential compound motor,* in which the series field opposes the shunt field in producing flux, has limited applications. Each type of motor will exhibit different speed and torque characteristics. For example, the DC series motor exhibits large torque at low speeds and is used to drive subway cars.

Figure 13-25 illustrates the general configuration of a compound motor. R_S refers to the resistance of the series field. R_{SH} refers to the resistance of the parallel or shunt field. When a motor has both a series field and a shunt field, it is called a **compound motor.** R_A represents the resistance of the armature coils and V_A represents the counter EMF induced in the armature. V_L refers to the input line voltage.

When a motor has both a series field and a shunt field, it is called a compound motor.

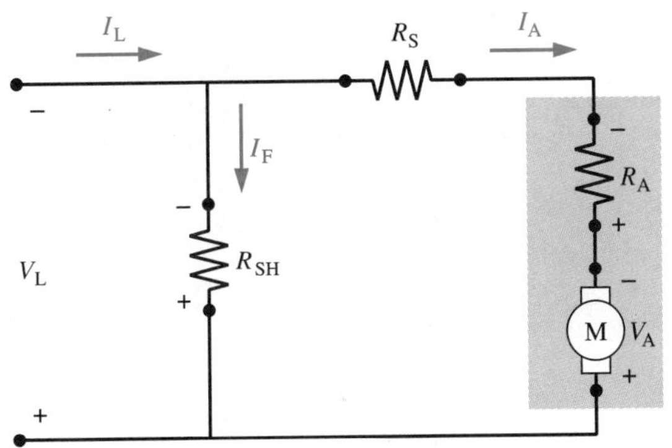

$I_L = I_A + I_F$

$V_L = V_A + I_A(R_S + R_A) = I_F R_{SH}$

Figure 13-25 Compound DC motor

EXAMPLE 13-15

A 2-horsepower, 220-volt DC shunt motor at full load draws 10 amperes (line current) when rotating at 1,800 revolutions per minute. The armature resistance is 0.2 ohms and the field resistance is 220 ohms. Find (A) the armature current, (B) the developed counter EMF, V_A, and (C) the output torque of the motor. See Figure 13-26.

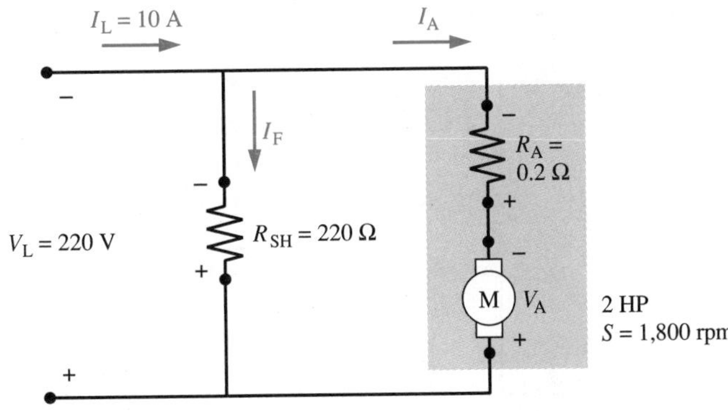

Figure 13-26

Solution:

(A) $I_L = I_A + I_F$ (Kirchhoff's current law)

$I_L = 10\text{ A}$

$$I_F = \frac{V_L}{R_{SH}}$$

$$= \frac{220\text{ V}}{220\ \Omega}$$

$$= 1\text{ A}$$

$$I_A = I_L - I_F$$

$$= 10\text{ A} - 1\text{ A}$$

$$= 9\text{ A}$$

(B) $V_L = V_F = V_A + I_A R_A$ (Kirchhoff's voltage law)

$$V_A = V_L - I_A R_A$$

$$= 220\text{ V} - (9\text{ A})(0.2\ \Omega)$$

$$= 220\text{ V} - 1.8\text{ V}$$

$$= 218.2\text{ V}$$

(C)
$$T = \frac{5{,}252\ HP}{S}$$

$$= \frac{(5{,}252)(2\ HP)}{1{,}800\text{ rpm}}$$

$$= 5.84\text{ lb-ft}$$

EXAMPLE 13-16

A 4.5-horsepower, 200-volt DC shunt motor at full load draws 20 amperes when rotating at 1,200 revolutions per minute. The armature resistance is 0.18 ohms and the field resistance is 180 ohms. Find (A) the developed counter EMF, (B) V_A, (C) the output torque, (D) the efficiency, and (E) the power loss due to copper and rotational losses. See Figure 13-27.

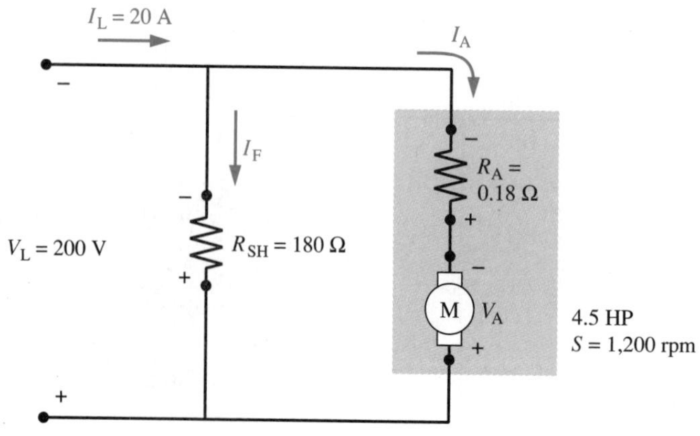

Figure 13-27

Solution:

(A)
$$I_L = I_F + I_A$$
$$I_L = 20\text{ A}$$
$$I_F = \frac{V_L}{R_{SH}} = \frac{200\text{ V}}{180\ \Omega} = 1.11\text{ A}$$
$$I_A = I_L - I_F = (20\text{ A}) - (1.11\text{ A}) = 18.89\text{ A}$$

(B)
$$V_L = V_F = V_A + I_A R_A$$
$$V_A = V_L - I_A R_A = 200\text{ V} - (18.89\text{ A})(0.18\ \Omega) = 200\text{ V} - 3.4\text{ V} = 196.6\text{ V}$$

(C)
$$T = \frac{5{,}252\ HP}{S} = \frac{(5{,}252)(4.5\text{ HP})}{1{,}200\text{ rpm}} = 19.7\text{ lb-ft}$$

(D)
$$\%\text{ efficiency} = \frac{P_{out}}{P_{in}} \times 100\% = \frac{(4.5\text{ HP})(746\text{ W/HP})}{(200\text{ V})(20\text{ A})} \times 100\% = \frac{3{,}357\text{ W}}{4{,}000\text{ W}} \times 100\% = 84\%$$

(E) POWER LOSSES

Field copper losses
$$P_F = I_F^2 R_{SH} = (1.11\text{ A})^2(180\ \Omega) = 222\text{ W}$$

or

$$P_F = V_F I_F = (200\text{ V})(1.11\text{ A}) = 222\text{ W}$$

Armature copper losses

$$P_A = I_A^2 R_A$$

$$= (18.89\text{ A})^2(0.18\ \Omega)$$

$$= 64\text{ W}$$

Total power losses

$$P_{tot} = P_{in} - P_{out}$$

$$= 4{,}000\text{ W} - 3{,}357\text{ W}$$

$$= 643\text{ W}$$

Rotational losses

$$P_{tot} = P_F + P_A + P_{rot}$$

$$P_{rot} = P_{tot} - P_F - P_A$$

$$= 643\text{ W} - 222\text{ W} - 64\text{ W}$$

$$= 357\text{ W}$$

The value of V_A, the induced counter EMF, is a function of the speed of the motor. The faster the speed, the greater the value of V_A, the slower the speed, the smaller the value of V_A. V_A is developed in the opposite direction to the applied voltage, so it subtracts from the applied voltage. This is an important point when considering the armature current. The armature current can be calculated from the following equation referring to Figure 13-25:

Equation 13-18 ▶

$$I_A = \frac{V_L - V_A}{R_A}$$

Generators convert mechanical energy into electrical energy.

Note that if the motor stops completely, V_A, which depends on speed, will equal 0 volts and I_A will become very large because R_A is a low resistance value. I_A will increase to a point that will burn out the motor.

In theory, an **electric generator** can be considered as a motor operating in the opposite mode. That is, it takes mechanical energy of rotation as its input and transforms it into electric energy for its output. The generated voltage is always a function of the speed and the strength of the magnetic field. Mathematically,

Equation 13-19 ▶

$$V_g = K\phi S$$

where K = constant

ϕ = strength of the magnetic field

S = speed of the motor

V_g = the generated voltage

The general configuration for a DC compound generator is shown in Figure 13-28. R_S is the resistance of the series field, R_{SH} is the resistance of the parallel or shunt field, R_A represents the resistance of the armature, V_A represents the generated voltage in the armature, and V_L represents the output or line voltage of the generator.

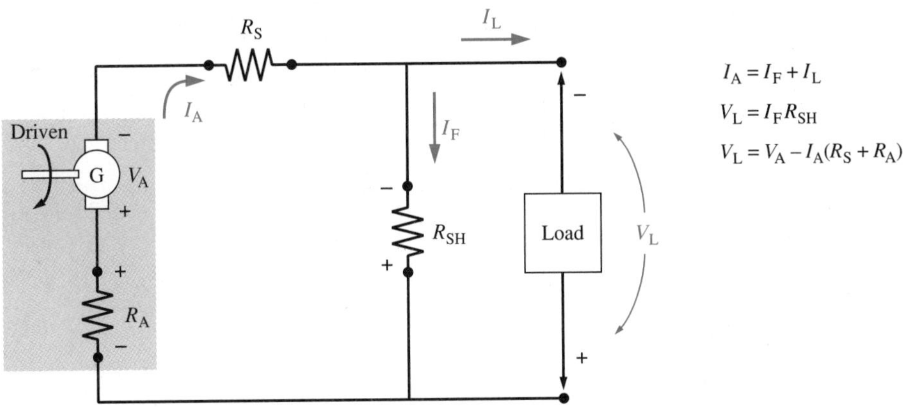

Figure 13-28 **DC compound generator**

EXAMPLE 13-17

A 10-kilowatt, 200-volt DC shunt generator at full load has rotational losses of 480 watts at 1,000 revoltions per minute. Assuming an armature resistance of 0.15 ohms, a shunt current of 1.5 amperes, and a line voltage of 200 volts, calculate the efficiency of the generator. See Figure 13-29.

Solution:

$$P_L = V_L I_L$$

$$I_L = \frac{P_L}{V_L}$$

$$= \frac{10{,}000 \text{ W}}{200 \text{ V}}$$

$$= 50 \text{ A}$$

$$I_A = I_F + I_L$$

$$= 1.5 \text{ A} + 50 \text{ A}$$

$$= 51.5 \text{ A}$$

POWER LOSSES

Field copper losses

$$P_F = V_F I_F$$

$$= (200 \text{ V})(1.5 \text{ A})$$

$$= 300 \text{ W}$$

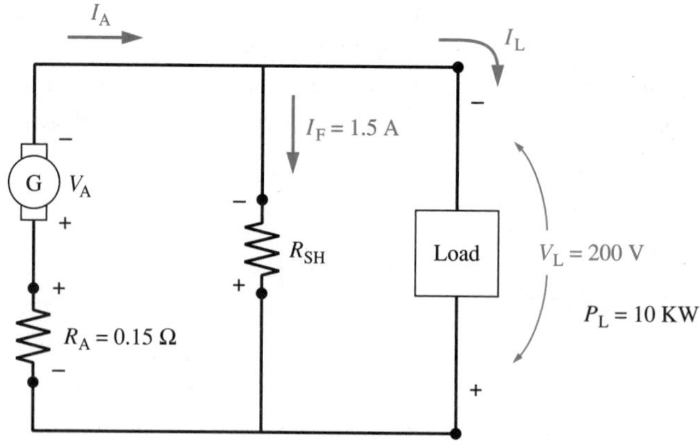

Figure 13-29

Armature copper losses

$$P_A = I_A^2 R_A$$
$$= (51.5\text{ A})^2(0.15\ \Omega)$$
$$= 398\text{ W}$$

Rotational losses

$$P_{rot} = 480\text{ W}$$

Total power losses

$$P_{tot} = P_F + P_A + P_{rot}$$
$$= 300\text{ W} + 398\text{ W} + 480\text{ W}$$
$$= 1{,}178\text{ W}$$

$$\%\ \text{efficiency} = \frac{P_{out}}{P_{out} + P_{loss}} \times 100\%$$
$$= \frac{10{,}000\text{ W}}{10{,}000\text{ W} + 1{,}178\text{ W}} \times 100\%$$
$$= 89\%$$

✓ Self-Test 4

1. Two main parts of a motor are ______________ and ______________.
2. Electricity is fed into a DC motor through the ______________ and ______________.
3. Motors are rated in units of ______________.
4. The ______________ of a motor is the ratio of the power output to the power input.
5. Three types of motor losses are ______________, ______________, and ______________.

13.8 AC Machines

The AC **induction motor** is one of the most widely used motors today. Its characteristics of good speed regulation and high starting torque make it particularly desirable, but its simplicity of construction is probably the biggest reason for its widespread use. The rotor may have windings or it may have solid brass bars (like a cage). The former would be called a wound rotor induction motor and the latter a squirrel cage induction motor. The stator is composed of windings and does not have physical field poles. Figure 13-30 and Figure 13-31 are photographs of AC induction motors.

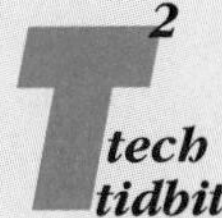

tech tidbit

Induction motors have no brushes.

The name, induction motor, comes from its construction and theory of operation. Energy is received in the rotor by means of magnetic induction rather than by direct connection as is the case in the DC motor. There is no need for brushes on a commutator or slip rings. The absence of the commutator and

Figure 13-30 **Induction motor**

Figure 13-31 **Induction motor cut open**

brushes, as well as the simplicity of the rotor construction, makes the induction motor more economical than an equally rated DC motor.

The induction motor may be considered as a rotating transformer. The voltage is induced into the secondary, the rotor, by a rotating magnetic field from the stator and then torque is developed in this same secondary or rotor. Figure 13-32 illustrates this principle. If a permanent magnet is rotated by some mechanical means or even by hand, the field moving past the single conductor or disk will induce a voltage and, therefore, a current into the disk. The final result will be rotation of the disk due to the interaction of the permanent magnet's field and the magnetic field created by the current in the disk. Thus, the disk will follow the magnet and rotate. Note that if the disk were to rotate at exactly the same speed as the magnet, there would be no cutting of flux lines, no voltage would be induced in the disk, and therefore no current or torque developed. Thus the disk can never rotate at the same speed as the magnetic field. The effect of rotating the permanent magnet is created by the phase relationships of the current in the stator conductors or windings. The induction motor creates the rotating magnetic field by the placement of stator windings into groups located out of phase with each other.

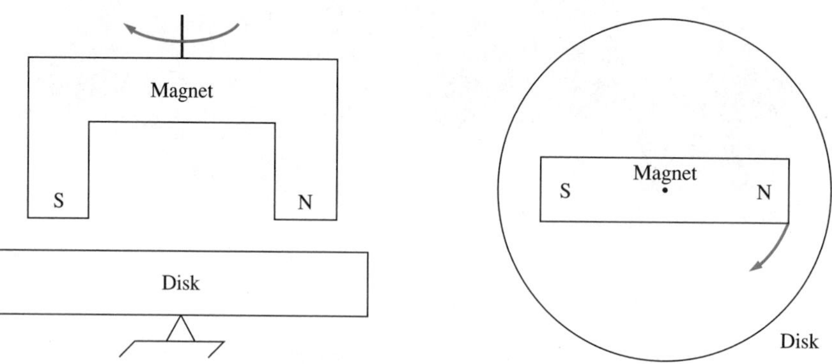

Figure 13-32 **Induction motor—principle of operation**

T^2 tech tidbit

The speed of the rotating field is called the synchronous speed.

The speed of the rotating field around the stator is called the **synchronous speed** (N_S). The synchronous speed is directly proportional to the supply frequency and inversely proportional to the number of stator poles or groupings of windings. Mathematically,

Equation 13-20 ▶

$$N_S = \frac{120f}{P}$$

where N_S = the synchronous speed in revolutions per minute

f = the input frequency in hertz

P = the number of magnetic poles

EXAMPLE 13-18

What is the synchronous speed of the rotating field of a 6-pole induction motor operating at a frequency of 60 hertz.

Solution:

$$N_S = \frac{120f}{P}$$

$$= \frac{(120)(60\ \text{Hz})}{6}$$

$$= 1{,}200\ \text{rpm}$$

Slip is the difference between the actual speed and the synchronous speed.

When loaded, all induction motors will run at a speed less than its synchronous speed and is therefore asynchronous. The difference between the actual motor speed and the synchronous speed of the rotating field is called the **slip** speed. Most large induction motors will operate between 90% and 98% of synchronous speed. The slip of a motor is often expressed as a percentage of the synchronous speed. Mathematically,

$$\%\ \text{slip} = \frac{N_S - N}{N_S} \times 100\%$$

◀ *Equation 13-21*

When expressed in terms of RPM,

$$\text{slip}_{(\text{RPM})} = N_S - N$$

◀ *Equation 13-22*

where N_S = the synchronous speed of the rotating field in revolutions per minute

N = the rotor speed in revolutions per minute

EXAMPLE 13-19

A 4-pole, 60-hertz induction motor has a full-load speed of 1,740 revolutions per minute. Find the full-load slip in revolutions per minute and the percent slip.

Solution:

$$N_S = \frac{120f}{P}$$

$$= \frac{(120)(60\ \text{Hz})}{4}$$

$$= 1{,}800\ \text{rpm}$$

$$\text{slip}_{(\text{RPM})} = N_S - N$$

$$= 1{,}800\ \text{rpm} - 1{,}740\ \text{rpm}$$

$$= 60\ \text{rpm}$$

$$\%\ \text{slip} = \frac{N_S - N}{N_S} \times 100\%$$

$$= \frac{1{,}800\ \text{rpm} - 1{,}740\ \text{rpm}}{1{,}800\ \text{rpm}} \times 100\%$$

$$= 3.33\%$$

Note: This means the motor is turning at a speed, *N,* which is 3.33% less than the synchronous speed.

Polyphase motors (more than one phase) produce a rotating magnetic field (N_S = 120 *f/P*), which cuts the rotor conductors to produce a torque. Industrial plants that receive three-phase (3ϕ) power from the power company use three-phase motors. Three-phase motors are more efficient than single-phase motors. Two-phase (2ϕ) motors have a widespread use as servo motors in feedback control systems.

Single phase motors are designated by the method of starting.

Single-phase motors, which in general have no inherent starting torque, require that a technique be incorporated to produce the rotating magnetic field. That is, the basic single-phase motor is not self-starting. Normally, single-phase motors are designated by the specific method of starting. Single-phase motors are used in homes and offices and are normally less than 1 horsepower (also known as fractional horsepower motors). The total power of all these fractional horsepower single-phase motors is greater than the total power of all other motors (AC and DC) combined because of the tremendous number of fractional horsepower motors in use today. Single-phase power is what is delivered to our homes. Single-phase motors may be either synchronous or asynchronous.

Synchronous Motors

Synchronous motors are used whenever a constant speed is required. These motors run at one speed regardless of the load. They can be used for power factor correction in industrial plants and in electric clocks, turntables, recorders, timers, and other constant speed requirements. The two basic types of synchronous motors are the reluctance motor and the hysteresis motor. The reluctance motor uses a noncylindrical motor, which is basically like a squirrel cage induction motor, to create its starting torque. The hysteresis motor has a rotor made of a special steel alloy. The starting torque is developed due to the large hysteresis in the steel.

Asynchronous Motors

Asynchronous motors run at varying speeds according to the load. When an AC voltage is applied to the stator windings, an alternating magnetic field will be produced. At one instant in time a north pole is created and then a south pole is created resulting in no rotation. Unless the field can be made to rotate the motor will never start to turn. Two common techniques for causing the field to rotate are the split phase and the shaded pole technique.

The **split phase motor** uses two windings. The main winding has many turns of heavy gauge wire. The auxiliary winding has fewer turns of fine wire. The main winding is largely inductive and the auxiliary winding is largely resistive, so a phase shift between the current and voltage causes the motor to start.

To develop a large starting torque, a capacitor is placed in series with the auxiliary winding. This type of starting technique is called a **capacitor start motor.** Many variations exist:

1. Capacitor Start/Induction Run—the capacitor is removed from the circuit after starting.

2. Capacitor Start/Capacitor Run—to get the best starting and running characteristics, one capacitor is used for starting conditions and a second capacitor is used for running conditions.
3. Capacitor (fixed)—the same capacitor is used to start and run the motor.

Figure 13-33 is a photograph of a capacitor start motor. Notice the top of the housing. This is where the capacitor is mounted.

Figure 13-33 **Capacitor start motor**

Split phase motors like those described are used as fans, pumps, oil burner motors, washing machine motors, refrigerator motors, and blower motors.

The **shaded pole motor** has a small coil of wire around a portion of each pole. The coil is opposed to the build up of flux in that portion of the pole. This creates two flux components displaced in time and space and results in a rotating magnetic field.

Shaded pole motors are very inefficient but very reliable. They are generally used for very low power applications like toys, hair dryers, and fans.

Another type of asynchronous motor is the universal motor. Universal motors are basically series motors that are modified to run on DC or AC. They are used in electric drills, vacuum cleaners, and food blenders.

✓ Self-Test 5

1. The ______________ motor is one of the most widely used motors today.
2. The synchronous speed is ______________ proportional to the input frequency and ______________ proportional to the number of stator poles.
3. The ______________ of an AC motor is often expressed as a percentage of the synchronous speed of the rotating field.

T3 tech tips & trouble-shooting

T3 NAMEPLATE DATA

The National Electric Code requires that the electrical information about a motor be written on the motor. The label describing the electrical data about a motor is called the **nameplate**. Figure 13-34 shows a typical motor nameplate.

Figure 13-34 **Motor nameplate**

Typical motor information includes:

1. The name of the manufacturer.
2. The style or model number of the motor.
3. The serial number of the motor.
4. The size or output power of the motor at rated speed. This is usually given in horsepower.
5. The type of motor such as a split phase, induction, squirrel cage, and so forth. This data is sometimes given as a coded number from the manufacturer.
6. The number of phases required from the supply voltage. AC motors can be one-, two-, or three-phase. A DC motor would specify DC in this box.
7. The operating voltage of the motor.
8. The required frequency of the motor.
9. The full load current of the motor when operating at rated voltage and frequency.
10. The speed of the motor at rated horsepower.
11. The temperature rise of the motor. This is the difference between the operating temperature of the motor and the ambient temperature (usually 20°C).
12. The time rating of the motor. Normally the time rating of a motor is continuous. When a value is given, it represents the time that the motor may be operated without overheating.
13. The frame size of the motor. This is a standard number manufacturers use to represent the dimensions of a motor. When motors are manufactured according to the same frame standards, they may be replaced by another manufacturers motor of the same frame type. For the same frame type, the mechanical construction and dimensions along with the electrical ratings and operating characteristics will be the same.

SUMMARY

Real power is the power transformed by a resistor into heat. The unit of real power is the watt. The rate of change of energy in a reactive element like an inductor or capacitor is called reactive power. The unit of reactive power is the VAR. The product of voltage times current in an AC circuit is known as the apparent power. Apparent power is the power produced by the power company to supply us with electricity. Real power is the power we pay for. It is usually less than the apparent power. The three types of power are related using the power triangle. When the power factor is 1 (0 degrees), the real power will equal the apparent power. Power companies apply power factor correction techniques to maintain the power factor as close to 1 as possible.

Power plants produce electricity in the form of three-phase power. Three-phase AC generators are connected in either a wye or a delta configuration. The phase voltage on a three-phase system is measured from the neutral point to each phase. The line voltage is the voltage between any two line leads.

An electric motor changes electric energy to mechanical energy of rotation. An asynchronous motor changes speed depending on its load. A synchronous motor is a constant speed motor regardless of the load. A universal motor operates on DC or AC. Motor power losses are due to copper losses, core losses, and mechanical losses. The efficiency of a motor is the ratio of the power out of the motor to the power into the motor.

DC motors are classified by the connection of the field windings to the armature winding. The general configurations are the shunt motor, the series motor, and the compound motor. An electric generator can be considered as a motor operating in the opposite mode. That is, it takes mechanical energy of rotation as its input and transforms it into electric energy for its output.

The AC induction motor is one of the most widely used motors today. This is largely due to its simplicity of construction. The speed of the rotating magnetic field around the stator is called the synchronous speed. The difference between the actual motor speed and the synchronous speed is called the slip speed.

Single-phase AC motors have no inherent starting torque and are therefore not self-starting. The method of starting is usually used to name the motor. For example, split-phase motor, capacitor motor, shaded pole motor, and so forth. The National Electric Code requires that the electrical information about a motor be written on the motor. The label describing the electrical data is called the nameplate.

FORMULAS

$$P = VI \qquad P = I^2R \qquad P = \frac{V^2}{R}$$

$$P_{Q_L} = I^2X_L \qquad P_{Q_C} = I^2X_C$$

$$P = VI\cos\theta$$ (real power—watts)

$$P_Q = VI\sin\theta$$ (reactive power—VARS)

$$P_a = S = VI$$ (apparent power—VA)

$$P_a^2 = P^2 + P_Q^2$$

$$PF = \frac{P}{P_a} = \frac{VI\cos\theta}{VI} = \cos\theta$$

$$V_L = \sqrt{3}\,V_\phi \quad \text{(wye)}$$

$$I_L = I_\phi \quad \text{(wye)}$$

$$I_L = \sqrt{3}\,I_\phi \quad \text{(delta)}$$

$$V_L = V_\phi \quad \text{(delta)}$$

$$\%\text{ efficiency} = \frac{P_{out}}{P_{in}} \times 100\% = \frac{P_{in} - P_{loss}}{P_{in}} \times 100\% = \frac{P_{out}}{P_{out} + P_{loss}} \times 100\%$$

$$HP = \frac{TS}{5{,}252} \qquad T = \frac{5{,}252\,HP}{S}$$

$$V_L = V_A + I_A R_A \quad \text{(KVL—shunt motor)}$$

$$V_L = V_A - I_A R_A \quad \text{(KVL—shunt generator)}$$

$$V_g = K\phi S$$

$$I_A = \frac{V_L - V_A}{R_A}$$

$$N_S = \frac{120f}{P}$$

$$\text{slip} = N_S - N$$

$$\%\text{ slip} = \frac{N_S - N}{N_S} \times 100\%$$

EXERCISES

Section 13.1

1. A 10-ohm resistor has a current of 2 amperes passing through it. Calculate the power dissipated.
2. A 10-ohm resistor has 24 volts across it. Calculate the power dissipated.
3. A resistor has 30 volts across it and 3 amperes passing through it. Calculate the power dissipated.
4. A 5-kilohm resistor has 3 milliamperes passing through it. Calculate the power dissipated.
5. A 5-kilohm has 100 volts across it. Calculate the power dissipated.

6. A 200-ohm resistor, a 400-ohm resistor, and a 120-volt AC source are connected in series. Calculate the power dissipated in each resistor and the total power used by the circuit.
7. A 5-ohm resistor, an inductor with a reactance of 12 ohms, and a 26-volt AC source are connected in series. Calculate the power dissipated by the resistor and the total power used by the circuit.
8. A 6-ohm resistor, a capacitor with a reactance of 8 ohms, and a 50-volt AC source are connected in series. Calculate the power dissipated by the resistor and the total power consumed by the circuit.
9. A 10-ohm resistor, an inductor with a reactance of 24 ohms, and a 52-volt AC source are all connected in parallel. Calculate the power dissipated by the resistor and the total power consumed by the circuit.

Section 13.2

10. A 3-ohm resistor, an inductor with a reactance of 4 ohms, and a 15-volt AC source are connected in series. Calculate the real power and the reactive power of the circuit.
11. A 15-ohm resistor, a capacitor with a reactance of 36 ohms, and a 117-volt AC source are connected in series. Calculate the real power and the reactive power of the circuit.
12. A 5-ohm resistor, an inductor with a reactance of 42 ohms, a capacitor with a reactance of 30 ohms, and a 52-volt AC source are connected in series. Calculate the real power and the reactive power of the circuit.
13. A 40-ohm resistor, an inductor with a reactance of 20 ohms, a capacitor with a reactance of 60 ohms, and a 120-volt AC source are connected in parallel. Calculate the real power and the reactive power of the circuit.

Section 13.3

14. A 10-ohm resistor, an inductor with a reactance of 15 ohms, and a 72-volt AC source are connected in series. Calculate the real power, the reactive power, and the apparent power of the circuit.
15. A 30-ohm resistor, a capacitor with a reactance of 45 ohms, and a 135-volt AC source are connected in series. Calculate the real power, the reactive power, and the apparent power of the circuit.
16. A 6-ohm resistor, an inductor with a reactance of 12 ohms, a capacitor with a reactance of 4 ohms, and a 120-volt AC source are connected in series. Calculate the real power, the reactive power, and the apparent power of the circuit.

Section 13.4

17. A 220-volt AC motor draws 4.8 amperes. A wattmeter connected in the circuit indicates 1,000 watts. Calculate the power factor of the motor.
18. A 120-volt AC motor draws 5 amperes. A wattmeter connected in the circuit indicates 500 watts. Calculate the power factor of the motor.
19. A 50-ohm resistor, an inductor with the reactance of 120 ohms, and 120-volt AC source are connected in series. Calculate the power factor of the circuit.

20. A 12-ohm resistor, an inductor with the reactance of 20 ohms, a capacitor with a reactance of 15 ohms, and 120-volt AC source are connected in series. Calculate the power factor of the circuit.

21. A 2.2-kilohm resistor and an inductor with a reactance of 1.5 kilohms are connected in series with a 100-volt AC source.

    (A) Calculate the power factor of the circuit.

    (B) If a capacitor with a reactance of 1 kilohm is added in series to improve the power factor, calculate the improved power factor.

Section 13.5

22. A three-phase wye generator with a line voltage of 381 volts is connected to a wye-type resistive load of 22 ohms for each resistor. Calculate the voltage across each resistor and the current through each resistor.

23. A three-phase delta generator with a line voltage of 480 volts is connected to a delta-type resistive load with each resistor equaling 10 ohms. Calculate the voltage across each phase of the load and the current through each phase of the load.

24. For the circuit of Figure 13-35, find the current through each phase of the load.

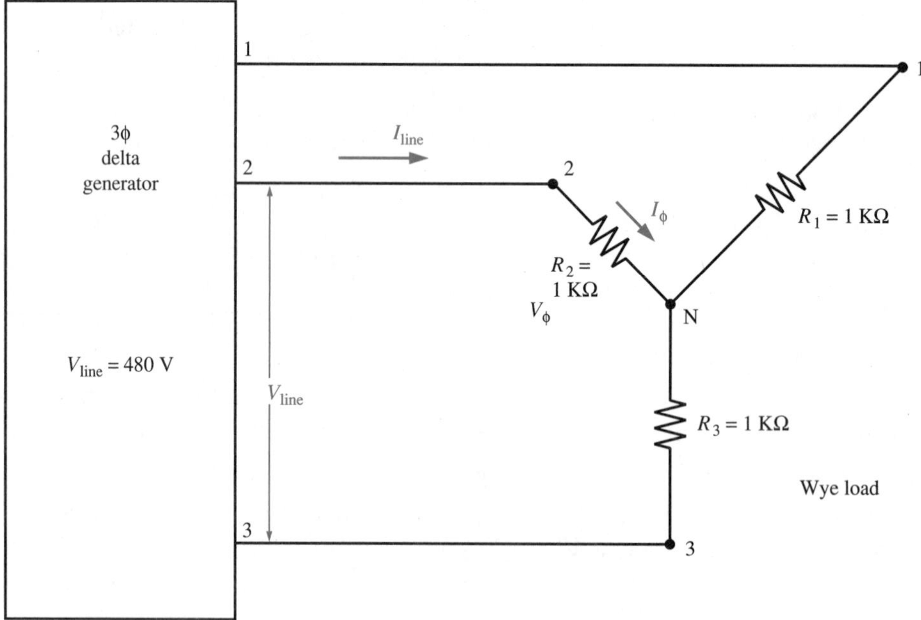

Figure 13-35

25. For the circuit of Figure 13-36, find the current through each phase of the load.

Section 13.6

26. A 2-horsepower DC motor operates at 120 volts and draws 15 amperes. Calculate the efficiency of the motor.

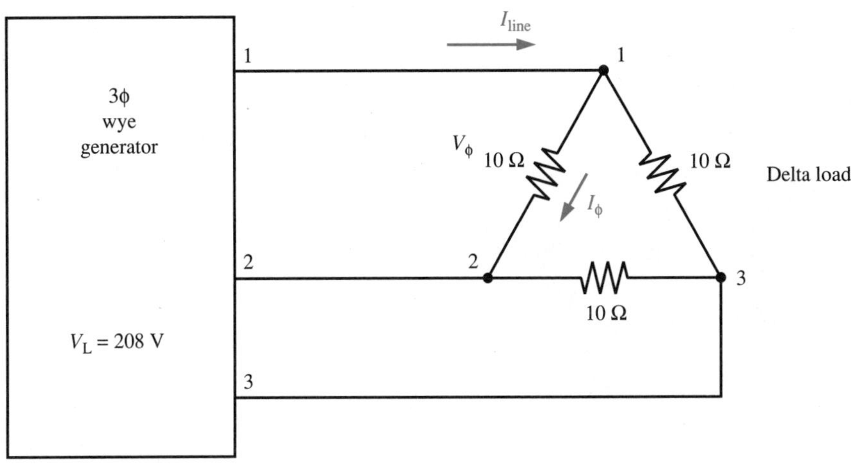

Figure 13-36

27. A ¼-horsepower DC motor operates at 12 volts and draws 20 amperes. Calculate the efficiency of the motor.
28. A motor generates an output torque of 15 pound-feet at 1,200 revolutions per minute. Calculate the horsepower of the motor.
29. Calculate the output torque of a 2-horsepower, 1,200-revolutions-per-minute motor.

Section 13.7

30. A 5-horsepower, 480-volt DC shunt motor draws 10 amperes while rotating at 1,800 revolutions per minute. The armature resistance is 0.4 ohms and the shunt field resistance is 200 ohms. Find the output torque and the counter EMF developed by the motor.
31. A 2-horsepower, 120-volt DC shunt motor draws 15 amperes when rotating at 1,200 revolutions per minute. The armature resistance is 0.2 ohms and the shunt field resistance is 150 ohms. Find the developed counter EMF, the output torque, the efficiency, and the power lost due to copper and rotational losses.
32. A 5-kilowatt, 100-volt DC shunt generator has rotational losses of 200 watts at 1,500 revolutions per minute. Assuming an armature resistance of 0.25 ohms and a field current of 2 amperes, calculate the efficiency of the generator at full load.

Section 13.8

33. Find the synchronous speed of the rotating magnetic field of a 4-pole induction motor operating at 60 hertz.
34. Find the synchronous speed of the rotating magnetic field of a 6-pole induction motor operating at 120 hertz.
35. A 60-hertz induction motor has a rotating magnetic field with a synchronous speed of 3,600 revolutions per minute. Find the number of poles in the motor.

36. A 6-pole, 60-hertz induction motor has a full load speed of 1,140 revolutions per minute. Find the full load slip in revolutions per minute and the percent slip of the motor.
37. A 2-pole, 60-hertz induction motor has a full load speed of 3,480 revolutions per minute. Find the full load slip in revolutions per minute and the percent slip of the motor.
38. See Lab Manual for the crossword puzzle for Chapter 13.

CHAPTER 14

COMPUTER-AIDED ANALYSIS

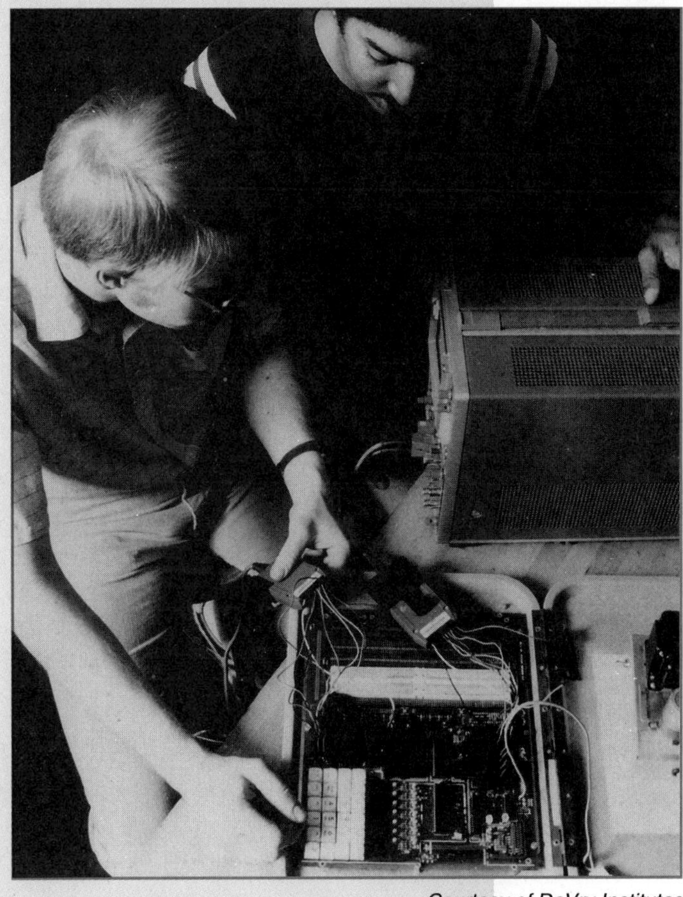

Courtesy of DeVry Institutes

OBJECTIVES

After completing this chapter, the student should be able to:

- define the application of circuit simulators.
- use circuit simulators to analyze DC circuits.
- use circuit simulators to analyze AC circuits.
- print circuit simulation outputs.
- plot circuit simulation outputs.

KEY TERMS

CircuitMaker
modeled
PSpice
simulator
SPICE

14.1 Introduction

SPICE is an acronym for Simulation Program with Integrated Circuit Emphasis.

The use of the computer as a problem-solving tool has been growing geometrically over the past number of years. At first, engineers and scientists used computer programming languages such as FORTRAN and BASIC to help in the analysis of circuits. In time, many types of specialized computer programs emerged to aid in the solution of complex circuit analysis. These programs are known as circuit **simulators.** Today, the de facto industry standard computer circuit simulation program for circuit analysis is **SPICE.** SPICE is an acronym for Simulation Program with Integrated Circuit Emphasis. SPICE was first developed at the University of California at Berkeley during the 1970s to run on large mainframe computers. MicroSim Corporation of Irvine, California, developed a version of SPICE called **PSpice,** which runs on personal computers and engineering workstations. MicroSim provides student versions of PSpice that are more than adequate to solve any circuit problem in this textbook at a nominal fee. MicroSim Corporation can be reached by telephone by dialing (800) 245-3022.

14.2 PSpice Installation

PSpice requires 512K of available memory.

PSpice may be installed on an IBM PC with 512K of RAM memory and one hard and one floppy disk drive. Make sure you are running a DOS 3.3 or higher operating system. Begin by creating a directory called PSPICE. To do this, at the DOS prompt (C:), type the following:

C:MD\PSPICE <ENTER>

Next, change to the PSPICE directory by typing the following:

C:CD\PSPICE <ENTER>

Copy all files from the PSPICE diskettes by invoking the following:

C:COPY A:*.* <ENTER>

Repeat the copy command for each of the PSPICE diskettes supplied. This completes the basic installation. You can begin PSpice by typing PS from the PSPICE directory.

C:PS <ENTER>

Before you run PSpice, verify that your CONFIG.SYS file in the root directory has the following two lines:

BUFFERS=20

FILES=20

If you do not have these two lines, add them by using your word processor in ASCII mode, EDLIN, or by any other method you prefer. DOS 5.0 and higher users can use the new EDIT command instead of EDLIN. If you do not have a CONFIG.SYS file in the root directory either create one using the DOS COPY CON command or simply copy the CONFIG.SYS file from the PSpice supplied diskette into your root directory. In any case reboot your system one more time before you try to start PSpice.

14.3 Getting Started With PSpice

PSpice comes with drop down menus and is not difficult to use. Once you get the basic idea, you'll find circuit simulation interesting and fun. The best way to learn PSpice is to do a circuit simulation. Basically using PSpice involves a three-step process:

1. Create an input file or circuit model.
2. Run the PSpice simulator (without errors).
3. Print the output file to obtain the results.

PSpice contains a circuit simulator that allows a circuit to be described or **modeled** by its connection nodes and circuit components. It calculates a circuit's voltages and currents and can display the results in tabular or graphical form. It can analyze circuits for DC voltages and currents, AC voltages and currents over a range of frequencies, and transient circuit responses.

PSpice models a circuit using its connection nodes and circuit components.

To start PSpice, enter the PSpice directory using the following DOS commands at the DOS prompt:

C:CD\PSPICE <ENTER>

Next type:

C:PS <ENTER>

The PSpice control shell will be displayed on the screen, as shown in Figure 14-1.

```
PSpice Control Shell - ver 4.05
Files    Circuit    StmEd    Analysis    Display    Probe    Quit

                Current File:                          (New)
F1=Help  F2=Move  F3=Manual  F4=Choices  F5=Calc  F6=Errors   ESC=Cancel
```

Figure 14-1 **PSpice control shell**

All PSpice programs contain three sections:

1. Title
2. Circuit Model
3. End

Files in PSpice use a standard required format as follows:

INPUT FILE NAME	.CIR
OUTPUT FILE NAME	.OUT

You will choose the input file name using the .CIR extension. PSpice will create the output file that has the results for printing with the .OUT extension. For example:

INPUT FILE NAME	TEST1.CIR
OUTPUT FILE NAME	TEST1.OUT

Let's begin by examining the circuit of Figure 14-2. Notice that the connection points or nodes have been numbered consecutively.

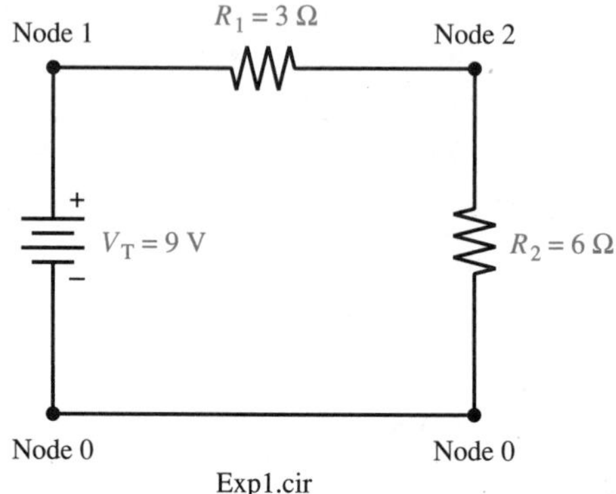

Figure 14-2

From the PSpice Control Shell, use the cursor control keys and the ENTER key to select the FILES drop down menu as follows:

FILES <ENTER>

From the drop down menus, Figure 14-3, select:

CURRENT <ENTER>

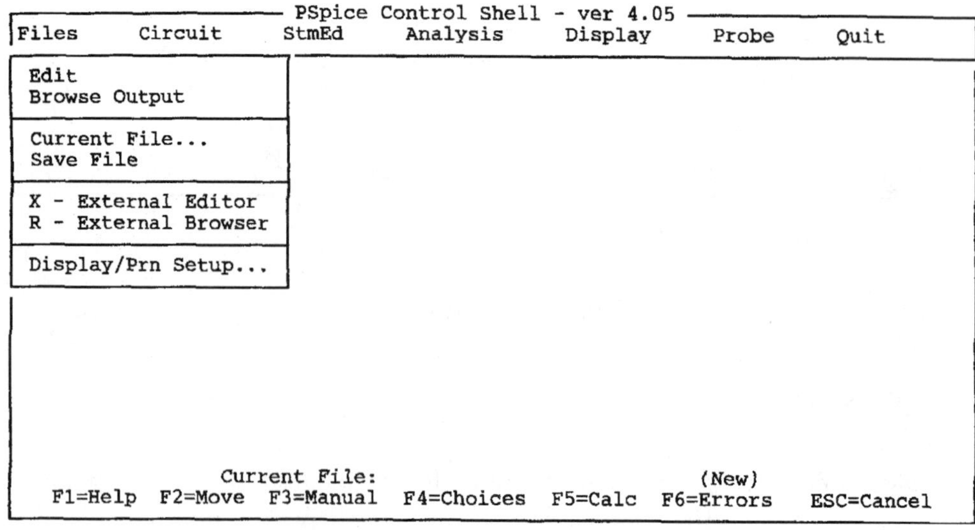

Figure 14-3 PSpice files drop-down menu

At the Define Input File window (Figure 14-4), type the input file name as follows:

Circuit File Name: EXP1.CIR <ENTER>

```
------------------------ PSpice Control Shell - ver 4.05 ------------------------
Files     Circuit     StmEd     Analysis     Display     Probe     Quit

 Edit
 Browse Output

 Current File...
 S ---------------Define Input File------------------------
 X
 R      Circuit File Name: EXP1.CIR

 D


              Current File:                         (New)
  F1=Help  F2=Move  F3=Manual  F4=Choices  F5=Calc  F6=Errors   ESC=Cancel
```

Figure 14-4 ***Define input file window***

Note that the lower menu bar now states:

Current File: EXP1.CIR

Next select the FILES drop down menu again as follows:

FILES <ENTER>

From the drop down menu select:

EDIT <ENTER>

The circuit editor window will appear on the screen. Next type in the circuit model for Figure 14-2, as shown in the following. It is a good idea to use all capital letters and to organize the lines into neat columns. The TAB key will be very helpful for making columns.

```
EXP1.CIR                          <ENTER>
VT    1    0    DC    9V          <ENTER>
R1    1    2    3                 <ENTER>
R2    2    0    6                 <ENTER>
.END                              <ENTER>
```

All PSpice programs must begin with a title line and conclude with the .END command.

This PSpice program models the circuit of Figure 14-2. Figure 14-5 illustrates the circuit editor window with the program typed in.

```
                     PSpice Control Shell - ver 4.05
               Circuit Editor     line:   6 col:  1      [Insert]
EXP1.CIR
VT     1     0     DC     9V
R1     1     2     3
R2     2     0     6
.END

                   Current File: EXP1.CIR                   (New)
  F1=Help  F2=Move  F3=Manual  F4=Choices  F5=Calc  F6=Errors   ESC=Cancel
```

Figure 14-5 ***PSice circuit ender***

Each node is labeled for use by the PSpice program. We begin with the reference node (usually ground) labeled as node 0. Continuing clockwise around the circuit each node is labeled consecutively. Unless specified, PSpice will automatically calculate the voltage at each node with respect to the reference node. The first line of our program is the title line. It identifies our program. Lines 2, 3, and 4 model the circuit. They define the components used, their values, and their connections to each other. Column 1 defines the component, column 2 is the positive node, and column 3 is the negative node. Column 4 specifies the size or type of component. Line 5 defines the end of the program.

Next press the ESCAPE key to save the program and exit the editor. At the message, Save Changes, type S. Next use the cursor control keys to select:

ANALYSIS <ENTER>

At the drop down menu (Figure 14-6), select:

RUN PSpice <ENTER>

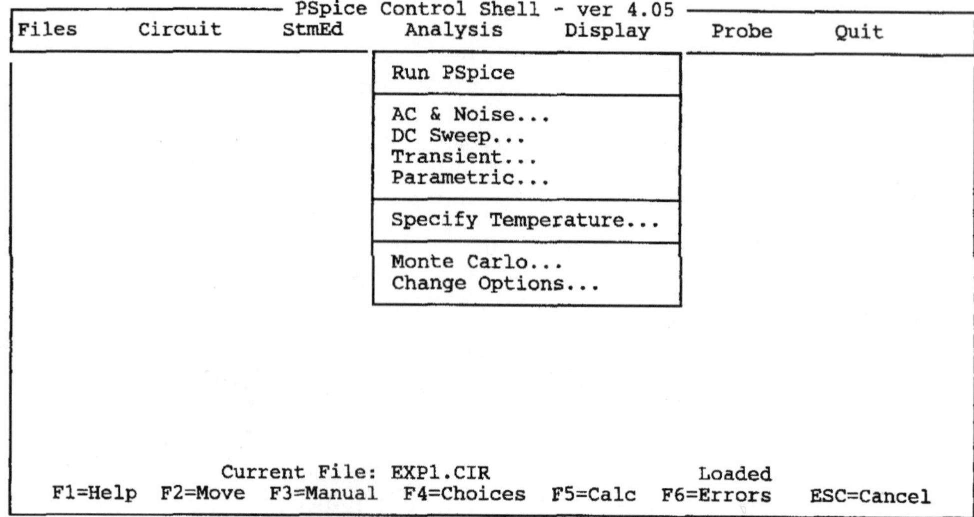

Figure 14-6 ***Analysis drop-down menu***

This tells PSpice to run the simulation and calculate our answers. When completed the lower menu bar will display:

CURRENT FILE: EXP1.CIR LOADED

To output the results use the cursor control keys to select:

QUIT <ENTER>

At the drop down menu (Figure 14-7), select:

DOS COMMAND <ENTER>

```
-------------------- PSpice Control Shell - ver 4.05 --------------------
Files     Circuit     StmEd     Analysis     Display     Probe     Quit
                                                              Exit to DOS
                                                              DOS Command...

                     Current File: EXP1.CIR                  Loaded
   F1=Help  F2=Move  F3=Manual  F4=Choices  F5=Calc  F6=Errors   ESC=Cancel
```

Figure 14-7 ***Quit drop-down menu***

At the Enter Command window (Figure 14-8), type:

PRINT EXP1.OUT <ENTER>

```
-------------------- PSpice Control Shell - ver 4.05 --------------------
Files     Circuit     StmEd     Analysis     Display     Probe     Quit
                                                              Exit to DOS
                                                              DOS Command...

   Enter Command:
   PRINT EXP1.OUT

                     Current File: EXP1.CIR                  Loaded
   F1=Help  F2=Move  F3=Manual  F4=Choices  F5=Calc  F6=Errors   ESC=Cancel
```

Figure 14-8 ***Enter command window***

At the Save Changes message, type S, the file name will appear on the screen, press the ENTER key and continue. Depending on your system configuration, you may get the message:

NAME OF LIST DEVICE <PRN>

If you see this message displayed on the screen, DOS is asking if the name of your printer is PRN. Press ENTER and your printer should start printing. Figure 14-9 illustrates the circuit description printer output and Figure 14-10 illustrates the printed results.

```
**** 03/21/91 16:29:59 ********* Evaluation PSpice (January 1991) ************
 EXP1.CIR

 ****     CIRCUIT DESCRIPTION

******************************************************************************


VT    1    0    DC    9V
R1    1    2    3
R2    2    0    6
.END
```

Figure 14-9 *PSpice circuit description output*

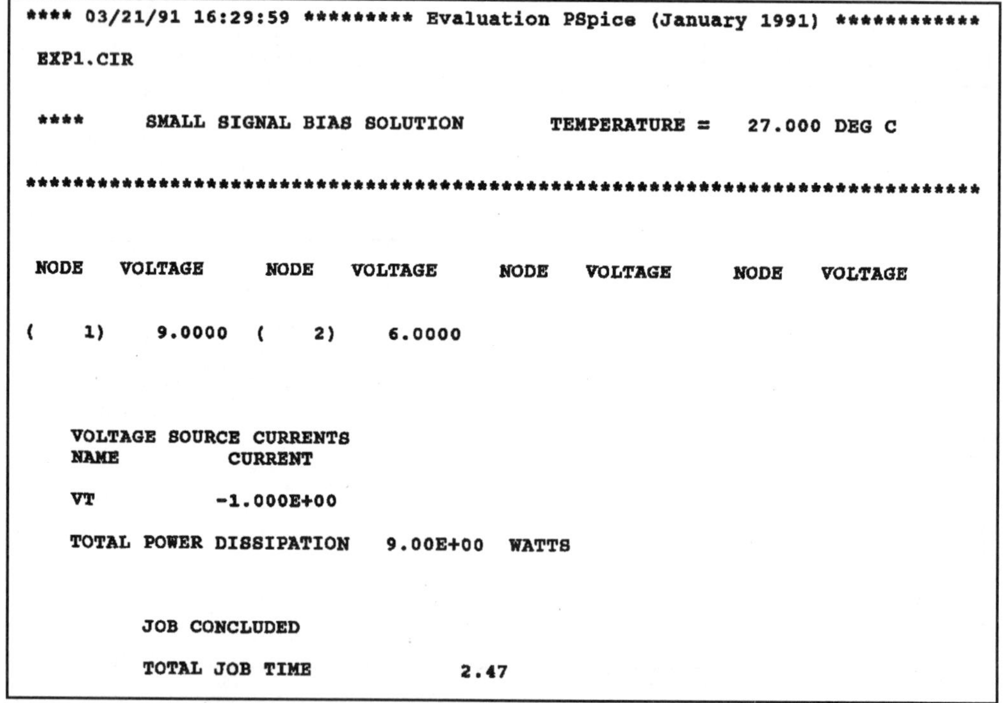

```
**** 03/21/91 16:29:59 ********* Evaluation PSpice (January 1991) ************
 EXP1.CIR

 ****     SMALL SIGNAL BIAS SOLUTION       TEMPERATURE =   27.000 DEG C

******************************************************************************


 NODE   VOLTAGE     NODE   VOLTAGE     NODE   VOLTAGE     NODE   VOLTAGE

(    1)    9.0000  (    2)    6.0000


    VOLTAGE SOURCE CURRENTS
    NAME            CURRENT

    VT             -1.000E+00

    TOTAL POWER DISSIPATION   9.00E+00  WATTS


          JOB CONCLUDED

          TOTAL JOB TIME             2.47
```

Figure 14-10 *PSpice solution output*

Let's modify our program in order to use and learn a few additional commands. First, bring up the program by selecting:

FILE <ENTER>

At the drop down menu, select:

EDIT <ENTER>

Our original program should now be on the screen. Edit the program to add the following lines as shown in the following:

```
EXP1.CIR                                 <ENTER>
VT        1    0    DC     9V            <ENTER>
R1        1    2    3                    <ENTER>
R2        2    0    6                    <ENTER>
.DC       VT   9    9      1             <ENTER>
.PRINT         DC   V(R1)  V(R2)  I(R1)  <ENTER>
.OPTIONS       NOPAGE                    <ENTER>
.END                                     <ENTER>
```

The .DC and the .PRINT command work together to specify exactly which voltages and currents you want printed in your output file. Without these commands, PSpice will calculate only the voltages at each node. The .DC command will allow you to sweep through a range of input voltages. This is demonstrated later. The .OPTIONS NOPAGE command simply saves you some paper. It tells PSpice to print the circuit description and the results on the same page. Let's run the modified program and compare the results. Begin by pressing the ESCAPE key. At the Save Changes message, type S to save the program and return to the PSpice Control Shell. Next select:

The .DC and the .PRINT commands work together to specify the output.

ANALYSIS <ENTER>

At the drop down menu select:

RUN PSpice <ENTER>

To print the results select:

QUIT <ENTER>

At the drop down menu select:

DOS COMMAND <ENTER>

At the enter command window type:

PRINT EXP1.OUT <ENTER>

At the Save Changes message, type S, the file name will appear on the screen, press the ENTER key and continue. The results are shown in Figure 14-11.

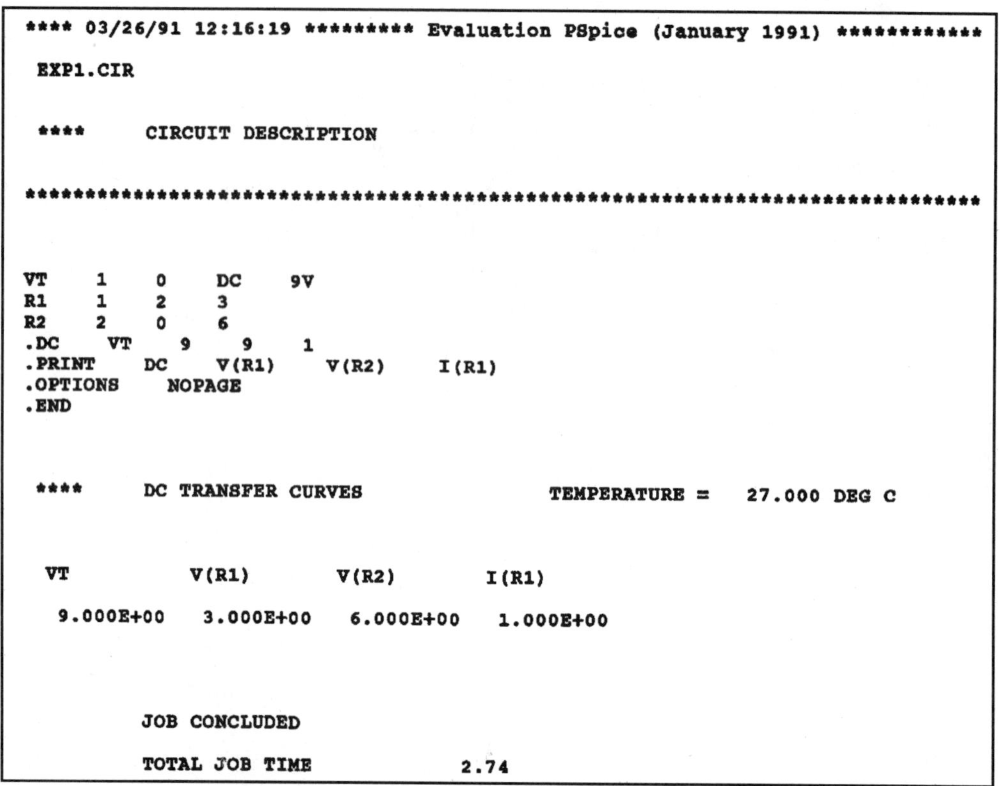

```
**** 03/26/91 12:16:19 ********* Evaluation PSpice (January 1991) ************

 EXP1.CIR

 ****     CIRCUIT DESCRIPTION

******************************************************************************

VT     1     0     DC     9V
R1     1     2     3
R2     2     0     6
.DC     VT     9     9     1
.PRINT     DC     V(R1)     V(R2)     I(R1)
.OPTIONS     NOPAGE
.END

 ****     DC TRANSFER CURVES               TEMPERATURE =   27.000 DEG C

  VT           V(R1)        V(R2)        I(R1)

   9.000E+00   3.000E+00   6.000E+00   1.000E+00

          JOB CONCLUDED

          TOTAL JOB TIME              2.74
```

Figure 14-11 *PSpice .print and .DC commands*

Notice that in this case the voltage across R_1 and R_2 is computed. In the previous example, the nodal voltages were calculated with respect to the reference point or ground. Note also in this example that the current I(R1) is a +1 ampere representing the current through R_1. In the previous example, the source current was calculated as −1 ampere, representing the current through or leaving the source. This is simply a result of the way PSpice works.

In either example, the answers are given in engineering notation where **E** represents the power of ten.

The .DC and the .PRINT commands added flexibility and allowed you to test your results for a range of input voltages. The general format for the .PRINT command is:

.PRINT   Desired Quantities

The general format for the .DC command is:

.DC   Source Name   Starting Value   Final Value   Increment

In EXP1.CIR the .DC command was as follows:

.DC   VT   9   9   1

Let's modify this line to select a range of voltage values for VT:

FILES   <ENTER>

At the drop down menu select:

EDIT <ENTER>

The EXP1.CIR program should appear on the screen. Modify the .DC line as follows:

.DC VT 1 9 1

Next press the ESCAPE key to exit the editor. At the Save Changes window, type S. Select:

ANALYSIS <ENTER>

At the drop down menu, select:

RUN PSpice <ENTER>

Select:

QUIT <ENTER>

At the drop down menu, select:

DOS COMMAND <ENTER>

At the Enter Command window, type:

PRINT EXP1.OUT <ENTER>

At the Save Changes window, type S. The file name will appear on the screen. Press the ENTER key and continue. The results are shown in Figure 14-12.

Note that we have swept the input voltage, VT, from 1 volt to 9 volts in 1-volt increments. The .PRINT command allowed us to calculate V(R1), V(R2), and I(R1) for the desired range of input voltages. The parenthesis is the command structure necessary for the proper output.

PSpice allows the user to include engineering prefixes. The following uppercase letters may be used with PSpice to represent engineering prefixes:

$$F = 10^{-15}$$

$$P = 10^{-12}$$

$$N = 10^{-9}$$

$$U = 10^{-6}$$

$$M = 10^{-3}$$

$$K = 10^{3}$$

$$MEG = 10^{6}$$

$$G = 10^{9}$$

$$T = 10^{12}$$

For example, 2200 may be written as 2.2K or 0.003 may be written as 3M.

```
**** 03/26/91 12:51:11 ********* Evaluation PSpice (January 1991) ************
 EXP1.CIR

 ****     CIRCUIT DESCRIPTION

******************************************************************************


VT    1    0    DC    9V
R1    1    2    3
R2    2    0    6
.DC     VT    1    9    1
.PRINT    DC    V(R1)    V(R2)    I(R1)
.OPTIONS    NOPAGE
.END


 ****     DC TRANSFER CURVES               TEMPERATURE =   27.000 DEG C


  VT          V(R1)        V(R2)        I(R1)

   1.000E+00   3.333E-01   6.667E-01   1.111E-01
   2.000E+00   6.667E-01   1.333E+00   2.222E-01
   3.000E+00   1.000E+00   2.000E+00   3.333E-01
   4.000E+00   1.333E+00   2.667E+00   4.444E-01
   5.000E+00   1.667E+00   3.333E+00   5.556E-01
   6.000E+00   2.000E+00   4.000E+00   6.667E-01
   7.000E+00   2.333E+00   4.667E+00   7.778E-01
   8.000E+00   2.667E+00   5.333E+00   8.889E-01
   9.000E+00   3.000E+00   6.000E+00   1.000E+00


          JOB CONCLUDED

          TOTAL JOB TIME              3.57
```

Figure 14-12 PSpice .print output

✓ Self-Test 1

1. Computer programs used to solve complex circuit analysis are called ________________.
2. State the three basic steps for modeling a circuit using PSpice.
3. The input file of a PSpice program always has a ________________ extension name. The output file always has a ________________ extension name.
4. The PSpice symbol for engineering prefixes is ________________.

14.4 DC Applications

When modeling a circuit, always begin with the reference node numbered as 0.

Figure 14-13 illustrates how to model a series-parallel circuit. We begin with the reference node 0 and continue numbering the nodes consecutively in a clockwise direction. The general format for the basic commands are:

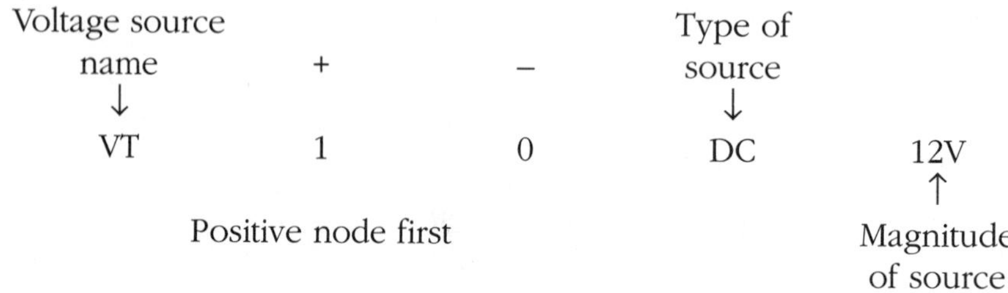

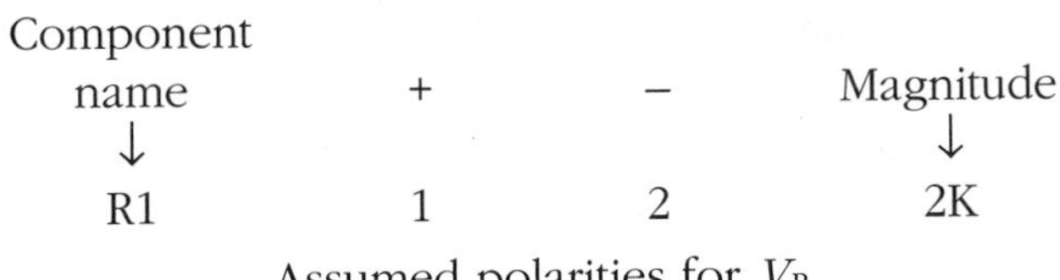

Figure 14-14 shows the modeled circuit program and the results.

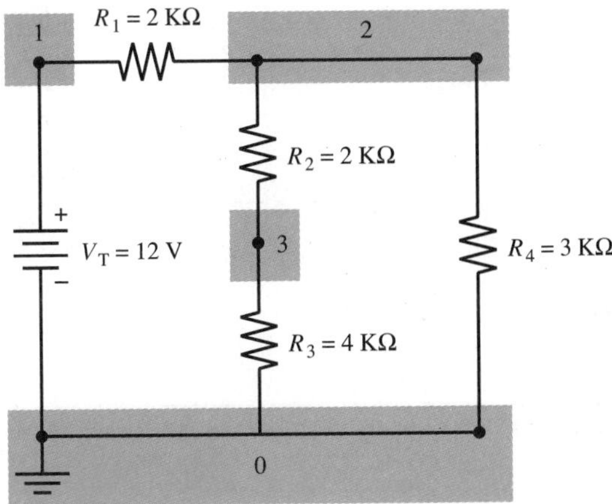

Figure 14-13 EXP2.CIR

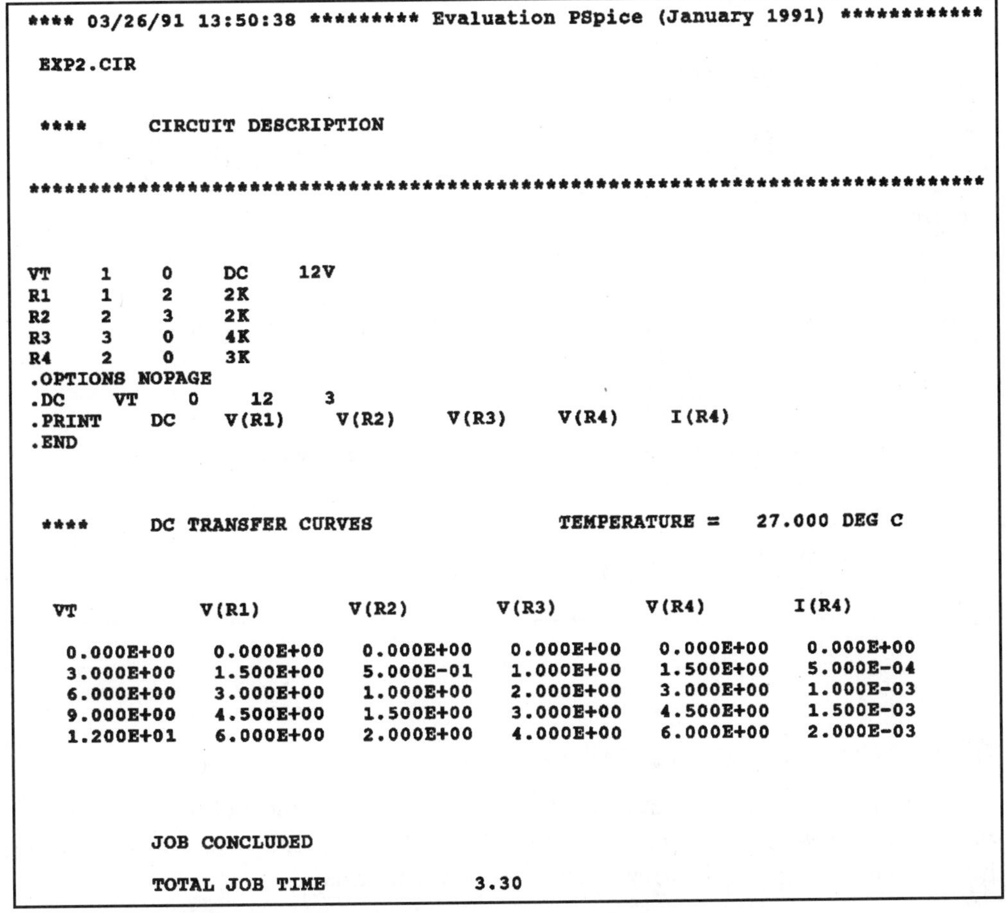

```
**** 03/26/91 13:50:38 ********* Evaluation PSpice (January 1991) ************

 EXP2.CIR

 ****      CIRCUIT DESCRIPTION

******************************************************************************


VT      1     0     DC      12V
R1      1     2     2K
R2      2     3     2K
R3      3     0     4K
R4      2     0     3K
.OPTIONS NOPAGE
.DC     VT     0      12      3
.PRINT     DC      V(R1)      V(R2)      V(R3)      V(R4)      I(R4)
.END


 ****      DC TRANSFER CURVES                    TEMPERATURE =    27.000 DEG C


  VT           V(R1)        V(R2)        V(R3)        V(R4)        I(R4)

  0.000E+00    0.000E+00    0.000E+00    0.000E+00    0.000E+00    0.000E+00
  3.000E+00    1.500E+00    5.000E-01    1.000E+00    1.500E+00    5.000E-04
  6.000E+00    3.000E+00    1.000E+00    2.000E+00    3.000E+00    1.000E-03
  9.000E+00    4.500E+00    1.500E+00    3.000E+00    4.500E+00    1.500E-03
  1.200E+01    6.000E+00    2.000E+00    4.000E+00    6.000E+00    2.000E-03


           JOB CONCLUDED

           TOTAL JOB TIME               3.30
```

Figure 14-14 PSpice EXP2.CIR output

Consider the multisource network of Figure 14-15. Modeling the circuit and running PSpice results in the output shown in Figure 14-16. Note that as the circuit gets more complex, we begin to appreciate the power of PSpice.

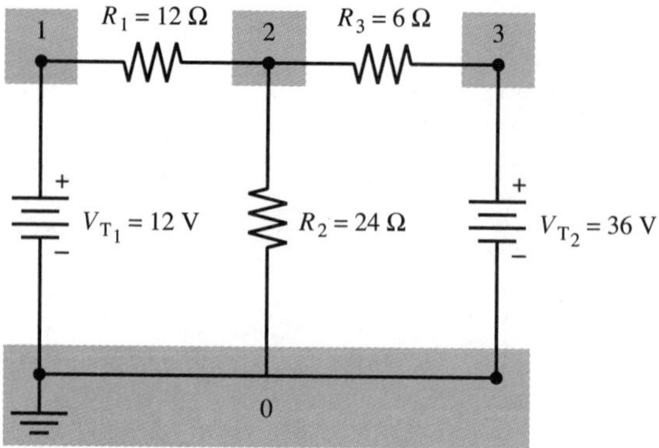

Figure 14-15 EXP3.CIR

```
**** 03/26/91 14:24:12 ********* Evaluation PSpice (January 1991) ************

 EXP3.CIR

 ****       CIRCUIT DESCRIPTION

******************************************************************************

VT1     1     0     DC      12V
VT2     3     0     DC      36V
R1      1     2     12
R2      2     0     24
R3      2     3     6
.DC     VT1      12       12      1
.PRINT     DC     V(1)      V(2)      V(3)      I(R3)
.OPTIONS NOPAGE
.END

 ****       DC TRANSFER CURVES                    TEMPERATURE =    27.000 DEG C

  VT1           V(1)          V(2)          V(3)          I(R3)

   1.200E+01    1.200E+01     2.400E+01     3.600E+01    -2.000E+00

            JOB CONCLUDED

            TOTAL JOB TIME                    2.75
```

Figure 14-16 PSpice EXP3.CIR output

The .AC command specifies the frequency value used during the analysis.

14.5 AC Applications

Using PSpice with AC analysis is similar to what we did before with the DC sweep command. The .AC command specifies the frequency value used during the analysis. The general format of the .AC command is:

.AC	LIN	1	5KH	5KH
		Number of data points	Starting frequency	Final frequency

Another example of the use of the .AC command is:

.AC	LIN	6	5KH	10KH

In the previous example, we begin our sweep at 5 kilohertz and increase in 1-kilohertz steps to 10 kilohertz. Note that step 1 is always the starting frequency.

Step	Frequency
1	5KHz
2	6KHz
3	7KHz
4	8KHz
5	9KHz
6	10KHz

To specify the value of an AC voltage source, we follow the following format.

VT	1	0	AC	120V	0
Name	Positive node	Negative node	Type of source	Magnitude	Phase angle

Note that for AC voltage sources the magnitude and phase must be specified. If no phase angle is specified, the default is 0 degrees. The frequency or range of frequencies is specified by using the .AC command.

The .PRINT command is modified for the .AC command as follows.

.PRINT	AC	VM(R1)	VP(R1)
Command	Type	AC magnitude	AC phase angle

The .PRINT command defines the value of VM(R1), the voltage magni– tude of resistor R1 to be printed. VP(R1) defines the voltage phase angle of resistor R1.

Consider the circuit of Figure 14-17.

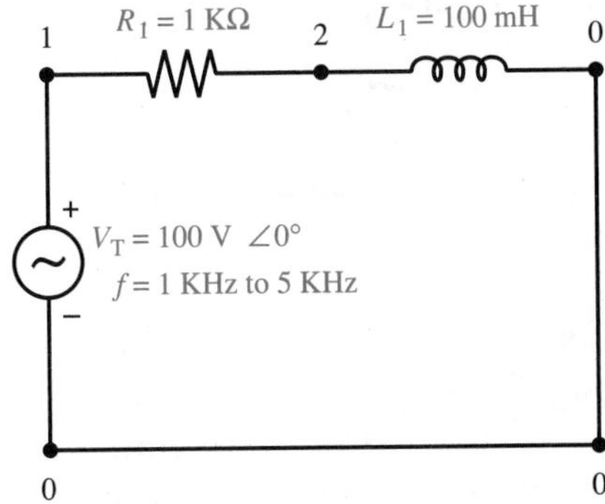

Figure 14-17 **EXP4.CIR**

Figure 14-18 illustrates how to model the circuit as well as the results. Note that inductors are prefixed using the letter, L, in a similar manner as resistors that use the prefix R.

```
**** 05/28/91 15:04:34 ********* Evaluation PSpice (January 1991) ************
 EXP4.CIR

 ****      CIRCUIT DESCRIPTION

******************************************************************************


VT     1     0     AC     100     0
R1     1     2     1K
L1     2     0     100M
.AC    LIN    6     1KH      5KH
.PRINT     AC     VM(L1)     VP(L1)     IM(L1)     IP(L1)
.OPTIONS NOPAGE
.END


 ****      SMALL SIGNAL BIAS SOLUTION          TEMPERATURE =   27.000 DEG C


 NODE   VOLTAGE     NODE   VOLTAGE     NODE   VOLTAGE     NODE   VOLTAGE

(    1)    0.0000  (    2)    0.0000



    VOLTAGE SOURCE CURRENTS
    NAME            CURRENT

    VT              0.000E+00

    TOTAL POWER DISSIPATION   0.00E+00  WATTS



 ****      AC ANALYSIS                         TEMPERATURE =   27.000 DEG C


  FREQ        VM(L1)       VP(L1)       IM(L1)       IP(L1)


   1.000E+03   5.320E+01   5.786E+01   8.467E-02   -3.214E+01
   1.800E+03   7.492E+01   4.148E+01   6.624E-02   -4.852E+01
   2.600E+03   8.529E+01   3.147E+01   5.221E-02   -5.853E+01
   3.400E+03   9.057E+01   2.508E+01   4.240E-02   -6.492E+01
   4.200E+03   9.351E+01   2.075E+01   3.544E-02   -6.925E+01
   5.000E+03   9.529E+01   1.766E+01   3.033E-02   -7.234E+01

          JOB CONCLUDED

          TOTAL JOB TIME              7.31
```

Figure 14-18 **EXP4.CIR output**

The .PLOT command is used to graph the results.

Another useful PSpice command is the .PLOT command. The .PLOT command is nearly the same as the .PRINT command. We must specify the type of analysis the .PLOT is for and which results we want plotted. The general format of the .PLOT command is:

.PLOT	AC	VM(L1)	VP(L1)
Command	Type	Magnitude of voltage V(L1)	Phase angle of voltage V(L1)

Note that when two output parameters are specified, two plots will result. In this case, VM(L1) and VP(L2) will be plotted for the frequency range specified in the .AC command. To illustrate the use of the .PLOT command, consider the circuit of Figure 14-19.

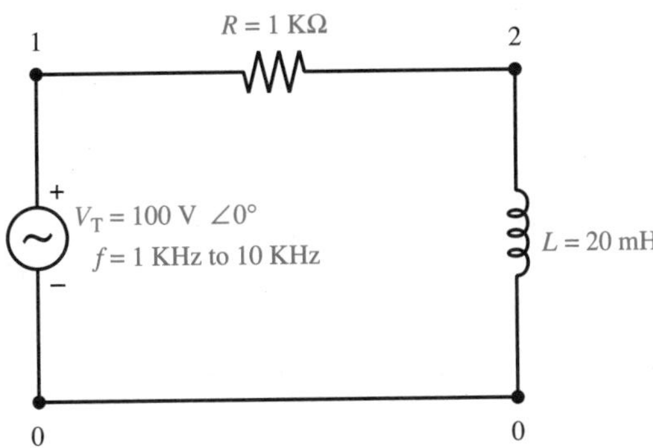

Figure 14-19 ***EXP5.CIR***

The PSpice model is shown in Figure 14-20A, 14-20B, and 14-20C. The .OPTIONS NOPAGE command was not used so that the entire plot would be on separate pages.

```
**** 05/28/91 15:37:47 ********* Evaluation PSpice (January 1991) ************

 EXP5.CIR

 ****      CIRCUIT DESCRIPTION

********************************************************************************


VT     1    0     AC     100     0
R1     1    2     1K
L1     2    0     20M
.AC     LIN     11     1KH     10KH
.PLOT       AC     VM(L1)     VP(L1)
.END
```

Figure 14-20A ***PSpice EXP5.CIR circuit model***

Figure 14-20A illustrates the PSpice circuit description using the .PLOT command. Figure 14-20B is the default print output page. If the .PRINT or .PLOT commands is not used, the nodal voltages and source currents would be calculated. In this example we are using the .PLOT command, so the nodal voltages and source currents are not calculated by default. Instead Figure 14-20C

```
**** 05/28/91 15:37:47 ********* Evaluation PSpice (January 1991) ************
 EXP5.CIR

 ****     SMALL SIGNAL BIAS SOLUTION      TEMPERATURE =   27.000 DEG C

******************************************************************************


 NODE   VOLTAGE     NODE   VOLTAGE     NODE   VOLTAGE     NODE   VOLTAGE

(    1)    0.0000  (    2)    0.0000



    VOLTAGE SOURCE CURRENTS
    NAME          CURRENT

    VT            0.000E+00

    TOTAL POWER DISSIPATION   0.00E+00  WATTS
```

Figure 14-20B *PSpice EXP5.CIR output*

```
**** 05/28/91 15:37:47 ********* Evaluation PSpice (January 1991) ************
 EXP5.CIR

 ****     AC ANALYSIS                        TEMPERATURE =   27.000 DEG C

******************************************************************************


 LEGEND:

*: VM(L1)
+: VP(L1)

 FREQ         VM(L1)

(*)----------    1.0000E+01   1.0000E+02   1.0000E+03   1.0000E+04   1.0000E+05
(+)----------    2.0000E+01   4.0000E+01   6.0000E+01   8.0000E+01   1.0000E+02

 1.000E+03  1.247E+01 . *  - - - - - - - - - - - - - - - - - - - - - - - -+- - - - - -
 1.900E+03  2.322E+01 .     *        .            .              . +       .
 2.800E+03  3.319E+01 .       *      .            .          +   .         .
 3.700E+03  4.216E+01 .         *    .            .     +        .         .
 4.600E+03  5.005E+01 .          *   .            +              .         .
 5.500E+03  5.686E+01 .          *   .         +  .              .         .
 6.400E+03  6.267E+01 .           *  .      +     .              .         .
 7.300E+03  6.760E+01 .           *  .   +        .              .         .
 8.200E+03  7.176E+01 .            *. +           .              .         .
 9.100E+03  7.528E+01 .            *.+            .              .         .
 1.000E+04  7.825E+01 .            X.             .              .         .
                       - - - - - - - - - - - - - - - - - - - - - - - - - - - - - - -



           JOB CONCLUDED

           TOTAL JOB TIME            8.41
```

Figure 14-20C *PSpce EXP5.CIR plot*

shows the plotted output of the AC analysis for this circuit. On the AC analysis plot, the legend defines which character is used to plot each curve. The voltage magnitude VM(L1) uses the "*" for the plot. The voltage phase angle VP(L1) uses the "+" for its plot. The first column on the left lists the range of frequencies specified in the .AC command. The second column lists the calculated voltage magnitude VM(L1) at the specified frequency. Note all values are given in engineering notation. The first column of data represents 10 volts or 20 degrees. The second column of data represents 100 volts or 40 degrees, and so forth. The point at which the two curves cross is indicated by a "X".

Let's consider one more practical example. For the RC filter circuit of Figure 14-21 we wish to plot the output of V(C1) when the input frequency changes from 0.5 KHz to 10 KHz.

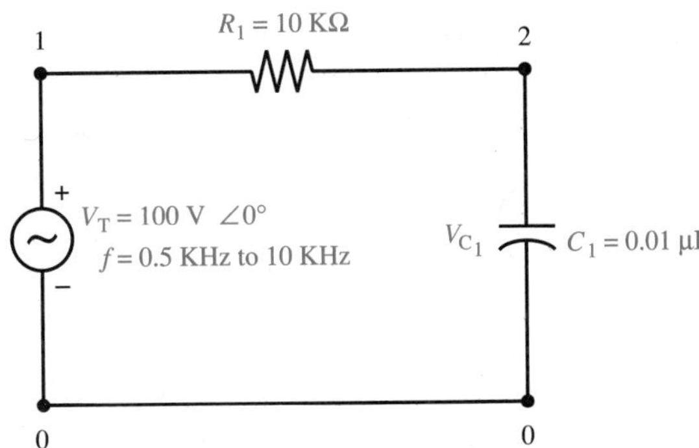

Figure 14-21 **EXP6.CIR**

Figures 14-22A, B, and C illustrate the circuit model and the results.

```
**** 05/28/91 16:41:28 ********* Evaluation PSpice (January 1991) ************

 EXP6.CIR

 ****     CIRCUIT DESCRIPTION

******************************************************************************


VT     1    0     AC      100V     0
R1     1    2     10K
C1     2    0     0.01U
.AC     LIN     20      0.5KH   10KH
.PLOT     AC     VM(C1)     VP(C1)
.END
```

Figure 14-22A **PSpice EXP6.CIR circuit model**

```
**** 05/28/91 16:41:28 ********* Evaluation PSpice (January 1991) ************
 EXP6.CIR

 ****     SMALL SIGNAL BIAS SOLUTION       TEMPERATURE =   27.000 DEG C

******************************************************************************

 NODE   VOLTAGE     NODE   VOLTAGE     NODE   VOLTAGE     NODE   VOLTAGE

(    1)    0.0000  (    2)    0.0000

    VOLTAGE SOURCE CURRENTS
    NAME          CURRENT

    VT            0.000E+00

    TOTAL POWER DISSIPATION   0.00E+00  WATTS
```

Figure 14-22B PSpice EXP6.CIR output

```
**** 05/28/91 16:41:28 ********* Evaluation PSpice (January 1991) ************
 EXP6.CIR

 ****     AC ANALYSIS                      TEMPERATURE =   27.000 DEG C

******************************************************************************

 LEGEND:

*: VM(C1)
+: VP(C1)

  FREQ         VM(C1)

(*)----------    1.0000E+01   1.0000E+02   1.0000E+03   1.0000E+04   1.0000E+05
(+)----------   -1.5000E+02  -1.0000E+02  -5.0000E+01  -7.1054E-15   5.0000E+01

  5.000E+02  9.540E+01 .  - - - - - - *- - - - - - - - - + - - .- - - - - - - -
  1.000E+03  8.467E+01 .              *             .     +      .             .
  1.500E+03  7.277E+01 .             *.             .  +         .             .
  2.000E+03  6.227E+01 .            * .             +            .             .
  2.500E+03  5.370E+01 .           *  .           + .            .             .
  3.000E+03  4.687E+01 .          *   .          +  .            .             .
  3.500E+03  4.139E+01 .          *   .         +   .            .             .
  4.000E+03  3.697E+01 .         *    .        +    .            .             .
  4.500E+03  3.334E+01 .        *     .        +    .            .             .
  5.000E+03  3.033E+01 .        *     .       +     .            .             .
  5.500E+03  2.780E+01 .       *      .       +     .            .             .
  6.000E+03  2.564E+01 .       *      .      +      .            .             .
  6.500E+03  2.378E+01 .      *       .      +      .            .             .
  7.000E+03  2.217E+01 .      *       .      +      .            .             .
  7.500E+03  2.076E+01 .      *       .      +      .            .             .
  8.000E+03  1.951E+01 .     *        .      +      .            .             .
  8.500E+03  1.840E+01 .     *        .     +       .            .             .
  9.000E+03  1.741E+01 .     *        .     +       .            .             .
  9.500E+03  1.652E+01 .    *         .     +       .            .             .
  1.000E+04  1.572E+01 .    *         .     +       .            .             .
                       - - - - - - - - - - - - - - - - - - - - - - - - - - - - -

         JOB CONCLUDED

         TOTAL JOB TIME            9.67
```

Figure 14-22C PSpice EXP6.CIR plot

14.6 CircuitMaker

CircuitMaker is a DC/AC tutorial and analysis program. It contains many practice exercises based on 100 preprogrammed circuit types. It can also perform many of the analysis functions of PSpice. CircuitMaker is graphical, fun, easy-to-use, and inexpensive. It makes circuit analysis a type of game by keeping score and awarding points for correct answers. In addition to basic drills, many troubleshooting exercises are incorporated. CircuitMaker is available from:

Coastal Computer Company
1609 Country Club Drive
Rocky Mount, NC 27804
(919) 442-7436

tech tidbit (T^2)

CircuitMaker contains many troubleshooting exercises.

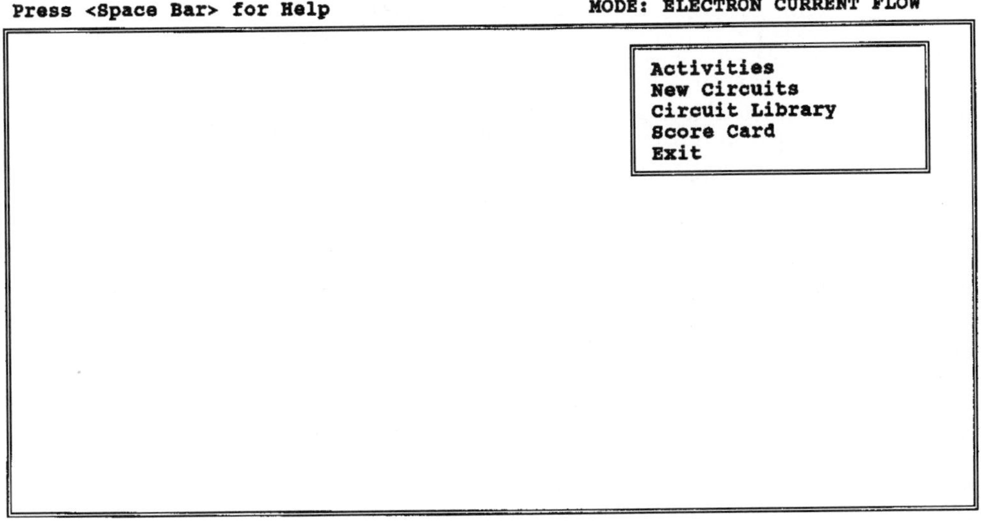

Figure 14-23 CircuitMaker main menu

```
Press <Space Bar> for Help                    MODE: ELECTRON CURRENT FLOW

HOW TO SELECT AN OPTION                       Activities
                                              New Circuits
This is the Main Menu.  No matter             Circuit Library
where you are in the program pressing         Score Card
<Esc> will return you to this point.          Exit

A list of options appears in the upper
right-hand corner.

To select an option press <↑> or <↓> to
move the cursor to that option.  Then
select that option by pressing <ENTER>.

Another way to select an option is to        <Space Bar>→Next page
press the first letter of its name.          <Pg Up>→Previous page
                                              <End>→Remove
     (Press <Space Bar> for more)
              1  of  6
```

Figure 14-24 CircuitMaker help screen

```
Activities              <Space Bar> For Help:<S> Score Card:<ESC> Main Menu

 ▸1-Resistor    Color Code -----------------------------------------------
  2-DC Series   Resistance Calculation ---------------------(3 Elements)
  3-DC Series   Resistance Calculation ---------------------(6 Elements)
  4-DC Series   Resistance Calculation ---------------------(9 Elements)
  5-DC Series   Resistance Calculation ---------------------(Random    )
  6-DC Series   Current    Calculation -------------------(3 Elements)
  7-DC Series   Voltage    Calculation -------------------(3 Elements)
  8-DC Series   Power      Calculation -------------------(3 Elements)
  9-DC Series   Unknown     Resistance -------------------(3 Elements)
 10-DC Series   Synthesis --------------------------------(3 Elements)
 11-DC Series   Troubleshooting Drill --------------------(3 Elements)
 12-DC Series   Current   Calculation   -------------------(6 Elements)
 13-DC Series   Voltage   Calculation   -------------------(6 Elements)
 14-DC Series   Power     Calculation   -------------------(6 Elements)
 15-DC Series   Unknown    Resistance   -------------------(6 Elements)
 16-DC Series   Synthesis ---------------------------------(6 Elements)
 17-DC Series   Troubleshooting Drill   -------------------(6 Elements)
 18-DC Series   Current Calculation     -----------------(9 Elements)
 19-DC Series   Voltage  Calculation    -----------------(9 Elements)
 20-DC Series   Power    Calculation    -----------------(9 Elements)

        Select activity - type # of problems (1-10) - Press <ENTER>
```

Figure 14-25 *CircuitMaker activities screen #1*

2 tech tidbit

CME command specifies electron current analysis.
CM command specifies conventional current analysis.

The entire program is contained on one disk. No installation is required. At the DOS prompt, A:, type the following:

A:CME <ENTER>

Follow the directions to enter your name, and so forth. You will only do this the first time you use the program. At the copyright screen, press any key to continue and the main menu will appear on the screen, as shown in Figure 14-23.

Pressing the space bar at any time will bring up the help screen, which gives you directions on how to continue, as shown in Figure 14-24.

Selecting the ACTIVITIES option from the main menu in Figure 14-23 will display a choice of 100 practice exercises, as shown in Figure 14-25. The up and down cursor control keys are used to scroll to the desired activity.

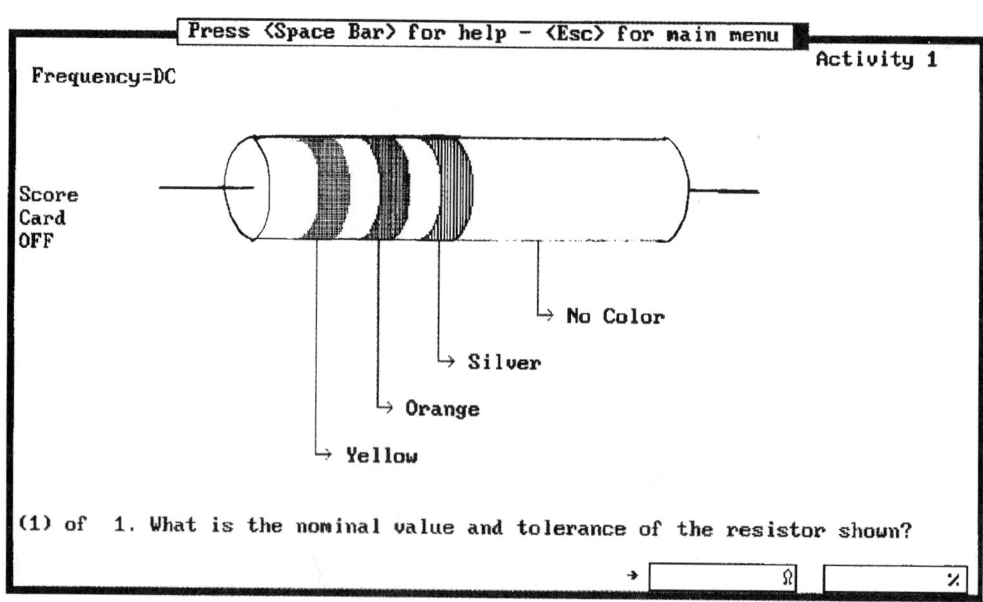

Figure 14-26 *Activity 1—code code screen*

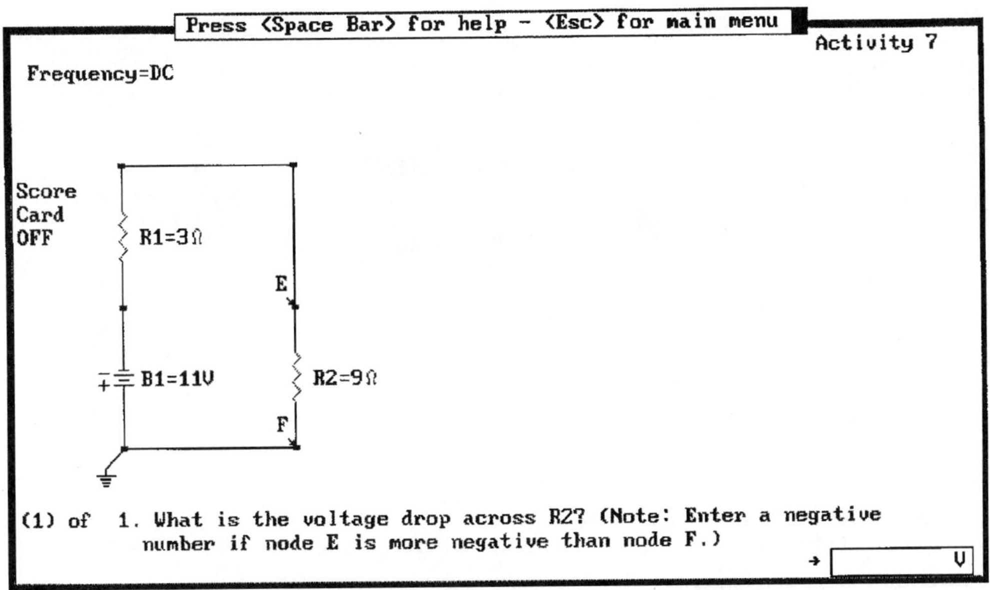

Figure 14-27 Activity 7—DC circuit example

From the ACTIVITY menu you can select color code problems, DC circuit analysis problems, AC circuit analysis problems, and troubleshooting problems. Each selected activity can be repeated with different circuit values up to a maximum of twenty times. For example, selecting Activity 1 will display the color code problem of Figure 14-26.

Selecting Activity 7 will display the DC circuit problem of Figure 14-27.

Selecting Activity 82 will display the AC circuit problem of Figure 14-28.

Selecting Activity 93 will display the troubleshooting example of Figure 14-29.

Selecting NEW CIRCUIT from the main menu allows you to draw your own circuit. The command ALT P will print your circuit. The command ALT O will

Selecting New Circuit allows you to draw, print, and analyze your own circuit diagrams.

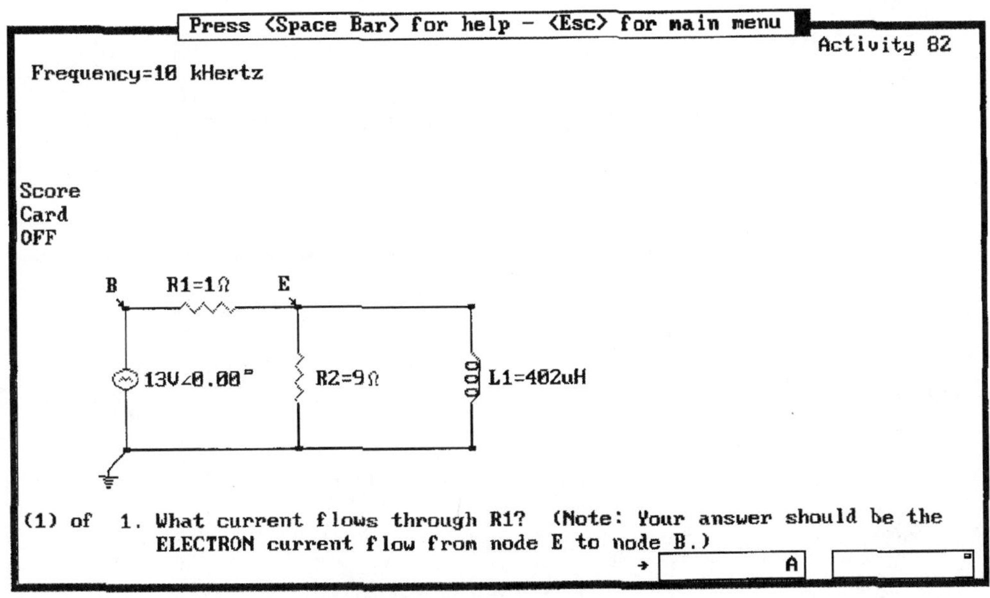

Figure 14-28 Activity 82—AC circuit example

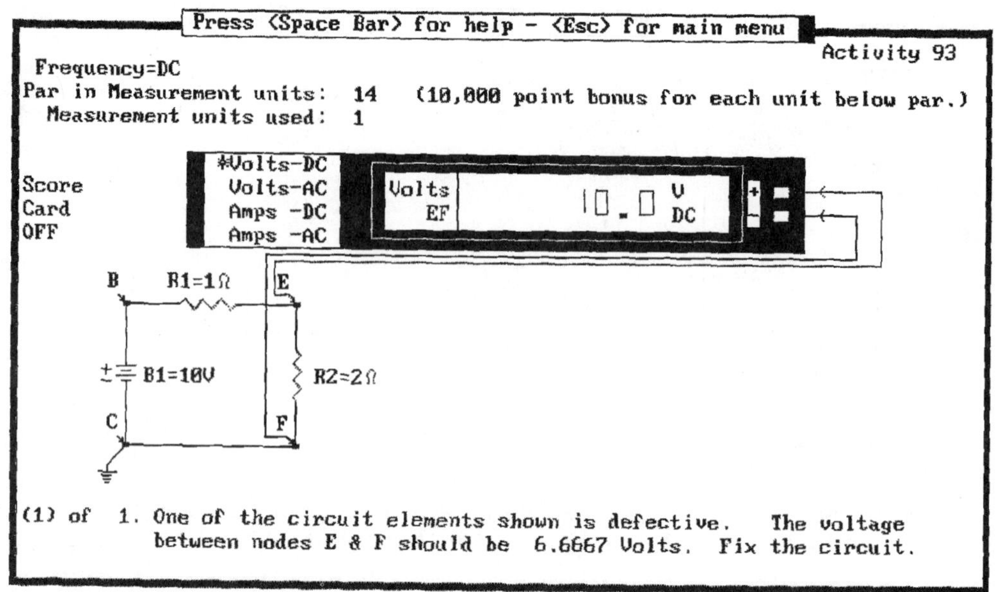

Figure 14-29 *Activity 93—Troubleshooting example*

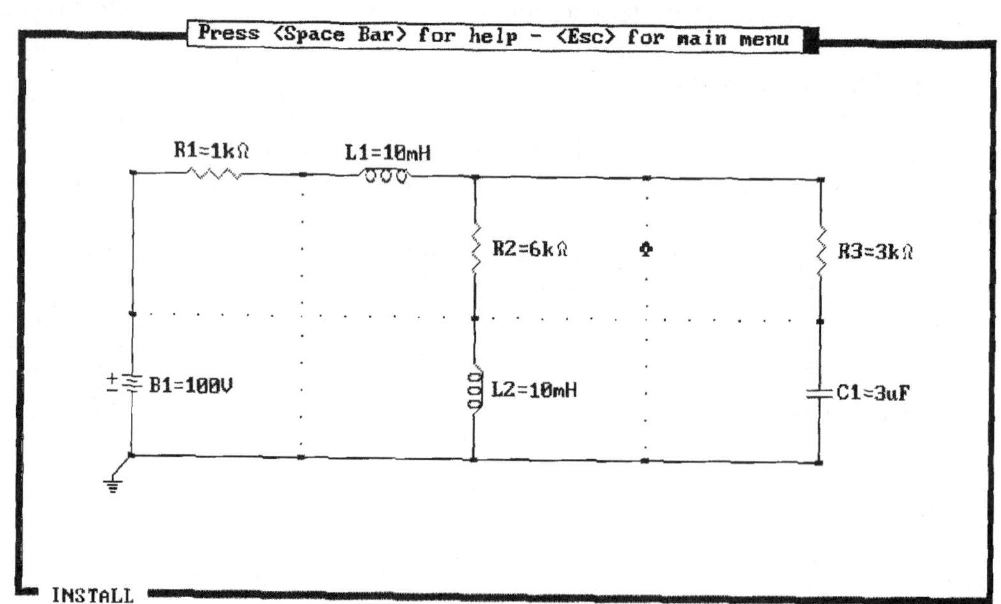

Figure 14-30A *New circuit design*

analyze the circuit and calculate all of the nodal voltages and currents, as shown in Figures 14-30A and 14-30B.

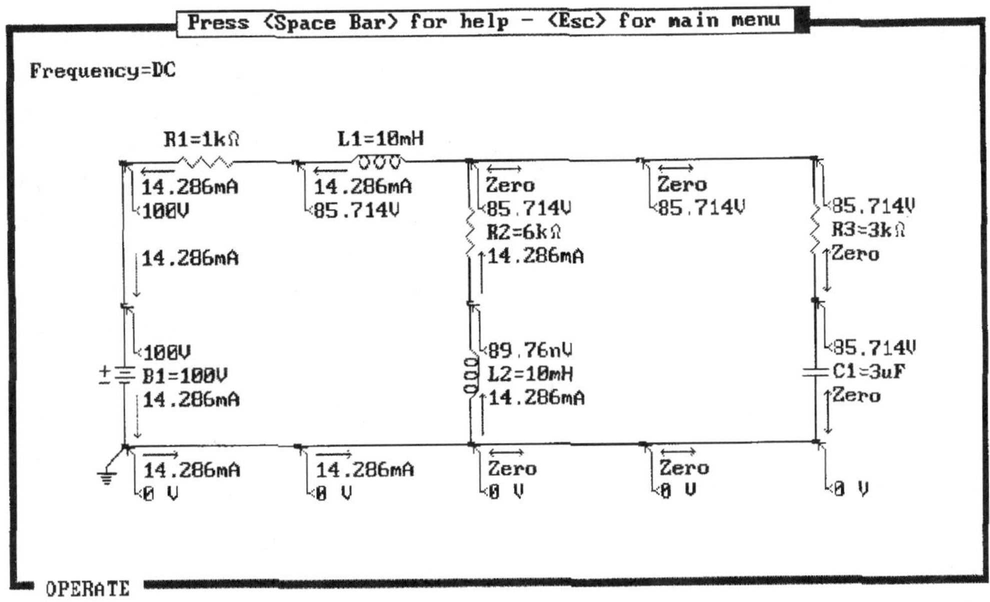

Figure 14-30B **New circuit analysis**

T³ SOFTWARE SETUP

Many versions of PSpice require a minimum of 512 KB of contiguous memory. For DOS versions 4.0 and earlier the operating system and device drivers use a portion of the 640 KB of available base memory. If you have a problem installing PSpice try booting your system without any device drivers. A device driver as small as MOUSE.COM, MOUSE.SYS, or ANSI.SYS could cause PSpice not to run. Terminate and Stay Resident (TSRs) programs should also be removed from the system. Hard disk users (running DOS 4.0 or earlier) should be aware of one other possible problem. Hard disk drives larger than 32 megabytes should be installed using the DOS SHARE.EXE program. There are at least three different ways to install SHARE.EXE. The preferred installation method for PSpice is to add the following lines to your AUTOEXEC.BAT and CONFIG.SYS files:

AUTOEXEC.BAT

SET COMSPEC = C:\DOS\COMMAND.COM

CONFIG.SYS

SHELL = C:\DOS\COMMAND.COM/P/E:256

Remember that the CONFIG.SYS file requires the following two lines:

FILES = 20

BUFFERS = 20

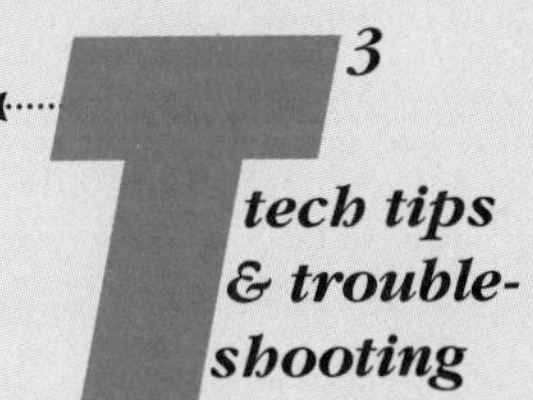

TSR programs once loaded take up memory.

SUMMARY

The industry standard computer circuit simulation program for circuit analysis is SPICE. There are many versions of SPICE. In this chapter, we discuss one of the more popular versions by MicroSim Corporation called PSpice. We also discuss a DC/AC tutorial and analysis program from Coastal Computer Company called CircuitMaker.

The use of the computer as a problem-solving tool has been growing in popularity. However, it is still necessary to know how to analyze electric circuits manually. It is the only way to develop confidence, depth, and understanding of electric theory. Computer simulation can be very handy and useful, especially when the circuit is complicated and requires lengthy mathematical solutions. Making a mistake and entering incorrect information will result in the wrong answer. Without a thorough understanding of circuit theory, you will not be able to sense that the computer has given you the wrong answer.

EXERCISES

Section 14.4

1. Use PSpice to solve for the nodal voltages in Figure 14-31.

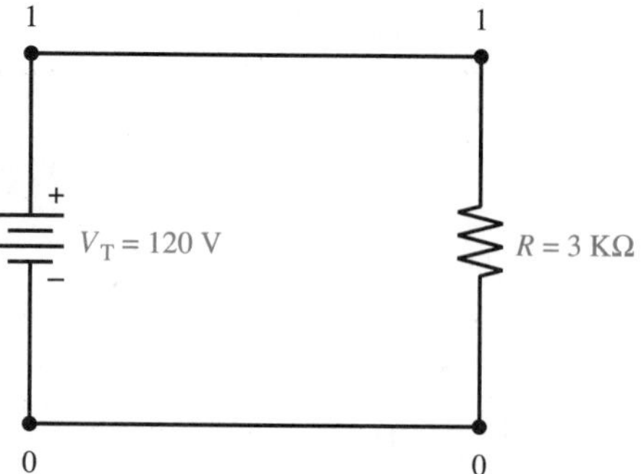

Figure 14-31

2. Use PSpice to solve for V(R1), V(R2), V(R3), and I(R1) in the circuit of Figure 14-32.

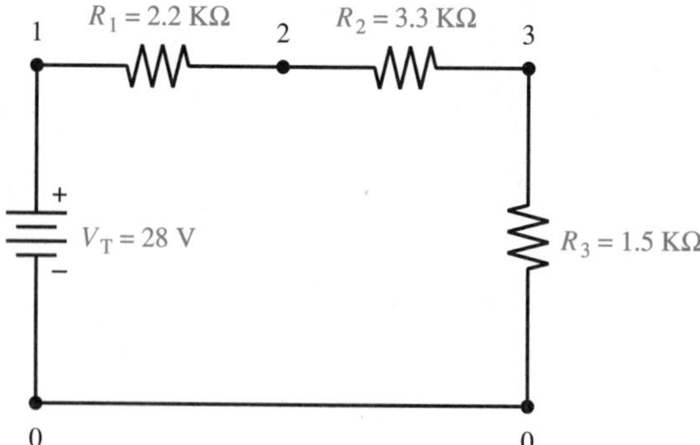

Figure 14-32

3. Repeat problem 2 for a supply voltage that sweeps from 1 to 28 volts in 1-volt steps.

4. Use PSpice to plot the results of problem 3.
5. Use PSpice to solve for I(R1), I(R2), and I(R3) in the circuit of Figure 14-33.

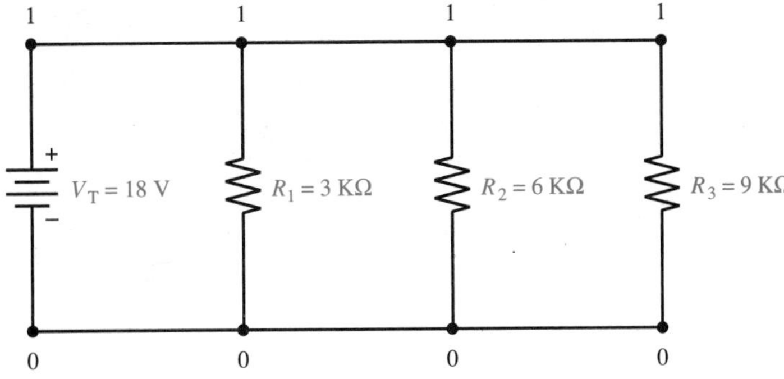

Figure 14-33

6. Repeat problem 5 for a supply voltage that sweeps from 1 to 18 volts in 1-volt steps.
7. Use PSpice to plot the results of problem 6.
8. Use PSpice to solve for the voltage and current through the load resistor, R_L, in Figure 14-34.

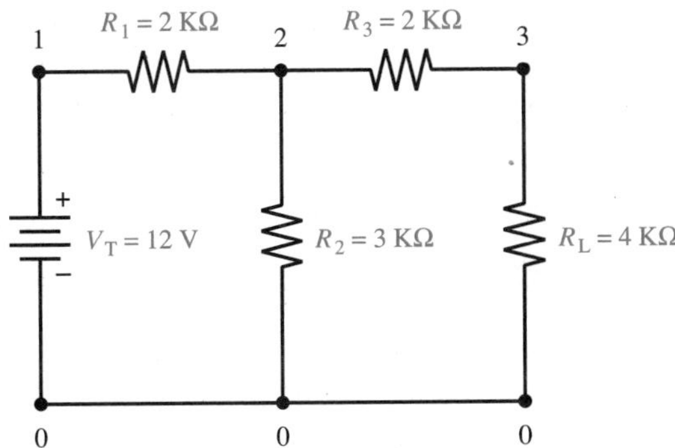

Figure 14-34

9. Repeat problem 8 for a supply voltage that sweeps from 1 to 12 volts in 1-volt steps.
10. Use PSpice to plot the results of problem 9.
11. Use PSpice to solve for the voltage across V(R1), V(R2), and V(R3) in the circuit of Figure 14-35.

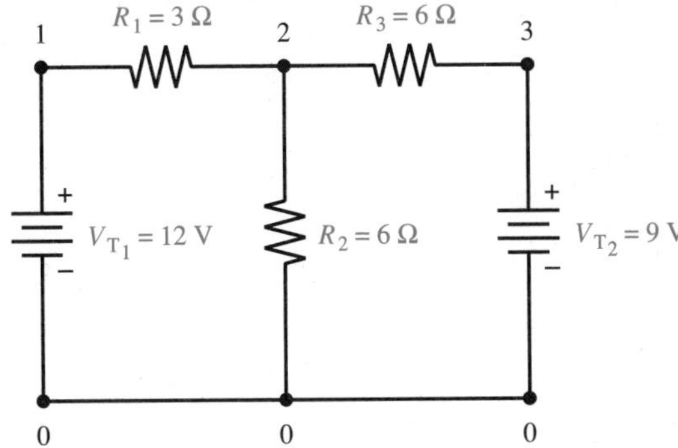

Figure 14-35

Section 14.5

12. Use PSpice to solve for the magnitude and phase of the voltage across the inductor in Figure 14-36.

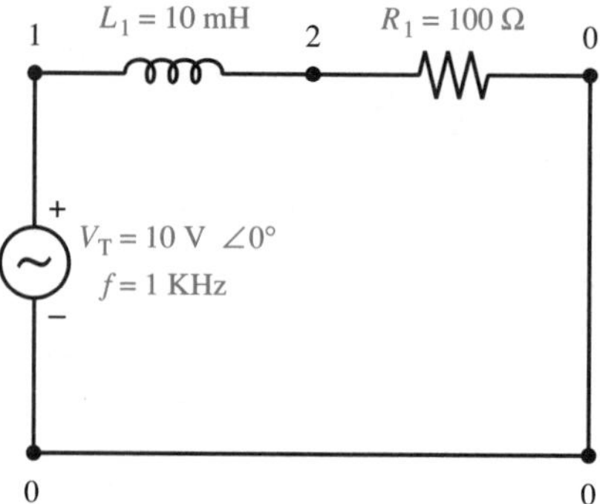

Figure 14-36

13. Use PSpice to solve for the magnitude and phase of the voltage across the capacitor in Figure 14-37.

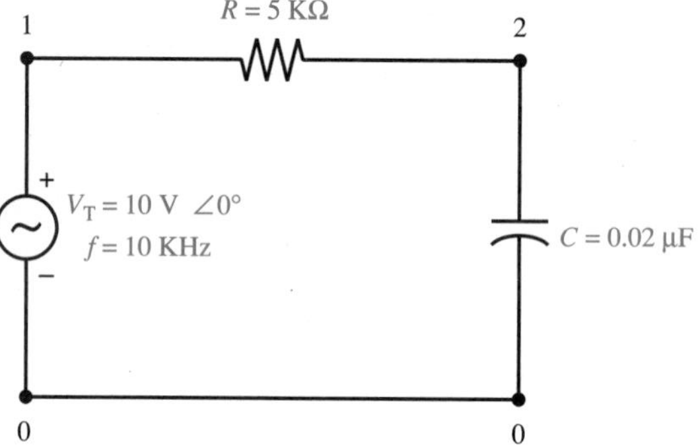

Figure 14-37

14. Use PSpice to solve for the magnitude and phase of the voltage across the capacitor in Figure 14-38.

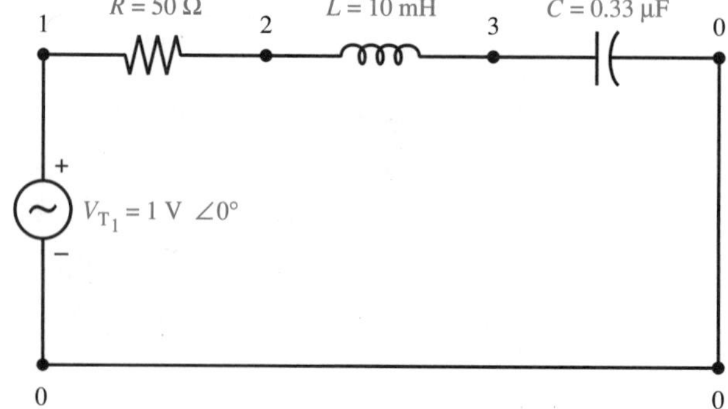

Figure 14-38

Section 14.6

15. Repeat problem 1 using CircuitMaker.
16. Repeat problem 2 using CircuitMaker.
17. Repeat problem 5 using CircuitMaker.
18. Repeat problem 8 using CircuitMaker.
19. Repeat problem 11 using CircuitMaker.
20. Repeat problem 12 using CircuitMaker.
21. Repeat problem 13 using CircuitMaker.
22. See Lab Manual for the crossword puzzle for Chapter 14.

Appendix A—Self-Test Answers

Chapter 1

Self-Test 1

1. amber rod
2. Leyden jar
3. voltaic cell

Self-Test 2

1. everyone's
2. 0.1
3. water
4. shut off the power, report the accident to the instructor

Self-Test 3

1. long-nose pliers, wire cutters, wire strippers, screwdrivers, nutdrivers
2. diagonal cutter
3. slotted, phillips, torx
4. overheat the component

Chapter 2

Self-Test 1

1. joule
2. atom
3. proton and neutron
4. negative
5. positive ion
6. insulators
7. 1/1,836

Self-Test 2

1. Lightning is caused by the movement of charge between the clouds and the earth's surface.
2. attract or move toward
3. couloumb
4. volt
5. 2 volts

Self-Test 3

1. current
2. 1 ampere
3. 40 amperes
4. resistance
5. siemens or mhos

Self-Test 4

1. rate of doing work
2. watts
3. TVs use more power.
4. 1,000
5. 746 watts
6. repel

Chapter 3

Self-Test 1

1. 10^2
2. 10^7
3. 10^{-2}
4. 10^{-5}
5. 10^0

Self-Test 2

1. 8.9×10^3
2. 4.7×10^4
3. 3.3×10^{-3}
4. 8.1×10^{-5}

Self-Test 3

1. (A) 2.4×10^4
   (B) $24 \times 10^3 = 24$ K
2. (A) 6.8×10^7
   (B) $68 \times 10^6 = 68$ M
3. (A) 4.4×10^{-2}
   (B) $44 \times 10^{-3} = 44$ m
4. (A) 1.5×10^{-5}
   (B) $15 \times 10^{-6} = 15$ μ

Self-Test 4

1. $30 \times 10^6 = 3.0 \times 107$
2. $7.7 \times 10^0 = 7.7$
3. $30 \times 10^3 = 3.0 \times 104$
4. 3×10^6

Self–Test 5

1. 10^{-9}
2. 16×10^6
3. 1,200 K
4. 500 μ

Chapter 4

Self-Test 1

1. energy source
2. charge
3. electron current
4. (A)
   (B) + −
   (C)

Self-Test 2

1. transducers
2. chemical, mechanical, light, heat, and pressure
3. battery
4. nickel-cadmium (Ni-Cad)
5. solar cells

Self-Test 3

1. ohm
2. watts
3. potentiometer
4. brown, red, orange, black
5. 1.588 ohms

Self-Test 4

1. coil or inductor
2. dielectric
3. SPDT
4. greater
5. (A) + V −
   (B)
   (C)
   (D)

Chapter 5

Self-Test 1

1. voltage
2. closed
3. 6 A
4. 200 Ω

Self-Test 2

1. sum
2. same
3. KVL
4. voltage divider rule
5. open circuit

Self-Test 3

1. same
2. KCL
3. $R_T = \dfrac{R_1 R_2}{R_1 + R_2}$
4. zero

Self-Test 4

1. sum
2. watts
3. $P = IV = I^2R = \dfrac{V^2}{R}$
4. 1 watt

Chapter 6

Self-Test 1

1. series-parallel
2. Thevenin's, Norton's, and Superposition
3. parallel branch

Self-Test 2

1. (A) $R_T = 5\ K\Omega$ (B) $I_T = 20$ mA
   (C) $I_2 = 10$ mA (D) $I_3 = 10$ mA
2. (A) $R_T = 3\ K\Omega$ (B) $I_T = 4$ mA
   (C) $I_2 = 1$ mA (D) $I_3 = 3$ A

Self-Test 3

1. constant voltage
2. constant current
3. $I_S = 24$ A, $R_S = 0.5\ \Omega$
4. $V_S = 50$ V, $R_S = 10\ M\Omega$

Self-Test 4

1. Thevenin's
2. Norton's
3. $R_L = 6\ K\Omega$
4. $P_{max} = 0.024$ W = 24 mW

Chapter 7

Self-Test 1

1. source conversion, superposition, loop, and nodal analysis
2. added
3. subtracted
4. 7 A

Self-Test 2

1. parallel
2. superposition
3. loop
4. nodal

Chapter 8

Self-Test 1

1. natural
2. artificial magnet
3. electromagnet
4. magnetic field
5. flux

Self-Test 2

1. Hans Christian Oersted
2. greater
3. Michael Faraday
4. rate of change
5. EMF

Chapter 9

Self-Test 1

1. pulsating DC
2. AC (alternating current)
3. 360°

Self-Test 2

1. peak
2. peak-to-peak
3. period
4. hertz

Self-Test 3

1. $V_{avg} = 0.637\ V_p$
2. $\text{avg} = \dfrac{\Sigma \text{ areas}}{\text{period}}$
3. 169.7 V_p
4. 155.5 V_{eff}

Self-Test 4

1. radians
2. square wave, rectangular wave, sawtooth wave, sine wave
3. (A) can be produced easier and more efficiently
   (B) can be distributed over long distances more economically
4. $P_{\text{supplied}} = P_{\text{generated}} - P_{\text{lost}}$

Chapter 10

Self-Test 1

1. greater
2. four
3. counter EMF
4. Lenz's law

Self-Test 2

1. $L_T = 50$ H
2. $L_T = 2$ H
3. current
4. inductance, resistance

Self-Test 3

1. ohm
2. $X_L = 12{,}650\ \Omega$
3. resistance
4. Q or quality

Self-Test 4

1. voltage
2. operating voltage, operating frequency, and power
3. primary, secondary
4. ohm
5. smaller

Self-Test 5

1. $P_{\text{loss}} = I^2R$
2. tapped
3. wiper
4. autotransformer
5. Eddy currents and hysteresis

Chapter 11

Self-Test 1

1. capacitance
2. capacitor
3. greater
4. less

Self-Test 2

1. increasing
2. $\frac{1}{C_T} = \frac{1}{C_1} + \frac{1}{C_2} + \frac{1}{C_3} + \ldots + \frac{1}{C_N}$
3. $C_T = C_1 + C_2 + C_3 + \ldots + C_N$
4. 63.2%
5. 5τ

Self-Test 3

1. equal
2. the sum of the charges
3. capacitive reactance
4. decrease

Chapter 12

Self-Test 1

1. short
2. open
3. magnitude, direction
4. in phase
5. lead
6. lead

Self-Test 2

1. impedance
2. $Z_T = 5\ \Omega$
3. $I_T = 20$ A
4. $\theta_T = 53.13°$

Self-Test 3

1. $Z_T = 13\ \Omega$
2. $I_T = 4$ A
3. $\theta_T = -67.4°$
4. $V_T = 48$ V

Self-Test 4

1. $Z_T = 2.5\ K\Omega$
2. $I_T = 40$ mA
3. current
4. voltage

Self-Test 5

1. resonance
2. reactance, resistance
3. narrower
4. low pass, high pass, band pass, band stop

Chapter 13

Self-Test 1

1. rate of doing work
2. real power
3. cos θ
4. $P = 100$ W

Self-Test 2

1. reactive power
2. real power, reactive power, and apparent power
3. the power factor
4. one

Self-Test 3

1. three-phase
2. wye, delta
3. greater
4. equals

Self-Test 4

1. rotor (armature), stator
2. brushes, commutator
3. horsepower
4. efficiency
5. mechanical, core, copper

Self-Test 5

1. induction
2. directly, inversely
3. slip

Chapter 14

Self-Test 1

1. simulators
2. creating an input file, running the simulation and printing the output f i l e results
3. .CIR, .OUT
4. E

Appendix B— Answers to Selected Exercises

Chapter 1

1. B
2. A
3. B
4. D
5. C
6. A
7. A
8. C
9. C
10. D

Chapter 2

1. 100 joules
3. 6 volts
5. 300 coulombs
7. 15 amperes
9. 5 milliamperes
11. 840 couloumbs
13. 0.01 siemens
15. 6.67 ohms
17. 2,400 watts
19. 15 seconds
21. 24 cents
23. 52 cents
25. 1658 watts

Chapter 3

1. (A) 10^2 (B) 10^4 (C) 10^{-2} (D) 10^{-4}
3. (A) 1,000 (B) 100,000 (C) 0.1 (D) 0.00001
5. (A) 5.6×10^4 (B) 9.1×10^7 (C) 9×10^{-6} (D) 3.9×10^{-5}
7. (A) 760 (B) 2,200 (C) 0.038 (D) 0.00046
9. (A) 1.2 K (B) 62 M (C) 15 m (D) 33 μ
11. (A) 15 K (B) 0.68 M or 680 K
    (C) 23 m (D) 560 μ or 0.56 m

13. (A) 8.0×10^5 (B) 9.66×10^6 (C) 2.87×10^8 (D) 1.824×10^{10}
15. (A) 3.0×10^1 (B) 2.5×10^4 (C) 4.0×10^7 (D) 2.0×10^{-1}
17. (A) 1×10^6 or 10^6 (B) 1×10^{-8} or 10^{-8}
    (C) 1×10^0 or 1.0 (D) 1×10^{30} or 10^{30}
19. (A) 5,000 μ (B) 0.2 M (C) 6,500 K (D) 0.5 m

Chapter 4

1. Energy source, load and conductor
3. Something that uses and converts energy.
5. Wood, rubber, and glass
7. Secondary cells can be recharged but primary cells cannot be recharged.
9. (A) 1.5 volts (B) 3 volts (C) 0 volts
11. 10 ampere-hours
13. 0.67 hours
15. Chemical, mechanical, light, temperature, pressure
17. Ratio of carbon to binding material
19. Very stable and rugged
21. (A) red, red, red
    (B) orange, white, red
    (C) orange, white, orange
    (D) brown, green, yellow
    (E) brown, green, green
23. 3.1925 ohms
25. 500 feet

Chapter 5

1. 4 A
3. 26.596 mA
5. 6.3 V
7. 80 Ω
9. 22 Ω
11. 10 KΩ
13. 23.53 mA
15. 4 mA
17. 80 Ω
19. 51 V
21. $R_T = 500\ \Omega$, $I_T = 0.4$ A, $V_1 = 60$ V, $V_2 = 140$ V
23. (A) $V_{AB} = -12$ V (B) $V_{AB} = -7.5$ V

25. $I_T = 1$ A

27. $V_1 = 3$ V, $V_2 = 6$ V

29. $I_T = 0$ A, $V_1 = 0$ V, $V_2 = 0$ V

31. $I_4 = 10$ A

33. $I_4 = 12$ A

35. $I_8 = 0$ A

37. $I_1 = 2$ mA, $I_2 = 1.5$ mA, $I_3 = 0.5$ mA, $I_T = 4$ mA, $R_T = 3\ \Omega$

39. $R_T = 2.66\ \Omega$, $G_T = 0.375$ S

41. $R_2 = 12\ \Omega$

43. $I_1 = 2$ A, $I_2 = 4$ A, $I_3 = 6$ A

45. $I_T = \infty$ A

47. (A) $P_T = 500$ W (B) $P_T = 500$ W

49. $P_1 = 40$ W, $P_2 = 80$ W, $P_3 = 120$ W

51. $V_T = 6$ V, $I_2 = 1$ A, $R_3 = 1\ \Omega$, $I_3 = 6$ A, $I_T = 9$ A

Chapter 6

1. (A) series (B) parallel (C) series

3. $R_T = 10\ \Omega$, $I_T = 1$ A, $V_1 = 4$ V, $V_2 = 6$ V, $V_3 = 6$ V

5. $R_T = 5\ \Omega$, $I_T = 2$ A, $I_2 = 0.667$ A, $I_4 = 1.333$ A

7. $R_T = 6$ KΩ, $I_T = 20$ mA, $I_2 = 10$ mA, $I_4 = 10$ mA

9. $R_T = 3$ KΩ, $I_T = 4$ mA, $I_1 = 2$ mA, $I_4 = 1$ mA, $V_{BD} = 8$ V, $V_{CD} = 9$ V

11. (A) $R_T = 1{,}000\ \Omega$ (B) $R_T = 8$ KΩ

13. (A) $V_S = 500$ V, $R_S = 50\ \Omega$ (B) $V_S = 25$ V, $R_S = 5$ KΩ

15. $V_L = 10.667$ V

17. $I_4 = 1$ A

19. $I_N = 4$ A, $R_N = 4\ \Omega$

21. $I_N = 1.35$ A, $R_N = 11.43\ \Omega$

23. $P_L = 14.22$ W

Chapter 7

1. (A) $V_{AB} = 12$ V (B) $V_{AB} = 10.5$ V

3. $I_T = 0.5$ A

5. $V_{AB} = 54$ V

7. $V_{AB} = 60$ V

9. $I_3 = 2$ A

11. $V_{AB} = 8$ V

13. $V_{AB} = 11$ V

15. $I_A = 2$ A, $I_B = 0$ A
17. $35\ I_A + 15\ I_B = 25$

    $15\ I_A + 20\ I_B = 35$
19. Node V_A: $I_A = I_1 + I_3$

    Node V_B: $I_B + I_3 = I_2$

Chapter 8

1. lodestone
3. relay, speakers, motors, compass
5. A permanent magnet retains its magnetic properties over a long period of time while a temporary magnet loses its magnetic properties in a short time.
7. magnetic field
9. north to south
11. The alignment of atoms caused by the spinning of electrons in the same direction.
13. permeability
15. reluctance
17. It intensifies or increases the magnetic field.
19. Lefthand Rule
21. The induced voltage in a conductor is directly proportional to the rate at which the conductor cuts the magnetic lines of force.
23. (A) generator (B) motor (C) speaker
25. By placing it in an alternating magnetic field.

Chapter 9

1. (A) $V_p = 25$ V, $V_{p-p} = 50$ V

   (B) $I_p = 10$ mA, $I_{p-p} = 20$ mA

   (C) $V_p = 311$ V, $V_{p-p} = 622$ V
3. (A) 0.25 Hz

   (B) 8.33 Hz

   (C) 50,000 Hz or 50 KHz
5. $V_p = 156.9\ V_{peak}$, $V_{p-p} = 313.9\ V_{peak\text{-}to\text{-}peak}$
7. $V_{avg} = 31.85$ V
9. $I_{avg} = 28.5$ mA
11. $V_{p-p} = 588.2\ V_{peak\text{-}to\text{-}peak}$
13. $I_{RMS} = 98.98$ mA
15. $I_{RMS} = 7.07$ mA

17. (A) V_A leads V_B by 90° or V_B lags V_A by 90°
    (B) V_A leads V_C by 240° or V_C lags V_A by 240°
    (C) V_B leads V_C by 150° or V_C lags V_B by 150°
19. (A) $\pi/4$ rad (B) $0.444\ \pi$ rad (C) $0.75\ \pi$ rad

Chapter 10

1. $C = 4.4275$ pF
3. $C_{new} = 10\ \mu F$
5. $C = 8\ \mu F$
7. $Q = 72\ \mu C$
9. $C_T = 2\ \mu F$
11. $C_T = 3\ \mu F$
13. $C_T = 600\ \mu F$
15. $C_T = 8\ \mu F$
17. 750 ms
19. $V_C = 72$ V
21. $W_C = 216\ \mu J$
23. (A) $C_T = 1\ \mu F$ (B) $Q = 60\ \mu C$
    (C) $V_1 = 10$ V, $V_2 = 20$ V, $V_3 = 30$ V
25. $V_C = 50$ V, $I_T = 8.33$ A
27. $X_{C_1} = 15\ \Omega$, $X_{C_2} = 7.5\ \Omega$, $X_{C_3} = 3.75\ \Omega$

Chapter 11

1. $V_L = 15{,}000$ V = 1.5 KV
3. $L = 20$ H
5. $L_T = 25$ mH
7. $L_T = 2.5$ H
9. $L_T = 2$ H
11. $L_T = 12$ H
13. $\tau = 8$ ms
15. Steady state = 0.5 s, $I_{SS} = 1.5$ A, $W_L = 11.25$ J
17. $\tau = 2$ ms, $I = 1.26$ A
19. (A) $X_L = 3.77\ \Omega$ (B) $X_L = 3.77\ K\Omega$
21. $Q = 251.2\ \Omega$
23. $L = 0.239$ mH, $R_L = 3\ \Omega$
25. $V_P = 110$ V
27. $I_P = 40$ A
29. $Z_P = 1{,}600\ \Omega$

31. V_{out} = 36 V
33. V_{out} = 200 V
35. efficiency = 90%
37. turns ratio = 110:1

Chapter 12

1. I_T = 0.5 A, V_1 = 25 V, V_2 = 0 V, V_3 = 75 V
3. I_T = 20 μA, V_1 = 60 V, V_2 = 40 V, V_3 = 0 V, V_4 = 0 V, V_5 = 20 V
5. I_T = 0.5 A
7. Z_T = 10 KΩ, I_T = 12 mA, V_{2K} = 24 V, V_{3K} = 36 V, V_{5K} = 60 V
9. Z_T = 1,265 Ω, θ_T = 71.57°, I_T = 94.86 mA, V_R = 37.94 V, V_L = 113.83 V
11. I_R = 4 A, I_L = 2 A, I_T = 4.47 A, Z_T = 22.37 Ω, θ_T = 26.5°
13. Z_T = 12.65 KΩ, θ_T = –18.45°, I_T = 3.79 mA, V_R = 45.48 V, V_C = 15.16 V
15. I_R = 4 mA, I_C = 3 mA, I_T = 5 mA, Z_T = 2.4 KΩ, θ_T = 36.87°
17. Z_T = 130 Ω
19. Z_T = 110 Ω, I_T = 1.09 A, V_R = 72 V, V_L = 130.8 V, V_C = 34.88 V, θ_T = 53.13°
21. I_R = 6 A, I_L = 3 A, I_C = 2 A, I_T = 6.08 A, Z_T = 49.32 Ω, θ_T = –9.3°
23. f_0 = 10,279 Hz
25. Z_T = 50 Ω, I_T = 2 A, V_R = 100 V, V_L = 2,000 V, V_C = 2,000 V
27. Q = 100, BW = 100 Hz
29. For the RC circuit: f_{cutoff} = 159.24 Hz
    For the RL circuit: f_{cutoff} = 250.37 Hz

Chapter 13

1. P = 40 W
3. P = 90 W
5. P = 2 W
7. P_R = 20 W, P_T = 20 W
9. P_R = 270.4 W, P_T = 270.4 W
11. P_R = 135 W, P_Q = 324 VARS (leading)
13. P_R = 1,000 W, P_Q = 1,000 W (leading)
15. P_R = 187.5 W, P_Q = 281.3 VARS (lagging), P_A = 338 VA
17. P_F = 0.947
19. P_F = 0.385
21. (A) P_F = 0.88 (B) P_F = 0.975
23. V_ϕ = 480 V, I_ϕ = 48 A
25. I_ϕ = 20.8 A
27. EFF = 77.7%
29. T = 8.75 lb-ft

31. V_A = 117.16 V, T = 8.76 lb-ft, EFF = 82.9%, P_f = 96 W, P_a = 40.33 W, P_R = 171.67 W

33. N_S = 1,800 rpm

35. P = 2 poles

37. slip = 120 rpm, % slip = 3.33%

Chapter 14

1. EXP14-31.CIR

```
VT       1     0     120V
R        1     0     3K
.OPTIONS NOPAGE
.END
```

3. EXP14-32M.CIR

```
VT       1     0       28V
R1       1     2       2.2K
R2       2     3       3.3K
R3       3     0       1.5K
.DC      VT    1       28      28
.PRINT   DC    V(R1)   V(R2)   V(R3)   I(R1)
.OPTIONS NOPAGE
.END
```

5. EXP14-33.CIR

```
VT       1     0       18V
R1       1     0       3K
R2       1     0       6K
R3       1     0       9K
.DC      VT    18      18      1
.PRINT   DC    I(R1)   I(R2)   I(R3)
.OPTIONS NOPAGE
.END
```

7. EXP14-33P.CIR

```
VT       1     0       18V
R1       1     0       3K
R2       1     0       6K
R3       1     0       9K
.DC      VT    1       18      1
.PLOT    DC    I(R1)   I(R2)   I(R3)
.OPTIONS NOPAGE
.END
```

9. EXP14-34M.CIR

```
VT      1      0      12V
R1      1      2      2K
R2      2      0      3K
R3      2      3      2K
RL      3      0      4K
.DC     VT     1      12     12
.PRINT  DC     V(RL)  I(RL)
.OPTIONS NOPAGE
.END
```

11. EXP14-35.CIR

```
VT1     1      0      12V
VT2     3      0      9V
R1      1      2      3
R2      2      0      6
R3      2      3      6
.DC     VT1    12     12     1
.PRINT  DC     V(R1)  V(R2)  V(R3)
.OPTIONS NOPAGE
.END
```

13. EXP14-37.CIR

```
VT      1      0      AC     10V    0
R       1      2      5K
C       2      0      0.02UF
.AC     LIN    1      10KH   10KH
.PRINT  AC     VM(C)  VP(C)
.OPTIONS NOPAGE
.END
```

Appendix C—Soldering

Solder Theory

The importance of good soldering habits cannot be overemphasized. All to frequently, the causes of electric failures are traced to poor solder connections. Some believe that soldering is an art. Actually it is a skill that takes practice and patience to master. Like learning to drive a car, the best method of learning to solder is to solder. Remember, poor solder connections are costly. It takes no longer to make a good solder connection than it does to make a poor solder connection. Perfect your soldering skills and you will insure your success in assembling and repairing electric circuits.

Soldering is the bonding together of metal parts with a tin-lead alloy (**solder**). In a good solder connection (**joint**), solder molecules actually mix with the molecules of the metal being soldered. This is called **wetting action**. Thus, when melted solder comes in contact with other metals, the molecules of the solder and the other metals actually mix together to form a bond. It should be noted that not all metals can be soldered. Not all metals have free electrons in their outer orbits that will accept wetting or bonding to a tin-lead alloy. Some metals like aluminum and stainless steel can only be soldered using other types of alloys (solders).

An **alloy** is a mixture of two or more different metals. Conventional solder is an alloy of tin and lead. The ratio of tin to lead determines the actual melting temperature of the alloy, as shown in Figure C-1.

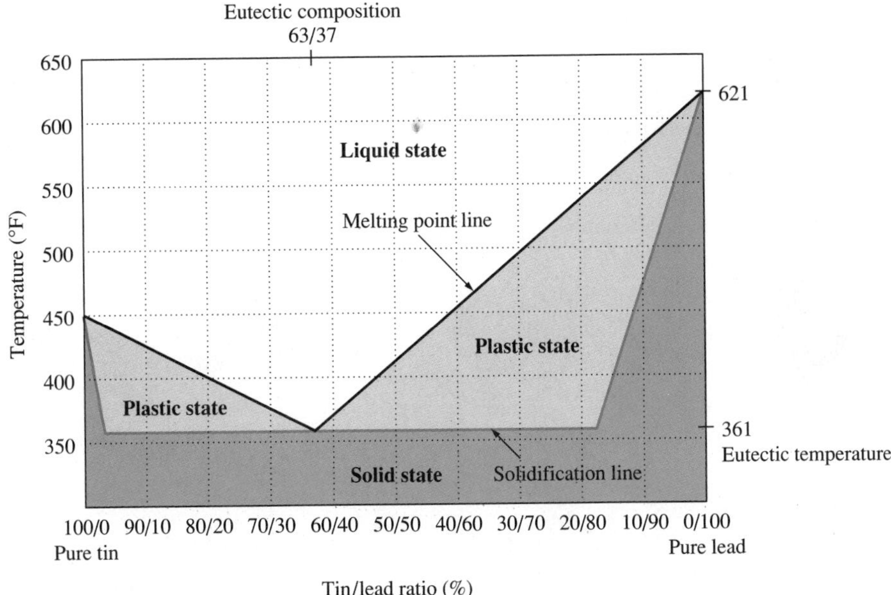

Figure C-1 **Melting point of tin/lead solders**

It is interesting to note that the melting temperature of a tin-lead alloy is less than the melting point of pure tin or pure lead. For example, referring to Figure C-1, we can see that pure tin melts at 450°F and that pure lead melts at 621°F. Solder that is 63% tin and 37% lead results in the lowest melting temperature. This ratio for tin and lead (63/37) is known as **eutectic** solder. Eutectic simply means the lowest possible melting temperature of an alloy. Furthermore, eutectic solder transitions from a solid state to a liquid state instantaneously. It never enters the plastic or pasty transition state. This means there is less need to worry about keeping the connection perfectly still while the solder cools.

Solder is often classified by its tin-lead ratio. For example, 63/37. The first number always represents the percentage of tin in the alloy. The second number always represents the percentage of lead. Although eutectic solder is the best solder because of its lowest melting temperature (361°F) and its ability to transition instantaneously from a solid to a liquid or from a liquid to a solid, 60/40 solder is most commonly available. It is far less expensive than eutectic solder and very close in characteristics. For example, 60/40 solder melts at 370°F and is in the plastic and the pasty state for a very short time.

The general procedure for soldering is to heat the connection to be soldered until its temperature is above the melting point of the solder. Next you apply the solder and allow it to flow over and *wet* the connection. When the heat is removed, the solder hardens. However, oxidation on the surfaces of the metal being soldered can cause an insulating layer to form on the surface that prevents the solder from wetting and adhering to the metal. When exposed to air most metals will oxidize. This results in the solder not sticking to the metal (no wetting). The solder beads up on the connection like a water bead on a polished car. In fact, dirt, grease, and other contaminants will prevent proper bonding, so the surfaces to be soldered must be clean.

To assist in the cleaning of oxides, solder manufacturers add a cleaning substance called **flux.** Flux comes as a powder, paste, or liquid, which can be brushed on to the connection. When flux is heated, it cleans away oxides permitting the solder to bond properly. Figure C-2 illustrates a cross-sectional view of commercial solder with a rosin core flux. Remember that flux takes care of the oxide only. You must still clean away dirt, grease, or other contaminants yourself.

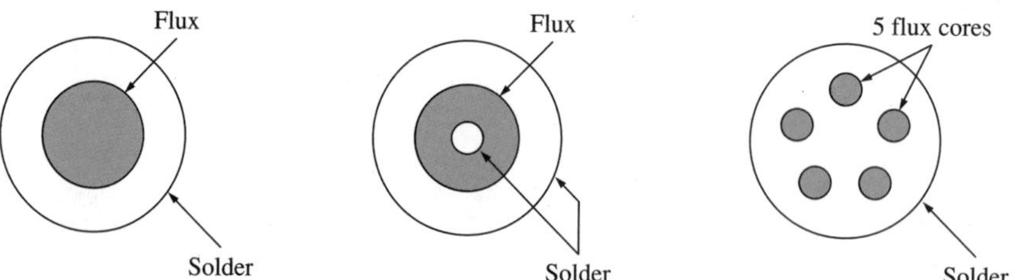

Figure C-2 **Flux core solder**

Cleanliness is the key to reliable soldering. Remember this rule, "When in doubt, reclean it!" Presolder and postsolder cleaning is evident in all *better* soldering. A point too frequently overlooked is the importance of cleanliness. A contact or wire contaminated with dirt, oil, grease, and so forth, cannot be soldered with any degree of success. This problem of contamination is too often left for the flux to correct. All parts should be cleaned with an approved solvent before soldering. The entire area to be soldered must be free of foreign particles or impurities such as oil and fingerprints. The soldering function should begin immediately after the parts have been cleaned.

Postsolder cleaning is also extremely important in order to remove all traces of flux residue. Flux residue is tacky or sticky, so foreign metallic particles can get stuck to the flux residue and produce a short by bridging two connections. Flux residue also attracts dust and dirt, which is somewhat conductive and can cause a high resistance short. Flux is not a substitute for the precleaning operation, because it only cleans oxidation. Generally, flux serves no value where precleaning of the area was bypassed.

Soldering Safety

The workbench and surrounding workspace should be maintained in a clean and orderly fashion. Enough light, properly directed, is a prime requirement for quality work. Adjust lights to illuminate the work. A neat and orderly work area keeps confusion to a minimum. Arrange tools within easy reach and have a specific place for each tool. Delays caused by misplaced or inaccessible tools can lead to poor workmanship and excessive rework.

The safety of a work area depends essentially on the development and maintenance of safety practices by everyone in the area. A general list of these practices appears here. Other practices for special areas should also be developed and maintained.

1. Keep the soldering iron where you don't have to reach across or around it.
2. Keep the soldering iron in an upper corner of the work area when not in use.
3. Safety glasses should be worn during all soldering operations. Wear safety glasses when cutting wires, soldering, and using solvents or an air hose.
4. When unsoldering a wire or component, make certain there is no tension or spring to the wire or component that could cause the solder to splash.
5. Plug the soldering iron into a properly grounded three-prong receptacle and keep the plug free from strands of wire or metal.
6. When cutting wires, keep the open side of the cutter away from your body. Keep wire ends from flying out of your area.
7. Disconnect all power before working on a chassis, printed circuit board, and so forth.
8. Be certain the tips of screwdrivers are square and sharp. Only use screwdrivers to screw and unscrew screws.
9. When cleaning with an air hose or pressurized can of cleaning solvent, make certain the air stream is kept away from people. If possible, use a vacuum hose instead of an air hose for cleaning.
10. Be certain there are no open flames and there is no smoking in the work area when working with solvents.
11. Always wear appropriate clothing when working. Long hair should be tied back in a safe manner.

Soldering irons are the cause of burns and illness from inhalation of fumes. Insulated noncombustible soldering iron holders help to eliminate the fire hazard and the danger of burns from accidental contact. Soldering iron holders should be inclined so the weight of the iron prevents it from falling out.

Harmful quantities of fumes from soldering should not be allowed to accumulate. Exhaust fans should be used if a lot of soldering is being performed in an enclosed area. Eye and face protection should always be used where there is a danger of solder or flux splattering.

Soldering Tools

The soldering iron is the most common tool used to supply heat to the surface of metals being joined together by solder. Soldering irons come in a variety of shapes, sizes, and styles. Figure C-3 is a photograph of a typical soldering iron and soldering iron holder used for electric circuit work. Proper use of the soldering iron determines how easily and how well soldering can be performed.

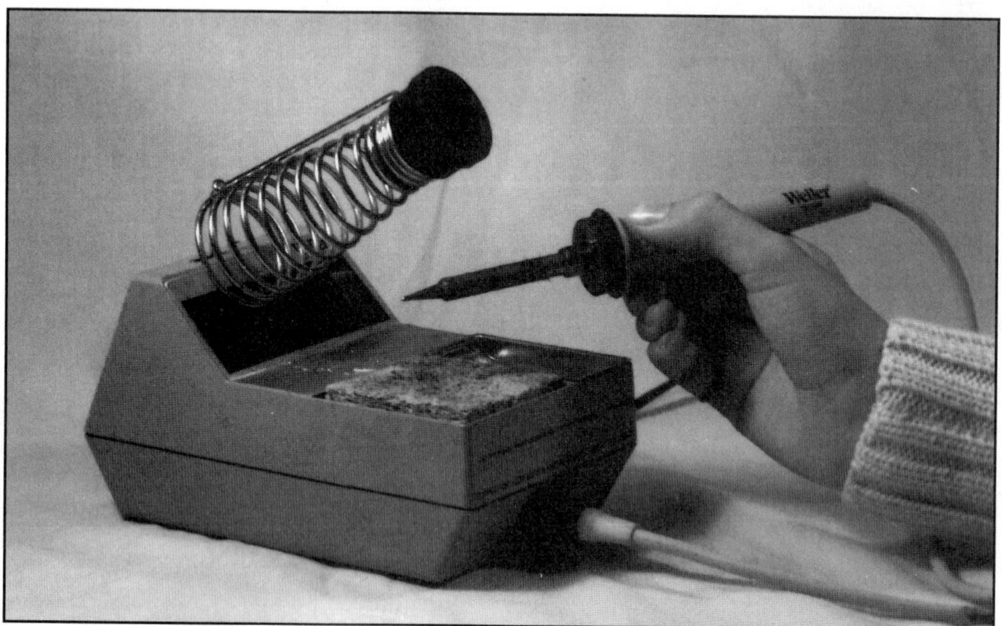

Figure C-3 **Soldering iron and holder**

Holding the soldering iron is up to the individual. The soldering iron should be held in the manner most comfortable to the user. Care should be taken to prevent the creation of dangerous situations in the area or adjacent area. Figure C-4 illustrates some optional methods of holding the soldering iron.

The size of the soldering iron is often rated by the size of the heating element in watts. Heavy duty soldering irons are generally 100 watts or greater.

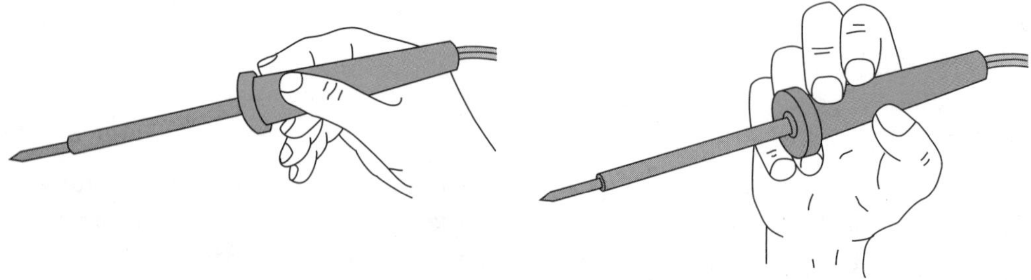

Figure C-4 **Optional methods of holding a soldering iron**

Medium duty soldering irons range between 45 and 75 watts. Light duty soldering irons are best suited for most electric work and printed circuit board work and range between 25 and 35 watts.

The tip is the part of the soldering iron that transfers the heat to the connection. The size, shape, cleanliness, and care of the tip is very important. Tips come in a variety of sizes and shapes for different soldering jobs, as shown in Figure C-5. Soldering iron tips are usually removable and interchangeable. When soldering, the tip should always be at least 100°F above the melting temperature of the solder. Generally, a minimum temperature of 650°F is recommended for printed circuit board work, while as much as 1,000°F may be required for soldering larger terminals. The larger the connection, the more heat required. Furthermore, the larger the connection, the longer it will take to raise the temperature of the connection to the melting point of the solder. Many soldering irons have digital readouts that monitor the temperature of the tip. Thus selecting the right size soldering iron and the shape and size of the tip is important for good soldering results. Generally, for electric circuit work, a 25- to 35-watt soldering iron with a pencil-shaped tip will perform most work adequately.

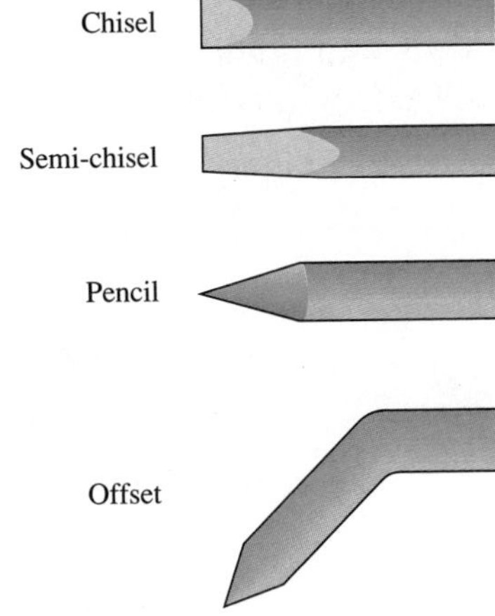

Figure C-5 **Soldering tips**

The long-nose pliers is one of the most important soldering tools. It is used for the holding and connection of wire and components where accessibility is difficult. Never use long-nose pliers for loosening or tightening nuts, screws, terminals or removing solder from an iron tip. The correct way to hold long-nose pliers is shown in Figure C-6.

Figure C-6 **Holding long-nose pliers**

The diagonal-cutter pliers is mainly used for cutting wire. Never use diagonal-cutter pliers to cut bolts, screws, or wire of such a large size that the cutting edges might be dulled or damaged. The correct way to hold a diagonal-cutter pliers is shown in Figure C-7.

Figure C-7 **Holding diagonal-cutter pliers**

A fiberglass solder probe is an aid for inspecting, probing, unsoldering, and assembling printed circuit boards and other components. Fiberglass probes are nonconductive and will not scratch printed circuit boards. Figure C-8 illustrates a typical fiberglass probe.

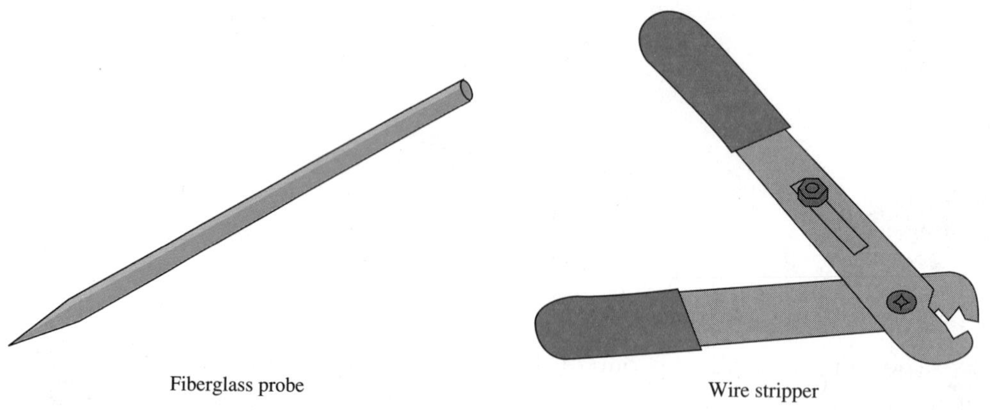

Figure C-8 **Fiberglass probe and wire stripper**

Wire strippers, like the one shown in Figure C-8, are used to remove the insulation from a wire. Care must be taken not to nick or cut any strands of a wire. Carefully examine the conductor in the area of the strip for cuts and nicks, as shown in Figure C-9.

Furthermore, when using stranded wire, do not allow the strands of the wire to **birdcage,** as shown in Figure C-10. Strands should be lightly twisted in their natural direction of rotation, as shown in Figure C-11. One final note is to be careful not to damage the insulation when holding the wire with pliers or by burning when soldering, as shown in Figure C-12.

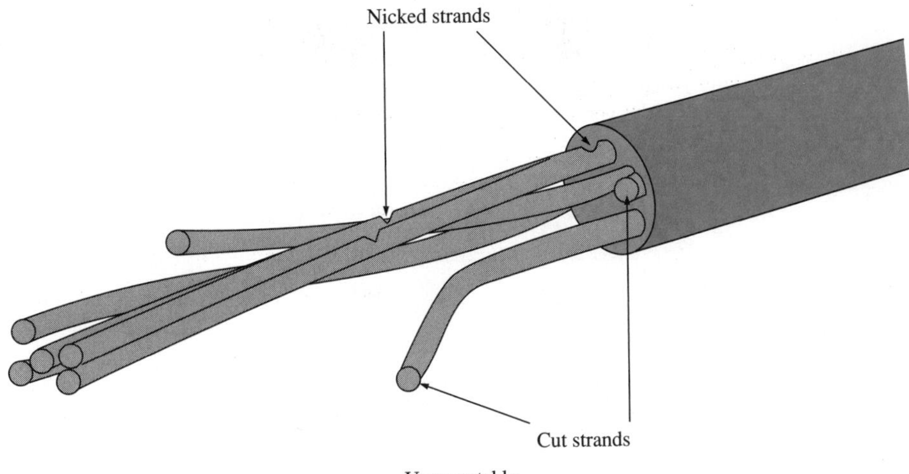

Figure C-9 **Cuts and nicks when stripping wire**

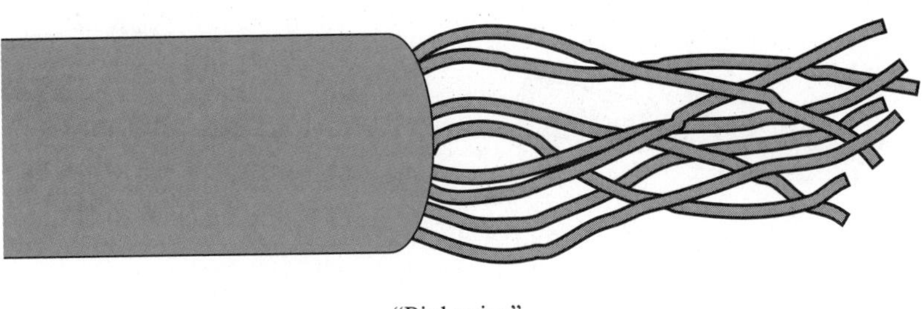

Figure C-10 **Birdcaged strands**

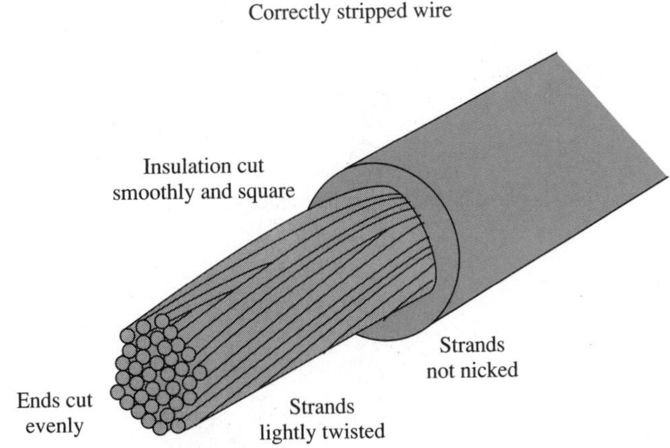

Figure C-11 **Correctly stripped wire**

Other tools include heat sinks, which protect sensitive components such as transistors and diodes from thermal damage, lead cleaners for cleaning component leads, suede brushes for cleaning hot soldering iron tips and soldering iron stands and tip-cleaning sponges, as shown in Figure C-13.

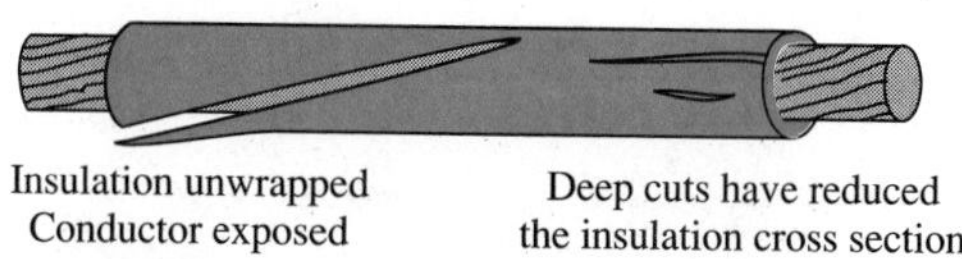

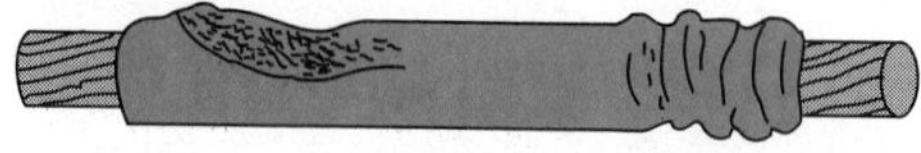

Figure C-12 **Insulation damage**

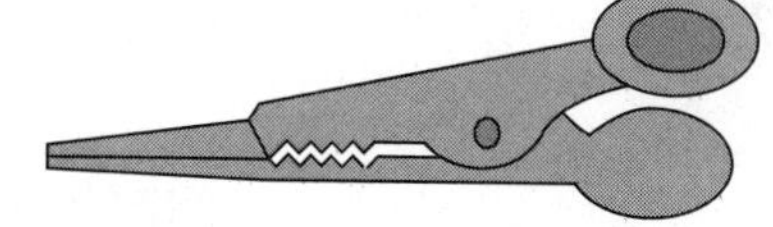

Heat Sinks
Protect heat sensitive components such as transistors and diodes from thermal damage

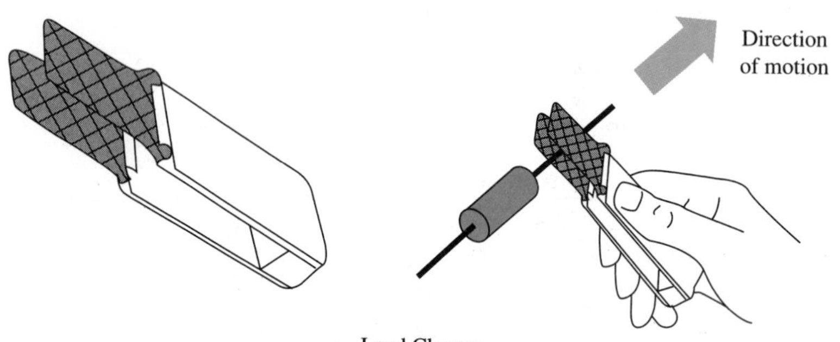

Lead Cleaner
A tool for cleaning component leads

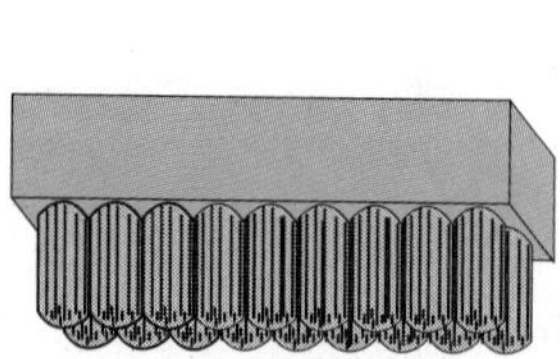

Suede Brush
Clean hot soldering iron plated tips

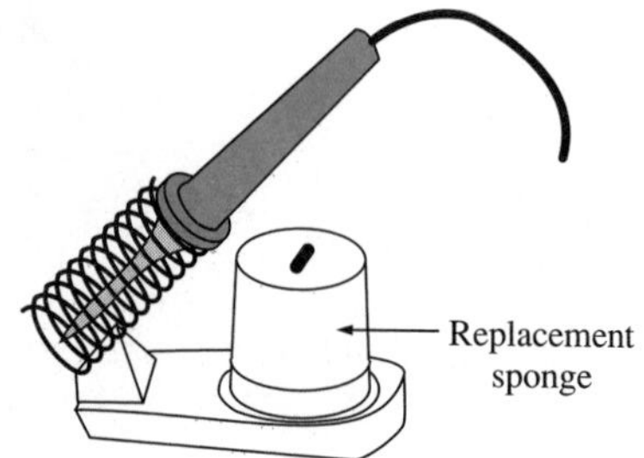

Iron Stand and Sponge
Tip cleaning sponge is used to ensure reliability of soldered joints and extends tip life

Figure C-13 **Other soldering tools**

Soldering

The first step is called "tinning the tip." Start by allowing the soldering iron to heat up to its operating temperature. Next, **tin** the tip by wiping it off on a wet sponge and applying solder to the clean tip. Tinning helps to resist oxidation and also provides a better transfer of heat to the connection. Now place the soldering iron tip on the connection to be soldered making certain it is touching both the component lead and the foil, pad, terminal, or wire to be soldered. Apply the solder to the component lead, foil, pad, terminal, or wire; not to the iron. Apply heat long enough to allow the solder to flow evenly and uniformly over the connection. Then remove the iron and be sure not to move the component until the solder cools. The entire process should not take more than 5 seconds. When soldering conventional components such as resistors and capacitors, it is a good idea to bend the leads to keep them from moving when the soldering iron is removed. You will find that it is well worth the extra time. The last step is to clean the connection with a flux remover or other noncorrosive solvent. Figure C-14 illustrates the soldering process.

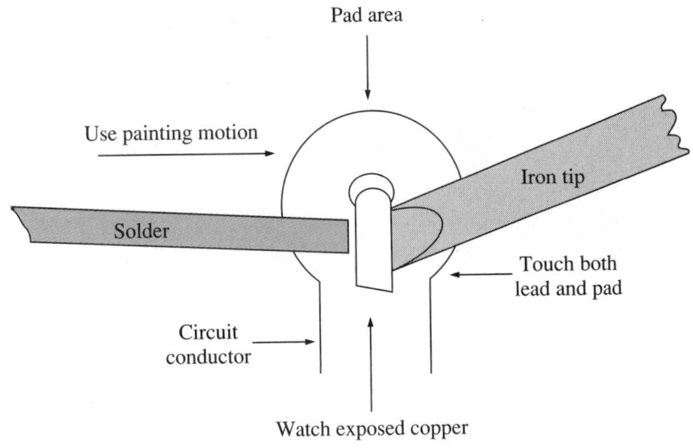

Figure C-14 **Soldering process**

If you have soldered the connection correctly, it will be smooth and shiny like the one shown in Figure C-15. If not, it will resemble the **cold** solder joint shown in Figure C-16. Cold solder joints are poor conductors. They are characterized by either a crystalline grainy texture or by blobs and uneven solder flow.

Figure C-15 **Good solder connection**

Figure C-16 Poor solder connection

A good solder connection should have just enough solder to cover the connection leads. The leads should be slightly visible in the solder, as shown in Figure C-17. When soldering small connection leads like those on an integrated circuit (IC), you can use extra thin solder. Remember that too much heat will cause the solder to boil and result in oxidation and other contaminants that will damage the connection. Too little heat will result in poor flow and wetting action. Practice is the best way to get everything in the proper proportions and to perfect your soldering skills.

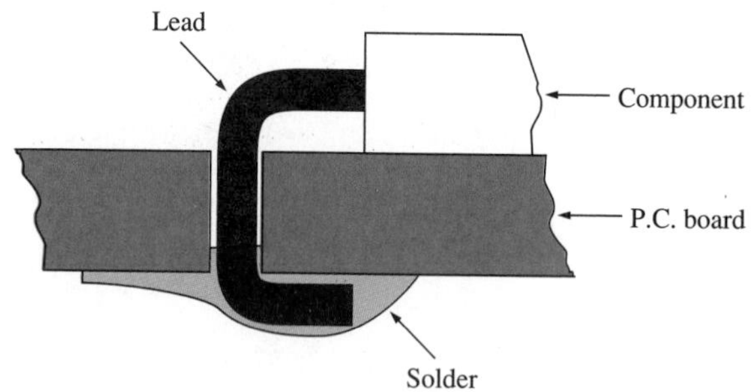

Figure C-17 Soldered component

Solder Removal

There are a number of ways to remove unwanted solder for component removal. You can use a product called solder removal braid, which is available under several trade names. Solder braid is a rosin flux coated braid of copper wire that absorbs **molten** solder. Solder braid comes in different widths for dif-

Figure C-18 Solder braid

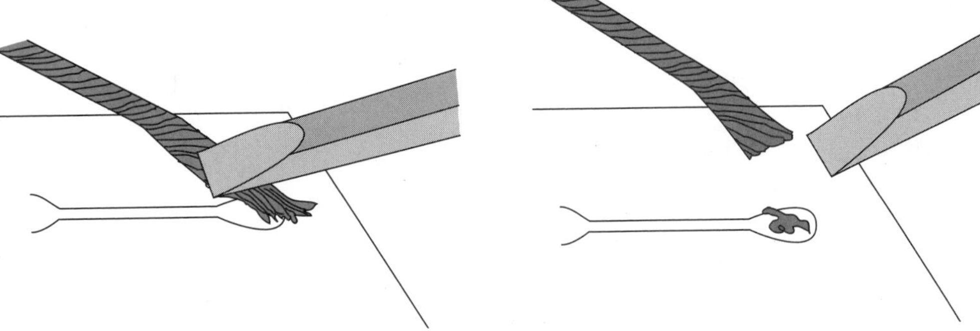

Figure C-19 **Use of solder braid**

ferent applications, as shown in Figure C-18. Figure C-19 demonstrates the use of solder braid.

Another technique is to use a solder sucker. Solder suckers come in a variety of sizes and shapes. Basically, a solder sucker uses a vacuum to literally suck solder off of a connection. Solder suckers come in both manual and automatic types, as shown in Figure C-20.

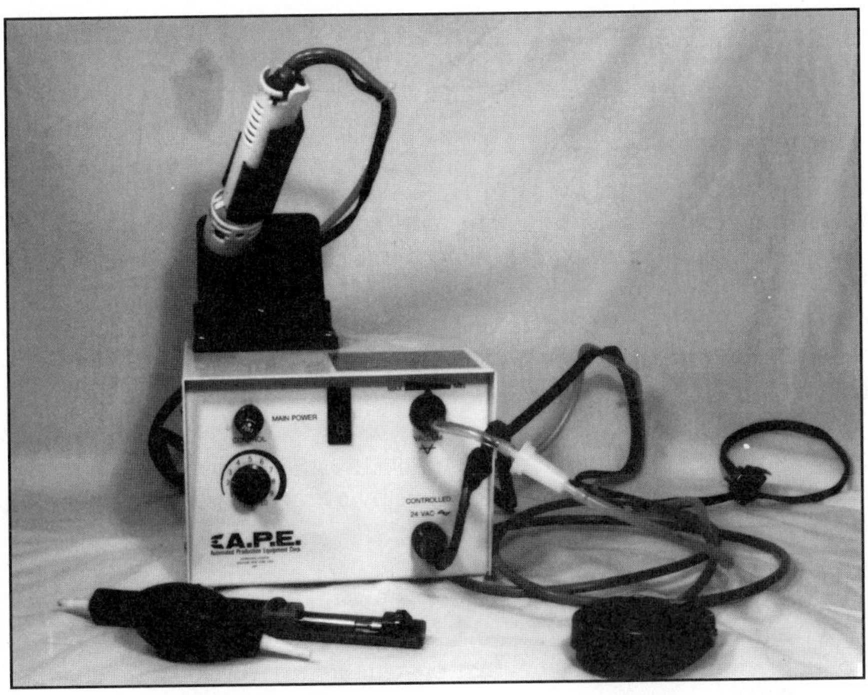

Figure C-20 **Solder sucker**

GLOSSARY

AC—alternating current; the flow of charge (current) that periodically changes direction

admittance—an AC measure of conductance

alternator—an AC generator

ammeter—a device designed to measure current

ampere—a unit for the measurement of charge flow

ampere-hour rating—a measure usually applied to batteries that indicates how long an amount of current can be drawn from the battery

apparent power—the power delivered to a load without considering the phase angle between the voltage and the current (VA rating)

armature—the part of a generator that produces a voltage; also the rotating part of a motor

artificial magnet—a magnet created by rubbing a piece of soft iron with a natural magnet

atom—the smallest amount of an element that still has all of the properties of the element

autotransformer—a transformer with a common primary and secondary winding

band pass filter—a filter designed to pass a range of frequencies and reject all others

band stop filter—a filter designed to reject a range of frequencies and pass all others

battery—two or more cells electrically connected together

bimetallic strip—a sandwich of two different metals with different coefficients of expansion

branch—a portion of a circuit that has one or more components connected in series

bridge circuit—a circuit configuration in the shape of a diamond

capacitance—a measure of the ability to store electric charge

capacitive reactance—the opposition to current by a capacitor

capacitor—a device used to store electric charge

cell—a fundamental source of electric energy

cemf—counter electromotive force; the voltage offered by self-inductance

choke—another name for an inductor

circuit—a configuration of components that provide a closed path for charge to flow (current)

circuit breaker—a resettable, protective device that opens a circuit when the current exceeds a certain amount

CircuitMaker—a computer program designed to perform electric circuit analysis

circular mil—a unit for the measurement of the cross-sectional area of wire; one circular mil is equivalent to the area of a circular wire whose diameter is one mil (1/1,000th of an inch)

color code—a system that uses bands of colors to indicate the resistance and tolerance of a resistor

commutator—a mechanical rectifier in a motor
compound motor—a motor with a series and a shunt field winding
condenser—obsolete name for a capacitor
conductance—the ability of a material to allow charge to flow (current); the opposite of resistance
conductor—a material with many free electrons that allows charge to flow (current) with little resistance
continuity—a low resistance path or connection for current
conventional current—a system that assumes the flow of charge (current) from positive to negative
copper losses—power losses due to heat or resistance
core losses—power losses in the iron core of a transformer; see also eddy currents and hysteresis
Coulomb's law—a mathematical relationship that determines the force of attraction or repulsion between two charges
coulomb—a unit for the measurement of charge; one coulomb is equal to the charge carried by 6.242×10^{18} electrons
current divider rule—a method of calculating the current through a parallel component without calculating the voltage
current source—an energy source that supplies a fixed amount of current to a circuit and whose terminal voltage varies with the load

DC—direct current; the flow of charge (current) in one direction
delta circuit—a three-phase circuit configuration in the shape of a triangle or the Greek letter delta; in a delta configuration the line and phase voltages are equal
determinant—a mathematical technique for the solution of linear equations with two or more unknown variables
dielectric—the insulating material between the plates of a capacitor
dielectric strength—the breakdown voltage of a capacitor; the voltage that will cause conduction across the plates of a capacitor
diode—a semiconductor device that allows charge flow (current) in one direction
DPDT—double-pole, double-throw switch
DPST—double-pole, single-throw switch

eddy currents—small circulating currents induced into an iron core of a magnetic device that cause power losses
effective value—an AC value with the same heating effect as a similar value of DC; equals 0.707 of the peak value
efficiency—the ratio of the output power to the input power in a device or circuit
electric field—invisible field of force that exists between electric charges
electrical degree—1/360th of an AC cycle
electromagnet—a magnet made by the flow of charge (current)
electron—a subatomic particle with a negative electric charge that orbits the nucleus of an atom
electron current—a system that assumes the flow of charge (current) from negative to positive
element—matter composed entirely of one type of atom

Faraday's law—the induced voltage in a conductor is directly proportional to the rate at which the conductor cuts the magnetic lines of force
filter—a network designed to pass or reject either high or low frequencies
flux—magnetic lines of force
flux density—the number of magnetic lines of force per unit area
free electrons—electrons that are not attached to any atom
frequency—a measure of how often a periodic waveform repeats itself
fuse—a protective device used to ensure that current does not exceed a certain level; fuses are usually nonresettable and operate by the melting of a metal link that opens the circuit

generator—a device that converts mechanical energy of rotation into electrical energy

henry—the unit of inductance; one henry is equal to one volt of cemf for a current change of one ampere in one second
hertz—the unit of frequency; one hertz is equal to one cycle per second
high pass filter—a filter designed to pass high frequencies and reject low frequencies
horsepower—a unit of power equal to 746 electrical watts
hysteresis—the lagging magnetic effect in time behind the applied magnetizing force

impedance—the total opposition to current in an AC circuit
inductance—the ability to produce an induced voltage when cut by magnetic lines of flux
induction motor—an AC motor in which energy is received by the rotor by induction rather than by direct connection
inductive reactance—the opposition to current by an inductor
inductor—a coil of wire
insulator—materials with few free electrons and a very high resistance to the flow of charge (current)
ion—an atom with more or less electrons than it has in its neutral state

joule—a unit for the measurement of energy; one joule is equal to 0.7378 foot pounds

kilowatt-hour—a unit of energy; one kilowatt-hour is equal to 3,600,000 joules
kilowatt-hour meter—a device that measures the energy supplied to a home or business
kinetic energy—energy due to motion
Kirchhoff's current law—the algebraic sum of the currents entering a node minus the currents leaving a node is equal to zero
Kirchhoff's voltage law—the algebraic sum of all voltages around a closed loop is equal to zero

Lenz's law—the induced voltage in a conductor is always in the opposite direction from the source voltage that produced it
Leyden jar—one of the first devices used to store charge; a capacitor
line current—the current that exists in a line connecting a three-phase energy source to a load

line voltage—the voltage between two lines of a three-phase system
loop analysis—a technique for determining the currents in a circuit
low pass filter—a filter designed to pass low frequencies and reject high frequencies

maximum power transfer theorem—a theorem used to determine the load resistance or impedance necessary to deliver maximum power to the load
mechanical losses—power losses due to windage and friction
meter—a unit for the measurement of length; one meter is equal to 1.094 yards
mho—an obsolete unit for the measurement of conductance (ohm spelled backward); replaced by the title siemens
motor—a device that converts electrical energy into mechanical energy of rotation

nameplate data—specifications about a motor that are written on a plate attached to the casing
natural magnet—a material, such as lodestone or magnetite, which exhibits magnetic properties
nucleus—the center of an atom that contains protons and neutrons
neutron—a subatomic particle found in the nucleus of an atom having no electrical charge
newton—a unit for the measurement of force; one newton is equal to 0.225 pounds
node—a connection point between at least two elements of a network
Norton's theorem—a method of equating a complicated network into a simpler network that consists of a current source and a parallel resistor
notch filter—see band stop filter

ohmmeter—a device designed to measure resistance
ohms—a unit for the measurement of resistance
Ohm's law—the fundamental relationship between current, voltage, and resistance in a circuit; current is equal to voltage divided by resistance
open circuit—no connection between two points in a circuit; infinite resistance

parallel circuit—a circuit configuration in which the components are connected together at two common points
period—the time required to complete one cycle
permanent magnet—a magnet that retains its magnetism for a long time
permeability—the ease in which a material can be magnetized
phase angle—a time relationship between voltage and/or current in a circuit
phase current—the current that exists in a three-phase generator or load
phase sequence—the order in which the generated voltage of a three-phase generator will affect the load
phase voltage—the voltage that appears between the line and neutral of a three-phase system
phasor—a vector that represents the voltage or current in an AC circuit at some instant in time
phasor diagram—a vector display of the magnitude and phase relationship of voltages and currents in a circuit
photovoltaic—producing a voltage from light energy; a solar cell

piezoelectric—producing a voltage by applying pressure to a crystal
poles—the ends of a magnet
potential energy—energy due to position
potentiometer—a three-terminal variable resistor used to control voltage levels
power—the rate of doing work; power is measured in joules/second or watts
power factor—a value between zero and one that is equal to the cosine of the phase angle of a circuit; a measure of the reactance in a circuit
power factor correction—the addition of inductance or capacitance to establish a power factor closer to unity
primary cell—a battery cell that cannot be recharged
primary winding—the input winding of a transformer
proton—a subatomic particle with a positive electrical charge that is found in the nucleus of an atom
PSpice—a computer program designed to perform electric circuit analysis

reactance—the opposition to current by an inductor or a capacitor
reactive power—the power associated with reactive elements of a circuit; VARS
real power—the power dissipated by a load; watts
rectifier—a device that changes AC to DC
reflected impedance—the impedance of the primary circuit of a transformer due to the secondary load impedance
reluctance—the opposition of a material to becoming magnetized
resistance—the opposition to current through a material
resistivity—a constant that relates the resistance of a material to its physical size; uses the Greek letter rho (ρ) as its symbol
resonance—a circuit condition where $X_L = X_C$
retentivity—the ability of a magnet to hold its magnetism
rheostat—a two-terminal, variable resistor used to control current
RMS value—route-mean-square; see effective value
rotary switch—a switch with multiple positions

scientific notation—a mathematical method for expressing very small and very large numbers as a power of ten and the multiplier between 1.00 and $9.\overline{99}$
secondary cell—a battery cell that can be recharged
secondary winding—the output winding of the transformer
selectivity—the ability of a circuit to separate signals (voltage) at different frequencies
series circuit—a configuration of components that has only one path for charge to flow (current)
series motor—a motor with its field winding in series with its armature
series-parallel circuit—a circuit consisting of both series and parallel branches
short circuit—a direct connection between two points; very low resistance
shunt motor—a motor with its field winding in parallel with its armature
SI system—Systems International; an international system of units adopted in the United States in 1967
siemens—a unit for the measurement of conductance; the reciprocal of ohms
slip speed—the difference between the actual motor speed and the synchronous speed of an AC motor
solder—a tin/lead alloy used for making electrical connections

source conversion—a method of changing a voltage source to a current source or vice versa
SPDT— single-pole, double-throw switch
SPST— single-pole, single-throw switch
static electricity—charge that is not in motion
stator—the stationary part of a generator or motor
step-down transformer—a transformer whose secondary voltage is less than the primary voltage
step-up transformer—a transformer whose secondary voltage is greater than the primary voltage
superposition theorem—a method of analyzing multisource circuits by considering the effects of each source separately
susceptance—a measure of AC conduction due to an inductor or a capacitor
synchronous speed—the speed of the rotating magnetic field around the stator of an AC motor

tank circuit—a parallel LC circuit
tapped transformer—a transformer winding with multiple connection points
tee circuit—see wye circuit
temperature coefficient of resistance—a constant that relates how the resistance of a material changes with changes in temperature; uses the symbol α
tesla—unit of flux density; one tesla is equal to 10^8 magnetic lines of force per square meter
thermistor—a semiconductor device that changes resistance with changes in temperature
Thevenin's theorem—a method of equating a complicated network into a simpler network consisting of a voltage source and a series resistor
three-phase system—an AC system with three sinusoidal voltages per rotation of the rotor; each 120 electrical degrees apart
time constant—a constant that relates how fast a capacitor or inductor will charge or discharge
torque—rotary motion produced in a motor
transformer—a device that changes voltage, current, or resistance levels
transient—an abrupt change in voltage or current in a circuit
transistor—a semiconductor device used for amplification and electronic switching
turns ratio—the ratio of the number of turns in the primary to the number of turns in the secondary of a transformer

valence electrons—electrons in the outermost shell of an atom
varistor—a semiconductor device that changes resistance with changes in voltage
volt—a unit for the measurement of the potential difference between two points
voltage—a difference in potential energy that causes charge to flow (current)
voltage divider—a series arrangement of components that provides a voltage less than the source voltage
voltage divider rule—a method of calculating the voltage across a series component without calculating the current

voltage source—an energy source that supplies a fixed amount of voltage to a circuit and whose terminal current varies with the load

voltaic cell—a device that a converts chemical energy into electrical energy

voltmeter—a device designed to measure voltage

VOM—Volt-Ohm-Milliammeter; a device designed to measure voltage, resistance, and current; also referred to as a multimeter

watt—the unit for electrical power

watt meter—a device that measures the power supplied to a circuit or a circuit element

Wheatstone bridge—a device used for very accurate current, voltage, or resistance measurements; see also, bridge circuit

working voltage—the maximum safe voltage that can be applied across the plates of a capacitor

wye circuit—a three-phase circuit configuration in the shape of the letter *Y* or *T;* in a wye configuration the line and phase currents are equal

INDEX

Note: Page number in **bold** type refer to non-text material.